计算机科学丛书

原书第12版

Java语言程序设计

[美] 梁勇（Y. Daniel Liang） 著
佐治亚南方大学

戴开宇 译
复旦大学

Introduction to Java Programming and Data Structures
Comprehensive Version, Twelfth Edition

进阶篇

机械工业出版社
China Machine Press

图书在版编目（CIP）数据

Java 语言程序设计. 进阶篇：原书第 12 版 /（美）梁勇（Y. Daniel Liang）著；戴开宇译. -- 北京：机械工业出版社，2021.9

（计算机科学丛书）

书名原文：Introduction to Java Programming and Data Structures, Comprehensive Version, Twelfth Edition

ISBN 978-7-111-68935-5

I. ① J… Ⅱ. ①梁… ②戴… Ⅲ. ① JAVA 语言 - 程序设计 Ⅳ. ① TP312.8

中国版本图书馆 CIP 数据核字（2021）第 165888 号

本书版权登记号：图字 01-2020-2395

Authorized translation from the English language edition, entitled *Introduction to Java Programming and Data Structures, Comprehensive Version, Twelfth Edition*, ISBN: 9780136520238, by Y. Daniel Liang, published by Pearson Education, Inc., Copyright © 2020, 2018, 2015 by Pearson Education, Inc. or its affiliates, 221 River Street, Hoboken, NJ 07030.

All rights reserved. No part of this book may be reproduced or transmitted in any form or by any means, electronic or mechanical, including photocopying, recording or by any information storage retrieval system, without permission from Pearson Education, Inc.

Chinese simplified language edition published by China Machine Press, Copyright © 2021.

本书中文简体字版由 Pearson Education（培生教育出版集团）授权机械工业出版社在中华人民共和国境内（不包括香港、澳门特别行政区及台湾地区）独家出版发行。未经出版者书面许可，不得以任何方式抄袭、复制或节录本书中的任何部分。

本书封底贴有 Pearson Education（培生教育出版集团）激光防伪标签，无标签者不得销售。

本书是 Java 语言的经典教材，中文版分为基础篇和进阶篇，主要介绍程序设计基础、面向对象程序设计、GUI 程序设计、数据结构和算法、高级 Java 程序设计等内容。本书通过示例讲解问题求解技巧，提供大量的程序清单，每章配有丰富的复习题和编程练习题，帮助读者掌握编程技术，并学会应用所学技术解决实际开发中遇到的问题。进阶篇主要讨论数据结构和算法，包括线性表、栈、队列、规则集、映射、排序、散列、树和图等内容。

本书可作为高等院校计算机相关专业程序设计课程的教材，也可作为 Java 语言及编程爱好者的参考资料。

出版发行：机械工业出版社（北京市西城区百万庄大街 22 号 邮政编码：100037）
责任编辑：曲 熠   责任校对：马荣敏
印　　刷：中国电影出版社印刷厂   版　　次：2021 年 9 月第 1 版第 1 次印刷
开　　本：185mm×260mm 1/16   印　　张：25
书　　号：ISBN 978-7-111-68935-5   定　　价：139.00 元

客服电话：(010) 88361066　88379833　68326294   投稿热线：(010) 88379604
华章网站：www.hzbook.com   读者信箱：hzjsj@hzbook.com

版权所有·侵权必究
封底无防伪标均为盗版
本书法律顾问：北京大成律师事务所　韩光 / 邹晓东

中文版序

Introduction to Java Programming and Data Structures, Comprehensive Version, Twelfth Edition

Welcome to the Chinese translation of *Introduction to Java Programming and Data Structure, Comprehensive Version, Twelfth Edition*. The first edition of the English version was published in 1998. Since then twelve editions of the book have been published in the last twenty-one years. Each new edition substantially improved the book in contents, presentation, organization, examples, and exercises. This book is now the #1 selling computer science textbook in the US. Hundreds and thousands of students around the world have learned programming and problem solving using this book.

I thank Dr. Kaiyu Dai of Fudan University for translating this latest edition. It is a great honor to reconnect with Fudan through this book. I personally benefited from teachings of many great professors at Fudan. Professor Meng Bin made Calculus easy with many insightful examples. Professor Liu Guangqi introduced multidimensional mathematic modeling in the Linear Algebra class. Professor Zhang Aizhu laid a solid mathematical foundation for computer science in the discrete mathematics class. Professor Xia Kuanli paid a great attention to small details in the PASCAL course. Professor Shi Bole showed many interesting sort algorithms in the data structures course. Professor Zhu Hong required an English text for the algorithm design and analysis course. Professor Lou Rongsheng taught the database course and later supervised my master's thesis.

My study at Fudan and teaching in the US prepared me to write the textbook. The Chinese teaching emphasizes on the fundamental concepts and basic skills, which is exactly I used to write this book. The book is fundamentals first by introducing basic programming concepts and techniques before designing custom classes. The fundamental-first approach is now widely adopted by the universities in the US. With the excellent translation from Dr. Dai, I hope more students will benefit from this book and excel in programming and problem solving.

欢迎阅读本书第12版的中文版。本书英文版的第1版于1998年出版。自那之后的21年中，本书共出版了12个版本。每个新的版本都在内容、表述、组织、示例以及练习题等方面进行了大量的改进。本书目前在美国计算机科学类教材中销量排名前列。全世界无数的学生通过本书学习程序设计以及问题求解。

感谢复旦大学的戴开宇博士翻译了这一最新版本。非常荣幸通过这本书和复旦大学重建联系，我本人曾经受益于复旦大学的许多杰出教授：孟斌教授采用许多富有洞察力的示例将微积分变得清晰易懂；刘光奇教授在线性代数课堂上介绍了多维数学建模；张霭珠教授的离散数学课程为我学习计算机科学打下了坚实的数学基础；夏宽理教授在Pascal课程中对许多小细节给予了极大的关注；施伯乐教授在数据结构课程中演示了许多有趣的排序算法；朱洪教授在算法设计和分析课程中使用了英文教材；楼荣生教授讲授了数据库课程，并且指导了我的硕士论文。

  我在复旦大学的学习经历以及在美国的授课经验为撰写本书奠定了基础。中国的教学重视基本概念和基础技能，这也是我写这本书所采用的方法。本书采用基础为先的方法，在介绍自定义类的设计之前先介绍基本的程序设计概念和方法。目前，基础为先的方法也被美国的大学广泛采用。我希望通过戴博士的优秀翻译，让更多的学生从中受益，并在程序设计和问题求解方面出类拔萃。

梁勇

y.daniel.liang@gmail.com

https://yongdanielliang.github.io/

译者序

Introduction to Java Programming and Data Structures, Comprehensive Version, Twelfth Edition

Java 是一门伟大的程序设计语言，同时，它还指基于 Java 语言的从嵌入式开发到企业级开发的平台。从 20 世纪 90 年代诞生至今，Java 凭借其优秀的语言和平台设计，以及适合互联网应用的"一次编译，到处运行"的跨平台特性，在 Web 应用、移动计算、云计算、大数据、人工智能、物联网及可穿戴设备等新兴技术领域得到了极其广泛的应用。除此之外，Java 还是一门设计优秀的教学语言。它是一门经典的面向对象编程语言，拥有优雅和简明的语法，体现了很多程序设计方面的理念和智慧，可帮助程序设计人员将精力尽可能地集中在业务领域的设计上。在版本迭代中，Java 还吸纳了其他程序设计语言的优点来进行完善，比如 Java 8 中 lambda 表达式的引入体现了函数式编程的特色。Java 还具有丰富且实用的类库，许多开源项目和科学研究的原型系统都是采用 Java 实现的。在针对编程语言流行趋势指标的 TIOBE 编程语言社区排行榜上，Java 多年来都居于前列。采用实际应用广泛的优秀程序设计语言进行教学，对学生今后进一步的科研和工作都有直接帮助。

在 14 年前机械工业出版社举办的一次教学研讨会上，我有幸认识了本书的作者梁勇（Y. Daniel Liang）教授并进行了交流。从那时起我开始在主讲的程序设计课程中采用本书英文版作为教材，并取得了很好的教学效果。本书知识点全面，体系结构清晰，重点突出，文字准确，内容组织循序渐进，并有大量精选的示例和配套素材，比如精心设计的大量练习题，甚至在配套网站中还有支持教学的大量动画演示。本书采用基础优先的方式，从编程基础开始，逐步引入面向对象思想，最后介绍应用框架。教学实践证明，这种方式很适合程序设计初学者。另外，强调问题求解和计算思维也是本书特色，这也是我在教授程序设计过程中遵循的教学理念。本书通过数学、经济、游戏等应用领域的生动实用的案例来引导学生学习程序设计，避免了单纯语法学习的枯燥，也让学生可以学以致用。程序设计教学中最重要的是培养学生的计算思维，通识教育和新工科建设背景下的教学理念都非常重视计算思维，这对于提升学生的综合素质并且将所学知识应用于生活中是很有裨益的。之前我翻译了本书第 10 版和第 11 版，得到了许多读者的好评，也收到了很多宝贵的建议。时隔 1 年，我很荣幸再次成为本书第 12 版的译者。这一版在上一版译文的基础上更加字斟句酌，修订了之前的一些问题，希望能对广大程序设计学习者有所帮助。

在本书的翻译过程中，我得到了本书作者梁勇教授的大力支持。非常感谢他不仅快速回复和详细解答了我在邮件中提出的一些问题，还拨冗写了中文版序，其一丝不苟的精神让人感动。感谢机械工业出版社的何方编辑，他在本书的整个翻译过程中提供了许多帮助。最后要感谢我的家人在翻译过程中给予的支持和鼓励。限于水平，书中难免还会存在问题，敬请大家指正。

戴开宇
kydai@fudan.edu.cn

前言

Introduction to Java Programming and Data Structures, Comprehensive Version, Twelfth Edition

许多读者就本书之前的版本给出了很多反馈，这些评论和建议极大地改进了本书。这一版在表述、组织、示例、练习题以及附录方面都有大幅改进。

本书采用基础优先的方法，在设计用户自定义类之前，首先介绍基本的程序设计概念和技术。选择语句、循环、方法和数组这样的基本概念与技术是程序设计的基础，打好这些基础将帮助学生为进一步学习面向对象程序设计和高级 Java 程序设计做好准备。

本书以问题驱动的方式来教授程序设计，将重点放在问题的解决而不是语法上。我们通过使用在各种应用场景中引发思考的问题，使程序设计的介绍变得更加有趣。前面章节的主线放在问题的解决上，引入合适的语法和库以支持编写解决问题的程序。为了支持以问题驱动的方式来教授程序设计，本书提供了大量不同难度的问题来激发学生的积极性。为了吸引各个专业的学生来学习，这些问题涵盖很多应用领域，包括数学、科学、商业、金融、游戏、动画以及多媒体等。

本书将程序设计、数据结构和算法无缝整合在一起，采用一种实用的方式来教授数据结构。首先介绍如何使用各种数据结构来开发高效的算法，然后演示如何实现这些数据结构。通过实现，学生可以深入理解数据结构的效率，以及如何和何时使用某种数据结构。最后，我们设计和实现了针对树和图的用户自定义数据结构。

本书广泛应用于全球众多大学的程序设计入门、数据结构和算法课程中。完全版⊖包括程序设计基础、面向对象程序设计、GUI 程序设计、数据结构、算法、并行、网络、数据库和 Web 程序设计。这个版本旨在把学生培养成精通 Java 的程序员。基础篇包含完全版的前 18 章内容，可用于程序设计的第一门课程（通常称为 CS1）。本书还有一个 AP 版本，适合学习 AP 计算机科学（AP Computer Science）课程的高中生使用。

教授编程的最好途径是通过示例，而学习编程的唯一途径是通过动手练习。本书通过示例对基本概念进行讲解，并提供大量不同难度的练习题供学生进行练习。在我们的程序设计课程中，每次课后都布置了编程练习。

我们的目标是编写一本可以通过各种应用场景中的有趣示例来教授问题求解和程序设计的教材。如果你有任何关于如何改进本书的意见或建议，请给我发邮件。

ACM/IEEE 课程体系 2013 版和 ABET 课程评价

新的 ACM/IEEE 计算机科学课程体系 2013 版将知识体系组织成 18 个知识领域。为了帮助教师基于本书设计课程，我们提供了示例教学大纲来确定知识领域和知识单元。作为一个常规的定制示例，示例教学大纲用于三学期的课程系列。示例教学大纲可以从教师资源配套网站获取。

许多读者来自 ABET 认证计划。ABET 认证的一个关键组成部分是，通过针对课程效果

⊖ 本书中文版将完全版分为基础篇和进阶篇出版，基础篇对应原书第 1~18 章，进阶篇对应原书第 19~30 章，你手中的这一本是进阶篇。——编辑注

的持续课程评价确定学习中的薄弱环节。我们在教师资源配套网站中提供了课程效果示例，以及用于检验课程效果的样卷。

本版新增内容

本版对各个细节都进行了全面修订，以更清晰地呈现知识、示例和练习题。本版的主要改进如下：

- 更新至 Java 9、10 和 11。使用 Java 9、10 和 11 版本中的新特性对示例进行了改进和简化。
- GUI 相关章节更新到 JavaFX 11，并改写了示例。示例和练习题中的用户界面现在可以改变尺寸并且居中显示。
- 数据结构相关章节中，更多的示例和练习题采用 lambda 表达式来简化编程。
- Comparable 和 Comparator 都被用于比较 Heap、PriorityQueue、BST 以及 AVLTree 中的元素。这样与 Java API 保持一致，更加实用、灵活。
- 第 22 章引入了字符串匹配算法。
- 添加了视频注解。
- 提供了没有出现在书中的额外习题，这些习题仅供教师使用。

可以访问本书配套网站 www.pearsonhighered.com/liang，了解这一版与前一版的关联以及全部的新特性。

教学特色

本书使用以下要素组织素材，以帮助读者高效学习：

- **教学目标**：在每章开始列出学生应该掌握的内容，学完这章后，学生能够判断自己是否达到这些目标。
- **引言**：提出引发思考的问题以展开讨论，激发读者深入探讨相关内容的积极性。
- **要点提示**：突出每节中涵盖的重要概念。
- **复习题**：帮助学生复习每节相关内容并评估掌握的程度。
- **问题和示例学习**：通过精心挑选示例，以易于理解的方式教授问题求解和程序设计概念。本书使用多个短小的、简单的、激发兴趣的例子来演示重要的概念。
- **本章小结**：回顾学生应该理解和记住的重要主题，有助于巩固所学的关键概念。
- **测试题**：可以在线访问，按章节组织，让学生可以就编程概念和技术进行自我测试。
- **编程练习题**：按章节组织，为学生提供自主应用所学新技能的机会。练习题的难度分为容易（没有星号）、适度（*）、难（**）和具有挑战性（***）四个级别。学习程序设计的窍门就是"实践，实践，再实践"。所以，本书提供了大量的编程练习题。教师资源网站还为教师提供了额外的 200 多道带有答案的编程练习题。
- **注意、提示、警告和设计指南**：贯穿全书，对程序开发的重要方面提供有价值的建议和见解。
  - ➢ **注意**：提供学习主题的附加信息，巩固重要概念。
  - ➢ **提示**：教授良好的程序设计风格和实践经验。
  - ➢ **警告**：帮助学生避开程序设计误区。
  - ➢ **设计指南**：提供设计程序的指南。

灵活的章节顺序

本书提供灵活的章节顺序，使 GUI、异常处理、递归、泛型和 Java 集合框架等内容可以或早或晚地讲解。下页的插图显示了各章之间的相关性。

本书的组织

本书章节分为五部分，共同构成了对 Java 程序设计、数据结构和算法、数据库以及 Web 程序设计的全面介绍。书中的知识是循序渐进的，前面的章节介绍程序设计的基本概念，并通过简单的例子和练习题引导学生，后续的章节逐步详细介绍 Java 程序设计，最后介绍开发综合的 Java 应用程序。附录包含数系、位操作符、正则表达式以及枚举类型等多个主题。

第一部分　程序设计基础（第 1～8 章）

第一部分是全书的基石，带你踏上 Java 学习之旅。你将了解 Java（第 1 章），还将学习像基本数据类型、变量、常量、赋值、表达式以及操作符这样的基本程序设计技术（第 2 章），选择语句（第 3 章），数学函数、字符和字符串（第 4 章），循环（第 5 章），方法（第 6 章），数组（第 7 和 8 章）。在第 7 章之后，可以跳到第 18 章学习如何编写递归方法来解决本身具有递归特性的问题。

第二部分　面向对象程序设计（第 9～13 章和第 17 章）

这一部分介绍面向对象程序设计。Java 是一种面向对象的程序设计语言，通过抽象、封装、继承和多态为软件开发提供了极大的灵活性、模块化和可重用性。你将学习如何使用对象和类（第 9 和 10 章）、类的继承（第 11 章）、多态（第 11 章）、异常处理（第 12 章）、抽象类（第 13 章）以及接口（第 13 章）进行程序设计。文本 I/O 将在第 12 章介绍，二进制 I/O 将在第 17 章介绍。

第三部分　GUI 程序设计（第 14～16 章和奖励章节第 31 章）

JavaFX 是一个用于开发 Java GUI 程序的新框架。它不仅对开发 GUI 程序有用，还是一个用于学习面向对象程序设计的优秀教学工具。第 14～16 章介绍使用 JavaFX 进行 Java GUI 程序设计。主要主题包括 GUI 基础（第 14 章）、容器面板（第 14 章）、绘制形状（第 14 章）、事件驱动编程（第 15 章）、动画（第 15 章）、GUI 控件（第 16 章），以及播放音频和视频（第 16 章）。你将学习采用 JavaFX 的 GUI 程序架构，并且使用控件、形状、面板、图像和视频来开发实用的应用程序。第 31 章讨论 JavaFX 的高级特性。

第四部分　数据结构和算法（第 18～30 章以及奖励章节第 42 和 43 章）

这一部分介绍典型的数据结构和算法课程中的主题。第 18 章介绍递归以编写解决本身具有递归特性的问题的方法。第 19 章介绍泛型如何提高软件的可靠性。第 20 和 21 章介绍 Java 集合框架，它为数据结构定义了一套有用的 API。第 22 章讨论算法效率的度量以便为应用程序选择合适的算法。第 23 章介绍经典的排序算法。你将在第 24 章中学到如何实现经典的数据结构，如线性表、队列和优先队列。第 25 和 26 章介绍二叉搜索树和 AVL 树。第 27 章介绍散列以及通过散列实现映射（map）和规则集（set）。第 28 和 29 章介绍图的应用。第 30 章介绍用于集合流的聚合操作。2-4 树、B 树以及红黑树在奖励章节第 42 和 43 章中介绍。

IX

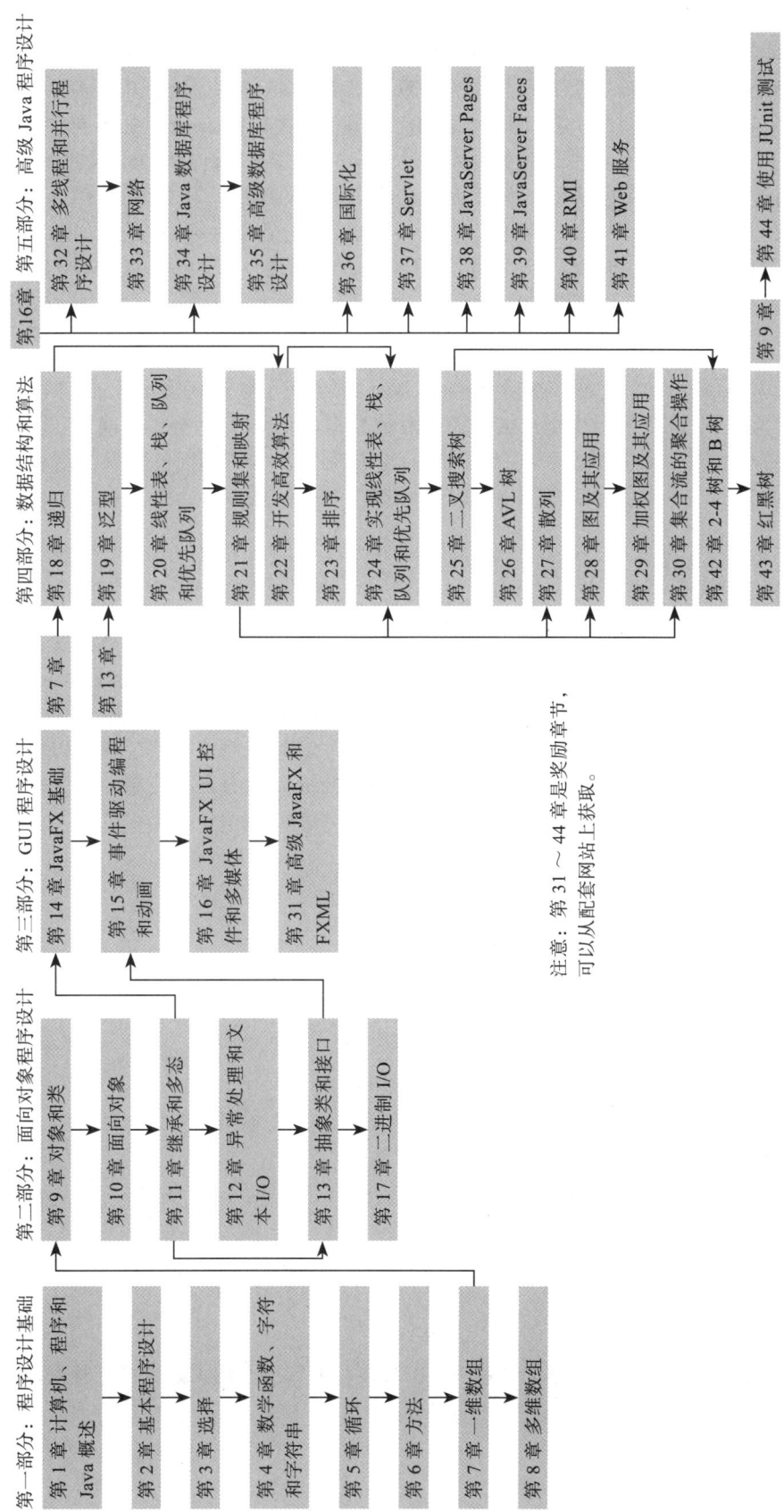

第五部分　高级 Java 程序设计（奖励章节第 32 ~ 41 章和第 44 章）

这一部分介绍高级 Java 程序设计。第 32 章介绍使用多线程使程序具有更好的响应性和交互性，并介绍并行程序设计。第 33 章讨论如何编写程序使得 Internet 上的不同主机能够相互对话。第 34 章介绍使用 Java 来开发数据库项目。第 35 章深入探讨高级 Java 数据库程序设计。第 36 章涵盖国际化支持的使用，以开发面向全球使用者的项目。第 37 和 38 章介绍如何使用 Java Servlet 和 JavaServer Pages 创建来自 Web 服务器的动态内容。第 39 章介绍使用 JavaServer Faces 进行现代 Web 应用开发。第 40 章介绍远程方法调用，第 41 章讨论 Web 服务。第 44 章介绍使用 JUnit 测试 Java 程序。

附录

附录部分涵盖多个主题。附录 A 列出 Java 关键字。附录 B 给出十进制和十六进制 ASCII 字符集。附录 C 给出操作符优先级。附录 D 总结 Java 修饰符及其使用。附录 E 讨论特殊的浮点值。附录 F 介绍数系以及二进制、十进制和十六进制间的转换。附录 G 介绍位操作符。附录 H 介绍正则表达式。附录 I 介绍枚举类型。附录 J 介绍大 O、大 Ω 和大 Θ 表示法。

Java 开发工具

可以使用 Windows 记事本（NotePad）或写字板（WordPad）这样的文本编辑器创建 Java 程序，然后从命令窗口编译、运行程序。也可以使用 Java 开发工具，例如 NetBeans 或者 Eclipse。这些工具是支持快速开发 Java 应用程序的集成开发环境（IDE）。编辑、编译、构建、运行和调试程序都集成在一个图形用户界面中。有效地使用这些工具可以极大地提高编写程序的效率。如果按照教程学习，NetBeans 和 Eclipse 也是易于使用的。关于 NetBeans 和 Eclipse 的教程，参见本书配套网站。

学生资源

本书配套网站上的学生资源包括：
- 复习题的答案。
- 绝大部分偶数编号编程练习题的答案。
- 书中示例的源代码。
- 交互式的自测题（按章节组织）。
- 补充材料。
- 调试技巧。
- 视频注解。
- 算法动画。
- 勘误表。

补充材料

教材涵盖了核心内容，补充材料进一步扩展教材内容，介绍了读者可能感兴趣的其他内容。补充材料可以从配套网站上获得。

教师资源[⊖]

本书配套网站上的教师资源包括：
- PowerPoint 教学幻灯片，通过交互性的按钮可以观看彩色、语法项高亮显示的源代码，并可以在幻灯片中直接运行程序。
- 绝大部分奇数编号编程练习题的答案。
- 按章节组织的 200 多道补充编程练习题和 300 道测试题。这些练习题和测试题仅对教师开放，并提供答案。
- 基于 Web 的测试题生成器（教师可以选择章节以从超过 2000 道题的大型题库中生成测试题）。
- 样卷。大多数试卷包含 4 个部分：
  - 多选题或者简答题。
  - 改正编程错误。
  - 跟踪程序。
  - 编写程序。
- 具有 ABET 课程评价的样卷。
- 课程项目。通常，每个项目给出一个描述，并且要求学生分析、设计和实现该项目。

使用 MyProgrammingLab 进行在线练习和评价

MyProgrammingLab 可帮助学生充分掌握编程的逻辑、语义和语法。通过实践性编程练习以及即时、个性化的反馈，MyProgrammingLab 还可帮助初学者提高编程能力。初学者经常受困于流行的高级编程语言的基本概念和范式。

作为一个自我学习和作业工具，MyProgrammingLab 课程由几百道小练习题组成，这些练习题是围绕本教材的结构进行组织的。对于学生，这套系统会自动检查他们所提交代码的逻辑和语法错误，并给出帮助学生理解哪里错了以及为何错了的针对性提示。对于教师，系统会提供一个综合的分数册，跟踪正确和非正确的答案，并保存学生输入的代码，以用于复习。

MyProgrammingLab 是和 Turing's Craft 合作提供给本书读者的。Turing's Craft 是 CodeLab 交互性编程练习系统的制作者。要得到该系统的完整演示，或者看到教师和学生的反馈，或者开始在你的课堂中使用 MyProgrammingLab，请访问 www.myprogramminglab.com。

视频注解

在新版中添加视频注解这一特色功能，我们感到很兴奋。这些视频针对关键内容提供了示例，并演示了从设计到编码的问题求解的完整过程。视频注解可以从本书配套网站上获取。

算法动画

我们提供了大量的算法演示动画，它们对于演示算法的机制是非常有价值的教学工具。

[⊖] 关于本书教辅资源，只有使用本书作为教材的教师才可以申请，需要的教师请联系机械工业出版社华章公司，电话 010-88378991，邮箱 wangguang@hzbook.com。此外，关于配套网站资源，有些内容需要访问码，访问码只有原英文版提供，中文版无法使用。——编辑注

可以从配套网站上获取算法的动画演示。

致谢

感谢佐治亚南方大学给我机会讲授我所写的内容，并支持我将所教的内容写成教材。教学是我持续改进本书的灵感之源。感谢使用本书的教师和学生提出的评价、建议、错误报告和赞扬。特别感谢拉玛尔大学的 Stefan Andrei、科罗拉多大学科罗拉多泉分校的 William Bahn，他们为改进本书数据结构部分的内容提供了帮助。

由于有了对本版和之前版本的富有见解的审阅，本书得到很大的改进。感谢以下审阅人员：Elizabeth Adams（James Madison University）, Syed Ahmed（North Georgia College and State University）, Omar Aldawud（Illinois Institute of Technology）, Stefan Andrei（Lamar University）, Yang Ang（University of Wollongong, Australia）, Kevin Bierre（Rochester Institute of Technology）, Aaron Braskin（Mira Costa High School）, David Champion（DeVry Institute）, James Chegwidden（Tarrant County College）, Anup Dargar（University of North Dakota）, Daryl Detrick（Warren Hills Regional High School）, Charles Dierbach（Towson University）, Frank Ducrest（University of Louisiana at Lafayette）, Erica Eddy（University of Wisconsin at Parkside）, Summer Ehresman（Center Grove High School）, Deena Engel（New Youk University）, Henry A. Etlinger（Rochester Institute of Technology）, James Ten Eyck（Marist College）, Myers Foreman（Lamar University）, Olac Fuentes（University of Texas at El Paso）, Edward F. Gehringer（North Carolina State University）, Harold Grossman（Clemson University）, Barbara Guillot（Louisiana State University）, Stuart Hansen（University of Wisconsin, Parkside）, Dan Harvey（Southern Oregon University）, Ron Hofman（Red River College, Canada）, Stephen Hughes（Roanoke College）, Vladan Jovanovic（Georgia Southern University）, Deborah Kabura Kariuki（Stony Point High School）, Edwin Kay（Lehigh University）, Larry King（University of Texas at Dallas）, Nana Kofi（Langara College, Canada）, George Koutsogiannakis（Illinois Institute of Technology）, Roger Kraft（Purdue University at Calumet）, Norman Krumpe（Miami University）, Hong Lin（DeVry Institute）, Dan Lipsa（Armstrong Atlantic State University）, James Madison（Rensselaer Polytechnic Institute）, Frank Malinowski（Darton College）, Tim Margush（University of Akron）, Debbie Masada（Sun Microsystems）, Blayne Mayfield（Oklahoma State University）, John McGrath（J.P. McGrath Consulting）, Hugh McGuire（Grand Valley State）, Shyamal Mitra（University of Texas at Austin）, Michel Mitri（James Madison University）, Kenrick Mock（University of Alaska Anchorage）, Frank Murgolo（California State University, Long Beach）, Jun Ni（University of Iowa）, Benjamin Nystuen（University of Colorado at Colorado Springs）, Maureen Opkins（CA State University, Long Beach）, Gavin Osborne（University of Saskatchewan）, Kevin Parker（Idaho State University）, Dale Parson（Kutztown University）, Mark Pendergast（Florida Gulf Coast University）, Richard Povinelli（Marquette University）, Roger Priebe（University of Texas at Austin）, Mary Ann Pumphrey（De Anza Junior College）, Pat Roth（Southern Polytechnic State University）, Amr Sabry（Indiana University）, Ben Setzer（Kennesaw State University）, Carolyn Schauble（Colorado State University）, David Scuse（University of Manitoba）, Ashraf Shirani（San Jose State University）, Daniel Spiegel（Kutztown University）, Joslyn A. Smith（Florida Atlantic University）, Lixin Tao

(Pace University)、Ronald F. Taylor(Wright State University)、Russ Tront(Simon Fraser University)、Deborah Trytten(University of Oklahoma)、Michael Verdicchio(Citadel)、Kent Vidrine(George Washington University)、Bahram Zartoshty(California State University at Northridge)。

能够与 Pearson 出版社一起工作，我感到非常愉快和荣幸。感谢 Tracy Johnson 和她的同事 Marcia Horton、Demetrius Hall、Yvonne Vannatta、Kristy Alaura、Carole Snyder、Scott Disanno、Bob Engelhardt、Shylaja Gattupalli，感谢他们组织、开展和积极促进本项目。

一如既往，感谢妻子 Samantha 的爱、支持和鼓励。

Y. Daniel Liang
y.daniel.liang@gmail.com
www.pearsonhighered.com/liang

目 录

Introduction to Java Programming and Data Structures, Comprehensive Version, Twelfth Edition

中文版序
译者序
前言

第 19 章　泛型 ································ 1

19.1　引言 ································ 1
19.2　动机和优点 ························ 1
19.3　定义泛型类和接口 ················ 4
19.4　泛型方法 ···························· 5
19.5　示例学习：对一个对象数组
　　　进行排序 ···························· 7
19.6　原生类型和向后兼容 ············· 8
19.7　通配泛型 ·························· 10
19.8　泛型的擦除和限制 ·············· 12
19.9　示例学习：泛型矩阵类 ········ 15
关键术语 ····································· 19
本章小结 ····································· 19
测试题 ······································· 20
编程练习题 ································· 20

第 20 章　线性表、栈、队列和
　　　　　　优先队列 ···················· 22

20.1　引言 ······························ 22
20.2　集合 ······························ 23
20.3　迭代器 ··························· 26
20.4　使用 forEach 方法 ············· 27
20.5　线性表 ··························· 28
　　20.5.1　List 接口中的通用方法 ··· 28
　　20.5.2　ArrayList 和 LinkedList 类 ··· 30
20.6　Comparator 接口 ·············· 32
20.7　用于线性表和集合的静态方法 ··· 36
20.8　示例学习：弹球 ················ 40
20.9　向量类和栈类 ··················· 43
20.10　队列和优先队列 ·············· 44

　　20.10.1　Queue 接口 ················ 44
　　20.10.2　双端队列 Deque 和链表
　　　　　　LinkedList ··············· 45
20.11　示例学习：表达式求值 ······ 47
关键术语 ····································· 51
本章小结 ····································· 51
测试题 ······································· 52
编程练习题 ································· 52

第 21 章　规则集和映射 ················ 57

21.1　引言 ······························ 57
21.2　规则集 ··························· 57
　　21.2.1　HashSet ···················· 58
　　21.2.2　LinkedHashSet ············ 61
　　21.2.3　TreeSet ····················· 62
21.3　比较规则集和线性表的性能 ··· 65
21.4　示例学习：关键字计数 ······· 67
21.5　映射 ······························ 69
21.6　示例学习：单词的出现次数 ·· 73
21.7　单例与不可变的集合和映射 ·· 75
关键术语 ····································· 76
本章小结 ····································· 76
测试题 ······································· 77
编程练习题 ································· 77

第 22 章　开发高效算法 ················ 79

22.1　引言 ······························ 79
22.2　使用大 O 表示法来衡量算法效率 ··· 79
22.3　示例：确定大 O ················ 81
22.4　分析算法的时间复杂度 ······· 85
　　22.4.1　分析二分查找算法 ········ 85
　　22.4.2　分析选择排序算法 ········ 85
　　22.4.3　分析汉诺塔问题 ··········· 85
　　22.4.4　常用的递推关系 ··········· 86
　　22.4.5　比较常用的增长函数 ····· 86

22.5 使用动态编程求斐波那契数 …… 87
22.6 使用欧几里得算法求
最大公约数 …… 89
22.7 求素数的高效算法 …… 93
22.8 使用分治法寻找最近点对 …… 98
22.9 使用回溯法解决
八皇后问题 …… 101
22.10 计算几何：寻找凸包 …… 103
　22.10.1 卷包裹算法 …… 104
　22.10.2 格雷厄姆算法 …… 105
22.11 字符串匹配 …… 106
　22.11.1 暴力算法 …… 106
　22.11.2 Boyer-Moore 算法 …… 107
　22.11.3 Knuth-Morris-Pratt 算法 …… 109
关键术语 …… 112
本章小结 …… 112
测试题 …… 113
编程练习题 …… 113

第23章 排序 …… 120

23.1 引言 …… 120
23.2 插入排序 …… 121
23.3 冒泡排序 …… 123
23.4 归并排序 …… 125
23.5 快速排序 …… 128
23.6 堆排序 …… 132
　23.6.1 堆的存储 …… 133
　23.6.2 添加一个新结点 …… 133
　23.6.3 删除根结点 …… 134
　23.6.4 Heap 类 …… 135
　23.6.5 使用 Heap 类进行排序 …… 137
　23.6.6 堆排序的时间复杂度 …… 138
23.7 桶排序和基数排序 …… 139
23.8 外部排序 …… 141
　23.8.1 实现阶段 I …… 143
　23.8.2 实现阶段 II …… 143
　23.8.3 结合两个阶段 …… 145
　23.8.4 外部排序复杂度 …… 148
关键术语 …… 148
本章小结 …… 148

测试题 …… 149
编程练习题 …… 149

第24章 实现线性表、栈、队列和优先队列 …… 153

24.1 引言 …… 153
24.2 线性表的通用操作 …… 153
24.3 数组线性表 …… 156
24.4 链表 …… 163
　24.4.1 结点 …… 163
　24.4.2 MyLinkedList 类 …… 165
　24.4.3 实现 MyLinkedList …… 166
　24.4.4 MyArrayList 和
　　　　 MyLinkedList …… 174
　24.4.5 链表的变体 …… 174
24.5 栈和队列 …… 176
24.6 优先队列 …… 179
本章小结 …… 180
测试题 …… 181
编程练习题 …… 181

第25章 二叉搜索树 …… 183

25.1 引言 …… 183
25.2 二叉搜索树基础 …… 183
25.3 表示二叉搜索树 …… 184
25.4 查找一个元素 …… 185
25.5 在 BST 中插入一个元素 …… 185
25.6 树的遍历 …… 187
25.7 BST 类 …… 188
25.8 删除 BST 中的一个元素 …… 197
25.9 树的可视化和 MVC …… 202
25.10 迭代器 …… 205
25.11 示例学习：数据压缩 …… 207
关键术语 …… 211
本章小结 …… 211
测试题 …… 212
编程练习题 …… 212

第26章 AVL 树 …… 216

26.1 引言 …… 216

26.2 重新平衡树 ……………………… 217
26.3 为 AVL 树设计类 ………………… 219
26.4 重写 insert 方法 ………………… 220
26.5 实现旋转 ………………………… 221
26.6 实现 delete 方法 ………………… 221
26.7 AVLTree 类 ……………………… 222
26.8 测试 AVLTree 类 ………………… 227
26.9 AVL 树的时间复杂度分析 ……… 230
关键术语 ……………………………… 231
本章小结 ……………………………… 231
测试题 ………………………………… 231
编程练习题 …………………………… 231

第 27 章 散列 ……………………… 233
27.1 引言 ……………………………… 233
27.2 什么是散列 ……………………… 233
27.3 散列函数和散列码 ……………… 234
    27.3.1 基本数据类型的散列码 …… 234
    27.3.2 字符串的散列码 …………… 235
    27.3.3 压缩散列码 ………………… 235
27.4 使用开放地址法处理冲突 ……… 236
    27.4.1 线性探测法 ………………… 236
    27.4.2 二次探测法 ………………… 237
    27.4.3 双重散列法 ………………… 238
27.5 使用分离链接法处理冲突 ……… 240
27.6 装填因子和再散列 ……………… 241
27.7 使用散列实现映射 ……………… 242
27.8 使用散列实现规则集 …………… 250
关键术语 ……………………………… 256
本章小结 ……………………………… 257
测试题 ………………………………… 257
编程练习题 …………………………… 257

第 28 章 图及其应用 ……………… 259
28.1 引言 ……………………………… 259
28.2 基本的图术语 …………………… 260
28.3 表示图 …………………………… 262
    28.3.1 表示顶点 …………………… 262
    28.3.2 表示边：边数组 …………… 263
    28.3.3 表示边：Edge 对象 ……… 264

    28.3.4 表示边：邻接矩阵 ………… 264
    28.3.5 表示边：邻接线性表 ……… 265
28.4 图的建模 ………………………… 267
28.5 图的可视化 ……………………… 276
28.6 图的遍历 ………………………… 279
28.7 深度优先搜索 …………………… 280
    28.7.1 DFS 算法 …………………… 280
    28.7.2 DFS 的实现 ………………… 281
    28.7.3 DFS 的应用 ………………… 283
28.8 示例学习：连通圆问题 ………… 283
28.9 广度优先搜索 …………………… 285
    28.9.1 BFS 算法 …………………… 286
    28.9.2 BFS 的实现 ………………… 286
    28.9.3 BFS 的应用 ………………… 288
28.10 示例学习：9 枚硬币
     反面问题 ………………… 288
关键术语 ……………………………… 294
本章小结 ……………………………… 294
测试题 ………………………………… 294
编程练习题 …………………………… 294

第 29 章 加权图及其应用 ………… 299
29.1 引言 ……………………………… 299
29.2 加权图的表示 …………………… 300
    29.2.1 加权边的表示：边数组 …… 300
    29.2.2 加权邻接矩阵 ……………… 301
    29.2.3 邻接线性表 ………………… 301
29.3 WeightedGraph 类 ……………… 302
29.4 最小生成树 ……………………… 309
    29.4.1 最小生成树算法 …………… 310
    29.4.2 完善 Prim 的 MST 算法 …… 311
    29.4.3 MST 算法的实现 …………… 312
29.5 寻找最短路径 …………………… 315
29.6 示例学习：加权的 9 枚硬币
    反面问题 …………………… 323
关键术语 ……………………………… 326
本章小结 ……………………………… 326
测试题 ………………………………… 326
编程练习题 …………………………… 327

第 30 章　集合流的聚合操作 ……… 333

- 30.1　引言 ………………………… 333
- 30.2　流管道 ……………………… 334
  - 30.2.1　Stream.of、limit 和 forEach 方法 ……………… 336
  - 30.2.2　sorted 方法 ……………… 336
  - 30.2.3　filter 方法 ……………… 337
  - 30.2.4　max 和 min 方法 ………… 337
  - 30.2.5　anyMatch、allMatch 和 noneMatch 方法 …………… 337
  - 30.2.6　map、distinct 和 count 方法 ………………… 337
  - 30.2.7　findFirst、findAny 和 toArray 方法 ……………… 338
- 30.3　IntStream、LongStream 和 DoubleStream ………………… 339
- 30.4　并行流 ……………………… 341
- 30.5　使用 reduce 方法进行流的归约 … 344
- 30.6　使用 collect 方法进行流的归约 … 346
- 30.7　使用 groupingBy 收集器进行元素分组 ………………… 349
- 30.8　示例学习 …………………… 352
  - 30.8.1　示例学习：数字分析 ……… 352
  - 30.8.2　示例学习：计算字母的出现次数 ………………… 352
  - 30.8.3　示例学习：计算字符串中每个字母的出现次数 ……… 353
  - 30.8.4　示例学习：处理二维数组中的所有元素 ………… 354
  - 30.8.5　示例学习：得到目录大小 … 355
  - 30.8.6　示例学习：关键字计数 …… 356
  - 30.8.7　示例学习：单词出现次数 … 357
- 本章小结 ………………………… 358
- 测试题 …………………………… 359
- 编程练习题 ……………………… 359

附录 A　Java 关键字和保留字 ……… 360

附录 B　ASCII 字符集 ………………… 361

附录 C　操作符优先级表 ……………… 362

附录 D　Java 修饰符 …………………… 363

附录 E　特殊浮点值 …………………… 364

附录 F　数系 …………………………… 365

附录 G　位操作符 ……………………… 369

附录 H　正则表达式 …………………… 370

附录 I　枚举类型 ……………………… 376

附录 J　大 O、大 Ω 和大 Θ 表示法 …… 380

第 19 章

Introduction to Java Programming and Data Structures, Comprehensive Version, Twelfth Edition

泛 型

教学目标
- 描述泛型的优点（19.2 节）。
- 使用泛型类和接口（19.2 节）。
- 定义泛型类和接口（19.3 节）。
- 解释为什么泛型类型可以提高可靠性和可读性（19.3 节）。
- 定义并使用泛型方法和受限泛型类型（19.4 节）。
- 开发一个泛型排序方法来对任意一个 Comparable 对象数组排序（19.5 节）。
- 使用原生类型以向后兼容（19.6 节）。
- 解释为什么需要通配泛型类型（19.7 节）。
- 描述泛型类型消除，并列出一些由类型消除引起的泛型类型的限制和局限性（19.8 节）。
- 设计并实现泛型矩阵类（19.9 节）。

19.1 引言

☞ **要点提示**：泛型可以让我们在编译时而不是在运行时检测出错误。

你已经在第 11 章使用了一个泛型类 ArrayList，而第 13 章的泛型接口 Comparable 可以将类型参数化。有了这个功能，我们可以定义带泛型类型的类或方法，随后编译器会用具体的类型来替换它。例如，Java 定义了一个泛型类 ArrayList 用于存储泛型类型的元素。基于这个泛型类，可以创建用于保存字符串的 ArrayList 对象，以及保存数字的 ArrayList 对象。这里，字符串和数字是取代泛型类型的具体类型。

使用泛型的主要优点是能够在编译时而不是在运行时检测出错误。泛型类或方法允许指定可以和这些类或方法一起工作的对象类型。如果试图使用一个不相容的对象，编译器就会检测出这个错误。

本章介绍如何定义和使用泛型类、接口和方法，并且展示如何使用泛型来提高软件的可靠性和可读性。本章可以和第 13 章一起学习。

19.2 动机和优点

☞ **要点提示**：使用 Java 泛型的动机是在编译时就检测出错误。

从 JDK 1.5 开始，Java 允许定义泛型类、泛型接口和泛型方法。Java API 中的一些接口和类使用泛型也进行了修改。例如，在 JDK 1.5 之前，java.lang.Comparable 接口被定义为如图 19-1a 所示，但是，在 JDK 1.5 以后它被修改为如图 19-1b 所示。

这里的 <T> 表示形式泛型类型（formal generic type），随后可以用一个实际具体类型（actual concrete type）来替换它。替换泛型类型称为泛型实例化（generic instantiation）。按照惯例，使用像 E 或 T 这样的单个大写字母来表示形式泛型类型。

```
package java.lang;
public interface Comparable {
  public int compareTo(Object o)
}
```
a) JDK 1.5 之前

```
package java.lang;
public interface Comparable<T> {
  public int compareTo(T o)
}
```
b) JDK 1.5

图 19-1　从 JDK 1.5 开始，使用泛型类型重新定义了 java.lang.Comparable 接口

为了理解使用泛型的好处，我们来检查图 19-2 中的代码。图 19-2a 中的语句将 c 声明为一个类型为 Comparable 的引用变量，然后调用 compareTo 方法来比较 Date 对象和一个字符串。这样的代码可以成功编译，但是它会产生一个运行时错误，因为字符串不能与 Date 对象进行比较。

```
Comparable c = new Date();
System.out.println(c.compareTo("red"));
```
a) JDK 1.5 之前

```
Comparable<Date> c = new Date();
System.out.println(c.compareTo("red"));
```
b) JDK 1.5

图 19-2　新的泛型类型在编译时检测到可能的错误

图 19-2b 中的语句将 c 声明为一个类型为 Comparable<Date> 的引用变量，然后调用 compareTo 方法来比较 Date 对象和一个字符串。这样的代码会产生编译错误，因为传递给 compareTo 方法的参数必须是 Date 类型的。由于这个错误可以在编译时而不是运行时被检测到，因而泛型类型使程序更加可靠。

在 11.11 节中介绍过 ArrayList 类。从 JDK 1.5 开始，该类是一个泛型类。图 19-3 分别给出 ArrayList 类在 JDK 1.5 之前和从 JDK 1.5 开始的类图。

```
java.util.ArrayList
+ArrayList()
+add(o: Object): void
+add(index: int, o: Object): void
+clear(): void
+contains(o: Object): boolean
+get(index:int): Object
+indexOf(o: Object): int
+isEmpty(): boolean
+lastIndexOf(o: Object): int
+remove(o: Object): boolean
+size(): int
+remove(index: int): boolean
+set(index: int, o: Object): Object
```
a) JDK 1.5 之前的 ArrayList

```
java.util.ArrayList<E>
+ArrayList()
+add(o: E): void
+add(index: int, o: E): void
+clear(): void
+contains(o: Object): boolean
+get(index: int): E
+indexOf(o: Object): int
+isEmpty(): boolean
+lastIndexOf(o: Object): int
+remove(o: Object): boolean
+size(): int
+remove(index: int): boolean
+set(index: int, o: E): E
```
b) 从 JDK 1.5 开始的 ArrayList

图 19-3　从 JDK 1.5 开始，ArrayList 是一个泛型类

例如，下面的语句创建一个字符串的线性表：

```
ArrayList<String> list = new ArrayList<>();
```

现在，就只能向该线性表中添加字符串。例如，

```
list.add("Red");
```

如果试图向其中添加非字符串，就会产生编译错误。例如，下面的语句就是不合法的，

因为 list 只能包含字符串：

```
list.add(Integer.valueOf(1));
```

泛型类型必须是引用类型。不能使用 int、double 或 char 这样的基本类型来替换泛型类型。例如，下面的语句是错误的：

```
ArrayList<int> intList = new ArrayList<>();
```

为了给 int 值创建一个 ArrayList 对象，只能使用

```
ArrayList<Integer> intList = new ArrayList<>();
```

可以在 intList 中加入一个 int 值。例如，

```
intList.add(5);
```

Java 会自动地将 5 包装为 new Integer(5)。这个过程称为自动装箱（autoboxing），这在 10.8 节中介绍过。

无须类型转换就可以从一个已指定元素类型的线性表中获取一个值，因为编译器已经知道了这个元素的类型。例如，下面的语句创建了一个包含字符串的线性表，然后将字符串加入这个线性表，最后从这个线性表中获取该字符串。

```
1  ArrayList<String> list = new ArrayList<>();
2  list.add("Red");
3  list.add("White");
4  String s = list.get(0); // No casting is needed
```

在 JDK 1.5 之前，由于没有使用泛型，所以必须把返回值的类型转换为 String，如下所示：

```
String s = (String)(list.get(0)); // Casting needed prior to JDK 1.5
```

如果元素是包装类型，例如，Integer、Double 或 Character，那么可以直接将这个元素赋给一个基本类型的变量。这个过程称为自动拆箱（autounboxing），这在 10.8 节中介绍过。例如，请看下面的代码：

```
1  ArrayList<Double> list = new ArrayList<>();
2  list.add(5.5); // 5.5 is automatically converted to a Double object
3  list.add(3.0); // 3.0 is automatically converted to a Double object
4  Double doubleObject = list.get(0); // No casting is needed
5  double d = list.get(1); // Automatically converted to double
```

在第 2 行和第 3 行，5.5 和 3.0 自动转换为 Double 对象并添加到 list 中。在第 4 行，list 中的第一个元素被赋给一个 Double 变量。在此无须类型转换，因为 list 被声明为用于 Double 对象。在第 5 行，list 中的第二个元素被赋给一个 double 变量。list.get(1) 中的对象自动转换为一个基本类型的值。

复习题

19.2.1 图 a 和图 b 中有编译错误吗？

```
ArrayList dates = new ArrayList();
dates.add(new Date());
dates.add(new String());
```

```
ArrayList<Date> dates =
  new ArrayList<>();
dates.add(new Date());
dates.add(new String());
```

a) JDK 1.5 之前

b) JDK 1.5 之后

19.2.2 图 a 中有什么错误？图 b 中的代码正确吗？

```
ArrayList dates = new ArrayList();
dates.add(new Date());
Date date = dates.get(0);
```
a) JDK 1.5 之前

```
ArrayList<Date> dates =
  new ArrayList<>();
dates.add(new Date());
Date date = dates.get(0);
```
b) JDK 1.5 之后

19.2.3 使用泛型类型的优点是什么？

19.3 定义泛型类和接口

要点提示：可以为类或者接口定义泛型。当使用类来创建对象，或者使用类或接口来声明引用变量时，必须指定具体的类型。

我们修改 11.13 节中的栈类，使用泛型将元素类型通用化。新的名为 GenericStack 的栈类如图 19-4 所示，在程序清单 19-1 中实现它。

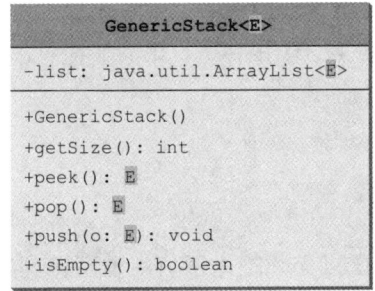

图 19-4 GenericStack 类封装了栈的存储，并提供对该栈的操作

程序清单 19-1 GenericStack.java

```
1  public class GenericStack<E> {
2    private java.util.ArrayList<E> list = new java.util.ArrayList<>();
3
4    public int getSize() {
5      return list.size();
6    }
7
8    public E peek() {
9      return list.get(getSize() - 1);
10   }
11
12   public void push(E o) {
13     list.add(o);
14   }
15
16   public E pop() {
17     E o = list.get(getSize() - 1);
18     list.remove(getSize() - 1);
19     return o;
20   }
21
22   public boolean isEmpty() {
23     return list.isEmpty();
24   }
25
26   @Override
```

```
27    public String toString() {
28      return "stack: " + list.toString();
29    }
30  }
```

下面的例子中，先创建一个存储字符串的栈，然后向这个栈添加三个字符串：

```
GenericStack<String> stack1 = new GenericStack<>();
stack1.push("London");
stack1.push("Paris");
stack1.push("Berlin");
```

该示例创建一个存储整数的栈，然后向这个栈添加三个整数：

```
GenericStack<Integer> stack2 = new GenericStack<>();
stack2.push(1); // autoboxing 1 to an Integer object
stack2.push(2);
stack2.push(3);
```

可以不使用泛型，而将元素类型设置为 Object，也可以容纳任何对象类型。但是，使用泛型能够提高软件的可靠性和可读性，因为某些错误能在编译时而不是运行时被检测到。例如，由于 stack1 被声明为 GenericStack<String>，所以，只可以将字符串添加到这个栈中。如果试图向 stack1 中添加整数就会产生编译错误。

> **警告**：为了创建一个字符串堆栈，可以使用 new GenericStack<String>() 或 new GenericStack<>()。这可能会误导你认为 GenericStack 的构造方法应该定义为
>
> **public** GenericStack<E>()
>
> 这是错误的。它应该被定义为
>
> **public** GenericStack()

> **注意**：有时候，泛型类可能会有多个参数。在这种情况下，应将所有参数一起放在尖括号中，并用逗号分隔开，比如 <E1,E2,E3>。

> **注意**：可以定义一个类或接口作为泛型类或者泛型接口的子类型。例如，在 Java API 中，java.lang.String 类被定义为实现 Comparable 接口，如下所示：
>
> **public class** String **implements** Comparable<String>

✓ 复习题

19.3.1 Java API 中，java.lang.Comparable 的泛型定义是什么？

19.3.2 既然使用 new ArrayList<String>() 创建了字符串的 ArrayList 的一个实例，那么应该将 ArrayList 类的构造方法定义为如下所示吗？

    public ArrayList<E>()

19.3.3 泛型类可以拥有多个泛型参数吗？

19.3.4 如何在类中声明一个泛型类型？

19.4 泛型方法

> **要点提示**：可以为静态方法定义泛型类型。

可以定义泛型接口（例如，图 19-1b 中的 Comparable 接口）和泛型类（例如，程序清单 19-1 中的 GenericStack 类），也可以使用泛型类型来定义泛型方法。例如，程序清单 19-2

定义了一个泛型方法 print（第 10～14 行）来打印一个对象数组。第 6 行传递一个整数对象的数组来调用泛型方法 print。第 7 行用字符串数组调用 print。

程序清单 19-2 GenericMethodDemo.java

```java
 1  public class GenericMethodDemo {
 2    public static void main(String[] args ) {
 3      Integer[] integers = {1, 2, 3, 4, 5};
 4      String[] strings = {"London", "Paris", "New York", "Austin"};
 5
 6      GenericMethodDemo.<Integer>print(integers);
 7      GenericMethodDemo.<String>print(strings);
 8    }
 9
10    public static <E> void print(E[] list) {
11      for (int i = 0; i < list.length; i++)
12        System.out.print(list[i] + " ");
13      System.out.println();
14    }
15  }
```

要声明泛型方法，将泛型类型 <E> 置于方法头中紧跟在关键字 static 之后。例如，

```
public static <E> void print(E[] list)
```

要调用泛型方法，需要将实际类型放在尖括号内作为方法名的前缀。例如，

```
GenericMethodDemo.<Integer>print(integers);
GenericMethodDemo.<String>print(strings);
```

或者如下简单调用：

```
print(integers);
print(strings);
```

在后一种情况中，没有明确指定实际类型。编译器自动发现实际类型。

可以将泛型指定为另外一种类型的子类型。这样的泛型类型称为受限的（bounded）。例如，程序清单 19-3 修改了程序清单 13-4 中的 equalArea 方法，以测试两个几何对象是否具有相同的面积。受限的泛型类型 <E extends GeometricObject>（第 10 行）将 E 指定为 GeometricObject 的泛型子类型。必须传递两个 GeometricObject 的实例来调用 equalArea。

程序清单 19-3 BoundedTypeDemo.java

```java
 1  public class BoundedTypeDemo {
 2    public static void main(String[] args ) {
 3      Rectangle rectangle = new Rectangle(2, 2);
 4      Circle circle = new Circle(2);
 5
 6      System.out.println("Same area? " +
 7        equalArea(rectangle, circle));
 8    }
 9
10    public static <E extends GeometricObject> boolean equalArea(
11        E object1, E object2) {
12      return object1.getArea() == object2.getArea();
13    }
14  }
```

☞ 注意：非受限泛型类型 <E> 等同于 <E extends Object>。

☞ 注意：为一个类定义泛型类型，需要将泛型类型放在类名之后，例如 GenericStack<E>。为

一个方法定义泛型类型，要将泛型类型放在方法返回类型之前，例如 <E> void max (E o1, E o2)。

✓ **复习题**

19.4.1 如何声明一个泛型方法？如何调用一个泛型方法？
19.4.2 什么是受限泛型类型？

19.5 示例学习：对一个对象数组进行排序

要点提示：可以开发一个泛型方法，对一个 Comparable 对象数组进行排序。

本节提供一个泛型方法，对一个 Comparable 对象数组进行排序。这些对象是 Comparable 接口的实例，它们使用 compareTo 方法进行比较。为了测试该方法，程序对一个整数数组、一个双精度数字数组、一个字符数组以及一个字符串数组分别进行了排序。程序如程序清单 19-4 所示。

程序清单 19-4 GenericSort.java

```java
public class GenericSort {
  public static void main(String[] args) {
    // Create an Integer array
    Integer[] intArray = {Integer.valueOf(2), Integer.valueOf(4),
      Integer.valueOf(3)};

    // Create a Double array
    Double[] doubleArray = {Double.valueOf(3.4), Double.valueOf(1.3),
      Double.valueOf(-22.1)};

    // Create a Character array
    Character[] charArray = {Character.valueOf('a'),
      Character.valueOf('J'), Character.valueOf('r')};

    // Create a String array
    String[] stringArray = {"Tom", "Susan", "Kim"};

    // Sort the arrays
    sort(intArray);
    sort(doubleArray);
    sort(charArray);
    sort(stringArray);

    // Display the sorted arrays
    System.out.print("Sorted Integer objects: ");
    printList(intArray);
    System.out.print("Sorted Double objects: ");
    printList(doubleArray);
    System.out.print("Sorted Character objects: ");
    printList(charArray);
    System.out.print("Sorted String objects: ");
    printList(stringArray);
  }

  /** Sort an array of comparable objects */
  public static <E extends Comparable<E>> void sort(E[] list) {
    E currentMin;
    int currentMinIndex;

    for (int i = 0; i < list.length - 1; i++) {
      // Find the minimum in the list[i+1..list.length-2]
      currentMin = list[i];
      currentMinIndex = i;
```

```
44        for (int j = i + 1; j < list.length; j++) {
45          if (currentMin.compareTo(list[j]) > 0) {
46            currentMin = list[j];
47            currentMinIndex = j;
48          }
49        }
50
51        // Swap list[i] with list[currentMinIndex] if necessary;
52        if (currentMinIndex != i) {
53          list[currentMinIndex] = list[i];
54          list[i] = currentMin;
55        }
56      }
57    }
58
59    /** Print an array of objects */
60    public static void printList(Object[] list) {
61      for (int i = 0; i < list.length; i++)
62        System.out.print(list[i] + " ");
63      System.out.println();
64    }
65  }
66
```

```
Sorted Integer objects: 2 3 4
Sorted Double objects: -22.1 1.3 3.4
Sorted Character objects: J a r
Sorted String objects: Kim Susan Tom
```

sort 方法的算法和程序清单 7-8 中的一样。那个程序中的 sort 方法对一个 double 数值的数组进行了排序。本例中的 sort 方法可以对任意对象类型的数组进行排序，只要这些对象是 Comparable 接口的实例。泛型类型定义为 <E extends Comparable <E>>（第 36 行）。这具有两个含义：首先，它指定 E 是 Comparable 的子类型；其次，它还指定进行比较的元素是 E 类型的。

sort 方法使用 compareTo 方法来确定数组中对象的排序（第 46 行）。Integer、Double、Character 以及 String 实现了 Comparable，因此这些类的对象可以使用 compareTo 方法进行比较。该程序创建一个 Integer 对象数组、一个 Double 对象数组、一个 Character 对象数组以及一个 String 对象数组（第 4～16 行），然后调用 sort 方法来对这些数组进行排序（第 19～22 行）。

✓ 复习题

19.5.1 给定 int[] list = {1, 2, -1}，可以使用程序清单 19-4 中的 sort 方法调用 sort(list) 吗？

19.5.2 给定 int[] list = {new Integer(1), new Integer(2), new Integer(-1)}，可以使用程序清单 19-4 中的 sort 方法调用 sort(list) 吗？

19.6 原生类型和向后兼容

🞂 **要点提示**：没有指定具体类型的泛型类和泛型接口被称为原生类型，用于让早期的 Java 版本向后兼容。

可以使用泛型类而不指定具体类型，如下所示：

```
GenericStack stack = new GenericStack(); // raw type
```

它大体等价于下面的语句：

```
GenericStack<Object> stack = new GenericStack<Object>();
```

像 GenericStack 和 ArrayList 这样不带类型参数的泛型类称为原生类型（raw type）。使用原生类型可以让 Java 的早期版本向后兼容。例如，从 JDK 1.5 开始，在 java.lang.Comparable 中使用了泛型类型，但许多代码仍然使用原生类型 Comparable，如程序清单 19-5 所示。

程序清单 19-5 Max.java

```
1  public class Max {
2    /** Return the maximum of two objects */
3    public static Comparable max(Comparable o1, Comparable o2) {
4      if (o1.compareTo(o2) > 0)
5        return o1;
6      else
7        return o2;
8    }
9  }
```

Comparable o1 和 Comparable o2 都是原生类型声明。但是小心：原生类型是不安全的。例如，你可能会使用下面的语句调用 max 方法：

```
Max.max("Welcome", 23); // 23 is autoboxed into an Integer object
```

这会引起一个运行时错误，因为不能将字符串与整数对象进行比较。如果在编译时使用了选项 -Xlint:unchecked，Java 编译器就会对第 3 行显示一条警告，如图 19-5 所示。

```
Command Prompt                                                    —  □  ×
c:\book>javac -Xlint:unchecked Max.java
Max.java:4: warning: [unchecked] unchecked call to compareTo(T) as a member of the raw type Comparable
    if (o1.compareTo(o2) > 0)
          ^
  where T is a type-variable:
    T extends Object declared in interface Comparable
1 warning

c:\book>
```

图 19-5 使用编译器选项 -Xlint:unchecked 会显示一条免检警告（来源：Oracle 或其附属公司版权所有 ©1995 ~ 2016，经授权使用）

编写 max 方法的更好方式是使用泛型类型，如程序清单 19-6 所示。

程序清单 19-6 MaxUsingGenericType.java

```
1  public class MaxUsingGenericType {
2    /** Return the maximum of two objects */
3    public static <E extends Comparable<E>> E max(E o1, E o2) {
4      if (o1.compareTo(o2) > 0)
5        return o1;
6      else
7        return o2;
8    }
9  }
```

如果使用下面的语句调用 max 方法：

```
// 23 is autoboxed into an Integer object
MaxUsingGenericType.max("Welcome", 23);
```

就会显示一个编译错误，因为 MaxUsingGenericType 中的 max 方法的两个参数必须是相同的类型（例如，两个字符串或两个整数对象）。此外，类型 E 必须是 Comparable<E> 的子类型。

下面的代码是另外一个例子，可以在第 1 行声明一个原生类型 stack，在第 2 行将 new GenericStack<String> 赋给它，然后在第 3 行和第 4 行将一个字符串和一个整数对象压入栈中。

```
1  GenericStack stack;
2  stack = new GenericStack<String>();
3  stack.push("Welcome to Java");
4  stack.push(Integer.valueOf(2));
```

然而，第 4 行是不安全的，因为该栈原本是用于存储字符串的，但是一个 Integer 对象被添加到该栈中。第 3 行应该是可行的，但是编译器会在第 3 行和第 4 行都显示警告，因为它不能理解程序的语义。编译器所知道的就是该栈是一个原生类型，并且在执行某些操作时会不安全。因此，它会显示警告以提醒潜在的问题。

☞ 提示：由于原生类型是不安全的，所以，本书后面不再使用原生类型。

✔ 复习题

19.6.1 什么是原生类型？为什么原生类型是不安全的？为什么 Java 中允许使用原生类型？

19.6.2 可以使用什么样的语法来声明一个使用原生类型的 ArrayList 引用变量，以及将一个原生类型的 ArrayList 对象赋值给该变量？

19.7 通配泛型

☞ 要点提示：可以使用非受限通配（unbounded wildcard）、受限通配（bounded wildcard）或者下限通配（lower-bound wildcard）来对一个泛型类型指定范围。

什么是通配泛型？为什么需要通配泛型？程序清单 19-7 给出了一个例子，以展示为什么需要通配泛型。该例子定义了一个泛型方法 max，该方法可以找出数字栈中的最大数（第 12～22 行）。main 方法创建了一个整数对象栈，然后向该栈添加三个整数，最后调用 max 方法找出该栈中的最大数字。

程序清单 19-7 WildCardNeedDemo.java

```
1   public class WildCardNeedDemo {
2     public static void main(String[] args ) {
3       GenericStack<Integer> intStack = new GenericStack<>();
4       intStack.push(1); // 1 is autoboxed into an Integer object
5       intStack.push(2);
6       intStack.push(-2);
7
8       System.out.print("The max number is " + max(intStack));
9     }
10
11    /** Find the maximum in a stack of numbers */
12    public static double max(GenericStack<Number> stack) {
13      double max = stack.pop().doubleValue(); // Initialize max
14
15      while (!stack.isEmpty()) {
16        double value = stack.pop().doubleValue();
17        if (value > max)
18          max = value;
19      }
20
21      return max;
```

程序清单 19-7 中的程序在第 8 行会出现编译错误，因为 intStack 不是 GenericStack<Number> 的实例。因此，不能调用 max(intStack)。

尽管事实上 Integer 是 Number 的子类型，但 GenericStack<Integer> 并不是 GenericStack<Number> 的子类型。为了避免这个问题，可以使用通配泛型类型。通配泛型类型有三种形式——?、? extends T 或者 ? super T，其中 T 是泛型类型。

第一种形式 ? 称为非受限通配，它和 ? extends Object 是一样的。第二种形式 ? extends T 称为受限通配，表示 T 或 T 的一个子类型。第三种形式 ? super T 称为下限通配，表示 T 或 T 的一个父类型。

可以使用下面的语句替换程序清单 19-7 中的第 12 行，从而修复上面的错误：

```
public static double max(GenericStack<? extends Number> stack) {
```

<? extends Number> 是一个表示 Number 或 Number 的子类型的通配类型。因此，调用 max(new GenericStack<Integer>()) 或 max(new GenericStack<Double>()) 都是合法的。

程序清单 19-8 给出一个例子，它在 print 方法中使用 ? 通配符打印栈中的对象以及清空栈。<?> 是一个通配符，表示任何一种对象类型。它等价于 <? extends Object>。如果用 GenericStack<Object> 替换 GenericStack<?>，会发生什么情况呢？这样调用 print(intStack) 会出错，因为 intStack 不是 GenericStack<Object> 的实例。注意，尽管 Integer 是 Object 的一个子类型，但是 GenericStack<Integer> 并不是 GenericStack<Object> 的子类型。

程序清单 19-8 AnyWildCardDemo.java

```
1  public class AnyWildCardDemo {
2    public static void main(String[] args) {
3      GenericStack<Integer> intStack = new GenericStack<>();
4      intStack.push(1); // 1 is autoboxed into an Integer object
5      intStack.push(2);
6      intStack.push(-2);
7
8      print(intStack);
9    }
10
11   /** Prints objects and empties the stack */
12   public static void print(GenericStack<?> stack) {
13     while (!stack.isEmpty()) {
14       System.out.print(stack.pop() + " ");
15     }
16   }
17 }
```

什么时候需要 <? super T> 通配符呢？考虑程序清单 19-9 中的例子。该示例创建了一个字符串栈 stack1（第 3 行）和一个对象栈 stack2（第 4 行），然后调用 add(stack1,stack2)（第 8 行）将 stack1 中的字符串添加到 stack2 中。在第 13 行使用 GenericStack<? super T> 来声明栈 stack2。如果用 <T> 代替 <? super T>，那么在第 8 行的 add(stack1,stack2) 上就会产生一个编译错误，因为 stack1 的类型为 GenericStack<String>，而 stack2 的类型为 GenericStack<Object>。<? super T> 表示类型 T 或 T 的父类型。Object 是 String 的父类型。

程序清单 19-9 SuperWildCardDemo.java

```
1  public class SuperWildCardDemo {
2    public static void main(String[] args) {
3      GenericStack<String> stack1 = new GenericStack<>();
4      GenericStack<Object> stack2 = new GenericStack<>();
5      stack2.push("Java");
6      stack2.push(2);
7      stack1.push("Sun");
8      add(stack1, stack2);
9      AnyWildCardDemo.print(stack2);
10    }
11
12    public static <T> void add(GenericStack<T> stack1,
13        GenericStack<? super T> stack2) {
14      while (!stack1.isEmpty())
15        stack2.push(stack1.pop());
16    }
17  }
```

如果第 12 ~ 13 行的方法头如下修改，该程序也可以运行：

```
public static <T> void add(GenericStack<? extends T> stack1,
    GenericStack<T> stack2)
```

涉及泛型类型和通配类型的继承关系在图 19-6 中进行了总结。在该图中，A 和 B 表示类或者接口，而 E 是泛型类型参数。

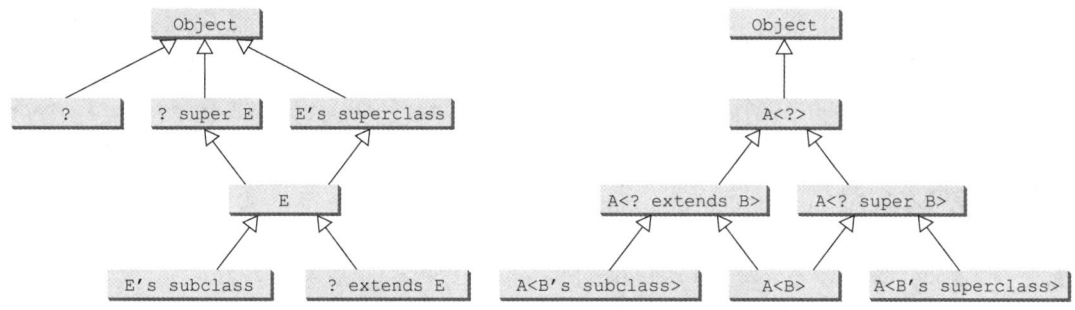

图 19-6 泛型类型和通配类型之间的关系

✓ 复习题

19.7.1 GenericStack 等同于 GenericStack<Object> 吗？

19.7.2 什么是非受限通配、受限通配、下限通配？

19.7.3 如果将程序清单 19-9 中的第 12 ~ 13 行改为如下所示，会发生什么情况？

```
public static <T> void add(GenericStack<T> stack1,
    GenericStack<T> stack2)
```

19.7.4 如果将程序清单 19-9 中的第 12 ~ 13 行改为如下所示，会发生什么情况？

```
public static <T> void add(GenericStack<? extends T> stack1,
    GenericStack<T> stack2)
```

19.8 泛型的擦除和限制

🔑 **要点提示**：泛型相关信息可被编译器使用，但这些信息在运行时是不可用的，这被称为类型擦除。

泛型是使用一种称为类型擦除（type erasure）的方法来实现的。编译器使用泛型类型信息来编译代码，但是随后会擦除它。因此，泛型信息在运行时是不可用的。这种方法可以使泛型代码向后兼容使用原生类型的遗留代码。

泛型存在于编译时。一旦编译器确认泛型类型是安全使用的，就会将它转换为原生类型。例如，编译器会检查以下图 a 的代码中泛型是否被正确使用，然后将它翻译成如图 b 所示的运行时使用的等价代码。图 b 中的代码使用的是原生类型。

```
ArrayList<String> list = new ArrayList<>();
list.add("Oklahoma");
String state = list.get(0);
```
a)

```
ArrayList list = new ArrayList();
list.add("Oklahoma");
String state = (String)(list.get(0));
```
b)

当编译泛型类、接口和方法时，编译器用 Object 类型代替泛型类型。例如，编译器会将以下图 a 中的方法转换为图 b 中的方法。

```
public static <E> void print(E[] list) {
  for (int i = 0; i < list.length; i++)
    System.out.print(list[i] + " ");
  System.out.println();
}
```
a)

```
public static void print(Object[] list) {
  for (int i = 0; i < list.length; i++)
    System.out.print(list[i] + " ");
  System.out.println();
}
```
b)

如果一个泛型类型是受限的，那么编译器就会用该受限类型来替换它。例如，编译器会将以下图 a 中的方法转换为图 b 中的方法。

```
public static <E extends GeometricObject>
    boolean equalArea(
    E object1,
    E object2) {
  return object1.getArea() ==
    object2.getArea();
}
```
a)

```
public static
    boolean equalArea(
    GeometricObject object1,
    GeometricObject object2) {
  return object1.getArea() ==
    object2.getArea();
}
```
b)

值得注意的是，不管实际的具体类型是什么，泛型类是被它的所有实例所共享的。假定按如下方式创建 list1 和 list2：

```
ArrayList<String> list1 = new ArrayList<>();
ArrayList<Integer> list2 = new ArrayList<>();
```

尽管在编译时 ArrayList<String> 和 ArrayList<Integer> 是两种类型，但是，在运行时只有一个 ArrayList 类会被加载到 JVM 中。list1 和 list2 都是 ArrayList 的实例，因此，下面两条语句的执行结果都为 true：

```
System.out.println(list1 instanceof ArrayList);
System.out.println(list2 instanceof ArrayList);
```

然而，表达式 list1 instanceof ArrayList<String> 是错误的。由于 ArrayList<String> 并没有在 JVM 中存储为单独一个类，所以，在运行时使用它是毫无意义的。

由于泛型类型在运行时被擦除，因此，对于如何使用泛型类型是有一定限制的。下面是其中的一些限制。

限制 1：不能使用 new E()

不能使用泛型类型参数创建实例。例如，下面的语句是错误的：

```
E object = new E();
```

出错的原因是运行时执行的是 new E()，但是运行时泛型类型 E 是不可用的。

限制 2：不能使用 new E[]

不能使用泛型类型参数创建数组。例如，下面的语句是错误的：

```
E[] elements = new E[capacity];
```

可以通过创建一个 Object 类型的数组，然后将它的类型转换为 E[] 来规避这个限制，如下所示：

```
E[] elements = (E[])new Object[capacity];
```

但是，类型转换到 (E[]) 会导致一个免检的编译警告。该警告会出现是因为编译器无法确保在运行时类型转换是否能成功。例如，如果 E 是 String，而 new Object[] 是 Integer 对象的数组，那么 (String[])(new Object[]) 将会导致 ClassCastException 异常。这种类型的编译警告是使用 Java 泛型的不足之处，也是无法避免的。

使用泛型类创建泛型数组也是不允许的。例如，下面的代码是错误的：

```
ArrayList<String>[] list = new ArrayList<String>[10];
```

可以使用下面的代码来规避这种限制：

```
ArrayList<String>[] list = (ArrayList<String>[])new
  ArrayList[10];
```

然而，你依然会得到一个编译警告。

限制 3：在静态上下文中不允许类的参数是泛型类型

由于泛型类的所有实例都有相同的运行时类，所以泛型类的静态变量和方法是被它的所有实例所共享的。因此，在静态方法、数据域或者初始化语句中，为类引用泛型类型参数是非法的。例如，下面的代码是非法的：

```
public class Test<E> {
  public static void m(E o1) {  // Illegal
  }
  public static E o1; // Illegal

  static {
    E o2; // Illegal
  }
}
```

限制 4：异常类不能是泛型的

泛型类可能没有继承 java.lang.Throwable，因此，下面的类声明是非法的：

```
public class MyException<T> extends Exception {
}
```

为什么呢？因为如果允许这样做，就应为 MyException<T> 添加一个 catch 子句，如下所示：

```
try {
  ...
}
catch (MyException<T> ex) {
  ...
}
```

JVM 必须检查这个从 try 子句中抛出的异常以确定它是否与 catch 子句中指定的类型

匹配。但这是不可能的，因为在运行时是没有该类型信息的。

✓ 复习题

19.8.1 什么是擦除？为什么使用擦除来实现 Java 泛型？
19.8.2 如果你的程序使用了 ArrayList<String> 和 ArrayList<Date>，JVM 会对它们都加载吗？
19.8.3 可以使用 new E() 为泛型类型 E 创建一个实例吗？为什么？
19.8.4 使用泛型类作为参数的方法可以是静态的吗？为什么？
19.8.5 可以定义一个自定义的泛型异常类吗？为什么？

19.9 示例学习：泛型矩阵类

o━ 要点提示：本节给出一个示例学习，使用泛型类型来设计用于矩阵运算的类。

除了元素类型不同以外，所有矩阵的加法和乘法操作都是类似的。因此，可以设计一个父类，描述所有类型的矩阵共享的通用操作，而不管它们的元素类型是什么，还可以创建若干个适用于指定矩阵类型的子类。这里的示例学习给出了两种类型 int 和 Rational 的实现。对于 int 类型而言，应该用包装类 Integer 将一个 int 类型的值包装到一个对象中，从而对象被传递给方法进行操作。

该类的类图如图 19-7 所示。addMatrix 和 multiplyMatrix 方法将两个泛型类型的矩阵 E[][] 进行相加和相乘。静态方法 printResult 显示矩阵、操作符以及它们的结果。方法 add、multiply 和 zero 都是抽象的，因为它们的实现依赖于数组元素的特定类型。例如，zero() 方法对于 Integer 类型返回 0，而对于 Rational 类型返回 0/1。这些方法将会在指定了矩阵元素类型的子类中实现。

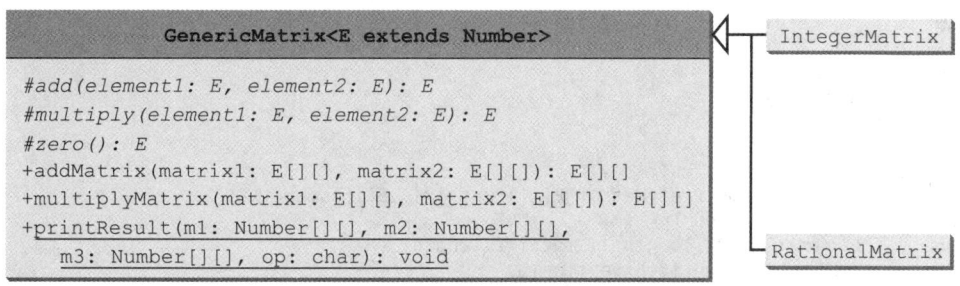

图 19-7 GenericMatrix 类是 IntegerMatrix 和 RationalMatrix 的抽象父类

IntegerMatrix 和 RationalMatrix 是 GenericMatrix 的具体子类。这两个类实现了在 GenericMatrix 类中定义的 add、multiply 和 zero 方法。

程序清单 19-10 实现了 GenericMatrix 类。第 1 行的 <E extends Number> 指明该泛型类型是 Number 的子类型。三个抽象方法 add、multiply 和 zero 在第 3、6 和 9 行定义。这些方法是抽象的，因为在不知道元素的确切类型时是无法实现它们的。addMatrix 方法（第 12～30 行）和 multiplyMatrix 方法（第 33～57 行）实现了两个矩阵的相加和相乘。所有这些方法都必须是非静态的，因为它们使用的是泛型类型 E 来表示类。printResult 方法（第 60～84 行）是静态的，因为它没有绑定到特定的实例。

矩阵元素的类型是 Number 的泛型子类型。这样就可以使用任意 Number 子类的对象，只要在子类中实现了抽象方法 add、multiply 和 zero 即可。

addMatrix 和 multiplyMatrix 方法（第 12 ～ 57 行）是具体的方法。只要在子类中实现了 add、multiply 和 zero 方法，就可以使用它们。

addMatrix 和 multiplyMatrix 方法在进行操作之前检查矩阵的边界。如果两个矩阵的边界不兼容，那么程序会抛出一个异常（第 16 和 36 行）。

程序清单 19-10 GenericMatrix.java

```java
 1  public abstract class GenericMatrix<E extends Number> {
 2    /** Abstract method for adding two elements of the matrices */
 3    protected abstract E add(E o1, E o2);
 4
 5    /** Abstract method for multiplying two elements of the matrices */
 6    protected abstract E multiply(E o1, E o2);
 7
 8    /** Abstract method for defining zero for the matrix element */
 9    protected abstract E zero();
10
11    /** Add two matrices */
12    public E[][] addMatrix(E[][] matrix1, E[][] matrix2) {
13      // Check bounds of the two matrices
14      if ((matrix1.length != matrix2.length) ||
15          (matrix1[0].length != matrix2[0].length)) {
16        throw new RuntimeException(
17          "The matrices do not have the same size");
18      }
19
20      E[][] result =
21        (E[][])new Number[matrix1.length][matrix1[0].length];
22
23      // Perform addition
24      for (int i = 0; i < result.length; i++)
25        for (int j = 0; j < result[i].length; j++) {
26          result[i][j] = add(matrix1[i][j], matrix2[i][j]);
27        }
28
29      return result;
30    }
31
32    /** Multiply two matrices */
33    public E[][] multiplyMatrix(E[][] matrix1, E[][] matrix2) {
34      // Check bounds
35      if (matrix1[0].length != matrix2.length) {
36        throw new RuntimeException(
37          "The matrices do not have compatible size");
38      }
39
40      // Create result matrix
41      E[][] result =
42        (E[][])new Number[matrix1.length][matrix2[0].length];
43
44      // Perform multiplication of two matrices
45      for (int i = 0; i < result.length; i++) {
46        for (int j = 0; j < result[0].length; j++) {
47          result[i][j] = zero();
48
49          for (int k = 0; k < matrix1[0].length; k++) {
50            result[i][j] = add(result[i][j],
51              multiply(matrix1[i][k], matrix2[k][j]));
52          }
53        }
54      }
55
56      return result;
57    }
```

```
58
59     /** Print matrices, the operator, and their operation result */
60     public static void printResult(
61         Number[][] m1, Number[][] m2, Number[][] m3, char op) {
62       for (int i = 0; i < m1.length; i++) {
63         for (int j = 0; j < m1[0].length; j++)
64           System.out.print(" " + m1[i][j]);
65
66         if (i == m1.length / 2)
67           System.out.print("  " + op + "  ");
68         else
69           System.out.print("      ");
70
71         for (int j = 0; j < m2.length; j++)
72           System.out.print(" " + m2[i][j]);
73
74         if (i == m1.length / 2)
75           System.out.print(" = ");
76         else
77           System.out.print("     ");
78
79         for (int j = 0; j < m3.length; j++)
80           System.out.print(m3[i][j] + " ");
81
82         System.out.println();
83       }
84     }
85   }
```

程序清单19-11实现了IntegerMatrix类。在第1行该类继承了GenericMatrix <Integer>。在泛型实例化之后，GenericMatrix<Integer>中的add方法就成为Integer add(Integer o1, Integer o2)。该程序为Integer对象实现了add、multiply和zero方法。因为这些方法只能被addMatrix和multiplyMatrix方法调用，所以，它们仍然是protected的。

程序清单19-11 IntegerMatrix.java

```
1   public class IntegerMatrix extends GenericMatrix<Integer> {
2     @Override /** Add two integers */
3     protected Integer add(Integer o1, Integer o2) {
4       return o1 + o2;
5     }
6
7     @Override /** Multiply two integers */
8     protected Integer multiply(Integer o1, Integer o2) {
9       return o1 * o2;
10    }
11
12    @Override /** Specify zero for an integer */
13    protected Integer zero() {
14      return 0;
15    }
16  }
```

设计模式提示：GenericMatrix类中的代码应用了模板方法模式，该模式使用抽象方法的方式来实现方法，其具体实现将在子类中提供。在GenericMatrix中，addMatrix和multiplyMatrix方法使用抽象的add、multiply以及zero方法实现，它们的具体实现在子类IntegerMatrix和RationalMatrix中给出。

程序清单19-12实现了RationalMatrix类。Rational类在程序清单13-13中介绍过。Rational是Number的子类型。在第1行RationalMatrix类继承了GenericMatrix<Rational>。在泛型实例化之后，GenericMatrix<Rational>中的add方法现在为Rational add(Rational r1,Rational r2)。该程序为 Rational 对象实现了 add、multiply 和 zero 方法。因为这些方法只能被 addMatrix 和 multiplyMatrix 方法调用，所以它们仍然是 protected 的。

程序清单 19-12 RationalMatrix.java

```java
1   public class RationalMatrix extends GenericMatrix<Rational> {
2     @Override /** Add two rational numbers */
3     protected Rational add(Rational r1, Rational r2) {
4       return r1.add(r2);
5     }
6
7     @Override /** Multiply two rational numbers */
8     protected Rational multiply(Rational r1, Rational r2) {
9       return r1.multiply(r2);
10    }
11
12    @Override /** Specify zero for a Rational number */
13    protected Rational zero() {
14      return new Rational(0, 1);
15    }
16  }
```

程序清单 19-13 给出了一个程序，该程序创建两个 Integer 矩阵（第 4～5 行）和一个 IntegerMatrix 对象（第 8 行），然后在第 12 行和第 16 行对这两个矩阵进行相加和相乘操作。

程序清单 19-13 TestIntegerMatrix.java

```java
1   public class TestIntegerMatrix {
2     public static void main(String[] args) {
3       // Create Integer arrays m1, m2
4       Integer[][] m1 = new Integer[][]{{1, 2, 3}, {4, 5, 6}, {1, 1, 1}};
5       Integer[][] m2 = new Integer[][]{{1, 1, 1}, {2, 2, 2}, {0, 0, 0}};
6
7       // Create an instance of IntegerMatrix
8       IntegerMatrix integerMatrix = new IntegerMatrix();
9
10      System.out.println("\nm1 + m2 is ");
11      GenericMatrix.printResult(
12        m1, m2, integerMatrix.addMatrix(m1, m2), '+');
13
14      System.out.println("\nm1 * m2 is ");
15      GenericMatrix.printResult(
16        m1, m2, integerMatrix.multiplyMatrix(m1, m2), '*');
17    }
18  }
```

```
m1 + m2 is
 1 2 3       1 1 1       2 3 4
 4 5 6   +   2 2 2   =   6 7 8
 1 1 1       0 0 0       1 1 1
m1 * m2 is
 1 2 3       1 1 1       5  5  5
 4 5 6   *   2 2 2   =   14 14 14
 1 1 1       0 0 0       3  3  3
```

程序清单 19-14 给出了一个程序，该程序创建两个 Rational 矩阵（第 4～10 行）和一个 RationalMatrix 对象（第 13 行），然后在第 17 行和第 19 行对这两个矩阵进行相加和相乘操作。

程序清单 19-14 TestRationalMatrix.java

```java
1   public class TestRationalMatrix {
2     public static void main(String[] args) {
3       // Create two Rational arrays m1 and m2
```

```
 4      Rational[][] m1 = new Rational[3][3];
 5      Rational[][] m2 = new Rational[3][3];
 6      for (int i = 0; i < m1.length; i++)
 7        for (int j = 0; j < m1[0].length; j++) {
 8          m1[i][j] = new Rational(i + 1, j + 5);
 9          m2[i][j] = new Rational(i + 1, j + 6);
10        }
11
12      // Create an instance of RationalMatrix
13      RationalMatrix rationalMatrix = new RationalMatrix();
14
15      System.out.println("\nm1 + m2 is ");
16      GenericMatrix.printResult(
17        m1, m2, rationalMatrix.addMatrix(m1, m2), '+');
18
19      System.out.println("\nm1 * m2 is ");
20      GenericMatrix.printResult(
21        m1, m2, rationalMatrix.multiplyMatrix(m1, m2), '*');
22    }
23 }
```

```
m1 + m2 is
 1/5 1/6 1/7     1/6 1/7 1/8        11/30 13/42 15/56
 2/5 1/3 2/7  +  1/3 2/7 1/4    =   11/15 13/21 15/28
 3/5 1/2 3/7     1/2 3/7 3/8        11/10 13/14 45/56
m1 * m2 is
 1/5 1/6 1/7     1/6 1/7 1/8        101/630 101/735 101/840
 2/5 1/3 2/7  *  1/3 2/7 1/4    =   101/315 202/735 101/420
 3/5 1/2 3/7     1/2 3/7 3/8        101/210 101/245 101/280
```

✓ 复习题

19.9.1 为什么 GenericMatrix 类中的 add、multiple 以及 zero 方法定义为抽象的？

19.9.2 IntegerMatrix 类中 add、multiple 以及 zero 方法是如何实现的？

19.9.3 RationalMatrix 类中 add、multiple 以及 zero 方法是如何实现的？

19.9.4 如果 printResult 方法如下定义，将会报什么错？

```
public static void printResult(
  E[][] m1, E[][] m2, E[][] m3, char op)
```

关键术语

actual concrete type（实际具体类型）
bounded generic type（受限泛型类型）
bounded wildcard(<? extends E>)（受限通配）
formal generic type（形式泛型类型）
generic instantiation（泛型实例化）

lower-bound wildcard(<? super E>)（下限通配）
raw type（原生类型）
unbounded wildcard(<?>)（非受限通配）
type erasure（类型擦除）

本章小结

1. 泛型使你能对类型参数化。可以定义使用泛型类型的类或方法，编译器会用具体类型来替换泛型类型。

2. 泛型的主要优势是能够在编译时而不是运行时检测错误。

3. 泛型类或方法允许指定这个类或方法可以具有的对象类型。如果试图使用具有不兼容对象的类或方法，编译器会检测出这个错误。

4. 定义在类、接口或者静态方法中的泛型称为形式泛型类型，之后可以用一个实际具体类型来替换它。替换泛型类型的过程称为泛型实例化。
5. 不使用类型参数的泛型类称为原生类型，例如 ArrayList。使用原生类型是为了向后兼容 Java 较早的版本。
6. 通配泛型类型有三种形式——?、? extends T 和 ? super T，这里的 T 代表一个泛型类型。第一种形式 ? 称为非受限通配，它等同于 ? extends Object。第二种形式 ? extends T 称为受限通配，代表 T 或者 T 的一个子类型。第三种类型 ? super T 称为下限通配，表示 T 或者 T 的一个父类型。
7. 泛型是使用称为类型擦除的方法来实现的。编译器使用泛型类型信息来编译代码，但是随后擦除它。因此，泛型信息在运行时是不可用的。这种做法能够使泛型代码向后兼容使用原生类型的遗留代码。
8. 不能使用泛型类型参数来创建实例，例如 new E()。
9. 不能使用泛型类型参数来创建数组，例如 new E[10]。
10. 不能在静态环境中使用类的泛型类型参数。
11. 在异常类中不能使用泛型类型参数。

测试题

回答位于本书配套网站上的本章测试题。

编程练习题

19.1 （修改程序清单 19-1）修改程序清单 19-1 中的 GenericStack 类，使用数组而不是 ArrayList 来实现它。你应该在给栈添加新元素之前检查数组的大小。如果数组满了，就创建一个新数组，该数组是当前数组大小的两倍，然后将当前数组的元素复制到新数组中。

19.2 （使用继承实现 GenericStack）程序清单 19-1 中，GenericStack 是使用组合实现的。定义一个新的继承自 ArrayList 的栈类。画出 UML 类图，然后实现 GenericStack。编写一个测试程序，提示用户输入 5 个字符串，然后以逆序显示它们。

19.3 （ArrayList 中的不同元素）编写以下方法，返回一个新的 ArrayList。该新列表中包含来自原列表中的不重复元素。

```
public static <E> ArrayList<E> removeDuplicates(ArrayList<E> list)
```

19.4 （泛型线性搜索）为线性搜索实现以下泛型方法。

```
public static <E extends Comparable<E>>
    int linearSearch(E[] list, E key)
```

19.5 （数组中的最大元素）实现下面的方法，返回数组中的最大元素。

```
public static <E extends Comparable<E>> E max(E[] list)
```

编写一个测试程序，提示用户输入 10 个整数，调用该方法找到最大数并显示。

19.6 （二维数组中的最大元素）编写一个泛型方法，返回二维数组中的最大元素。

```
public static <E extends Comparable<E>> E max(E[][] list)
```

19.7 （泛型二分查找法）使用二分查找法实现下面的方法。

```
public static <E extends Comparable<E>>
    int binarySearch(E[] list, E key)
```

19.8 （打乱 ArrayList）编写以下方法，打乱 ArrayList。

```
public static <E> void shuffle(ArrayList<E> list)
```

19.9 （对 ArrayList 排序）编写以下方法，对 ArrayList 排序。

```
public static <E extends Comparable<E>>
  void sort(ArrayList<E> list)
```

编写一个测试程序，提示用户输入 10 个整数，调用该方法对数字进行排序并以升序显示这些数字。

19.10 （ArrayList 中的最大元素）编写以下方法，返回 ArrayList 中的最大元素。

```
public static <E extends Comparable<E>> E max(ArrayList<E> list)
```

19.11 （ComplexMatrix）使用编程练习题 13.17 中所介绍的 Complex 类来开发 ComplexMatrix 类，用于执行涉及复数的矩阵运算。ComplexMatrix 类应该继承自 GenericMatrix 类并实现 add、multiple 以及 zero 方法。你需要修改 GenericMatrix 并将每个出现的 Number 替换为 Object，因为 Complex 不是 Number 的子类。编写一个测试程序，创建两个矩阵并且调用 printResult 方法显示它们相加和相乘的结果。

第 20 章

Introduction to Java Programming and Data Structures, Comprehensive Version, Twelfth Edition

线性表、栈、队列和优先队列

教学目标
- 探索 Java 集合框架层次结构中接口和类的关系（20.2 节）。
- 使用 Collection 接口中定义的通用方法来操作集合（20.2 节）。
- 使用 Iterator 接口来遍历一个集合（20.3 节）。
- 使用 foreach 循环遍历集合中的元素（20.3 节）。
- 使用 forEach 方法为集合中的每一个元素执行一个操作（20.4 节）。
- 探索如何以及何时使用 ArrayList 或 LinkedList 来存储元素线性表（20.5 节）。
- 使用 Comparable 接口和 Comparator 接口来比较元素（20.6 节）。
- 使用 Collections 类中的静态工具方法来排序、查找和打乱线性表，以及找出集合中的最大元素和最小元素（20.7 节）。
- 使用 ArrayList 开发一个多弹球的应用程序（20.8 节）。
- 区分 Vector 与 ArrayList，然后使用 Stack 类创建栈（20.9 节）。
- 探索 Collection、Queue、LinkedList 以及 PriorityQueue 之间的关系，然后使用 PriorityQueue 类创建优先队列（20.10 节）。
- 使用栈编写一个程序来对表达式求值（20.11 节）。

20.1 引言

要点提示：为一个特定的任务选择最好的数据结构和算法是开发高性能软件的关键。

第 18～29 章一般会在数据结构课程中教授。数据结构（data structure）是以某种形式将数据组织在一起的集合。数据结构不仅存储数据，还支持访问和处理数据的操作。即使不懂得数据结构，你也可以编写程序，但是你的程序可能不会很高效。如果具有数据结构的知识，你可以编写出更加高效的程序，这对实际的应用编程非常重要。

在面向对象思想里，数据结构也被认为是一种容器（container）或者容器对象（container object），是一个能存储其他对象的对象，这里的其他对象常被称为数据或者元素。定义一种数据结构从本质上讲就是定义一个类。数据结构类应该使用数据域存储数据，并提供方法支持查找、插入和删除等操作。因此，创建一个数据结构就是创建这个类的一个实例。然后，可以使用这个实例上的方法来操作这个数据结构，例如，向该数据结构中插入一个元素，或者从这个数据结构中删除一个元素。

11.11 节已经介绍了 ArrayList 类，它是一种将元素存储在线性表中的数据结构。Java 还提供了更多能有效地组织和操作数据的数据结构（线性表、向量、栈、队列、优先队列、规则集和映射）。这些数据结构通常称为 Java 集合框架（Java Collections Framework）。我们将在本章介绍线性表（list）、向量（vector）、栈（stack）、队列（queue）和优先队列（priority queue）的应用，在下一章介绍规则集（set）和映射（map）。这些数据结构的实现将在第 24～27 章讨论。通过实现这些数据结构，学生会对数据结构为何高效以及何时、如何使用数据结构有深入的理解。最后，我们将在第28和29章介绍图（graph）的数据结构和算法的设计与实现。

20.2 集合

☞ **要点提示**：Collection 接口为线性表、向量、栈、队列、优先队列以及规则集定义了通用的操作。

Java 集合框架支持以下两种类型的容器：
- 一种是为了存储一个元素集合，称为集合（collection）。
- 另一种是为了存储键值对，称为映射（map）。

映射是一种使用键（key）快速搜索元素的高效数据结构。我们将在下一章介绍映射。现在我们将注意力集中在以下集合上。
- Set 用于存储一组不重复的元素。
- List 用于存储一个有序元素的集合。
- Stack 用于存储采用后进先出方式处理的对象。
- Queue 用于存储采用先进先出方式处理的对象。
- PriorityQueue 用于存储按照优先级顺序处理的对象。

这些集合的通用操作在接口中定义，而实现是在具体类中提供的，如图 20-1 所示。

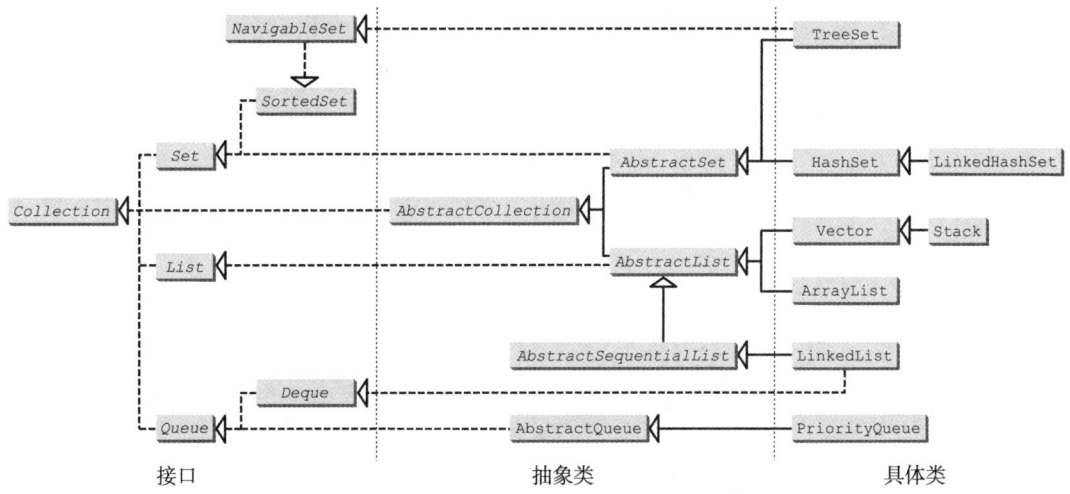

图 20-1 集合是存储对象的容器

☞ **注意**：在 Java 集合框架中定义的所有接口和类都组织在 java.util 包中。

☞ **设计指南**：Java 集合框架的设计是一个使用接口、抽象类和具体类的优秀示例。接口定义了通用的操作，抽象类提供部分实现，具体类用具体的数据结构实现接口。提供一个部分实现接口的抽象类为用户编写代码提供了方便。用户可以简单地定义一个继承自抽象类的具体类，而无须实现接口中的所有方法。为了方便，提供了如 AbstractCollection 这样的抽象类。因为这个原因，这些抽象类被称为便利抽象类（convenience abstract class）。

Collection 接口是处理对象集合的根接口，它的公共方法在图 20-2 中列出。AbstractCollection 类提供了 Collection 接口的部分实现，除了 add、size 和 iterator 方法之外，它实现了 Collection 接口中的其他所有方法。add、size 和 iterator 方法在具体的子类中实现。

　　Collection 接口提供了在集合中添加与删除元素的基本操作。add 方法添加一个元素到集合中。addAll 方法把指定集合中的所有元素添加到这个集合中。remove 方法从集合中删除一个元素。removeAll 方法从这个集合中删除位于指定集合中的所有元素。retainAll 方法保留既出现在这个集合中也出现在指定集合中的元素。所有这些方法都返回 boolean 值。如果执行方法改变了这个集合，那么返回值为 true。clear() 方法简单地移除集合中的所有元素。

　　注意：addAll、removeAll、retainAll 方法类似于规则集上的并、差、交运算。

　　Collection 接口提供了多种查询操作。size 方法返回集合中元素的个数。contains 方法检测集合中是否包含指定的元素。containsAll 方法检测这个集合中是否包含指定集合中的所有元素。isEmpty 方法在集合为空时返回 true。

　　Collection 接口提供的 toArray() 方法针对该集合返回一个 Object 数组，它还提供了 toArray(T[]) 方法，返回一个 T[] 类型的数组。

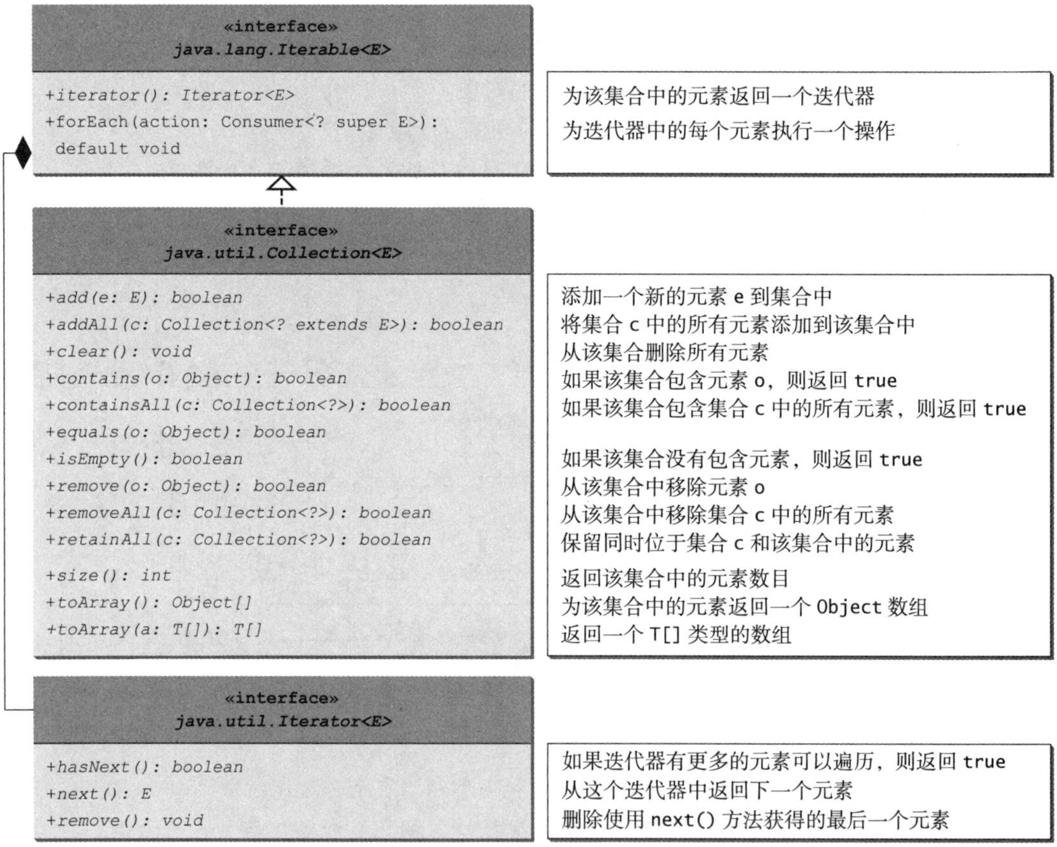

图 20-2　Collection 接口包含了处理集合中元素的方法，并且可以得到一个迭代器对象用于遍历集合中的元素

　　设计指南：Collection 接口中的有些方法是不能在具体子类中实现的。在这种情况下，这些方法会抛出异常 java.lang.UnsupportedOperationException，它是 RuntimeException 异常类的一个子类。这是一种很好的设计，推荐在项目中使用。如果一个方法在子类中没有意义，可以按如下方式实现它：

```java
public void someMethod() {
  throw new UnsupportedOperationException
    ("Method not supported");
}
```

程序清单 20-1 给出了一个使用 Collection 接口中所定义的方法的示例。

程序清单 20-1　TestCollection.java

```java
 1  import java.util.*;
 2
 3  public class TestCollection {
 4    public static void main(String[] args) {
 5      ArrayList<String> collection1 = new ArrayList<>();
 6      collection1.add("New York");
 7      collection1.add("Atlanta");
 8      collection1.add("Dallas");
 9      collection1.add("Madison");
10
11      System.out.println("A list of cities in collection1:");
12      System.out.println(collection1);
13
14      System.out.println("\nIs Dallas in collection1? "
15        + collection1.contains("Dallas"));
16
17      collection1.remove("Dallas");
18      System.out.println("\n" + collection1.size() +
19        " cities are in collection1 now");
20
21      Collection<String> collection2 = new ArrayList<>();
22      collection2.add("Seattle");
23      collection2.add("Portland");
24      collection2.add("Los Angeles");
25      collection2.add("Atlanta");
26
27      System.out.println("\nA list of cities in collection2:");
28      System.out.println(collection2);
29
30      ArrayList<String> c1 = (ArrayList<String>)(collection1.clone());
31      c1.addAll(collection2);
32      System.out.println("\nCities in collection1 or collection2: ");
33      System.out.println(c1);
34
35      c1 = (ArrayList<String>)(collection1.clone());
36      c1.retainAll(collection2);
37      System.out.print("\nCities in collection1 and collection2: ");
38      System.out.println(c1);
39
40      c1 = (ArrayList<String>)(collection1.clone());
41      c1.removeAll(collection2);
42      System.out.print("\nCities in collection1, but not in 2: ");
43      System.out.println(c1);
44    }
45  }
```

```
A list of cities in collection1:
[New York, Atlanta, Dallas, Madison]
Is Dallas in collection1? true
3 cities are in collection1 now
A list of cities in collection2:
[Seattle, Portland, Los Angeles, Atlanta]
Cities in collection1 or collection2:
[New York, Atlanta, Madison, Seattle, Portland, Los Angeles, Atlanta]
Cities in collection1 and collection2: [Atlanta]
Cities in collection1, but not in 2: [New York, Madison]
```

该程序使用 ArrayList 创建了一个具体的集合对象（第 5 行），然后调用 Collection 接口的 contains 方法（第 15 行）、remove 方法（第 17 行）、size 方法（第 18 行）、addAll 方法（第 31 行）、retainAll 方法（第 36 行）以及 removeAll 方法（第 41 行）。

对于该例子而言，我们使用了 ArrayList。你可以使用 Collection 的任意具体类，如 HashSet、LinkedList 替代 ArrayList 来测试这些定义在 Collection 接口中的方法。

程序创建了一个数组线性表的副本（第 30、35 和 40 行）。这样做的目的是保持原数组不被改变，而使用它的副本来执行 addAll、retainAll 以及 removeAll 操作。

☞ 注意：除了 java.util.PriorityQueue 没有实现 Cloneable 接口外，Java 集合框架中的其他所有具体类都实现了 java.lang.Cloneable 和 java.io.Serializable 接口。因此，除了优先队列外，所有 Collection 的实例都是可克隆的，并且所有 Collection 的实例都是可序列化的。

✓ 复习题

20.2.1 什么是数据结构？
20.2.2 描述 Java 集合框架。列出 Collection 接口下面的接口、便利抽象类以及具体类。
20.2.3 一个集合对象是否可以克隆以及序列化？
20.2.4 使用什么方法可以将一个集合中的所有元素添加到另一个集合中？
20.2.5 什么时候一个方法应该抛出 UnsupportedOperationException 异常？

20.3 迭代器

☞ **要点提示**：每种集合都是可迭代的（iterable）。可以获得集合的 Iterator 对象来遍历集合中的所有元素。

迭代器（Iterator）是一种经典的设计模式，用于在不需要暴露数据如何保存在数据结构中这一细节的情况下，遍历一个数据结构。

Collection 接口继承自 Iterable 接口。Iterable 接口中定义了 iterator 方法，该方法会返回一个迭代器。Iterator 接口为遍历各种类型的集合中的元素提供了一种统一的方法。Iterable 接口中的 iterator() 方法返回一个 Iterator 的实例，如图 20-2 所示，它使用 next() 方法提供对集合中元素的顺序访问。还可以使用 hasNext() 方法来检测迭代器中是否还有更多的元素，以及使用 remove() 方法来移除迭代器返回的最后一个元素。

程序清单 20-2 给出了一个在数组线性表中使用迭代器遍历元素的示例。

程序清单 20-2 TestIterator.java

```java
1  import java.util.*;
2
3  public class TestIterator {
4    public static void main(String[] args) {
5      Collection<String> collection = new ArrayList<>();
6      collection.add("New York");
7      collection.add("Atlanta");
8      collection.add("Dallas");
9      collection.add("Madison");
10
11     Iterator<String> iterator = collection.iterator();
12     while (iterator.hasNext()) {
13       System.out.print(iterator.next().toUpperCase() + " ");
14     }
```

```
15        System.out.println();
16      }
17  }
```

```
NEW YORK ATLANTA DALLAS MADISON
```

该程序使用ArrayList创建一个具体的集合对象（第5行），然后添加4个字符串到线性表中（第6～9行）。接下来该程序获得集合的一个迭代器（第11行），并使用该迭代器来遍历线性表中的所有字符串，然后以大写方式来显示该字符串（第12～14行）。

提示：可以使用foreach循环来简化第11～14行的代码，而不使用迭代器，如下所示：

```
for (String element: collection)
  System.out.print(element.toUpperCase() + " ");
```

可以将该循环理解为"对集合中的每个元素，做以下事情"。foreach循环可以用于数组（见7.2.7节），也可以用于Iterable的任何实例。

在Java 10中，可以使用var声明一个变量。编译器自动基于上下文识别出变量的类型。比如，

```
var x = 3;
var y = x;
```

编译器自动从赋给x的整数值3来确定变量x是int类型，而变量x赋给y，从而y也是int类型。这不是一个使用var类型的好例子。var类型的真正用途在于替换长的类型。例如，

```
Iterator<String> iterator = new ArrayList<String>().iterator();
```

可以被替换为

```
var iterator = new ArrayList<String>().iterator();
```

这样做简单便利，并且可以避免错误，因为有时程序员可能会忘记迭代器的正确类型。

警告：我们刻意延迟到现在才介绍var，是为了避免这一有用的特性被误用。不要对基本数据类型使用它。

注意：Java 10引入了var。理想情况下，它应该作为一个关键字。然而为了向后兼容，var不是关键字和保留字，而是作为一个对局部变量声明具有特别含义的标识符。

复习题

20.3.1 如何从一个集合对象获得迭代器？
20.3.2 使用什么方法从迭代器得到集合中的一个元素？
20.3.3 可以使用foreach循环来遍历任何Collection实例中的元素吗？
20.3.4 使用foreach循环遍历一个集合中的所有元素时，需要使用迭代器中的next()或者hasNext()方法吗？

20.4 使用forEach方法

要点提示：可以使用forEach方法对集合中的每个元素执行一个操作。

Java 8在Iterable接口中添加了一个新的默认方法forEach。该方法使用一个参数来指定动作，该动作是函数接口Consumer<? super E>的一个实例。Consumer接口定义了在元素e上执行操作的accept(E e)方法。可以使用程序清单20-3中的forEach方法重写前面的示例。

程序清单 20-3 TestForEach.Java

```java
1  import java.util.*;
2
3  public class TestForEach {
4    public static void main(String[] args) {
5      Collection<String> collection = new ArrayList<>();
6      collection.add("New York");
7      collection.add("Atlanta");
8      collection.add("Dallas");
9      collection.add("Madison");
10
11     collection.forEach(e -> System.out.print(e.toUpperCase() + " "));
12   }
13 }
```

```
NEW YORK ATLANTA DALLAS MADISON
```

第 11 行中的语句使用下面图 a 中的 lambda 表达式，这相当于使用图 b 中所示的匿名内部类。使用 lambda 表达式不仅简化了语法，而且简化了语义。

```
forEach(e ->
  System.out.print(e.toUppserCase() + " "))
```

```
forEach(
  new java.util.function.Consumer<String>() {
    public void accept(String e) {
      System.out.print(e.toUpperCase() + " ");
    }
  }
)
```

  a) 使用 lambda 表达式   b) 使用匿名内部类

可以使用 foreach 循环或者 forEach 方法来编写代码。在大多数情况下，使用 forEach 方法比较简单。

✔ 复习题

20.4.1 Collection 中的任何实例都可以用 forEach 方法吗？forEach 方法在哪里被定义？

20.4.2 假设 list 中的每个元素都是一个 StringBuilder，使用 forEach 方法编写语句，将 list 中每一个元素的首字母变为大写。

20.5 线性表

> **要点提示**：List 接口继承自 Collection 接口，定义了一个用于顺序存储元素的接口。可以使用它的两个具体类 ArrayList 或者 LinkedList 来创建一个线性表。

前面小节中我们使用 ArrayList 来测试 Collection 接口中的方法。现在我们将更深入地考察 ArrayList。本节中我们还将介绍另外一个有用的线性表：LinkedList。

20.5.1 List 接口中的通用方法

ArrayList 和 LinkedList 定义在 List 接口下。List 接口继承自 Collection 接口，定义了一个允许重复的有序集合。List 接口增加了面向位置（position-oriented）的操作，并且增加了一个能够双向遍历线性表的新线性表迭代器。List 接口中引入的方法如图 20-3 所示。

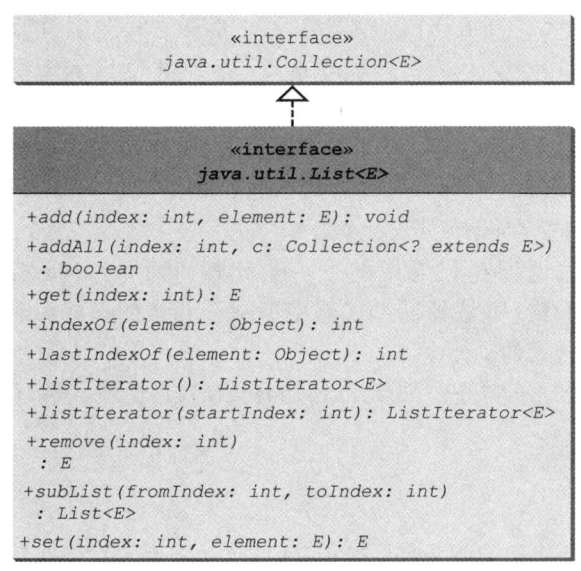

图 20-3　List 接口顺序存储元素并允许元素重复

add(index，element) 方法用于在指定下标处插入一个元素，而 addAll(index，collection) 方法用于在指定下标位置插入一个集合。remove(index) 方法用于从线性表中删除指定下标位置的元素，set(index，element) 方法用于在指定下标位置设置一个新元素。

indexOf(element) 方法用于获取指定元素在线性表中第一次出现时的下标，而 lastIndexOf(element) 方法用于获取指定元素在线性表中最后一次出现时的下标。使用 subList(fromIndex,toIndex) 方法可以获得一个子线性表。

listIterator() 或 listIterator(startIndex) 方法都会返回 ListIterator 的一个实例。ListIterator 接口继承了 Iterator 接口，以增加对线性表的双向遍历能力。ListIterator 接口中的方法如图 20-4 所示。

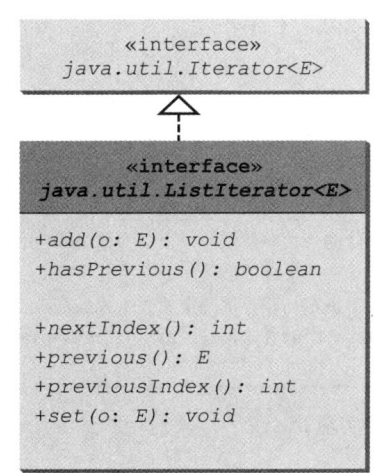

图 20-4　ListIterator 接口可以双向遍历线性表

add(element) 方法用于将指定元素插入线性表中。如果 Iterator 接口中定义的 next()

方法的返回值非空，则该元素将被插入 next() 方法返回的元素之前；如果 previous() 方法的返回值非空，则该元素将被插入 previous() 方法返回的元素之后。如果线性表中没有元素，这个新元素即成为线性表中唯一的元素。set(element) 方法用于将 next 方法或 previous 方法返回的最后一个元素替换为指定元素。

在 Iterator 接口中定义的 hasNext() 方法用于检测迭代器向前遍历时是否还有元素，而 hasPrevious() 方法用于检测迭代器往回遍历时是否还有元素。

在 Iterator 接口中定义的 next() 方法返回迭代器中的下一个元素，而 previous() 方法返回迭代器中的前一个元素。nextIndex() 方法返回迭代器中下一个元素的下标，而 previousIndex() 方法返回迭代器中前一个元素的下标。

AbstractList 类提供了 List 接口的部分实现。AbstractSequentialList 类扩展了 AbstractList 类，以提供对链表的支持。

20.5.2 ArrayList 和 LinkedList 类

ArrayList 和 LinkedList 类是实现 List 接口的两个具体类。ArrayList 将元素存储在数组中，这个数组是动态创建的。如果元素个数超过数组的容量，则创建一个更大的新数组，并将当前数组中的所有元素都复制到新数组中。LinkedList 将元素存储在一个链表中。选用这两种类中的哪一个依赖于特定需求。如果需要通过下标随机访问元素，而不会在线性表起始位置插入或删除元素，那么 ArrayList 是最高效的。但是，如果应用程序需要在线性表的起始位置插入或删除元素，就应该选择 LinkedList 类。线性表的大小是可以动态增大或减小的。然而数组一旦被创建，其大小就是固定的。如果应用程序不需要插入或删除元素，那么数组是效率最高的数据结构。

ArrayList 是 List 接口的一种可变大小数组实现。它还提供了一些方法，用于管理存储线性表的内部数组的大小，如图 20-5 所示。每个 ArrayList 实例都有一个容量，这个容量是指存储线性表中元素的数组的大小。它一定不会小于线性表的大小。向 ArrayList 中添加元素时，其容量会自动增大。ArrayList 不能自动减小。可以使用 trimToSize() 方法将数组容量减小到线性表的大小。ArrayList 可以用它的无参构造方法、ArrayList(Collection) 或 ArrayList(initialCapacity) 来创建。

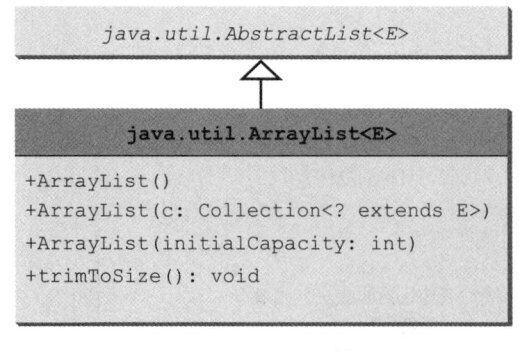

图 20-5　ArrayList 使用数组实现 List

LinkedList 是 List 接口的链表实现。除了实现 List 接口外，这个类还提供从线性表两端获取、插入和删除元素的方法，如图 20-6 所示。LinkedList 可以用它的无参构造方法或 LinkedList(Collection) 来创建。

线性表、栈、队列和优先队列 31

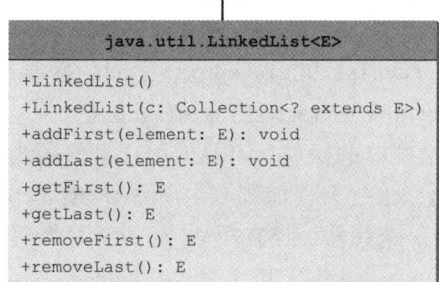

图20-6 LinkedList 提供从线性表两端添加和删除元素的方法

程序清单 20-4 给出一个程序，创建一个用数字填充的数组线性表，并且将新元素插入线性表的指定位置。本例还从数组线性表创建了一个链表，并且向该链表中插入元素或从中删除元素。最后，这个例子分别向前、向后遍历该链表。

程序清单 20-4 TestArrayAndLinked.java

```java
 1  import java.util.*;
 2
 3  public class TestArrayAndLinkedList {
 4    public static void main(String[] args) {
 5      List<Integer> arrayList = new ArrayList<>();
 6      arrayList.add(1); // 1 is autoboxed to an Integer object
 7      arrayList.add(2);
 8      arrayList.add(3);
 9      arrayList.add(1);
10      arrayList.add(4);
11      arrayList.add(0, 10);
12      arrayList.add(3, 30);
13
14      System.out.println("A list of integers in the array list:");
15      System.out.println(arrayList);
16
17      LinkedList<Object> linkedList = new LinkedList<Object>(arrayList);
18      linkedList.add(1, "red");
19      linkedList.removeLast();
20      linkedList.addFirst("green");
21
22      System.out.println("Display the linked list forward:");
23      ListIterator<Object> listIterator = linkedList.listIterator();
24      while (listIterator.hasNext()) {
25        System.out.print(listIterator.next() + " ");
26      }
27      System.out.println();
28
29      System.out.println("Display the linked list backward:");
30      listIterator = linkedList.listIterator(linkedList.size());
31      while (listIterator.hasPrevious()) {
32        System.out.print(listIterator.previous() + " ");
33      }
34    }
35  }
```

```
A list of integers in the array list:
[10, 1, 2, 30, 3, 1, 4]
Display the linked list forward:
green 10 red 1 2 30 3 1
Display the linked list backward:
1 3 30 2 1 red 10 green
```

线性表可以存储相同的元素。整数 1 就在线性表中存储了两次（第 6 和 9 行）。ArrayList 和 LinkedList 的操作类似，它们最主要的不同体现在内部实现上，这影响了它们的性能。在线性表的起始位置插入和删除元素，LinkedList 的效率会高一些；ArrayList 对于所有其他的操作效率会高一些。证明 ArrayList 和 LinkedList 之间的性能差异的示例，请参见 liveexample.pearsoncmg.com/supplement/ArrayListvsLinkedList.pdf。

链表可以使用 get(i) 方法，但这是一个耗时的操作。不要使用它来遍历线性表中的所有元素，如下面图 a 中所示。应该使用 foreach 循环（如图 b 中所示），或者 forEach 方法（如图 c 中所示）。注意图 b 和 c 隐式地使用了迭代器，当在第 24 章中学习如何实现链表的时候，你将知道原因。

```
for (int i = 0; i < list.size(); i++)
  process list.get(i);
}
```
a) 非常低效

```
for (listElementType e: list) {
  process e;
}
```
b) 高效的

```
list.forEach(e ->
  process e
)
```
c) 高效的

> **提示**：为了从可变长参数表中创建线性表，Java 提供了静态的 asList 方法。这样，就可以使用下面的代码创建一个字符串线性表和一个整数线性表：
>
> ```
> List<String> list1 = Arrays.asList("red", "green", "blue");
> List<Integer> list2 = Arrays.asList(10, 20, 30, 40, 50);
> ```
>
> 在 Java 9 中，可以使用静态方法 List.of 来取代 Arrays.asList，以从变长的参数列表中创建一个线性表。List.of 这个名字比 Arrays.asList 更直观和易于记忆。

✓ **复习题**

20.5.1 如何向线性表中添加元素和从线性表中删除元素？如何从两个方向遍历线性表？

20.5.2 假设 list1 是一个包含字符串 red、yellow、green 的线性表，list2 是一个包含字符串 red、yellow、blue 的线性表，回答下面的问题：

(a) 执行完 list1.addAll(list2) 方法之后，线性表 list1 和 list2 分别是什么？

(b) 执行完 list1.add(list2) 方法之后，线性表 list1 和 list2 分别是什么？

(c) 执行完 list1.removeAll(list2) 方法之后，线性表 list1 和 list2 分别是什么？

(d) 执行完 list1.remove(list2) 方法之后，线性表 list1 和 list2 分别是什么？

(e) 执行完 list1.retainAll(list2) 方法之后，线性表 list1 和 list2 分别是什么？

(f) 执行完 list1.clear() 方法之后，线性表 list1 是什么？

20.5.3 ArrayList 与 LinkedList 之间的区别是什么？应该使用哪种线性表在一个线性表的起始位置插入和删除元素。

20.5.4 LinkedList 是否包含 ArrayList 中的所有方法？哪些方法在 LinkedList 中有而在 ArrayList 中没有？

20.5.5 如何从一个对象数组创建一个线性表？

20.6 Comparator 接口

> **要点提示**：Comparator 可用于比较没有实现 Comparable 的类的对象，或者定义比较对象的新标准。

你已经学习了如何使用 Comparable 接口来比较元素（13.6 节中介绍）。Java API 的一些类，比如 String、Date、Calendar、BigInteger、BigDecimal 以及所有基本类型的数字包装类都实现了 Comparable 接口。Comparable 接口定义了 compareTo 方法，用于比较实现了 Comparable 接口的同一个类的两个元素。

如果元素的类没有实现 Comparable 接口又将如何呢？这些元素可以比较吗？可以定义一个 Comparator 来比较不同类的元素。要做到这一点，需要创建一个实现了 java.util.Comparator<T> 接口的类并重写它的 compare(a,b) 方法。

```
public int compare(T a, T b)
```

如果 element1 小于 element2，则返回一个负值；如果 element1 大于 element2，则返回一个正值；若两者相等，则返回 0。

13.2 节中介绍了 GeometricObject 类，该类没有实现 Comparable 接口。为了比较 GeometricObject 类的对象，可以定义一个比较器类，如程序清单 20-5 所示。

程序清单 20-5 GeometricObjectComparator.java

```java
1  import java.util.Comparator;
2
3  public class GeometricObjectComparator
4      implements Comparator<GeometricObject>, java.io.Serializable {
5    public int compare(GeometricObject o1, GeometricObject o2) {
6      double area1 = o1.getArea();
7      double area2 = o2.getArea();
8
9      if (area1 < area2)
10       return -1;
11     else if (area1 == area2)
12       return 0;
13     else
14       return 1;
15   }
16 }
```

第 4 行实现了 Comparator<GeometricObject>，第 5 行通过重写 compare 方法来比较两个几何对象。比较器类也实现了 Serializable 接口。对于比较器来说，实现 Serializable 通常是一个好主意，这样它们可以被序列化。如果 o1.getArea()>o2.getArea()，则 compare(o1, o2) 方法返回 1；如果 o1.getArea()==o2.getArea()，则返回 0；否则返回 -1。

程序清单 20-6 给出了一个方法，返回两个几何对象中较大的那个。两个对象使用 GeometricObjectComparator 进行比较。

程序清单 20-6 TestComparator.java

```java
1  import java.util.Comparator;
2
3  public class TestComparator {
4    public static void main(String[] args) {
5      GeometricObject g1 = new Rectangle(5, 5);
6      GeometricObject g2 = new Circle(5);
7
8      GeometricObject g =
9        max(g1, g2, new GeometricObjectComparator());
10
11     System.out.println("The area of the larger object is " +
12       g.getArea());
13   }
14
15   public static GeometricObject max(GeometricObject g1,
16       GeometricObject g2, Comparator<GeometricObject> c) {
17     if (c.compare(g1, g2) > 0)
```

```
18         return g1;
19       else
20         return g2;
21     }
22   }
```

```
The area of the larger object is 78.53981633974483
```

该程序在第 5 和 6 行创建了一个 Rectangle 对象和一个 Circle 对象（Rectangle 类和 Circle 类在 13.2 节中定义）。它们都是 GeometricObject 的子类。程序调用 max 方法得到具有较大面积的几何对象（第 8 和 9 行）。

比较器是 Comparator 类型的对象，其 compare(a,b) 方法被用于比较两个元素。GeometricObjectComparator 被创建并且传递给 max 方法（第 9 行），第 17 行中的 max 方法使用该比较器来比较几何对象。

由于 Comparator 接口是单一的抽象方法接口，所以可以使用 lambda 表达式将第 9 行替换为以下代码来简化程序：

```
max(g1, g2, (o1, o2) -> o1.getArea() > o2.getArea() ?
  1 : o1.getArea() == o2.getArea() ? 0 : -1);
```

这里，o1 和 o2 是 Comparator 接口中 compare 方法的两个参数。如果 o1.getArea()>o2.getArea()，那么方法返回 1；如果 o1.getArea()==o2.getArea()，那么方法返回 0；其他情况返回 -1。

> **注意**：使用 Comparable 接口比较元素称为使用自然顺序（natural order）进行比较，使用 Comparator 接口比较元素则称为使用比较器进行比较。

前面的例子定义了一个比较器来比较两个几何对象，因为 GeometricObject 类没有实现 Comparable 接口。有时类实现了 Comparable 接口，但是如果想要使用不同的标准来比较它们的对象，就可以自定义一个比较器。清单 20-7 给出了一个比较字符串长度的示例。

程序清单 20-7 SortStringByLength.java

```
 1  public class SortStringByLength {
 2    public static void main(String[] args) {
 3      String[] cities = {"Atlanta", "Savannah", "New York", "Dallas"};
 4      java.util.Arrays.sort(cities, new MyComparator());
 5
 6      for (String s : cities) {
 7        System.out.print(s + " ");
 8      }
 9    }
10
11    public static class MyComparator implements
12        java.util.Comparator<String> {
13      @Override
14      public int compare(String s1, String s2) {
15        return s1.length() - s2.length();
16      }
17    }
18  }
```

```
Dallas Atlanta Savannah New York
```

通过实现 Comparator 接口（第 11 和 12 行），该程序定义了一个比较器类。compare 方法通过比较两个字符串的长度来比较它们（第 14 ～ 16 行）。该程序调用 sort 方法，使用比较器（第 4 行）对字符串数组进行排序。

由于 Comparator 是一个函数式接口，所以可以使用 lambda 表达式来简化程序，如下所示：

```
java.util.Arrays.sort(cities,
  (s1, s2) -> {return s1.length() - s2.length();});
```

或者简化为：

```
java.util.Arrays.sort(cities,
  (s1, s2) -> s1.length() - s2.length());
```

List 接口定义了 sort(comparator) 方法，该方法可以使用指定的比较器对线性表中的元素进行排序。程序清单 20-8 给出了一个在线性表中使用比较器，忽略字母大小写对字符串进行排序的例子。

程序清单 20-8 SortStringIgnoreCase.java

```
1  public class SortStringIgnoreCase {
2    public static void main(String[] args) {
3      java.util.List<String> cities = java.util.Arrays.asList
4        ("Atlanta", "Savannah", "New York", "Dallas");
5      cities.sort((s1, s2) -> s1.compareToIgnoreCase(s2));
6  
7      for (String s: cities) {
8        System.out.print(s + " ");
9      }
10   }
11 }
```

```
Atlanta dallas new York Savannah
```

该程序使用比较器来对字符串线性表进行排序，比较字符串时忽略大小写（第 5 行）。如果调用 list.sort(null)，那么该线性表将按照它的自然顺序排序。

比较器是使用 lambda 表达式创建的。注意这里的 lambda 表达式只调用了 compareToIgnoreCase 方法。在这种情况下，可以使用更简单、更清晰的语法来代替 lambda 表达式，如下所示：

```
cities.sort(String::compareToIgnoreCase);
```

这里的 String::compareToIgnoreCase 称为方法引用，等价于一个 lambda 表达式。编译器自动将一个方法引用转换为等价的 lambda 表达式。

Comparator 接口还包含几个有用的静态方法和默认方法。可以使用静态方法 comparing (Function<? sup T,? sup R> keyExtracter) 来创建 Comparator<T>，该比较器使用从 Function 对象中提取的键来比较元素。Function 对象的 apply(T) 方法返回对象 T 的 R 类型的键。例如，下面图 a 中的代码创建了一个 Comparator，它使用 lambda 表达式根据字符串的长度来比较字符串，这相当于使用图 b 中的匿名内部类，以及图 c 中的方法引用。

```
Comparator.comparing(e -> e.length())
```
a) 使用 lambda 表达式

```
Comparator.comparing(
  new java.util.function.Function<String, Integer>() {
    public Integer apply(String s) {
      return s.length();
    }
})
```
b) 使用匿名内部类

```
Comparator.comparing(String::length)
```
c) 使用方法引用

对于前面的示例而言，Comparator 接口中的 comparing 方法本质上是按如下方式实现的：

```java
// comparing returns a Comparator
public static Comparator<String> comparing(Function<String, Integer> f) {
  return (s1, s2) -> f.apply(s1).compareTo(f.apply(s2));
}
```

可以使用以下代码替换程序清单 20-7 中的比较器：

```java
java.util.Arrays.sort(cities, Comparator.comparing(String::length));
```

Comparator.comparing 方法对于使用对象的属性创建 Comparator（比较器）特别有用。例如，下面的代码是根据 loanAmount 属性对 Loan 对象列表进行排序（参见程序清单 10-2）。

```java
Loan[] list = {new Loan(5.5, 10, 2323), new Loan(5, 10, 1000)};
Arrays.sort(list, Comparator.comparing(Loan::getLoanAmount));
```

可以使用 Comparator 的默认方法 thenComparing 来设置排序的首要标准、次要标准、次次要标准等。例如，下面的代码对 Loan 对象列表进行排序，首先基于它们的 loanAmount 属性，然后基于 annualInterestRate 属性。

```java
Loan[] list = {new Loan(5.5, 10, 100), new Loan(5, 10, 1000)};
Arrays.sort(list, Comparator.comparing(Loan::getLoanAmount)
  .thenComparing(Loan::getAnnualInterestRate));
```

默认的 reverse() 方法可以用于反转比较器的顺序。例如，下面的代码以递减的顺序基于 loanAmount 属性对 Load 对象列表进行排序。

```java
Arrays.sort(list, Comparator.comparing(Loan::getLoanAmount).
  reversed());
```

✓ 复习题

20.6.1 Comparable 接口与 Comparator 接口之间的不同之处是什么？它们分别属于哪一个包？

20.6.2 如何定义一个实现了 Comparable 接口的类 A？类 A 的两个实例可以比较吗？如何定义一个实现了 Comparator 接口的类 B，并且重写 compare 方法来比较 B1 类型的两个对象？如何使用比较器调用 sort 方法来对 B1 类型的对象列表进行排序？

20.6.3 编写 lambda 表达式来创建一个比较器，对两个 Loan 对象按照其 annualInterestRate 属性进行比较。使用 Comparator.comparing 方法创建一个比较器来基于 annualInterestRate 属性对 Loan 对象进行比较。创建一个比较器对 Loan 对象进行比较，首先基于 annualInterestRate 属性，然后基于 loanAmount 属性。

20.6.4 分别使用 lambda 表达式以及 Comparator.comparing 方法创建比较器，通过 Collection 对象的大小对它们进行比较。

20.6.5 编写一条语句，对一组 Point2D 对象进行排序，首先基于它们的 y 值，然后基于它们的 x 值。

20.6.6 编写一条语句，对名为 list 的字符串 ArrayList，根据它们的最后一个字符进行升序排序。

20.6.7 编写一条语句，根据第二列对二维数组 double[][] 按照递增顺序进行排序。例如，如果数组为 double[][] x = { {3,1}, {2,-1}, {2,0} }，排序后则为 { {2,-1}, {2,0}, {3,1} }。

20.6.8 编写一条语句，将二维数组的第二列作为首要标准，第一列作为次要标准，对二维数组 double[][] 按递增顺序排序。例如，如果数组 double[][] x = { {3,1}, {2,-1}, {2,0}, {1,-1} }，排序后为 { {1,-1}, {2,-1}, {2,0}, {3,1} }。

20.7 用于线性表和集合的静态方法

要点提示：Collections 类包含用于执行集合和线性表中通用操作的静态方法。

11.12节中介绍了Collections类中用于数组线性表的一些静态方法。Collections类包含用于线性表的sort、binarySearch、reverse、shuffle、copy和fill方法，以及用于集合的max、min、disjoint和frequency方法，如图20-7所示。

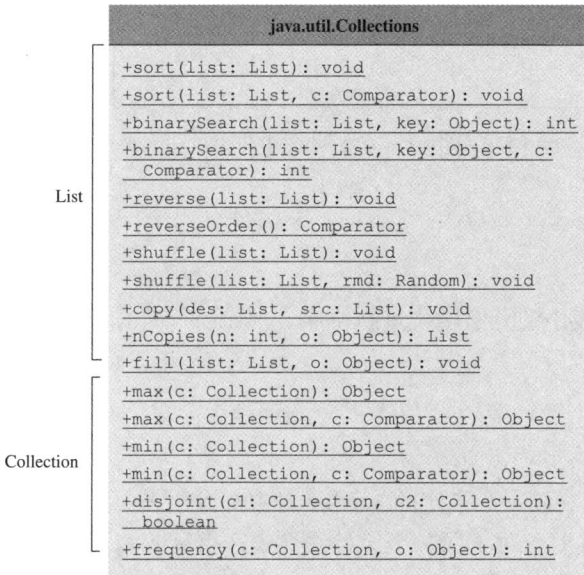

图20-7 Collections类包含用于操作线性表和集合的静态方法

可以使用Comparable接口中的compareTo方法，对线性表中可比较的元素以自然顺序排序。也可以指定比较器来对元素排序。例如，下面的代码对线性表中的字符串排序：

```
List<String> list = Arrays.asList("red", "green", "blue");
Collections.sort(list);
System.out.println(list);
```

输出为[blue, green, red]。

上面的代码以升序对线性表排序。要以降序排列，可以简单地使用Collections.reverseOrder()方法返回一个Comparator对象，该对象以自然顺序的逆序排列元素。例如，下面的代码对字符串线性表以降序排列：

```
List<String> list = Arrays.asList("yellow", "red", "green", "blue");
Collections.sort(list, Collections.reverseOrder());
System.out.println(list);
```

输出为[yellow, red, green, blue]。

可以使用binarySearch方法在线性表中查找一个键值。要使用该方法，线性表必须以升序排列好。如果键值没有在线性表中，那么该方法就会返回－(插入点+1)。回忆一下，如果存在该键值，插入点就是键值在线性表中的位置。例如，下面的代码在一个整数线性表和一个字符串线性表中查找键值：

```
List<Integer> list1 =
  Arrays.asList(2, 4, 7, 10, 11, 45, 50, 59, 60, 66);
System.out.println("(1) Index: " + Collections.binarySearch(list1, 7));
System.out.println("(2) Index: " + Collections.binarySearch(list1, 9));

List<String> list2 = Arrays.asList("blue", "green", "red");
System.out.println("(3) Index: " +
```

```
        Collections.binarySearch(list2, "red"));
System.out.println("(4) Index: " +
        Collections.binarySearch(list2, "cyan"));
```

上面代码的输出为：

```
(1) Index: 2
(2) Index: -4
(3) Index: 2
(4) Index: -2
```

可以使用 reverse 方法将线性表中的元素以逆序排列。例如，下面的代码显示 [blue, green, red, yellow]：

```
List<String> list = Arrays.asList("yellow", "red", "green", "blue");
Collections.reverse(list);
System.out.println(list);
```

可以使用 shuffle(List) 方法对线性表中的元素随机重新排序。例如，下面的代码打乱 list 中的元素：

```
List<String> list = Arrays.asList("yellow", "red", "green", "blue");
Collections.shuffle(list);
System.out.println(list);
```

也可以使用 shuffle(List, Random) 方法以一个指定的 Random 对象对线性表中的元素随机重新排序。使用指定的 Random 对象，有助于对于同一个原始线性表产生具有相同元素序列的线性表。例如，下面的代码打乱 list 中的元素：

```
List<String> list1 = Arrays.asList("yellow", "red", "green", "blue");
List<String> list2 = Arrays.asList("yellow", "red", "green", "blue");
Collections.shuffle(list1, new Random(20));
Collections.shuffle(list2, new Random(20));
System.out.println(list1);
System.out.println(list2);
```

你将看到 list1 和 list2 在打乱顺序之前和之后拥有相同的元素序列。

可以使用 copy(det,src) 方法将源线性表中的所有元素以同样的下标复制到目标线性表中。目标线性表必须和源线性表等长。如果目标线性表的长度大于源线性表，那么，目标线性表中的剩余元素不会受到影响。例如，下面的代码将 list2 复制到 list1 中：

```
List<String> list1 = Arrays.asList("yellow", "red", "green", "blue");
List<String> list2 = Arrays.asList("white", "black");
Collections.copy(list1, list2);
System.out.println(list1);
```

list1 的输出是 [white,black,green,blue]。copy 方法执行的是浅复制：复制的只是线性表中元素的引用。

可以使用 nCopies(int n,object o) 方法创建一个包含指定对象的 n 个副本的不可变线性表。例如，下面的代码创建一个具有 5 个 Calender 对象的线性表：

```
List<GregorianCalendar> list1 = Collections.nCopies
    (5, new GregorianCalendar(2005, 0, 1));
```

用 nCopies 方法创建的线性表是不可变的，因此，不能在该线性表中添加、删除或更新元素。所有的元素都有相同的引用。

可以使用 fill(List list,Object o) 方法以指定元素替换线性表中的所有元素。例如，下面的代码显示 [black,black,black]：

```
List<String> list = Arrays.asList("red", "green", "blue");
```

```java
Collections.fill(list, "black");
System.out.println(list);
```

可以使用 max 和 min 方法找出集合中的最大元素和最小元素。集合中的元素必须是可以使用 Comparable 接口或 Comparator 接口进行比较的。参见下面的示例代码：

```java
Collection<String> collection = Arrays.asList("red", "green", "blue");
System.out.println(Collections.max(collection));       // Use Comparable
System.out.println(Collections.min(collection,
          Comparator.comparing(String::length)));      // Use Comparator
```

如果两个集合没有共同的元素，那么 disjoint(collection1,collection2) 方法返回 true。例如，在下面的代码中，disjoint(collection1,collection2) 方法返回 false，但是 disjoint(collection1,collection3) 方法返回 true：

```java
Collection<String> collection1 = Arrays.asList("red", "cyan");
Collection<String> collection2 = Arrays.asList("red", "blue");
Collection<String> collection3 = Arrays.asList("pink", "tan");
System.out.println(Collections.disjoint(collection1, collection2));
System.out.println(Collections.disjoint(collection1, collection3));
```

使用 frequency(collection,element) 方法可以找出集合中某元素的出现次数。例如，下面代码中的 frequency(collection,"red") 返回 2。

```java
Collection<String> collection = Arrays.asList("red", "cyan", "red");
System.out.println(Collections.frequency(collection, "red"));
```

✓ 复习题

20.7.1 Collections 类中的所有方法都是静态的吗？

20.7.2 下面的 Collections 类中，哪些静态方法是用于线性表的？哪些是用于集合的？

sort, binarySearch, reverse, shuffle, max, min, disjoint, frequency

20.7.3 给出下面代码的输出结果：

```java
import java.util.*;

public class Test {
  public static void main(String[] args) {
    List<String> list =
      Arrays.asList("yellow", "red", "green", "blue");
    Collections.reverse(list);
    System.out.println(list);

    List<String> list1 =
      Arrays.asList("yellow", "red", "green", "blue");
    List<String> list2 = Arrays.asList("white", "black");
    Collections.copy(list1, list2);
    System.out.println(list1);

    Collection<String> c1 = Arrays.asList("red", "cyan");
    Collection<String> c2 = Arrays.asList("red", "blue");
    Collection<String> c3 = Arrays.asList("pink", "tan");
    System.out.println(Collections.disjoint(c1, c2));
    System.out.println(Collections.disjoint(c1, c3));

    Collection<String> collection =
      Arrays.asList("red", "cyan", "red");
    System.out.println(Collections.frequency(collection, "red"));
  }
}
```

20.7.4 可以使用哪个方法对 ArrayList 或 LinkedList 中的元素进行排序？可以使用哪个方法对字符串数组进行排序？

20.7.5 可以使用哪个方法对 ArrayList 或 LinkedList 中的元素进行二分查找？可以使用哪个方法对字符串数组中的元素进行二分查找？

20.7.6 编写一条语句，找出由可比较对象构成的数组中的最大元素。

20.8 示例学习：弹球

要点提示：本节给出一个显示弹球的程序，该程序支持用户添加和移除球。

15.12 节给出了一个显示弹球的程序，本节给出一个显示多个弹球的程序。可以使用两个按钮来暂停和恢复球的移动，用一个滚动条来控制球速，以及用"+"和"-"按钮来添加和移除一个球，如图 20-8 所示。

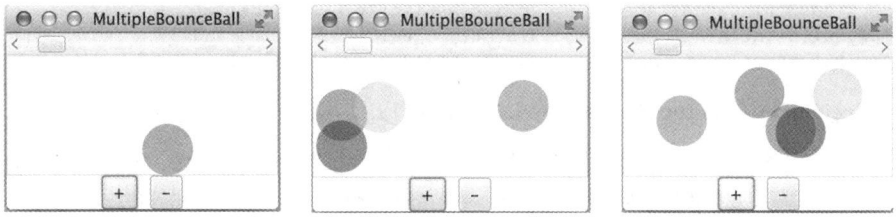

图 20-8 按"+"和"-"按钮来添加和移除球（来源：Oracle 或其附属公司版权所有 ©1995～2016，经授权使用）

15.12 节中的例子只需要保存一个球。如何在该例中保存多个球呢？Pane 的 getChildren() 方法返回一个 List<Node> 的子类型 ObservableList<Node>，用于存储面板中的结点。该线性表初始为空。当创建一个新球时，将其添加到线性表的末尾。要移除一个球，只需要简单地移除线性表的最后一个元素。

每个球的状态包括：x 坐标，y 坐标，颜色，以及移动的方向。可以定义一个继承自 javafx.scene.shape.Circle 的 Ball 类。Circle 中已经定义了 x、y 坐标以及颜色。当球被创建时，它将从左上角开始向右下角移动。一个随机颜色将被赋给一个新的球。

MultipleBallPane 类负责显示球，MultipleBounceBall 类放置控制组件并实现控制。这些类的关系显示在图 20-9 中。程序清单 20-9 给出该程序。

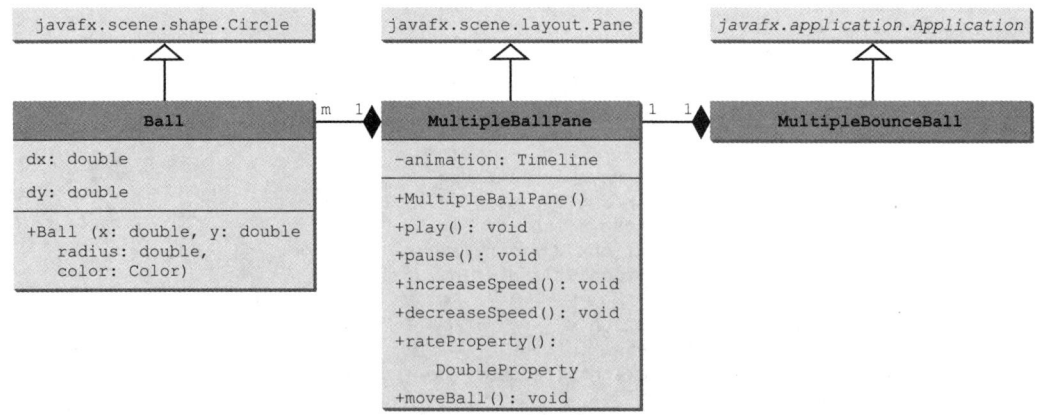

图 20-9 MultipleBounceBall 包含 MultipleBallPane，MultipleBallPane 包含 Ball

程序清单20-9 MultipleBounceBall.java

```java
1   import javafx.animation.KeyFrame;
2   import javafx.animation.Timeline;
3   import javafx.application.Application;
4   import javafx.beans.property.DoubleProperty;
5   import javafx.geometry.Pos;
6   import javafx.scene.Node;
7   import javafx.stage.Stage;
8   import javafx.scene.Scene;
9   import javafx.scene.control.Button;
10  import javafx.scene.control.ScrollBar;
11  import javafx.scene.layout.BorderPane;
12  import javafx.scene.layout.HBox;
13  import javafx.scene.layout.Pane;
14  import javafx.scene.paint.Color;
15  import javafx.scene.shape.Circle;
16  import javafx.util.Duration;
17
18  public class MultipleBounceBall extends Application {
19    @Override // Override the start method in the Application class
20    public void start(Stage primaryStage) {
21      MultipleBallPane ballPane = new MultipleBallPane();
22      ballPane.setStyle("-fx-border-color: yellow");
23
24      Button btAdd = new Button("+");
25      Button btSubtract = new Button("-");
26      HBox hBox = new HBox(10);
27      hBox.getChildren().addAll(btAdd, btSubtract);
28      hBox.setAlignment(Pos.CENTER);
29
30      // Add or remove a ball
31      btAdd.setOnAction(e -> ballPane.add());
32      btSubtract.setOnAction(e -> ballPane.subtract());
33
34      // Pause and resume animation
35      ballPane.setOnMousePressed(e -> ballPane.pause());
36      ballPane.setOnMouseReleased(e -> ballPane.play());
37
38      // Use a scroll bar to control animation speed
39      ScrollBar sbSpeed = new ScrollBar();
40      sbSpeed.setMax(20);
41      sbSpeed.setValue(10);
42      ballPane.rateProperty().bind(sbSpeed.valueProperty());
43
44      BorderPane pane = new BorderPane();
45      pane.setCenter(ballPane);
46      pane.setTop(sbSpeed);
47      pane.setBottom(hBox);
48
49      // Create a scene and place the pane in the stage
50      Scene scene = new Scene(pane, 250, 150);
51      primaryStage.setTitle("MultipleBounceBall"); // Set the stage title
52      primaryStage.setScene(scene); // Place the scene in the stage
53      primaryStage.show(); // Display the stage
54    }
55
56    private class MultipleBallPane extends Pane {
57      private Timeline animation;
58
59      public MultipleBallPane() {
60        // Create an animation for moving the ball
61        animation = new Timeline(
62          new KeyFrame(Duration.millis(50), e -> moveBall()));
63        animation.setCycleCount(Timeline.INDEFINITE);
64        animation.play(); // Start animation
```

```
65      }
66
67      public void add() {
68        Color color = new Color(Math.random(),
69          Math.random(), Math.random(), 0.5);
70        getChildren().add(new Ball(30, 30, 20,color));
71      }
72
73      public void subtract() {
74        if (getChildren().size() > 0) {
75          getChildren().remove(getChildren().size() - 1);
76        }
77      }
78
79      public void play() {
80        animation.play();
81      }
82
83      public void pause() {
84        animation.pause();
85      }
86
87      public void increaseSpeed() {
88        animation.setRate(animation.getRate() + 0.1);
89      }
90
91      public void decreaseSpeed() {
92        animation.setRate(
93          animation.getRate() > 0 ? animation.getRate() - 0.1 : 0);
94      }
95
96      public DoubleProperty rateProperty() {
97        return animation.rateProperty();
98      }
99
100     protected void moveBall() {
101       for (Node node:  this.getChildren()) {
102         Ball ball = (Ball)node;
103         // Check boundaries
104         if (ball.getCenterX() < ball.getRadius() ||
105             ball.getCenterX() > getWidth() - ball.getRadius()) {
106           ball.dx *= -1; // Change ball move direction
107         }
108         if (ball.getCenterY() < ball.getRadius() ||
109             ball.getCenterY() > getHeight() - ball.getRadius()) {
110           ball.dy *= -1; // Change ball move direction
111         }
112
113         // Adjust ball position
114         ball.setCenterX(ball.dx + ball.getCenterX());
115         ball.setCenterY(ball.dy + ball.getCenterY());
116       }
117     }
118   }
119
120   class Ball extends Circle {
121     private double dx = 1, dy = 1;
122
123     Ball(double x, double y, double radius, Color color) {
124       super(x, y, radius);
125       setFill(color); // Set ball color
126     }
127   }
128 }
```

add()方法使用一个随机颜色创建一个新球并将其加入面板中(第70行)。面板将所有的球存储在一个线性表中。subtract() 方法移除线性表中的最后一个球（第 75 行）。

当用户点击"+"按钮时，一个新球被加入面板中（第 31 行）。当用户点击"-"按钮时，数组线性表中的最后一个球被移除（第 32 行）。

MultipleBallPane 类中的 moveBall() 方法获取面板线性表中的每个球，并且调整球的位置（第 114 和 115 行）。

✓ 复习题

20.8.1 针对面板调用 pane.getChildren() 将返回什么值？
20.8.2 如何修改 MutipleBallApp 程序中的代码，使得当按钮被点击的时候移除线性表中的第一个球？
20.8.3 如何修改 MutipleBallApp 程序中的代码，使得每个球的半径为 10 和 20 之间的随机值？

20.9 向量类和栈类

要点提示：在 Java API 中，Vector 是 AbstractList 的子类，Stack 是 Vector 的子类。

Java 集合框架是在 Java 2 中引入的。一些数据结构在较早版本中得到支持，其中就有向量类 Vector 与栈类 Stack。为了融入 Java 集合框架，Java 2 对这些类进行了重新设计，但是为了兼容性，保留了旧的方法。

除了包含用于访问和修改向量的同步方法之外，Vector 类与 ArrayList 类是一样的。同步方法可以防止两个或多个线程同时访问某个向量时引起数据损坏。我们将在第 32 章中讨论同步问题。对于许多不需要同步的应用程序来说，使用 ArrayList 比使用 Vector 效率更高。

Vector 类继承了 AbstractList 类，它还包含 Java 2 以前的版本中原始 Vector 类中的方法，如图 20-10 所示。

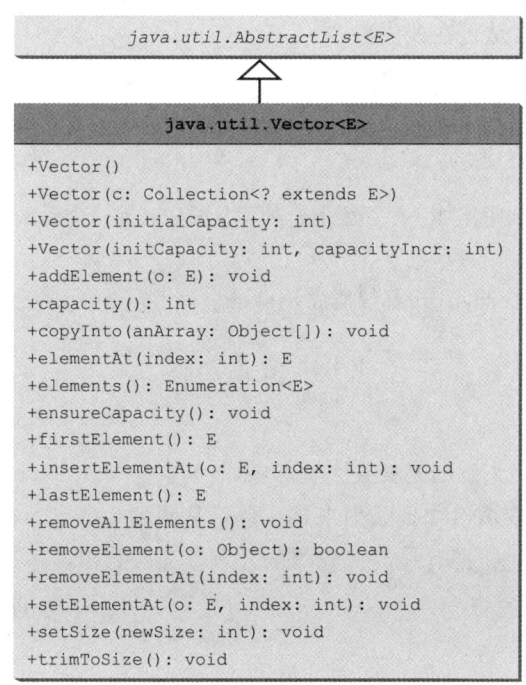

图 20-10 从 Java 2 开始，Vector 类继承了 AbstractList 类，并保留了原来 Vector 类中的所有方法

图 20-10 中 UML 图所列出的 Vector 类中的大多数方法都类似于 List 接口中的方法。这些方法都是在 Java 集合框架之前引入的。例如，addElement(Object element) 方法除了是同步的之外，它与 add(Object element) 方法是一样的。如果不需要同步，最好使用 ArrayList，因为它比 Vector 快得多。

注意：elements() 方法返回一个 Enumeration 对象。Enumeration 接口是在 Java 2 之前引入的，已经被 Iterator 接口所取代。

注意：Vector 类被广泛应用于 Java 的遗留代码中，因为它是 Java 2 之前可变大小数组的实现。在 Java 集合框架中，Stack 类是作为 Vector 类的扩展来实现的，如图 20-11 所示。

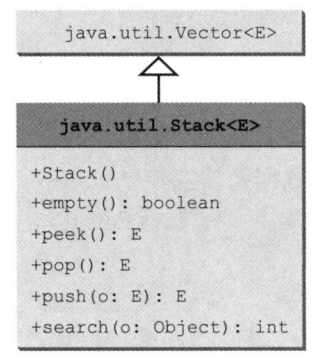

图 20-11　Stack 类继承自 Vector，提供了后进先出的数据结构

Stack 类是在 Java 2 之前引入的。图 20-11 给出的方法在 Java 2 之前已经使用。empty() 方法与 isEmpty() 方法的功能是一样的。peek() 方法可以返回栈顶元素而不删除它。pop() 方法返回栈顶元素并删除它。push(Object element) 方法将指定元素添加到栈中。search(Object element) 方法检测指定元素是否在栈内。

复习题

20.9.1　如何创建 Vector 的一个实例？如何在向量中添加或插入一个新元素？如何从向量中删除一个元素？如何获取向量的大小？

20.9.2　如何创建 Stack 的一个实例？如何向栈中添加一个新元素？如何从栈中删除一个元素？如何获取栈的大小？

20.9.3　在程序清单 20-1 中，如果所有出现的 ArrayList 都替换成 LinkedList、Vector 或者 Stack，那么可以编译运行吗？

20.10　队列和优先队列

要点提示：在优先队列中，具有最高优先级的元素最先被移除。

队列是一种先进先出的数据结构。元素被追加到队列末尾，然后从队列头移出。在优先队列中，元素被赋予优先级。当访问元素时，拥有最高优先级的元素首先被移出。本节将介绍 Java API 中的队列和优先队列。

20.10.1　Queue 接口

Queue 接口继承自 java.util.Collection，加入了插入、提取和检查等操作，如图 20-12 所示。

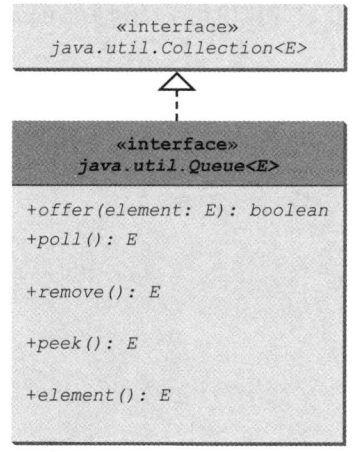

图 20-12 Queue 接口继承自 Collection,并提供额外的插入、提取和检验等操作

offer 方法用于向队列添加一个元素。该方法类似于 Collection 接口中的 add() 方法,但是 offer 方法更适用于队列。poll() 方法和 remove() 方法类似,但是如果队列为空,poll() 方法会返回 null,而 remove() 方法会抛出一个异常。peek() 方法和 element() 方法类似,但是如果队列为空,peek() 方法会返回 null,而 element() 方法会抛出一个异常。

20.10.2 双端队列 Deque 和链表 LinkedList

LinkedList 类实现了 Deque 接口,Deque 又继承自 Queue 接口,如图 20-13 所示。因此,可以使用 LinkedList 创建一个队列。LinkedList 很适合队列操作,因为它可以高效地在列表的两端插入和移除元素。

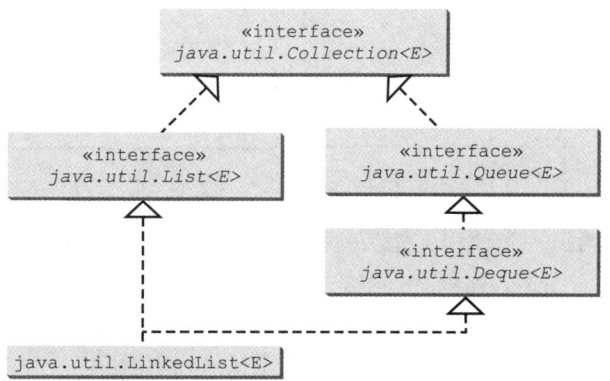

图 20-13 LinkedList 实现了 List 和 Deque

Deque 支持在两端插入和删除元素。deque 是"双端队列"(double-ended queue)的简称,通常的发音为"deck"。Deque 接口继承自 Queue,增加了从队列两端插入和删除元素的方法。方法 addFirst(e)、removeFirst()、addLast(e)、removeLast()、getFirst() 和 getLast() 都在 Deque 接口中进行了定义。

程序清单 20-10 给出一个使用队列存储字符串的例子。程序第 3 行使用 LinkedList 创建一个队列,第 4~7 行将 4 个字符串添加到队列中。在 Collection 接口中定义的方法

size() 返回队列中的元素数目（第 9 行）。remove() 方法获取并移出位于队列头部的元素（第 10 行）。

程序清单 20-10 TestQueue.java

```java
 1  public class TestQueue {
 2    public static void main(String[] args) {
 3      java.util.Queue<String> queue = new java.util.LinkedList<>();
 4      queue.offer("Oklahoma");
 5      queue.offer("Indiana");
 6      queue.offer("Georgia");
 7      queue.offer("Texas");
 8
 9      while (queue.size() > 0)
10        System.out.print(queue.remove() + " ");
11    }
12  }
```

```
Oklahoma Indiana Georgia Texas
```

PriorityQueue 类实现了一个优先队列，如图 20-14 所示。默认情况下，优先队列使用 Comparable 以元素的自然顺序进行排序。拥有最小数值的元素被赋予最高优先级，因此最先从队列中删除。如果几个元素具有相同的最高优先级，则任意选择一个。也可以使用构造方法 PriorityQueue(initialCapacity,comparator) 中的 Comparator 来指定一个顺序。

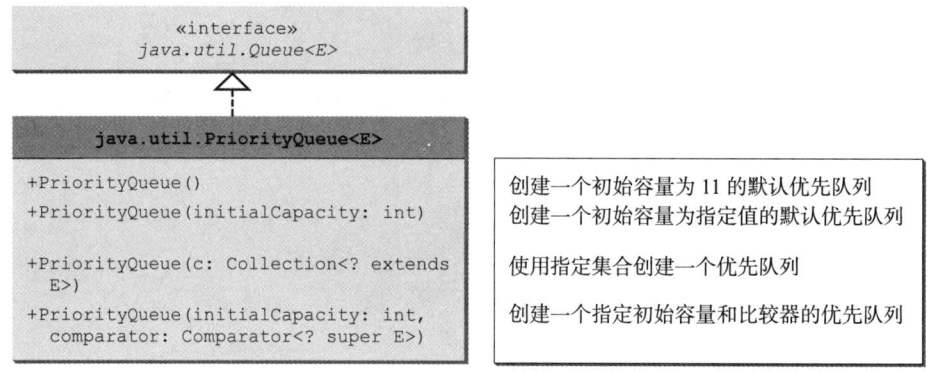

图 20-14 PriorityQueue 类实现了一个优先队列

程序清单 20-11 给出一个使用优先队列存储字符串的例子。第 5 行使用无参构造方法创建字符串优先队列。该优先队列以字符串的自然顺序进行排序，这样，字符串以升序从队列中删除。第 16 和 17 行使用从 Collections.reverseOrder() 中获得的比较器创建优先队列，该比较器以降序对元素排序，因此，字符串以降序从队列中删除。

程序清单 20-11 PriorityQueueDemo.java

```java
 1  import java.util.*;
 2
 3  public class PriorityQueueDemo {
 4    public static void main(String[] args) {
 5      PriorityQueue<String> queue1 = new PriorityQueue<>();
 6      queue1.offer("Oklahoma");
 7      queue1.offer("Indiana");
 8      queue1.offer("Georgia");
 9      queue1.offer("Texas");
10
```

```
11      System.out.println("Priority queue using Comparable:");
12      while (queue1.size() > 0) {
13        System.out.print(queue1.remove() + " ");
14      }
15
16      PriorityQueue<String> queue2 = new PriorityQueue<>(
17        4, Collections.reverseOrder());
18      queue2.offer("Oklahoma");
19      queue2.offer("Indiana");
20      queue2.offer("Georgia");
21      queue2.offer("Texas");
22
23      System.out.println("\nPriority queue using Comparator:");
24      while (queue2.size() > 0) {
25        System.out.print(queue2.remove() + " ");
26      }
27    }
28  }
```

```
Priority queue using Comparable:
Georgia Indiana Oklahoma Texas
Priority queue using Comparator:
Texas Oklahoma Indiana Georgia
```

✓ 复习题

20.10.1 java.util.Queue 是 java.util.Collection、java.util.Set 还是 java.util.List 的子接口？LinkedList 实现了 Queue 吗？

20.10.2 如何创建一个整数优先队列？默认情况下，优先队列的元素如何排序？在优先队列中，拥有最小数值的元素被赋予最高优先级吗？

20.10.3 如何创建一个逆转元素的自然顺序的优先队列？

20.11 示例学习：表达式求值

要点提示：栈可以用于进行表达式求值。

栈和队列具有许多应用。本节给出一个使用栈来对表达式求值的程序。你可以在 Google 页面上输入一个算术表达式来求值，如图 20-15 所示。

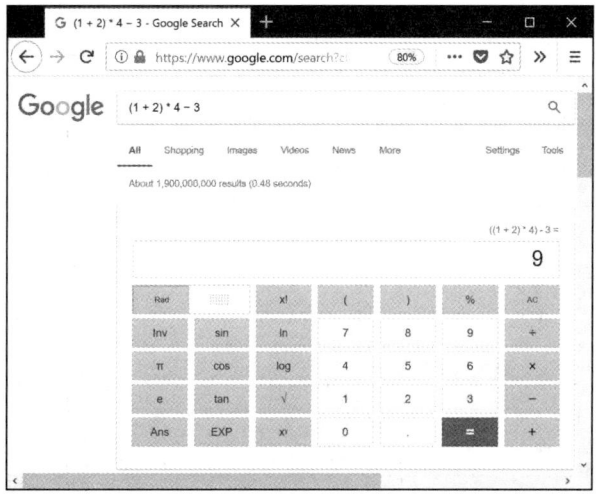

图 20-15 可以使用 Google 搜索引擎来对算术表达式求值（来源：Google 和 Google 标志是谷歌公司的注册商标，经授权使用）

Google 是如何对表达式求值的呢？本节给出一个程序，对具有多个操作符和括号的复合表达式求值（例如，`(1 + 2)*4 - 3`）。为简化起见，假设操作数都是整数，并且操作符是 +、-、* 和 / 这四种之一。

这个问题可以使用两个栈来解决，将这两个栈命名为 operandStack 和 operatorStack，分别用于存储操作数和操作符。操作数和操作符在被处理前先压入栈中。当一个操作符被处理时，它从 operatorStack 中弹出，并应用于 operandStack 的前面两个操作数（这两个操作数是从 operandStack 中弹出的）。结果数值被压回 operandStack。

该算法分两个阶段进行。

阶段 1：扫描表达式

程序从左到右扫描表达式，提取出操作数、操作符以及括号。

1.1　如果提取的项是操作数，则将其压入 operandStack。

1.2　如果提取的项是 + 或 - 操作符，处理在 operatorStack 栈顶的所有操作符，并将提取出的操作符压入 operatorStack。

1.3　如果提取的项是 * 或 / 操作符，处理在 operatorStack 栈顶的所有 * 和 / 操作符，将提取出的操作符压入 operatorStack。

1.4　如果提取的项是（符号，将它压入 operatorStack。

1.5　如果提取的项是）符号，重复处理来自 operatorStack 栈顶的操作符，直到看到栈上的（符号。

阶段 2：清除栈

重复处理来自 operatorStack 栈顶的操作符，直到 operatorStack 为空为止。

表 20-1 显示了如何应用该算法来计算表达式 `(1+2)*4-3`。

表 20-1　对一个表达式求值

表达式	扫描	动作	operandStack	operatorStack
(1 + 2) * 4 − 3　↑	(	阶段 1.4		(
(1 + 2) * 4 − 3　↑	1	阶段 1.1	1	(
(1 + 2) * 4 − 3　↑	+	阶段 1.2	1	+ (
(1 + 2) * 4 − 3　↑	2	阶段 1.1	2 1	(
(1 + 2) * 4 − 3　↑	)	阶段 1.5	3	
(1 + 2) * 4 − 3　↑	*	阶段 1.3	3	*
(1 + 2) * 4 − 3　↑	4	阶段 1.1	4 3	*
(1 + 2) * 4 − 3　↑	−	阶段 1.2	12	−
(1 + 2) * 4 − 3　↑	3	阶段 1.1	3 12	−
(1 + 2) * 4 − 3　↑	（无）	阶段 2	9	

程序清单 20-12 给出了这个程序。图 20-16 给出了一些示例输出。

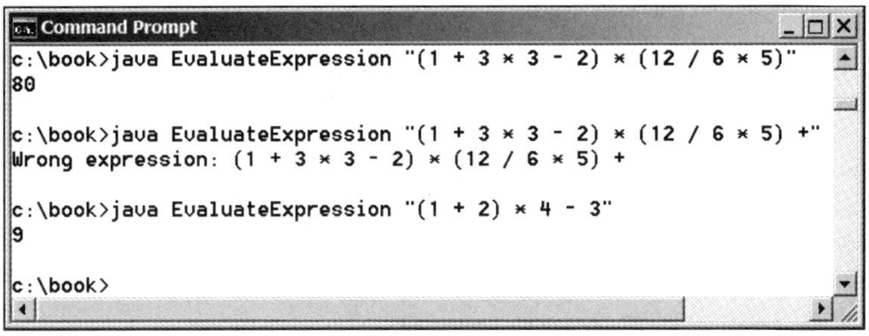

图 20-16 该程序将一个表达式作为命令行参数（来源：Oracle 或其附属公司版权所有 ©1995 ～ 2016，经授权使用）

程序清单 20-12 EvaluateExpression.java

```java
 1  import java.util.Stack;
 2
 3  public class EvaluateExpression {
 4    public static void main(String[] args) {
 5      // Check number of arguments passed
 6      if (args.length != 1) {
 7        System.out.println(
 8          "Usage: java EvaluateExpression \"expression\"");
 9        System.exit(1);
10      }
11
12      try {
13        System.out.println(evaluateExpression(args[0]));
14      }
15      catch (Exception ex) {
16        System.out.println("Wrong expression: " + args[0]);
17      }
18    }
19
20    /** Evaluate an expression */
21    public static int evaluateExpression(String expression) {
22      // Create operandStack to store operands
23      Stack<Integer> operandStack = new Stack<>();
24
25      // Create operatorStack to store operators
26      Stack<Character> operatorStack = new Stack<>();
27
28      // Insert blanks around (, ), +, -, /, and *
29      expression = insertBlanks(expression);
30
31      // Extract operands and operators
32      String[] tokens = expression.split(" ");
33
34      // Phase 1: Scan tokens
35      for (String token: tokens) {
36        if (token.length() == 0) // Blank space
37          continue; // Back to the while loop to extract the next token
38        else if (token.charAt(0) == '+' || token.charAt(0) == '-') {
39          // Process all +, -, *, / in the top of the operator stack
40          while (!operatorStack.isEmpty() &&
41            (operatorStack.peek() == '+' ||
42             operatorStack.peek() == '-' ||
43             operatorStack.peek() == '*' ||
44             operatorStack.peek() == '/')) {
45            processAnOperator(operandStack, operatorStack);
46          }
```

```java
47
48          // Push the + or - operator into the operator stack
49          operatorStack.push(token.charAt(0));
50        }
51        else if (token.charAt(0) == '*' || token.charAt(0) == '/') {
52          // Process all *, / in the top of the operator stack
53          while (!operatorStack.isEmpty() &&
54            (operatorStack.peek() == '*' ||
55            operatorStack.peek() == '/')) {
56            processAnOperator(operandStack, operatorStack);
57          }
58
59          // Push the * or / operator into the operator stack
60          operatorStack.push(token.charAt(0));
61        }
62        else if(token.trim().charAt(0) =='(') {
63          operatorStack.push('('); // Push '(' to stack
64        }
65        else if (token.trim().charAt(0) ==')') {
66          // Process all the operators in the stack until seeing '('
67          while (operatorStack.peek() != '(') {
68            processAnOperator(operandStack, operatorStack);
69          }
70
71          operatorStack.pop(); // Pop the '(' symbol from the stack
72        }
73        else { // An operand scanned
74          // Push an operand to the stack
75          operandStack.push(Integer.valueOf(token));
76        }
77      }
78
79      // Phase 2: Process all the remaining operators in the stack
80      while (!operatorStack.isEmpty()) {
81        processAnOperator(operandStack, operatorStack);
82      }
83
84      // Return the result
85      return operandStack.pop();
86    }
87
88    /** Process one operator: Take an operator from operatorStack and
89     *  apply it on the operands in the operandStack */
90    public static void processAnOperator(
91        Stack<Integer> operandStack, Stack<Character> operatorStack) {
92      char op = operatorStack.pop();
93      int op1 = operandStack.pop();
94      int op2 = operandStack.pop();
95      if (op == '+')
96        operandStack.push(op2 + op1);
97      else if (op == '-')
98        operandStack.push(op2 - op1);
99      else if (op == '*')
100        operandStack.push(op2 * op1);
101      else if (op == '/')
102        operandStack.push(op2 / op1);
103    }
104
105    public static String insertBlanks(String s) {
106      String result = "";
107
108      for (int i = 0; i < s.length(); i++) {
109        if (s.charAt(i) == '(' || s.charAt(i) == ')' ||
110          s.charAt(i) == '+' || s.charAt(i) == '-' ||
111          s.charAt(i) == '*' || s.charAt(i) == '/')
112          result += " " + s.charAt(i) + " ";
```

```
113         else
114             result += s.charAt(i);
115     }
116
117     return result;
118   }
119 }
```

可以使用本书提供的 GenericStack 或者定义在 Java API 中的 java.util.Stack 类来创建栈。本示例使用 java.util.Stack 类。如果替换成 GenericStack，程序依然可以运行。

该程序将一个表达式以一个字符串的形式作为命令行参数。

evaluateExpression 方法创建 operandStack 和 operatorStack 两个栈（第 23 和 26 行），并且提取被空格分隔的操作数、操作符以及括号（第 29 ~ 32 行）。insertBlanks 方法用于保证操作数、操作符以及括号至少被一个空格分隔（第 29 行）。

程序在 for 循环中扫描每个标记（第 35 ~ 77 行）。如果标记是空的，就跳过它（第 37 行）。如果标记是一个操作数，就将它压入 operandStack（第 75 行）。如果标记是一个 + 或 - 操作符（第 38 行），就处理在 operatorStack 栈顶的所有操作符（如果有的话）（第 40 ~ 46 行），并将新扫描到的操作符压入栈中（第 49 行）。如果标记是一个 * 或 / 操作符（第 51 行），就处理在 operatorStack 栈顶的所有 * 和 / 操作符（如果有的话）（第 53 ~ 57 行），并将新扫描到的操作符压入栈中（第 60 行）。如果标记是一个 "(" 符号（第 62 行），则将其压入 operatorStack。如果标记是一个 ")" 符号（第 65 行），则处理 operatorStack 栈顶的所有操作符，直到看到 "(" 符号为止（第 67 ~ 69 行），然后从栈中弹出 "(" 符号。

在考虑完所有的标记之后，程序处理 operatorStack 中剩余的操作符（第 80 ~ 82 行）。

processAnOperator 方法（第 90 ~ 103 行）用来处理一个操作符。该方法从 operatorStack 中弹出一个操作符（第 92 行），并且从 operandStack 中弹出两个操作数（第 93 和 94 行）。依据所弹出的操作符，该方法完成对应的操作，然后将操作结果压回 operandStack 中（第 96、98、100 和 102 行）。

✓ 复习题

20.11.1 EvauateExpression 程序可以对表达式 "1+2"、"1+2"、"(1)+2"、"((1))+2" 以及 "(1+2)" 求值吗？

20.11.2 使用 EvaluateExpression 程序对 "3+(4+5)*(3+5) +4*5" 求值时，给出栈中内容的变化。

20.11.3 如果输入表达式 "4+5 5 5"，程序将显示 10。如何修改这个问题？

关键术语

collection（集合）
comparator（比较器）
convenience abstract class（便利抽象类）
data structure（数据结构）
linked list（链表）
list（线性表）
priority queue（优先队列）
queue（队列）

本章小结

1. Collection 接口为线性表、向量、栈、队列、优先队列、规则集等定义了通用的操作。
2. 每一个集合都是 Iterable，可以获得其 Iterator 对象来遍历集合中的所有元素。

3. 除了 PriorityQueue 以外，Java 集合框架中的所有具体类都实现了 Cloneable 和 Serializable 接口。所以，它们的实例都是可复制和可序列化的。
4. 一个线性表存储一个有序的元素集合。若要在集合中存储重复的元素，就需要使用线性表。线性表不仅可以存储重复的元素，而且还允许用户指定存储位置。用户可以通过下标来访问线性表中的元素。
5. Java 集合框架支持两种类型的线性表：数组线性表 ArrayList 和链表 LinkedList。ArrayList 是 List 接口的一个可变大小数组实现。ArrayList 中的所有方法都在 List 接口中有定义。LinkedList 是 List 接口的链表实现。除实现了 List 接口外，该类还提供了可从线性表两端提取、插入以及删除元素的方法。
6. Comparator 可以用于比较没有实现 Comparable 接口的类的对象。
7. Vector 类继承了 AbstractList 类。从 Java 2 开始，Vector 类和 ArrayList 是一样的，所不同的是它所包含的访问和修改向量的方法是同步的。Stack 类继承了 Vector 类，并且提供了几种对栈进行操作的方法。
8. Queue 接口表示队列。PriorityQueue 类实现了 Queue 接口，用于优先队列。

测试题

回答位于本书配套网站上的本章测试题。

编程练习题

20.2 ～ 20.7 节

*20.1 （按字母序的升序显示单词）编写一个程序，从文本文件读取单词，并按字母的升序显示所有的单词（可以重复）。单词必须以字母开头。文本文件作为命令行参数传递。

*20.2 （在链表中存储数字）编写一个程序，让用户从图形用户界面中输入数字，然后在文本域显示它们，如图 20-17a 所示。使用链表存储这些数字，但不要存储重复的数值。添加按钮 Sort、Shuffle 和 Reverse，分别对该表进行排序、打乱顺序与逆序排列操作。

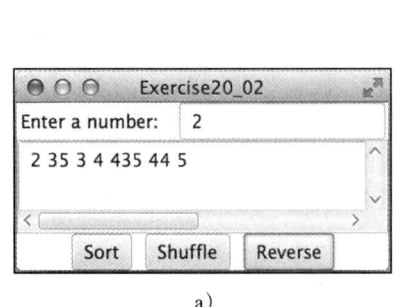
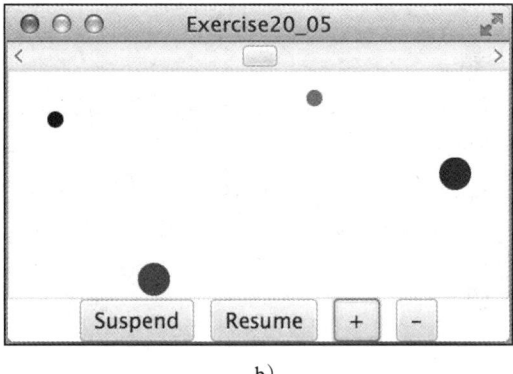

a) b)

图 20-17 a) 数字保存在线性表中并显示在一个文本区域内（来源：Oracle 或其附属公司版权所有 ©1995 ～ 2016，经授权使用）; b) 相撞的球结合在一起

*20.3 （猜首府）改写编程练习题 8.37，保存州和首府的匹配对，从而随机显示问题。

*20.4 （对面板上的点进行排序）编写一个程序，满足下面的要求：使用 Point2D 随机创建 100 个点，并且应用 Arrays.sort(list,Comparator) 方法进行排序，首先通过 y 坐标的升序对点进行排序，如果 y 相同，则按照 x 坐标的升序排序。显示前 5 个点的 x 和 y 坐标。

***20.5 （合并碰撞的弹球）20.8 节的示例中显示了多个弹球。扩充该例子来进行碰撞检测。一旦两个球相撞，移除后加入面板的那个球，并且将它的半径加到另外一个球上，如图 20-17b 所示。用 Suspend 按钮来暂停动画，用 Resume 按钮来继续动画。添加一个按下鼠标的处理器，从而在鼠标按在球上的时候移除这个球。

20.6 （在链表上使用迭代器）编写一个测试程序，在一个链表上存储 500 万个整数，测试分别使用 iterator 和 get(index) 方法的遍历时间。

***20.7 （游戏：猜字游戏）编程练习题 7.35 给出了流行的猜字游戏的控制台版本。编写一个 GUI 程序让用户来玩这个游戏。用户通过一次输入一个字母来猜单词，如图 20-18 所示。如果用户 7 次都没猜对，悬挂的小人就摆动起来。一旦完成一个单词，用户就可以按下 Enter 键继续猜另一个单词。

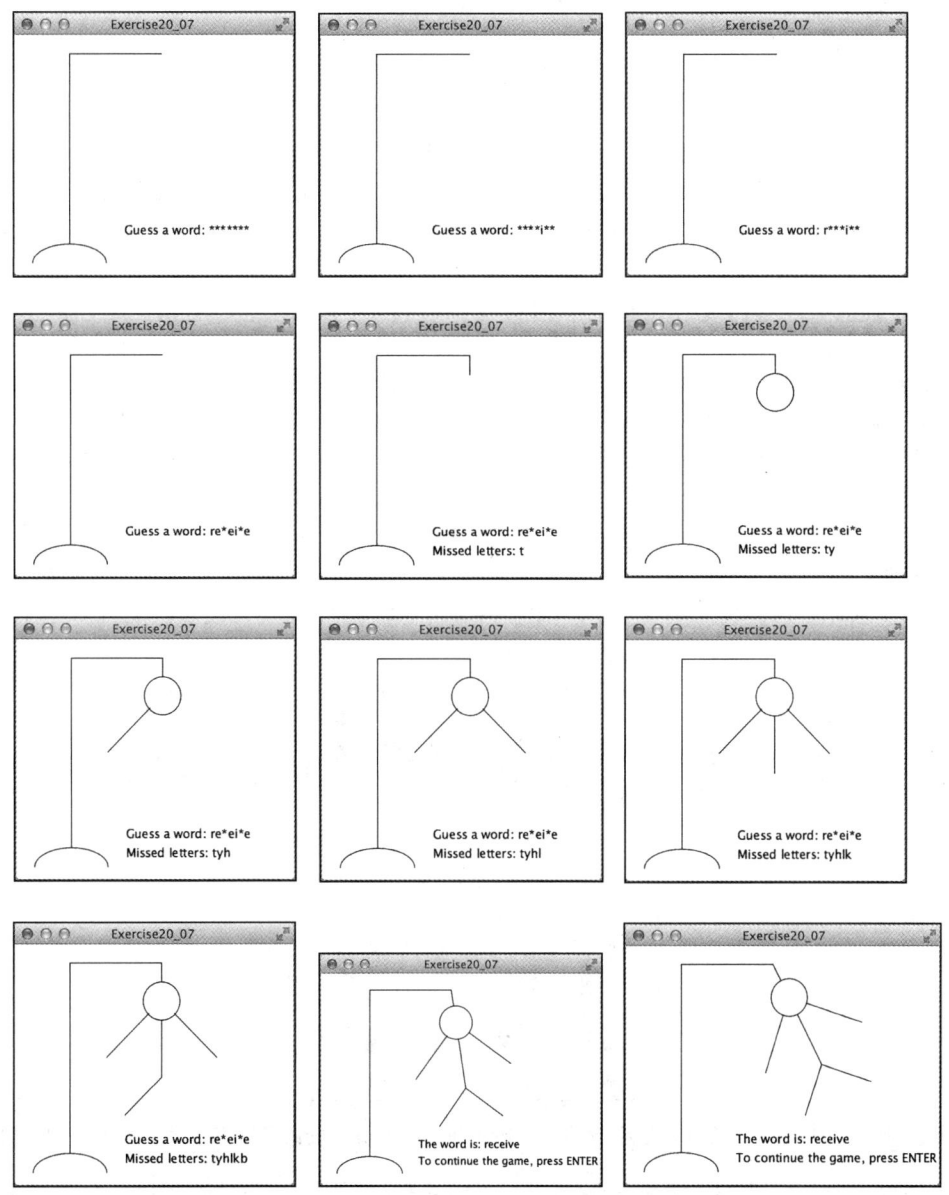

图 20-18 该程序显示猜字游戏（来源：Oracle 或其附属公司版权所有 ©1995～2016，经授权使用）

20.8 (游戏：彩票) 修改编程练习题 3.15，如果用户输入的两个数字在彩票号码之中，则增加额外的 2000 美元。(提示：对彩票中的三个数字和用户输入的三个数字进行排序，并分别存入两个线性表中，然后使用 Collection 的 containsAll 方法来检查用户输入的两个数字是否在彩票数字中。)

20.8～20.10 节

***20.9 (首先移除最大的球)** 修改程序清单 20-10，使得一个球在被创建的时候被赋予 2 到 20 的随机半径值。当点击 "-" 按钮时，则移除最大的球。

20.10 (在优先队列上进行集合操作) 编写一个程序创建两个优先队列 {"George", "Jim", "John", "Blake", "Kevin", "Michael"} 和 {"George", "Katie", "Kevin", "Michelle", "Ryan"}，求它们的并集、差集和交集。

*20.11 (匹配编组符号)** Java 程序包含各种编组符号对，例如：
- 圆括号：(和)
- 花括号：{ 和 }
- 方括号：[和]

请注意编组符号不能交叉。例如，(a{b}) 是不合法的。编写一个程序，检查一个 Java 源程序中的编组符号是否都正确匹配了。将源代码文件的名字作为命令行参数传递。

20.12 (克隆 PriorityQueue) 定义 MyPriorityQueue 类，继承自 PriorityQueue 并实现 Cloneable 接口和 clone() 方法，以克隆一个优先队列。

***20.13 (游戏：24 点扑克牌游戏)** 24 点扑克牌游戏是指从 52 张牌中任意选取 4 张，如图 20-19 所示。注意，将两个王排除在外。每张牌表示一个数字。A、K、Q 和 J 分别表示 1、13、12 和 11。你可以点击 Shuffle 按钮来获取 4 张新的扑克牌。输入这 4 张扑克牌面的 4 个数字构成的一个表达式。每个数字必须使用且只能使用一次。可以在表达式中使用运算符（加法、减法、乘法和除法）以及括号。表达式必须计算出 24。在输入表达式之后，单击 Verify 按钮来检查表达式中的数字是否是当前所选择的扑克牌面上的数，并检查表达式的结果是否正确。检查结果显示在 Shuffle 按钮前面的一个标签中。假设图像以黑桃、红心、方块和梅花的顺序存储在名为 1.png, 2.png, ⋯, 52.png 的文件中，这样前 13 个图像就是黑桃的 1, 2, 3, ⋯, 13。

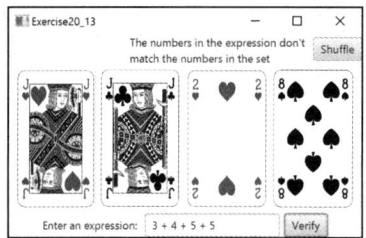
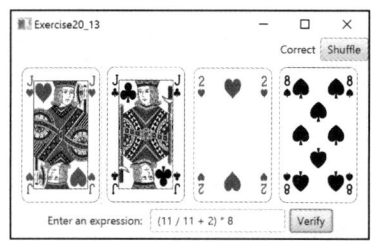
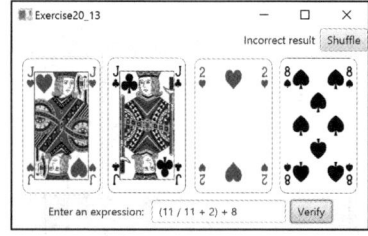

图 20-19 用户输入由牌面数字组成的表达式，并点击 Verify 按钮来检查结果（来源：Fotolia）

20.14 (后缀表示法) 后缀表示法是一种不使用括号编写表达式的方法。例如，表达式 (1 + 2) * 3 可以写为 1 2 + 3 *。后缀表达式是使用栈来计算的。从左到右扫描后缀表达式，将变量或常量压入栈内，当遇到运算符时，将该运算符应用在栈顶的两个操作数上，然后用运算结果替换这两个操作数。下图演示了如何计算 1 2 + 3 *。

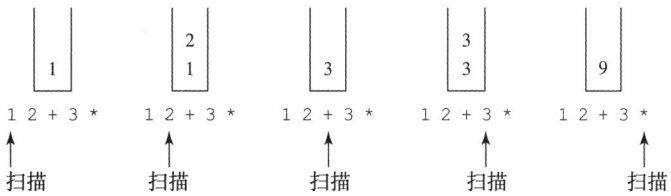

编写一个程序，计算后缀表达式，用一个字符串将后缀表达式作为命令行的参数传递。

***20.15 （游戏：24 点扑克牌游戏）改进编程练习题 20.13，如果表达式存在，则让计算机显示它，如图 20-20 所示；否则，报告这样的表达式不存在。将显示验证结果的标签置于 UI 的底部。表达式必须使用所有 4 张扑克牌并且值等于 24。

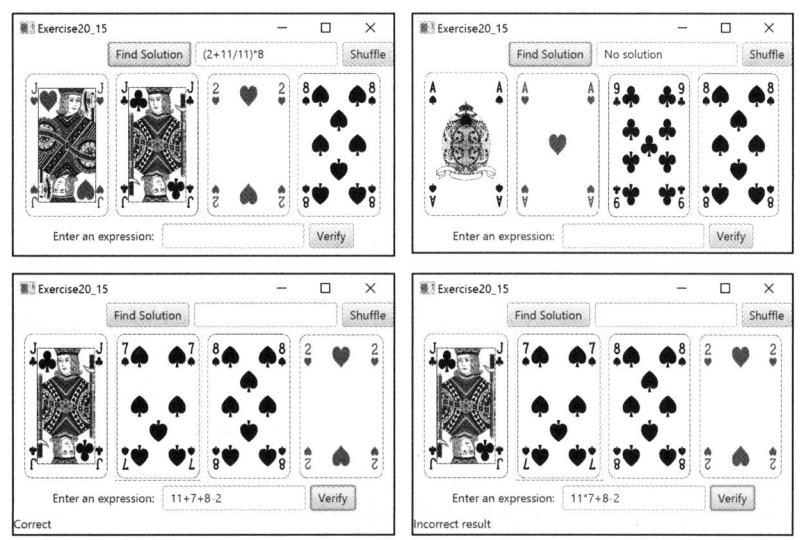

图 20-20　该程序可以自动找到一个解决方案，如果存在一个这样的方案的话（来源：Fotolia）

**20.16 （将中缀转换为后缀）使用下面的方法头编写方法，将中缀表达式转换为后缀表达式：

public static String infixToPostfix(String expression)

例如，该方法可以将中缀表达式 (1 + 2) *3 转换为 1 2 + 3 *，将 2 * (1 + 3) 转换为 2 1 3 + *。编写一个程序，接收命令行中作为一个参数的表达式，并显示相应的后缀表达式。

***20.17 （游戏：24 点扑克牌游戏）该题是编程练习题 20.13 中描述的 24 点扑克牌游戏的变体。编写一个程序，检查是否有这 4 个给定数的 24 点解决方案。该程序让用户输入在 1 到 13 之间的 4 个值，如图 20-21 所示。然后用户可以单击 Solve 按钮显示解决方案，若不存在解决方案，则提示"不存在解决方案"。

图 20-21　用户输入 4 个数字，然后该程序找出解决方案（来源：Oracle 或其附属公司版权所有 ©1995 ～ 2016，经授权使用）

*20.18 （目录大小）程序清单 18-10 使用递归方法来得到目录的大小。重写该方法，不使用递归。你的

程序应该使用队列来存储一个目录下的所有子目录。算法描述如下：

```
long getSize(File directory) {
  long size = 0;
  add directory to the queue;

  while (queue is not empty) {
    Remove an item from the queue into t;
    if (t is a file)
      size += t.length();
    else
      add all the files and subdirectories under t into the
        queue;
  }

  return size;
}
```

***20.19 （游戏：24点游戏有解的比例）回顾编程练习题20.13介绍的24点游戏，从52张牌中选择4张牌，这4张牌可能没有能得到24点的解决方案。从52张牌中选择4张牌的所有可能的挑选次数是多少？在这些所有可能的挑选中，有多少可以得到24点？成功的概率是多少（即（有解的挑选次数）/（四张牌的所有可能挑选次数））？编写一个程序，得出这些答案。

*20.20 （目录大小）重写编程练习题18.28，使用栈而不是队列来解决这个问题。

*20.21 （使用Comparator）使用选择排序和比较器，编写以下泛型方法：

```
public static <E> void selectionSort(E[] list,
    Comparator<? super E> comparator)
```

编写一个测试程序，提示用户输入6个字符串，调用sort方法根据它们的最后一个字符来对这6个字符串排序，并显示排好序的字符串。使用Scanner的next()方法来读取字符串。

*20.22 （非递归的汉诺塔实现）使用栈而不是递归实现程序清单18-8中的moveDisks方法。

**20.23 （表达式求值）修改程序清单20-12，增加指数运算符^和求模运算符%。例如，3 ^ 2等于9，3 % 2等于1。运算符^具有最高优先级，运算符%与*和/具有一样的优先级。你的程序应该提示用户输入一个表达式。下面是一个程序的运行示例：

```
Enter an expression: (5 * 2 ^ 3 + 2 * 3 % 2) * 4 ↵Enter
(5 * 2 ^ 3 + 2 * 3 % 2) * 4 = 160
```

第 21 章

规则集和映射

教学目标
- 使用规则集存储无序的、没有重复的元素（21.2 节）。
- 探索如何使用以及何时使用 HashSet（21.2.1 节）、LinkedHashSet（21.2.2 节）或者 TreeSet（21.2.3 节）来存储元素集。
- 比较规则集和线性表的性能（21.3 节）。
- 使用规则集开发一个计算 Java 源文件中关键字数目的程序（21.4 节）。
- 区分 Collection 与 Map，并描述何时及如何使用 HashMap、LinkedHashMap 或者 TreeMap 来存储与键关联的值（21.5 节）。
- 使用映射开发一个计算文本文件中单词出现次数的程序（21.6 节）。
- 使用 Collections 类中的静态方法来获得单例规则集、线性表和映射，以及不可修改的规则集、线性表和映射（21.7 节）。

21.1 引言

要点提示：规则集（set）是一个用于存储和处理无重复元素的高效数据结构。映射（map）类似于字典，提供了使用键快速查询以获取值。

"禁飞"列表是一个由美国政府的恐怖分子筛选检查中心创建和维护的一张表，列出了不允许搭乘商业飞机进出美国的人员名单。假设我们需要写一个程序，检验一个人是否在禁飞名单上，可以使用一个线性表来存储禁飞名单上面的名字。然而，针对这个应用的更有效的数据结构是规则集（set）。

假设你的程序还需要存储禁飞名单上恐怖分子的详细信息，可以使用名字作为键来获取诸如性别、身高、体重以及国籍等详细信息。映射（map）是针对这种任务的有效数据结构。

本章介绍 Java 集合框架中的规则集和映射。

21.2 规则集

要点提示：可以使用规则集的三个具体类 HashSet、LinkedHashSet 或者 TreeSet 来创建规则集。

Set 接口扩展了 Collection 接口，如图 20-1 所示。它没有引入新的方法或常量，只是规定了 Set 的实例不能包含重复的元素。实现 Set 的具体类必须确保不能向这个规则集添加重复的元素。

AbstractSet 类继承自 AbstractCollection 类并部分实现 Set 接口。AbstractSet 类提供 equals 方法和 hashCode 方法的具体实现。一个规则集的散列码是这个规则集中所有元素散列码的和。由于 AbstractSet 类没有实现 size 方法和 iterator 方法，所以 AbstractSet 类是一个抽象类。

Set 接口的三个具体类是 HashSet、LinkedHashSet 和 TreeSet，如图 21-1 所示。

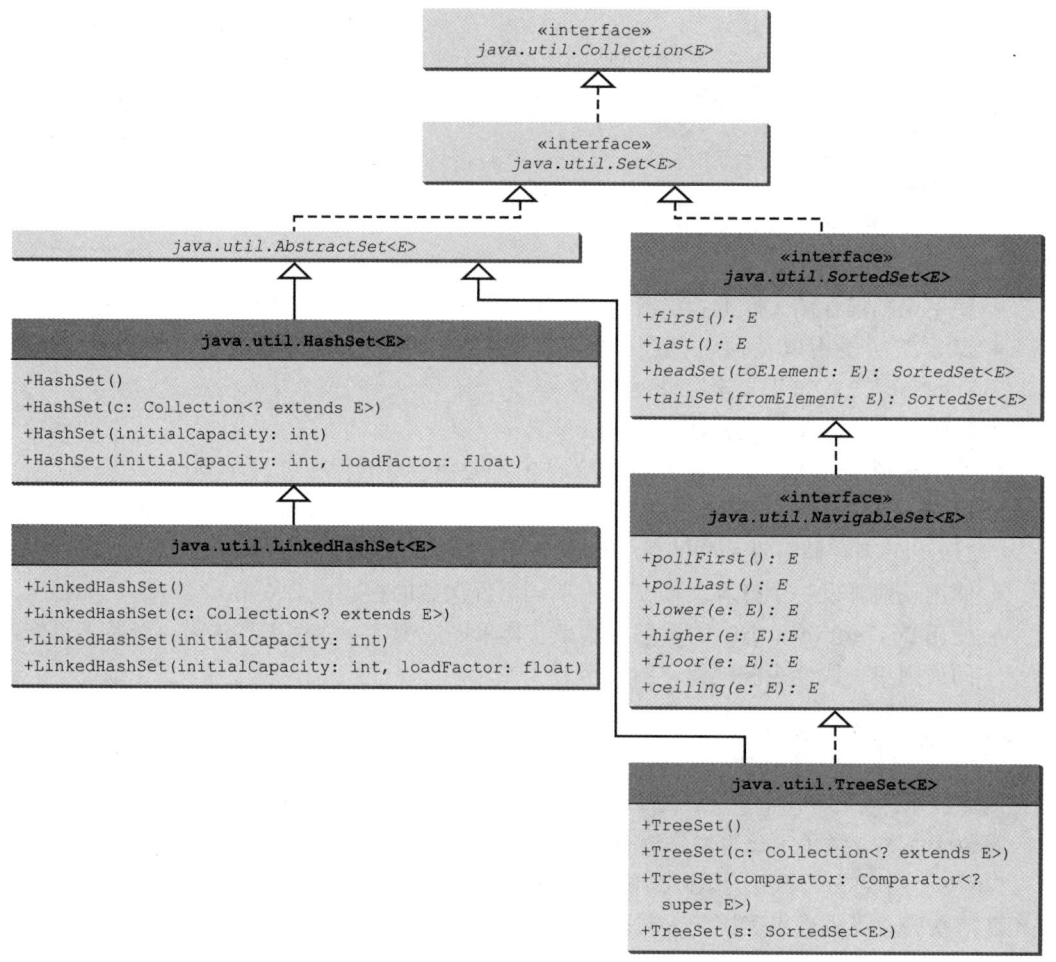

图 21-1 Java 集合框架提供了三个具体的规则集类

21.2.1 HashSet

　　HashSet 类是一个实现了 Set 接口的具体类。可以使用它的无参构造方法来创建空的散列集（hash set），也可以由一个现有的集合创建散列集。默认情况下，初始容量为 16 而负载系数是 0.75。如果知道规则集的大小，就可以在构造方法中指定初始容量和负载系数。否则，使用默认的设置，负载系数的值在 0.0 ~ 1.0 之间。

　　在增加规则集的容量之前，负载系数（load factor）测量该规则集允许填充多满。当元素个数超过了容量与负载系数的乘积，容量就会自动翻倍。例如，如果容量是 16 而负载系数是 0.75，那么当尺寸达到 12（16×0.75=12）时，容量将会翻倍到 32。比较高的负载系数会降低空间开销，但是会增加查找时间。通常情况下，默认的负载系数 0.75 是时间开销与空间开销之间的一个好的权衡。我们将在第 27 章更深入地讨论负载系数。

　　HashSet 类可以用来存储不重复的元素。考虑到效率，添加到散列集中的对象必须以一种正确分布散列码的方式来实现 hashCode 方法。hashCode 方法在 Object 类中定义。如果两个对象相等，那么这两个对象的散列码必须一样。两个不相等的对象可能会有相同的散列码，因此你应该实现 hashCode 方法以避免出现太多这样的情况。Java API 中的大多数类都实现了 hashCode 方法。例如，Integer 类中的 hashCode 方法返回其 int 值，Character 类中的 hashCode 方法返回这个字符的 Unicode 码，String 类中的 hashCode 方法返回 $s_0*31^{(n-1)}+s_1*31^{(n-2)}+\cdots+s_{n-1}$，其中 s_i 是 s.charAt(i)。

Set 不存储重复元素。如果两个元素 e1 和 e2 满足 e1.equals(e2) 为真，并且 e1.hashCode()==e2.hashCode()，那么对于一个 HashSet 而言来说认为 e1 和 e2 是重复的。注意，反过来说，如果两个元素相等，它们的 hashCode 必须一样。因此，当类中的 equals 方法被重写时，需要重写 hashCode() 方法。

程序清单 21-1 给出的程序创建了一个散列集来存储字符串，并且使用一个 foreach 循环和一个 forEach 方法来遍历这个规则集中的元素。

程序清单 21-1 TestHashSet.java

```
1  import java.util.*;
2
3  public class TestHashSet {
4    public static void main(String[] args) {
5      // Create a hash set
6      Set<String> set = new HashSet<>();
7
8      // Add strings to the set
9      set.add("London");
10     set.add("Paris");
11     set.add("New York");
12     set.add("San Francisco");
13     set.add("Beijing");
14     set.add("New York");
15
16     System.out.println(set);
17
18     // Display the elements in the hash set
19     for (String s: set) {
20       System.out.print(s.toUpperCase() + " ");
21     }
22
23     // Process the elements using a forEach method
24     System.out.println();
25     set.forEach(e -> System.out.print(e.toLowerCase() + " "));
26   }
27 }
```

```
[San Francisco, New York, Paris, Beijing, London]
SAN FRANCISCO NEW YORK PARIS BEIJING LONDON
san francisco beijing new york london pairs
```

该程序将多个字符串添加到规则集中（第 9 ~ 14 行）。New York 被添加多次，但是只有一个被存储，因为规则集不允许有重复元素。

如输出所示，字符串没有按照它们被插入规则集时的顺序存储。散列集中的元素没有特定的顺序。要强加给它们一个顺序，就需要使用 LinkedHashSet 类，这个类将在下一节中介绍。

回顾前面提到的，Collection 接口继承自 Iterable 接口，因此规则集中的元素是可遍历的。使用了 foreach 循环来遍历规则集中的所有元素（第 19 ~ 21 行），还可以使用 forEach 方法对规则集中的每个元素进行操作（第 25 行）。

由于规则集是 Collection 的实例，因此，定义在 Collection 中的所有方法都可以用于规则集。程序清单 21-2 给出一个应用 Collection 接口中方法的例子。

程序清单 21-2 TestMethodsInCollection.java

```
1  public class TestMethodsInCollection {
```

```java
 2  public static void main(String[] args) {
 3    // Create set1
 4    java.util.Set<String> set1 = new java.util.HashSet<>();
 5
 6    // Add strings to set1
 7    set1.add("London");
 8    set1.add("Paris");
 9    set1.add("New York");
10    set1.add("San Francisco");
11    set1.add("Beijing");
12
13    System.out.println("set1 is " + set1);
14    System.out.println(set1.size() + " elements in set1");
15
16    // Delete a string from set1
17    set1.remove("London");
18    System.out.println("\nset1 is " + set1);
19    System.out.println(set1.size() + " elements in set1");
20
21    // Create set2
22    java.util.Set<String> set2 = new java.util.HashSet<>();
23
24    // Add strings to set2
25    set2.add("London");
26    set2.add("Shanghai");
27    set2.add("Paris");
28    System.out.println("\nset2 is " + set2);
29    System.out.println(set2.size() + " elements in set2");
30
31    System.out.println("\nIs Taipei in set2? "
32      + set2.contains("Taipei"));
33
34    set1.addAll(set2);
35    System.out.println("\nAfter adding set2 to set1, set1 is "
36      + set1);
37
38    set1.removeAll(set2);
39    System.out.println("After removing set2 from set1, set1 is "
40      + set1);
41
42    set1.retainAll(set2);
43    System.out.println("After retaining common elements in set2 "
44      + "and set2, set1 is " + set1);
45  }
46 }
```

```
set1 is [San Francisco, New York, Paris, Beijing, London]

5 elements in set1

set1 is [San Francisco, New York, Paris, Beijing]
4 elements in set1

set2 is [Shanghai, Paris, London]
3 elements in set2

Is Taipei in set2? false

After adding set2 to set1, set1 is
  [San Francisco, New York, Shanghai, Paris, Beijing, London]

After removing set2 from set1, set1 is
  [San Francisco, New York, Beijing]

After retaining common elements in set1 and set2, set1 is []
```

该程序创建了两个规则集（第 4 和 22 行）。size() 方法返回规则集中的元素个数（第 14

行）。第 17 行

```
set1.remove("London");
```

从 set1 中删除 London。

contains 方法（第 32 行）检查一个元素是否在规则集中。

第 34 行

```
set1.addAll(set2);
```

将 set2 添加到 set1 中。这样，set1 就变成 [San Francisco, New York, Shanghai, Paris, Beijing, London]。

第 38 行

```
set1.removeAll(set2);
```

从 set1 中删除 set2。这样，set1 就变成 [San Francisco, New York, Beijing]。

第 42 行

```
set1.retainAll(set2);
```

保留 set1 和 set2 共有的元素。因为 set1 和 set2 没有公共的元素，所以 set1 变为空。

21.2.2 LinkedHashSet

LinkedHashSet 扩展了 HashSet 类，用一个链表实现支持对规则集内的元素排序。HashSet 中的元素是没有顺序的，而 LinkedHashSet 中的元素可以按照它们插入规则集的顺序获取。可以使用 4 个构造方法之一来创建 LinkedHashSet，如图 21-1 所示。这些构造方法类似于 HashSet 的构造方法。

程序清单 21-3 了给出一个 LinkedHashSet 的测试程序。该程序只是简单地使用 LinkedHashSet 来替换程序清单 21-1 中的 HashSet。

程序清单 21-3　TestLinkedHashSet.java

```java
 1  import java.util.*;
 2
 3  public class TestLinkedHashSet {
 4    public static void main(String[] args) {
 5      // Create a hash set
 6      Set<String> set = new LinkedHashSet<>();
 7
 8      // Add strings to the set
 9      set.add("London");
10      set.add("Paris");
11      set.add("New York");
12      set.add("San Francisco");
13      set.add("Beijing");
14      set.add("New York");
15
16      System.out.println(set);
17
18      // Display the elements in the hash set
19      for (String element: set)
20        System.out.print(element.toLowerCase() + " ");
21    }
22  }
```

```
[London, Paris, New York, San Francisco, Beijing]
london paris new york san francisco beijing
```

第 6 行创建了一个 `LinkedHashSet`。如输出中所示,字符串按照它们插入规则集的顺序存储。由于 `LinkedHashSet` 是一个规则集,所以它不会存储重复的元素。

`LinkedHashSet` 保持了元素插入时的顺序。要强加一个不同的顺序(例如,升序或降序),可以使用下一节介绍的 `TreeSet` 类。

☞ **提示**:如果不需要维护元素插入时的顺序,则使用 `HashSet`,它比 `LinkedHashSet` 更加高效。

21.2.3 TreeSet

如图 21-1 所示,`SortedSet` 是 `Set` 的一个子接口,它可以确保规则集中的元素是有序的。另外,它还提供了 `first()` 和 `last()` 方法以分别返回规则集中的第一个元素和最后一个元素,以及 `headSet(toElement)` 和 `tailSet(fromElement)` 方法以分别返回规则集中元素小于 `toElement` 和大于或等于 `fromElement` 的那一部分。

`NavigableSet` 扩展了 `SortedSet`,并提供导航方法 `lower(e)`、`floor(e)`、`ceiling(e)` 和 `higher(e)` 以分别返回小于、小于或等于、大于或等于以及大于给定元素的元素。如果没有这样的元素,则返回 `null`。`pollFirst()` 和 `pollLast()` 方法会分别删除并返回树形集中的第一个元素和最后一个元素。

`TreeSet` 实现了 `SortedSet` 接口。为了创建 `TreeSet` 对象,可以使用如图 21-1 所示的构造方法。只要对象是可以互相比较的,就可以将它们添加到一个树形集(tree set)中。

如 20.5 节所讨论的,元素可以有两种方法进行比较:使用 `Comparable` 接口或者 `Comparator` 接口。

程序清单 21-4 给出使用 `Comparable` 接口对元素进行排序的例子。前面程序清单 21-3 中的例子以插入的顺序显示所有的字符串。该示例重写前面的例子,使用 `TreeSet` 类按照字母顺序来显示这些字符串。

程序清单 21-4 TestTreeSet.java

```
1  import java.util.*;
2
3  public class TestTreeSet {
4    public static void main(String[] args) {
5      // Create a hash set
6      Set<String> set = new HashSet<>();
7
8      // Add strings to the set
9      set.add("London");
10     set.add("Paris");
11     set.add("New York");
12     set.add("San Francisco");
13     set.add("Beijing");
14     set.add("New York");
15
16     TreeSet<String> treeSet = new TreeSet<>(set);
17     System.out.println("Sorted tree set: " + treeSet);
18
19     // Use the methods in SortedSet interface
20     System.out.println("first(): " + treeSet.first());
21     System.out.println("last(): " + treeSet.last());
22     System.out.println("headSet(\"New York\"): " +
23       treeSet.headSet("New York"));
24     System.out.println("tailSet(\"New York\"): " +
25       treeSet.tailSet("New York"));
26
```

```
27          // Use the methods in NavigableSet interface
28          System.out.println("lower(\"P\"): " + treeSet.lower("P"));
29          System.out.println("higher(\"P\"): " + treeSet.higher("P"));
30          System.out.println("floor(\"P\"): " + treeSet.floor("P"));
31          System.out.println("ceiling(\"P\"): " + treeSet.ceiling("P"));
32          System.out.println("pollFirst(): " + treeSet.pollFirst());
33          System.out.println("pollLast(): " + treeSet.pollLast());
34          System.out.println("New tree set: " + treeSet);
35      }
36  }
```

```
Sorted tree set: [Beijing, London, New York, Paris, San Francisco]
first(): Beijing
last(): San Francisco
headSet("New York"): [Beijing, London]
tailSet("New York"): [New York, Paris, San Francisco]
lower("P"): New York
higher("P"): Paris
floor("P"): New York
ceiling("P"): Paris
pollFirst(): Beijing
pollLast(): San Francisco
New tree set: [London, New York, Paris]
```

该示例创建了一个由字符串填充的散列集，然后创建一个由相同字符串构成的树形集，使用 Comparable 接口中的 compareTo 方法对树形集中的字符串进行排序。对于 TreeSet 而言，如果其中两个元素 e1 和 e2 满足使用 Comparable 的话 e1.compareTo(e2) 为 0，以及使用 Comparator 的话 e1.compare(e2) 为 0，则认为 e1 和 e2 是重复元素。

当使用语句 new TreeSet<>(Set)（第 16 行）从一个 HashSet 对象创建一个 TreeSet 对象时，规则集中的元素被排序。可以改写这个程序，使用 TreeSet 的无参构造方法来创建一个 TreeSet 的实例，然后将字符串添加到这个 TreeSet 对象中。

treeSet.first() 返回 treeSet 中的第一个元素（第 20 行）。treeSet.last() 返回 treeSet 中的最后一个元素（第 21 行）。treeSet.headSet("New York") 返回 treeSet 中 New York 之前的那些元素（第 22～23 行）。treeSet.tailSet("New York") 返回 treeSet 中 New York 之后的那些元素，包括 New York（第 24～25 行）。

treeSet.lower("P") 返回 treeSet 中小于 P 的最大元素（第 28 行）。treeSet.higher("P") 返回 treeSet 中大于 P 的最小元素（第 29 行）。treeSet.floor("P") 返回 treeSet 中小于或等于 P 的最大元素（第 30 行）。treeSet.ceiling("P") 返回 treeSet 中大于或等于 P 的最小元素（第 31 行）。treeSet.pollFirst() 删除 treeSet 中的第一个元素，并返回被删除的元素（第 32 行）。treeSet.pollLast() 删除 treeSet 中的最后一个元素，并返回被删除的元素（第 33 行）。

- 注意：Java 集合框架中的所有具体类（参见图 20-1）都至少有两个构造方法：一个是创建空集合的无参构造方法，另一个是从某个集合来创建实例的构造方法。这样，TreeSet 类中含有从集合 c 创建 TreeSet 对象的构造方法 TreeSet(Collection c)。在这个例子中，new TreeSet<>(set) 方法从集合 set 创建了 TreeSet 的一个实例。

- 提示：当更新一个规则集时，如果不需要保持元素的排序关系，就应该使用散列集，因为在散列集中插入和删除元素所花的时间较少。当需要一个排序的规则集时，可以从该散列集创建一个树形集。

如果使用无参构造方法创建一个 TreeSet，则会假定元素的类实现了 Comparable 接口，并使用 compareTo 方法来比较规则集中的元素。要使用 comparator，则必须用构造方法 TreeSet(Comparator comparator)，使用比较器中的 compare 方法来创建一个排好序的规则集。

程序清单 21-5 给出了一个程序，演示了如何使用 Comparator 接口来对树形集中的元素进行排序。

程序清单 21-5 TestTreeSetWithComparator.java

```java
import java.util.*;

public class TestTreeSetWithComparator {
  public static void main(String[] args) {
    // Create a tree set for geometric objects using a comparator
    Set<GeometricObject> set =
      new TreeSet<>(new GeometricObjectComparator());
    set.add(new Rectangle(4, 5));
    set.add(new Circle(40));
    set.add(new Circle(40));
    set.add(new Rectangle(4, 1));

    // Display geometric objects in the tree set
    System.out.println("A sorted set of geometric objects");
    for (GeometricObject element: set)
      System.out.println("area = " + element.getArea());
  }
}
```

```
A sorted set of geometric objects
area = 4.0
area = 20.0
area = 5021.548245743669
```

GeometricObjectComparator 类在程序清单 20-4 中定义。该程序创建了一个几何对象的树形集，并使用 GeometricObjectComparator 来比较规则集中的元素（第 6 ~ 7 行）。

Circle 类和 Rectangle 类已经在 13.2 节中定义，它们都是 GeometricObject 的子类，被加入规则集中（第 8 ~ 11 行）。

两个半径相同的圆被添加到树形集中（第 9 ~ 10 行），但是只存储了一个，因为这两个圆是相等的（在本例中由比较器决定），而规则集中不允许有重复的元素。

✓ **复习题**

21.2.1 如何创建一个 Set 的实例？如何在规则集内插入一个新元素？如何从规则集中删除一个元素？如何获取一个规则集的大小？

21.2.2 如果两个对象 o1 和 o2 相等，那么 o1.equals(o2) 和 o1.hashCode() == o2.hashCode() 分别为多少？

21.2.3 如何遍历规则集中的元素？

21.2.4 假设 set1 是包含字符串 red、yellow、green 的规则集，而 set2 是包含字符串 red、yellow、blue 的规则集，回答下面的问题：

- 执行完 set1.addAll(set2) 方法之后，规则集 set1 和 set2 中分别是什么？
- 执行完 set1.add(set2) 方法之后，规则集 set1 和 set2 中分别是什么？
- 执行完 set1.removeAll(set2) 方法之后，规则集 set1 和 set2 中分别是什么？
- 执行完 set1.remove(set2) 方法之后，规则集 set1 和 set2 中分别是什么？
- 执行完 set1.retainAll(set2) 方法之后，规则集 set1 和 set2 中分别是什么？
- 执行完 set1.clear() 方法之后，规则集 set1 中是什么？

21.2.5 给出下面代码的输出结果：

```java
import java.util.*;

public class Test {
  public static void main(String[] args) {
    LinkedHashSet<String> set1 = new LinkedHashSet<>();
    set1.add("New York");
    LinkedHashSet<String> set2 = set1;
    LinkedHashSet<String> set3 =
      (LinkedHashSet<String>)(set1.clone());
    set1.add("Atlanta");
    System.out.println("set1 is " + set1);
    System.out.println("set2 is " + set2);
    System.out.println("set3 is " + set3);
    set1.forEach(e -> System.out.print(e + " "));
  }
}
```

21.2.6 给出下面代码的输出结果：

```java
Set<String> set = new LinkedHashSet<>();
set.add("ABC");
set.add("ABD");
System.out.println(set);
```

21.2.7 HashSet、LinkedHashSet 和 TreeSet 之间的区别是什么？

21.2.8 如何使用 Comparable 接口中的 compareTo 方法对规则集内的元素进行排序？如何使用 Comparator 接口对规则集内的元素进行排序？如果向树形集内添加一个不能与已有元素进行比较的元素，会发生什么情况？

21.2.9 如果程序清单 21-5 中的第 6～7 行被下面的代码所替换，输出将会是什么？

```java
Set<GeometricObject> set = new HashSet<>();
```

21.2.10 给出下列代码的输出结果：

```java
Set<String> set = new TreeSet<>(
  Comparator.comparing(String::length));
set.add("ABC");
set.add("ABD");
System.out.println(set);
```

21.3 比较规则集和线性表的性能

☛ 要点提示：在存储无重复元素方面，规则集比线性表更加高效。线性表在通过下标来访问元素方面非常有用。

线性表中的元素可以通过下标来访问。而规则集不支持下标，因为规则集中的元素是无序的。要遍历规则集中的所有元素，使用 foreach 循环。现在，我们来做一个有趣的试验，以测试规则集和线性表的性能。程序清单 21-6 给出一个程序，该程序显示了以下任务的执行时间：(1) 测试一个元素是否在一个散列集、链式散列集、树形集、数组线性表以及链表中；(2) 从一个散列集、链式散列集、树形集、数组线性表以及链表中删除元素。

程序清单 21-6 SetListPerformanceTest.java

```java
1  import java.util.*;
2
3  public class SetListPerformanceTest {
```

```java
4    static final int N = 50000;
5
6    public static void main(String[] args) {
7      // Add numbers 0, 1, 2, ..., N - 1 to the array list
8      List<Integer> list = new ArrayList<>();
9      for (int i = 0; i < N; i++)
10       list.add(i);
11     Collections.shuffle(list); // Shuffle the array list
12
13     // Create a hash set, and test its performance
14     Collection<Integer> set1 = new HashSet<>(list);
15     System.out.println("Member test time for hash set is " +
16       getTestTime(set1) + " milliseconds");
17     System.out.println("Remove element time for hash set is " +
18       getRemoveTime(set1) + " milliseconds");
19
20     // Create a linked hash set, and test its performance
21     Collection<Integer> set2 = new LinkedHashSet<>(list);
22     System.out.println("Member test time for linked hash set is " +
23       getTestTime(set2) + " milliseconds");
24     System.out.println("Remove element time for linked hash set is "
25       + getRemoveTime(set2) + " milliseconds");
26
27     // Create a tree set, and test its performance
28     Collection<Integer> set3 = new TreeSet<>(list);
29     System.out.println("Member test time for tree set is " +
30       getTestTime(set3) + " milliseconds");
31     System.out.println("Remove element time for tree set is " +
32       getRemoveTime(set3) + " milliseconds");
33
34     // Create an array list, and test its performance
35     Collection<Integer> list1 = new ArrayList<>(list);
36     System.out.println("Member test time for array list is " +
37       getTestTime(list1) + " milliseconds");
38     System.out.println("Remove element time for array list is " +
39       getRemoveTime(list1) + " milliseconds");
40
41     // Create a linked list, and test its performance
42     Collection<Integer> list2 = new LinkedList<>(list);
43     System.out.println("Member test time for linked list is " +
44       getTestTime(list2) + " milliseconds");
45     System.out.println("Remove element time for linked list is " +
46       getRemoveTime(list2) + " milliseconds");
47   }
48
49   public static long getTestTime(Collection<> c) {
50     long startTime = System.currentTimeMillis();
51
52     // Test if a number is in the collection
53     for (int i = 0; i < N; i++)
54       c.contains((int)(Math.random() * 2 * N));
55
56     return System.currentTimeMillis() - startTime;
57   }
58
59   public static long getRemoveTime(Collection<Integer> c) {
60     long startTime = System.currentTimeMillis();
61
62     for (int i = 0; i < N; i++)
63       c.remove(i);
64
65     return System.currentTimeMillis() - startTime;
66   }
67 }
```

```
Member test time for hash set is 20 milliseconds
Remove element time for hash set is 27 milliseconds
Member test time for linked hash set is 27 milliseconds
Remove element time for linked hash set is 26 milliseconds
Member test time for tree set is 47 milliseconds
Remove element time for tree set is 34 milliseconds
Member test time for array list is 39802 milliseconds
Remove element time for array list is 16196 milliseconds
Member test time for linked list is 52197 milliseconds
Remove element time for linked list is 14870 milliseconds
```

该程序创建了一个包含数字 0 到 N-1（N=50000）的线性表（第 8～10 行），并打乱该表（第 11 行）。程序然后基于该线性表创建一个散列集（第 14 行）、一个链式散列集（第 21 行）、一个树形集（第 28 行）、一个数组线性表（第 35 行）以及一个链表（第 42 行）。该程序获得测试一个数字是否在散列集中（第 16 行）、链式散列集中（第 23 行）、树形集中（第 30 行）、数组线性表中（第 37 行）以及链表中（第 44 行）的执行时间；然后获得将元素从散列集中（第 18 行）、链式散列集中（第 25 行）、树形集中（第 32 行）、数组线性表中（第 39 行）以及链表中（第 46 行）删除的执行时间。

getTestTime 方法调用 contains 方法测试一个数字是否在容器中（第 54 行），getRemoveTime 方法调用 remove 方法将一个元素从容器中移除（第 63 行）。

如这些运行时间所展示的，在测试一个元素是否在规则集或者线性表中，规则集比线性表更加高效。因此，前述的禁飞名单应该使用散列集实现，而不要采用线性表，因为测试一个元素是否在一个散列集中比测试它是否在一个线性表中要快得多。

你可能会困惑为什么规则集比线性表要更加高效。这些问题将在第 24 章和第 27 章介绍线性表和规则集的实现的时候得到回答。

✓ 复习题

21.3.1 假定你需要编写一个存储无序并且无重复元素的程序，应该使用什么数据结构？
21.3.2 假定你需要编写一个按照插入顺序来存储无重复元素的程序，应该使用什么数据结构？
21.3.3 假定你需要编写一个以元素值升序存储并且无重复元素的程序，应该使用什么数据结构？
21.3.4 假定你需要编写一个存储固定个数元素（可能有重复元素）的程序，应该使用什么数据结构？
21.3.5 假定你需要编写一个程序，将元素存储在一个表中并且需要经常在该表的末尾进行添加和删除元素的操作，应该使用什么数据结构？
21.3.6 假定你需要编写一个程序，将元素存储在一个表中并且需要经常在该表的开始处进行插入和删除元素的操作，应该使用什么数据结构？

21.4 示例学习：关键字计数

요点提示：本节给出一个应用，对一个 Java 源文件中的关键字进行计数。

对于 Java 源文件中的每个单词，需要确定该单词是否是一个关键字。为了高效处理这个问题，将所有的关键字保存在一个 HashSet 中，并且使用 contains 方法来测试一个单词是否在关键字规则集中。程序清单 21-7 给出了这个程序。

程序清单 21-7 CountKeywords.java

```
1  import java.util.*;
2  import java.io.*;
3
```

```
4   public class CountKeywords {
5     public static void main(String[] args) throws Exception {
6       Scanner input = new Scanner(System.in);
7       System.out.print("Enter a Java source file: ");
8       String filename = input.nextLine();
9
10      File file = new File(filename);
11      if (file.exists()) {
12        System.out.println("The number of keywords in " + filename
13          + " is " + countKeywords(file));
14      }
15      else {
16        System.out.println("File " + filename + " does not exist");
17      }
18    }
19
20    public static int countKeywords(File file) throws Exception {
21      // Array of all Java keywords + true, false and null
22      String[] keywordString = {"abstract", "assert", "boolean",
23        "break", "byte", "case", "catch", "char", "class", "const",
24        "continue", "default", "do", "double", "else", "enum",
25        "extends", "for", "final", "finally", "float", "goto",
26        "if", "implements", "import", "instanceof", "int",
27        "interface", "long", "native", "new", "package", "private",
28        "protected", "public", "return", "short", "static",
29        "strictfp", "super", "switch", "synchronized", "this",
30        "throw", "throws", "transient", "try", "void", "volatile",
31        "while", "true", "false", "null"};
32
33      Set<String> keywordSet =
34        new HashSet<>(Arrays.asList(keywordString));
35      int count = 0;
36
37      Scanner input = new Scanner(file);
38
39      while (input.hasNext()) {
40        String word = input.next();
41        if (keywordSet.contains(word))
42          count++;
43      }
44
45      return count;
46    }
47  }
```

```
Enter a Java source file: c:\ Welcome.java  ↵Enter
The number of keywords in c:\ Welcome.java is 5
```

```
Enter a Java source file: c:\ TTT.java  ↵Enter
File c:\ TTT.java does not exist
```

该程序提示用户输入一个Java源文件名（第7行）并且读取文件名（第8行）。如果文件存在，则调用countKeywords方法来统计文件中出现的关键字（第13行）。

countKeywords方法创建了一个包含所有关键字的字符串数组（第22～31行），并且从该数组创建一个散列规则集（第33～34行）。然后从文件中读取每个单词，并且测试这个单词是否在规则集中（第41行）。如果在，程序增加1个计数（第42行）。

也可以使用LinkedHashSet、TreeSet、ArrayList或者LinkedList来存储关键字。然而，对这个程序来说，使用HashSet是最高效的。

✓ 复习题

21.4.1 如果第33～34行改为以下语句，CountKeywords程序还能工作吗？

```
Set<String> keywordSet =
    new LinkedHashSet<>(Arrays.asList(keywordString));
```

21.4.2 如果第 33 ～ 34 行改为以下语句，CountKeywords 程序还能工作吗？

```
List<String> keywordSet =
    new ArrayList<>(Arrays.asList(keywordString));
```

21.5 映射

要点提示：可以使用三个具体的类来创建一个映射：HashMap、LinkedHashMap、TreeMap。

映射（map）是一个存储"键/值对"集合的容器对象。它提供了通过键快速获取、删除和更新键/值对的功能。映射将值和键一起保存。键很像下标。在 List 中，下标是整数；而在 Map 中，键可以是任意类型的对象。映射中不能有重复的键，每个键都对应一个值。键和它的对应值构成一个保存在映射中的条目，如图 21-2a 所示。图 21-2b 展示了一个映射，其中每个条目由作为键的社会安全号以及作为值的姓名所组成。

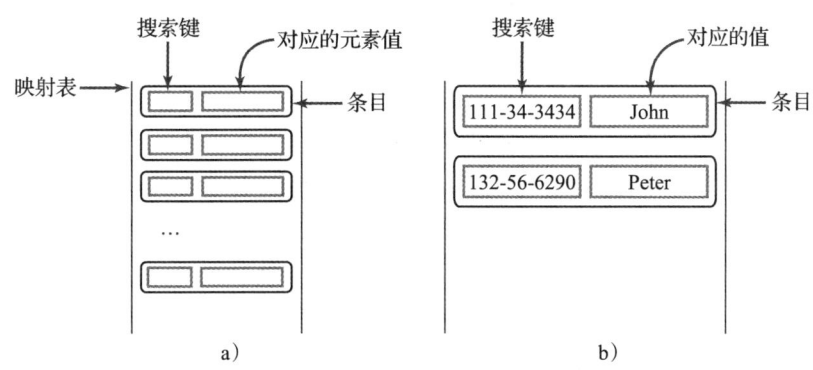

图 21-2 由键/值对组成的条目存储在映射中

有三种映射类型：散列映射 HashMap、链式散列映射 LinkedHashMap 和树形映射 TreeMap。这些映射的通用特性都在 Map 接口中定义，它们的关系如图 21-3 所示。

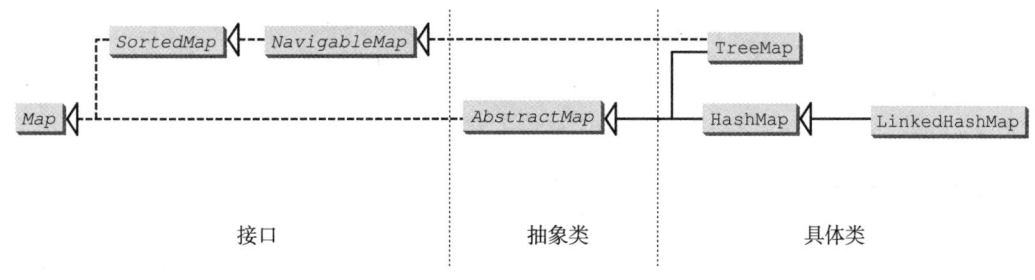

图 21-3 映射存储键/值对

Map 接口提供了查询、更新和获取值的集合和键的规则集的方法，如图 21-4 所示。

更新方法（update method）包括 clear、put、putAll 和 remove。clear() 方法从映射中删除所有的条目。put(K key,V value) 方法将一个指定的键和值作为一个条目添加到映射中。如果这个映射原来就包含该键的一个条目，则旧值将被新值所替代，并且返回与这个键相关联的旧值。putAll(Map m) 方法将 m 中的所有条目添加到这个映射中。remove(Object key)

方法将指定键对应的条目从映射中删除。

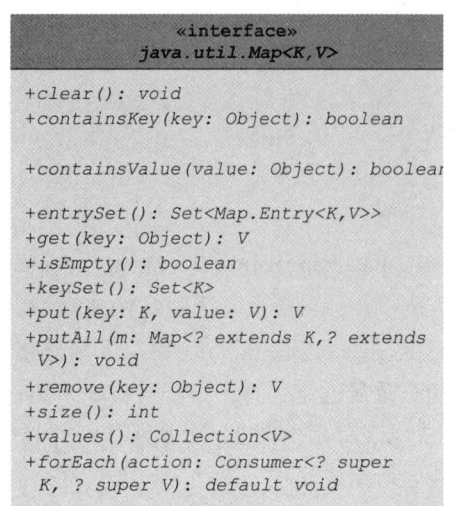

图 21-4　Map 接口将键映射到值

查询方法（query method）包括 containsKey、containsValue、isEmpty 和 size。containsKey(Object key) 方法检测映射中是否包含指定键的条目。containsValue (Object value) 方法检测映射中是否包含指定值的条目。isEmpty() 方法检测映射中是否包含任何条目。size() 方法返回映射中的条目数。

可以使用 keySet() 方法来获得一个映射中键的规则集，并可以使用 values() 方法获得一个映射中值的集合。entrySet() 方法返回一个条目的规则集。这些条目是 Map.Entry<K,V> 接口的实例，这里 Entry 是 Map 接口的一个内部接口，如图 21-5 所示。该规则集中的每个条目都是所在映射中一个键/值对。

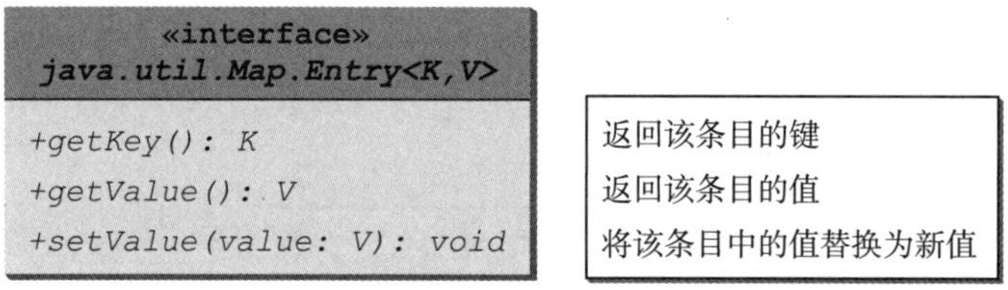

图 21-5　Map.Entry 接口对映射中条目的操作

Java 8 在 Map 接口中添加了一个默认的 forEach 方法，来对映射中的每一个条目执行操作。这个方法可以像一个迭代器一样，用于遍历映射中的条目。

AbstractMap 类是一个便利抽象类，它实现了 Map 接口中除了 entrySet() 方法之外的所有方法。

HashMap、LinkedHashMap 和 TreeMap 类是 Map 接口的三个具体实现（concrete implementation），如图 21-6 所示。

HashMap 类对于定位一个值、插入一个条目以及删除一个条目而言是高效的。

LinkedHashMap 类用链表实现来扩展 HashMap 类，它支持映射中条目的排序。HashMap 类中的条目是没有顺序的，但是在 LinkedHashMap 中，元素既可以按照它们插入映射的顺序排序

(称为插入顺序（insertion order）），也可以按它们被最后一次访问时的顺序，从最早到最晚（称为访问顺序（access order））排序。无参构造方法以插入顺序来创建 LinkedHashMap。要按访问顺序创建 LinkedHashMap 对象，可以使用构造方法 LinkedHashMap (initialCapacity,loadFactor, true)。

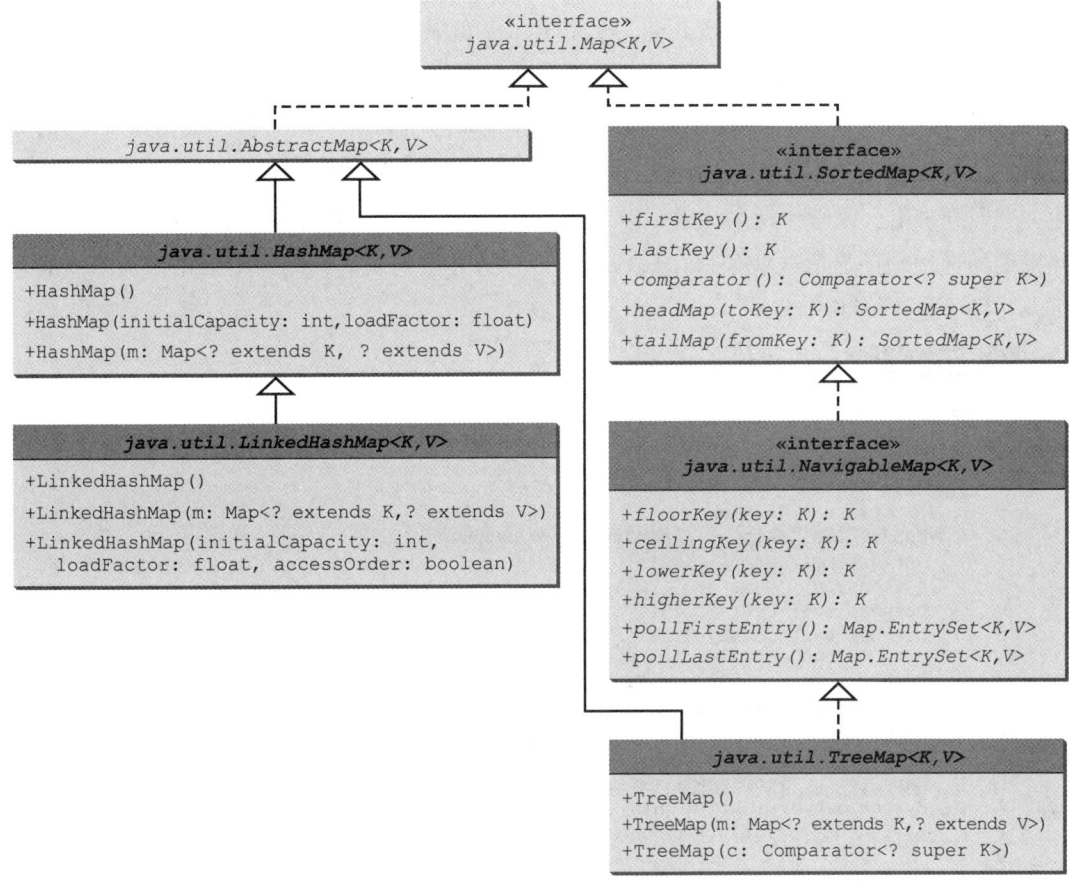

图 21-6　Java 集合框架提供三个具体映射类

TreeMap 类对于遍历排好顺序的键很高效。键可以使用 Comparable 接口或 Comparator 接口来排序。如果使用其无参构造方法创建一个 TreeMap 对象，假定键的类实现了 Comparable 接口，则可以使用 Comparable 接口中的 compareTo 方法来对映射内的键进行比较。要使用比较器，必须使用构造方法 TreeMap(Comparator comparator) 来创建一个有序映射，这样，该映射中的条目就能使用比较器中的 compare 方法基于键进行排序。

SortedMap 是 Map 的一个子接口，它可确保映射中的条目是排好序的。除此之外，它还提供了 firstKey() 和 lastKey() 方法来分别返回映射中的第一个和最后一个键，而 headMap(toKey) 和 tailMap(fromKey) 方法分别返回键小于 toKey 的那部分映射和键大于或等于 fromKey 的那部分映射。

NavigableMap 继承了 SortedMap，它提供导航方法 lowerKey(key)、floorKey(key)、 ceilingKey(key) 和 higherKey(key) 来分别返回小于、小于或等于、大于或等于、大于某个给定键的键，如果没有这样的键，则会返回 null。pollFirstEntry() 和 pollLastEntry() 方法分别删除并返回树形映射中的第一个和最后一个条目。

> **注意**：在 Java 2 以前，一般使用 java.util.Hashtable 来映射键和值。为了融入 Java 集合框架，Hashtable 被重新设计，但为了兼容性保留了所有的方法。Hashtable 实现了 Map 接口，除了 Hashtable 中的更新方法是同步的以外，它与 HashMap 的用法是一样的。

程序清单 21-8 给出的例子创建了一个散列映射（hash msp）、一个链式散列映射（linked hash map）和一个树形映射（tree map），以建立学生与年龄之间的映射关系。该程序首先创建一个散列映射，以学生姓名为键，以年龄为值，然后从这个散列映射创建一个树形映射，并按键的递增顺序显示这些条目，最后创建一个链式散列映射，向该映射中添加相同的条目，并显示这些条目。

程序清单 21-8 TestMap.java

```java
 1  import java.util.*;
 2
 3  public class TestMap {
 4    public static void main(String[] args) {
 5      // Create a HashMap
 6      Map<String, Integer> hashMap = new HashMap<>();
 7      hashMap.put("Smith", 30);
 8      hashMap.put("Anderson", 31);
 9      hashMap.put("Lewis", 29);
10      hashMap.put("Cook", 29);
11
12      System.out.println("Display entries in HashMap");
13      System.out.println(hashMap + "\n");
14
15      // Create a TreeMap from the preceding HashMap
16      Map<String, Integer> treeMap = new TreeMap<>(hashMap);
17      System.out.println("Display entries in ascending order of key");
18      System.out.println(treeMap);
19
20      // Create a LinkedHashMap
21      Map<String, Integer> linkedHashMap =
22        new LinkedHashMap<>(16, 0.75f, true);
23      linkedHashMap.put("Smith", 30);
24      linkedHashMap.put("Anderson", 31);
25      linkedHashMap.put("Lewis", 29);
26      linkedHashMap.put("Cook", 29);
27
28      // Display the age for Lewis
29      System.out.println("\nThe age for " + "Lewis is " +
30        linkedHashMap.get("Lewis"));
31
32      System.out.println("Display entries in LinkedHashMap");
33      System.out.println(linkedHashMap);
34
35      // Display each entry with name and age
36      System.out.print("\nNames and ages are ");
37      treeMap.forEach(
38        (name, age) -> System.out.print(name + ": " + age + " "));
39    }
40  }
```

```
Display entries in HashMap
{Cook=29, Smith=30, Lewis=29, Anderson=31}

Display entries in ascending order of key
{Anderson=31, Cook=29, Lewis=29, Smith=30}
The age for Lewis is 29
Display entries in LinkedHashMap
{Smith=30, Anderson=31, Cook=29, Lewis=29}

Names and ages are Anderson: 31 Cook: 29 Lewis: 29 Smith: 30
```

如输出所示，HashMap 中条目的顺序是随机的，而 TreeMap 中的条目是按键的升序排列的，LinkedHashMap 中的条目则是按元素最后一次被访问的时间从最早到最晚排序的。

实现 Map 接口的所有具体类至少有两种构造方法：一种是无参构造方法，它可用来创建一个空映射，而另一种构造方法是从 Map 的一个实例来创建映射。因此，语句 new TreeMap <>(hashMap)（第 16 行）是从一个散列映射来创建一个树形映射。

可以创建一个按插入顺序或访问顺序排序的链式散列映射。程序第 21~22 行创建一个按访问顺序排序的链式散列映射，最晚被访问的条目被放在映射的末尾。第 30 行中拥有键 Lewis 的条目最后被访问，因此，在第 33 行它最后被显示。

使用 forEach 方法处理映射中的所有条目是很方便的。该程序使用 forEach 方法来显示姓名和年龄（第 37~38 行）。

> **提示**：如果更新映射时不需要保持映射中元素的顺序，就使用 HashMap；如果需要保持映射中元素的插入顺序或访问顺序，就使用 LinkedHashMap；如果需要使映射按照键排序，就使用 TreeMap。

✓ 复习题

21.5.1 如何创建 Map 的一个实例？如何向映射中添加一个由键和值组成的条目？如何从映射中删除一个条目？如何获取映射的大小？如何遍历映射中的条目？

21.5.2 描述并比较 HashMap、LinkedHashMap 和 TreeMap。

21.5.3 给出下面代码的输出结果：

```
import java.util.*;
public class Test {
  public static void main(String[] args) {
    Map<String, String> map = new LinkedHashMap<>();
    map.put("123", "John Smith");
    map.put("111", "George Smith");
    map.put("123", "Steve Yao");
    map.put("222", "Steve Yao");
    System.out.println("(1) " + map);
    System.out.println("(2) " + new TreeMap<String, String>(map));
    map.forEach((k, v) -> {
      if (k.equals("123")) System.out.println(v);});
  }
}
```

21.6 示例学习：单词的出现次数

> **要点提示**：该示例学习编写一个程序，以统计一个文本中单词的出现次数，然后按照字母顺序显示这些单词以及它们的出现次数。

本程序使用一个 TreeMap 来存储包含单词及其次数的条目。对于每一个单词来说，都要判断它是否已经是映射中的一个键。如果不是，将由这个单词作为键而 1 作为值构成的条目存入该映射中。否则，将映射中该单词（键）对应的值加 1。假定单词不区分大小写，例如，Good 被认为是和 good 一样的。

程序清单 21-9 给出了该问题的解决方案。

程序清单 21-9 CountOccurrenceOfWords.java

```java
1   import java.util.*;
2
3   public class CountOccurrenceOfWords {
4     public static void main(String[] args) {
5       // Set text in a string
6       String text = "Good morning. Have a good class. " +
7         "Have a good visit. Have fun!";
8
9       // Create a TreeMap to hold words as key and count as value
10      Map<String, Integer> map = new TreeMap<>();
11
12      String[] words = text.split("[\\s+\\p{P}]");
13      for (int i = 0; i < words.length; i++) {
14        String key = words[i].toLowerCase();
15
16        if (key.length() > 0) {
17          if (!map.containsKey(key)) {
18            map.put(key, 1);
19          }
20          else {
21            int value = map.get(key);
22            value++;
23            map.put(key, value);
24          }
25        }
26      }
27
28      // Display key and value for each entry
29      map.forEach((k, v) -> System.out.println(k + "\t" + v));
30    }
31  }
```

```
a         2
class     1
fun       1
good      3
have      3
morning   1
visit     1
```

该程序创建了一个 TreeMap（第 10 行）来存储包含单词和它们的出现次数的键值对。单词作为键。因为映射中的所有值必须存储为对象，所以统计次数被包装在一个 Integer 对象中。

该程序使用 String 类中的 split 方法（第 12 行）从文本中提取单词（参见 10.10.4 节和附录 H）。文本使用空格 \s 或者标点 \p{P} 作为分隔符。对于每个被提取出的单词，程序都会检测它是否已经被存储为映射中的键（第 17 行）。如果没有，就将这个单词和它的初始统计次数（1）构成一个新键值对存储到映射中（第 18 行）。否则，将该单词的计数器加 1（第 21 ~ 23 行）。

程序使用 Map 类中的 forEach 方法（第 29 行），显示每个条目中的计数和键。

因为这个映射是一个树形映射，所以条目是以单词的升序显示的。要以出现次数的升序显示它们,参见编程练习题21.8。

现在回过头思考一下在不使用映射的情况下如何编写这个程序。新程序将会更长,也更复杂,由此可发现映射是解决此类问题的非常高效且功能强大的数据结构。

Java集合框架提供了组织和操作数据的全面支持。如果你希望按照出现次数的递增顺序来显示单词,应该如何修改程序?可以创建一个映射条目的线性表,并且创建一个Comparator来根据它们的值对条目进行排序,如下所示:

```
List<Map.Entry<String, Integer>> entries =
  new ArrayList<>(map.entrySet());
Collections.sort(entries, (entry1, entry2) -> {
  return entry1.getValue().compareTo(entry2.getValue()); });
for (Map.Entry<String, Integer> entry: entries) {
  System.out.println(entry.getKey() + "\t" + entry.getValue());
}
```

✔ 复习题

21.6.1 如果第10行改成下面语句,程序CountOccurrenceOfWords还能工作吗?

```
Map<String, int> map = new TreeMap<>();
```

21.6.2 如果第17行改成下面语句,程序CountOccurrenceOfWords还能工作吗?

```
if (map.get(key) == null) {
```

21.6.3 如果第29行改成下面语句,程序CountOccurrenceOfWords还能工作吗?

```
for (String key: map)
  System.out.println(key + "\t" + map.getValue(key));
```

21.6.4 如何使用条件表达式来简化程序清单21-9中的第17～24行代码?

21.7 单例与不可变的集合和映射

要点提示:可以使用Collections类中的静态方法来创建单例规则集、线性表和映射,以及不可变的规则集、线性表和映射。

Collections类包含了用于线性表和集合的静态方法。它还包含用于创建不可修改的单例规则集、线性表和映射的方法,以及用于创建只读规则集、线性表和映射的方法,如图21-7所示。

java.util.Collections	
+singleton(o: Object): Set	返回一个包含指定对象的不可变的集合
+singletonList(o: Object): List	返回一个包含指定对象的不可变的线性表
+singletonMap(key: Object, value: Object): Map	返回一个具有键值对的不可变的映射表
+unmodifiableCollection(c: Collection): Collection	返回一个集合的只读视图
+unmodifiableList(list: List): List	返回一个线性表的只读视图
+unmodifiableMap(m: Map): Map	返回一个映射的只读视图
+unmodifiableSet(s: Set): Set	返回一个规则集的只读视图
+unmodifiableSortedMap(s: SortedMap): SortedMap	返回一个排好序的映射的只读视图
+unmodifiableSortedSet(s: SortedSet): SortedSet	返回一个排好序的规则集的只读视图

图21-7 Collections类包含了用于创建单例以及只读的规则集、线性表和映射的静态方法

Collections类中为空规则集、空线性表、空映射定义了三个常量:EMPTY_SET、EMPTY_LIST和EMPTY_MAP。这些集合是不可修改的。该类还定义了如下几个方法:singleton(Object

o) 方法用于创建包含单一条目的不可变线性表；singletonList(Object o) 方法用于创建包含单一条目的不可变线性表；singletonMap(Object key,Object value) 方法用于创建包含单一条目的不可变映射。

  Collections 类还提供了 6 个用于返回集合的只读视图的静态方法：unmodifiableCollection (Collection c)、unmodifiableList(List list)、unmodifiableMap(Map m)、unmodifiableSet (Set set)、unmodifiableSortedMap(SortedMap m) 和 unmodifiableSortedSet(SortedSet s)。这种类型的视图类似于真实集合的引用。然而，不能通过一个只读的视图来修改集合。尝试通过只读视图修改集合将引发 UnsupportedOperationException 异常。

  JDK 9 中，可以使用静态的 Ser.of(e1, e2, ...) 方法来创建一个不可变规则集，使用 Map.of(key1, value1, key2, value2, ...) 方法来创建一个不可变映射。

✔ 复习题

21.7.1 下面代码中有什么错误？

```
Set<String> set = Collections.singleton("Chicago");
set.add("Dallas");
```

21.7.2 运行下面代码时将发生什么？

```
List list = Collections.unmodifiableList(Arrays.asList("Chicago",
  "Boston"));
list.remove("Dallas");
```

关键术语

hash map（散列映射）
hash set（散列集）
linked hash map（链式散列映射）
linked hash set（链式散列集）
map（映射）

read-only view（只读视图）
set（规则集）
tree map（树形映射）
tree set（树形集）

本章小结

1. 规则集存储的是无重复的元素。若要在集合中存储重复的元素，需要使用线性表。
2. 映射中存储的是键 / 值对。它提供使用键快速查询一个值。
3. Java 集合框架支持三种类型的规则集：散列集 HashSet、链式散列集 LinkedHashSet 和树形集 TreeSet。HashSet 以一个不可预知的顺序存储元素；LinkedHashSet 以元素被插入的顺序存储元素；TreeSet 存储已排好序的元素。HashSet、LinkedHashSet 和 TreeSet 都是 Collection 的子类型。
4. Map 接口将键映射到元素上。键类似于下标。List 中，下标为整数。Map 中，键可以为任何对象。映射不能包含相同的键。每个键可以映射最多一个值。Map 接口提供了查询、更新以及获取值的集合以及键的规则集的方法。
5. Java 集合框架支持三种类型的映射：散列映射 HashMap、链式散列映射 LinkedHashMap 和树形映射 TreeMap。对于定位一个值、插入一个条目和删除一个条目而言，HashMap 是高效的。LinkedHashMap 支持映射中的条目排序。HashMap 类中的条目是没有顺序的，但 LinkedHashMap 中的条目可以按某种顺序来获取，该顺序既可以是它们被插入映射中的顺序（称为插入顺序），也可以是根据最后一次被访问时间从最早到最晚的排序（称为访问顺序），从最早到最晚（称为访问顺序）。对于遍历排好序的键，TreeMap 是高效的。键可以使用 Comparable 接口来排序，也可以使用 Comparator 接口来排序。

测试题

回答位于本书配套网站上的本章测试题。

编程练习题

21.2～21.4 节

21.1 （在散列集上进行规则集操作）创建两个链接散列规则集 {"George","Jim","John","Blake","Kevin", "Michael"} 和 {"George","Katie","Kevin","Michelle","Ryan"}，然后求它们的并集、差集和交集。（可以先备份这些规则集，以保护初始规则集不被这些方法改变。）

21.2 （按升序显示不重复的单词）编写一个程序，从文本文件中读取单词，并将所有不重复的单词按升序显示。文本文件被作为命令行参数传递。

**21.3 （统计 Java 源代码中的关键字）修改程序清单 21-7 中的程序。如果关键字在注释或者字符串中，则不进行统计。将 Java 文件名从命令行传递。假设 Java 源代码是正确的，行注释和段落注释不会交叉。

*21.4 （对元音和辅音计数）编写一个程序，提示用户输入一个文本文件名，然后显示文件中的元音和辅音的数目。使用一个规则集存储元音 A、E、I、O 和 U。

***21.5 （突出显示语法）编写一个程序，将一个 Java 文件转换为一个 HTML 文件。在 HTML 文件中，关键字、注释和字面量分别用粗体的深蓝色、绿色和蓝色显示。使用命令行传递 Java 文件和 HTML 文件。例如，下面的命令

```
java Exercise21_05 Welcome.java Welcome.html
```

将 Welcome.java 转换为 Welcome.html。图 21-8a 显示了一个 Java 文件，它对应的 HTML 文件如图 21-8b 所示。

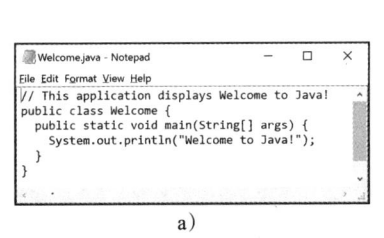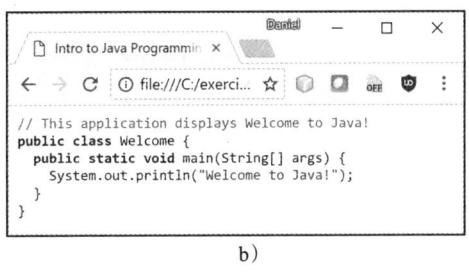

图 21-8 图 a 中纯文本形式的 Java 代码以 HTML 显示在图 b 中，其中语法被突出显示
（来源：Oracle 或其附属公司版权所有 ©1995～2016，经授权使用）

21.5～21.7 节

*21.6 （统计输入数字的出现次数）编写一个程序，读取不确定个数的整数，然后查找其中出现频率最高的数字。当输入为 0 时，表示结束输入。例如，如果输入的数据是 2 3 40 3 5 4 -3 3 3 2 0，那么数字 3 的出现频率是最高的。如果出现频率最高的数字不是一个而是多个，则应该将它们全部报告。例如，在线性表 9 30 3 9 3 2 4 中，3 和 9 都出现了两次，所以 3 和 9 都应该被报告。

**21.7 （改写程序清单 21-9）改写程序清单 21-9，将单词按出现次数的升序显示。

**21.8 （统计文本文件中单词的出现频率）改写程序清单 21-9，从文本文件中读取文本，文本文件名作为命令行参数传递。单词由空格、标点符号（,;.:?）、引号（'"）以及括号分隔。单词计数时不区分大小写（例如，认为 Good 和 good 是一样的单词）。单词必须以字母开头。以单词的字母顺序显示输出，每个单词前面显示其出现次数。

**21.9 (使用映射猜首府) 改写编程练习题 8.37,在映射中存储州和其首府的条目。你的程序应该提示用户输入一个州,然后显示该州的首府。

*21.10 (统计每个关键字的出现次数) 重写程序清单 21-7,读入一个 Java 源代码文件并且统计文件中每个关键字的出现次数。如果关键字是在注释中或者字符串字面值中,则不要进行统计。

**21.11 (婴儿姓名流行度排名) 使用编程练习题 12.31 中的数据文件编写一个程序,使用户可以选择一个年份、性别,输入一个姓名,然后显示在选择的年份和性别条件下该姓名的排名,如图 21-9 所示。为了获得最好的效率,创建两个数组分别用于男孩名字和女孩名字。每个数组包含 10 个元素代表 10 个年份。每个元素是一个映射,以键值对的方式存储了姓名和相应的排名,并将姓名作为键。

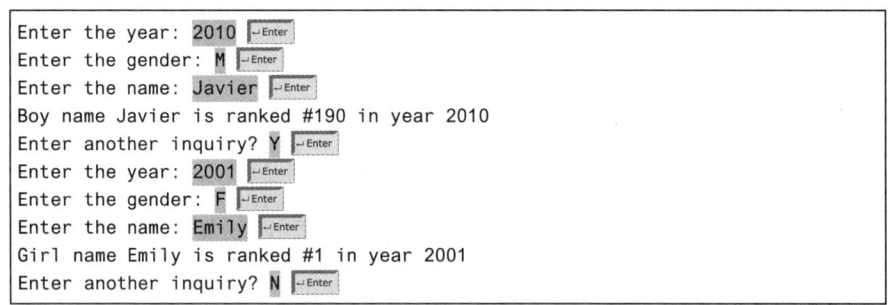

图 21-9 用户选择一个年份和性别,输入年份,单击 Find Ranking 按钮显示排名(来源:Oracle 或其附属公司版权所有 ©1995 ~ 2016,经授权使用)

**21.12 (两个性别都使用的姓名) 编写一个程序,提示用户输入编程练习题 12.31 中描述的文件名,然后显示文件中可以同时用于两种性别的姓名。使用规则集存储姓名并找到两个规则集中的共同姓名。下面是一个运行示例:

```
Enter a file name for baby name ranking: babynamesranking2001.txt  ↵Enter
69 names used for both genders
They are Tyler Ryan Christian ...
```

**21.13 (婴儿姓名流行度排名) 修改编程练习题 21.11,提示用户输入年份、性别和姓名,然后显示该名字的排名。提示用户输入另一个查询或者退出程序。下面是一个运行示例:

```
Enter the year: 2010  ↵Enter
Enter the gender: M  ↵Enter
Enter the name: Javier  ↵Enter
Boy name Javier is ranked #190 in year 2010
Enter another inquiry? Y  ↵Enter
Enter the year: 2001  ↵Enter
Enter the gender: F  ↵Enter
Enter the name: Emily  ↵Enter
Girl name Emily is ranked #1 in year 2001
Enter another inquiry? N  ↵Enter
```

**21.14 (Web 爬虫) 重写编程练习题 12.18,为 ListOfPendingURLs 和 listofTraversedURLs 采用合适的新数据结构以提高性能。

**21.15 (加法测试题) 重写编程练习题 11.16,将答案保存在一个规则集中,而不是线性表中。

第 22 章

Introduction to Java Programming and Data Structures, Comprehensive Version, Twelfth Edition

开发高效算法

教学目标

- 使用大 O 表示法判断算法效率（22.2 节）。
- 解释增长率以及为什么在分析时可以忽略常量和非主导项（22.2 节）。
- 确定各种类型算法的复杂度（22.3 节）。
- 分析二分查找算法（22.4.1 节）。
- 分析选择排序算法（22.4.2 节）。
- 分析汉诺塔算法（22.4.3 节）。
- 描述常用的增长函数（常量、对数、对数 – 线性、二次、三次和指数）（22.4.5 节）。
- 使用动态编程求斐波那契数（22.5 节）。
- 使用欧几里得算法求最大公约数（22.6 节）。
- 使用埃拉托色尼筛选法求素数（22.7 节）。
- 使用分治法寻找最近点对（22.8 节）。
- 使用回溯法解决八皇后问题（22.9 节）。
- 设计高效算法，为一个点集寻找凸包（22.10 节）。
- 使用暴力算法 Boyer-Moore 算法和 KMP 算法设计高效的字符串匹配算法（22.11 节）。

22.1 引言

要点提示：算法设计是指为解决某个问题开发一个数学流程。算法分析是指预测一个算法的性能。

前面两章介绍了经典的数据结构（线性表、栈、队列、优先队列、规则集和映射），并将它们应用于解决问题。本章将采用各种示例来介绍如何用常用的算法技术（动态编程、分治以及回溯）来开发高效的算法。本书后面的第 23 ~ 29 章将介绍一些高效的算法。在介绍高效算法的开发之前，我们需要讨论如何衡量算法效率的问题。

22.2 使用大 O 表示法来衡量算法效率

要点提示：大 O 表示法可以基于输入的大小得到一种衡量算法时间复杂度的函数。可以忽略函数中的倍乘常量和非主导项。

假定两个算法执行相同的任务，比如查找（线性查找与二分查找），哪个算法更好呢？为了回答这个问题，我们可以实现这两个算法，并运行程序得到执行时间。但是这种方法存在以下两个问题：

- 首先，计算机上同时运行着多个任务，一个特定程序的实际执行时间是依赖于系统负荷的。
- 其次，实际执行时间依赖于特定的输入。例如，考虑线性查找和二分查找。如果要查找的元素恰巧是线性表中的第一个元素，那么线性查找会比二分查找更快找到该元素。

通过测量执行时间来比较算法是非常困难的。为了克服这些问题，计算机科学家研发了一个独立于计算机和特定输入的理论方法来分析算法。该方法大致估计了输入大小的改变而产生的影响。通过这个方法可以看到随着输入大小的增长算法执行时间增长得有多快，因此可以通过检查两个算法的增长率（growth rate）来比较它们。

考虑线性查找的问题。线性查找算法顺序比较数组中的元素与键，直到找到键或者数组已搜索完毕。如果该键不在数组中，那么对于一个大小为 n 的数组需要进行 n 次比较。如果该键在数组中，那么平均需要进行 $n/2$ 次比较。该算法的执行时间与数组的大小成正比。如果将数组大小加倍，那么比较次数也会加倍。该算法是呈线性增长的，增长率为 n 的数量级。计算机科学家使用大 O 表示法（big O notation）表示数量级。使用该表示法，线性查找算法的复杂度就是 $O(n)$，读为"n 阶"。我们将时间复杂度为 $O(n)$ 的算法称为线性算法，它体现了线性的增长率。

注意：算法的时间复杂度（也就是运行时间）是使用大 O 表示法测量的算法所花费的时间。

对于相同的输入大小，算法的执行时间可能会有所不同。具有最短执行时间的输入称为最佳情况输入（best-case input），而具有最长执行时间的输入称为最差情况输入（worst-case input）。最佳和最差情况分析分别是在最佳和最差输入情况下对算法进行分析。最佳和最差情况分析都不具有代表性，但是最差情况分析却是非常有用的。我们可以确定的是自己的算法永远不会比最差情况还慢。平均情况分析（average-case analysis）试图确定所有相同大小的可能输入情形下的平均时间。平均情况分析是比较理想的，但是很难实现，这是因为对于许多问题而言，要确定各种输入实例的相对概率和分布是相当困难的。由于最差情况分析比较容易完成，所以分析通常针对最差情况进行。

线性查找算法在最差情况下需要进行 n 次比较，而如果你几乎总是在线性表中查找一个已知存在于其中的元素，平均情况下需要进行 $n/2$ 次比较。使用大 O 表示法，这两种情况需要的时间都为 $O(n)$。倍乘常量（1/2）可以忽略。算法分析的重点在于增长率，而倍乘常量对增长率没有影响。对于 $n/2$ 或 $100n$ 而言，增长率都和 n 一样，如表 22-1 所示。因此，$O(n)=O(n/2)=O(100n)$。

表 22-1 增长率

n \ $f(n)$	n	$n/2$	$100n$	
100	100	50	10000	
200	200	100	20000	
	2	2	2	$f(200)/f(100)$

考虑在一个包含 n 个元素的数组中找出最大数的算法。如果 n 为 2，找到最大数需要一次比较；如果 n 为 3，找到最大数需要两次比较。一般来说，在拥有 n 个元素的线性表中找到最大数需要 $n-1$ 次比较。算法分析主要用于大的输入规模。如果输入规模较小，那么估计算法效率是没有意义的。随着 n 的增大，表达式 $n-1$ 中的 n 就主导了复杂度。大 O 表示法允许忽略非主导部分（例如，表达式 $n-1$ 中的 -1），并强调重要部分（例如，表达式 $n-1$ 中的 n）。因此，该算法的复杂度为 $O(n)$。

大 O 表示法根据输入大小估算算法的执行时间。如果执行时间与输入规模无关，就称该算法耗费了常量时间（constant time），用符号 $O(1)$ 表示。例如，在数组中从给定下标处获取元素的方法耗费的时间即为常量时间，这是因为该时间不会随数组规模的增大而增加。

在算法分析中经常会用到下面的数学求和公式：

$$1+2+3+\cdots+(n-2)+(n-1) = \frac{n(n-1)}{2} = O(n^2)$$

$$1+2+3+\cdots+(n-1)+n = \frac{n(n+1)}{2} = O(n^2)$$

$$a^0 + a^1 + a^2 + a^3 + \cdots + a^{(n-1)} + a^n = \frac{a^{n+1}-1}{a-1} = O(a^n)$$

$$2^0 + 2^1 + 2^2 + 2^3 + \cdots + 2^{(n-1)} + 2^n = \frac{2^{n+1}-1}{2-1} = 2^{n+1}-1 = O(2^n)$$

> **注意**：时间复杂度是使用大 O 表示法对运行时间进行估算。类似地，也可以使用大 O 表示法对空间复杂度进行估算。空间复杂度衡量一个算法所使用的内存空间量。本书中大多数算法的空间复杂度为 $O(n)$。也就是说，相对于输入问题大小，它们体现出线性的增长率。例如，线性查找的空间复杂度为 $O(n)$。

> **注意**：我们以针对入门者的方式介绍了大 O 表示法，附录 J 给出了大 O 表示法以及大 Ω 和大 I 估算的精确数学定义。

复习题

22.2.1 为什么大 O 表示法中忽略掉常量因子？为什么大 O 表示法中忽略掉非主导项？

22.2.2 下面各个函数分别为多少阶？

$$\frac{(n^2+1)^2}{n}, \quad \frac{n^2+\log^2 n}{n}, \quad n^3+100n^2+n, \quad 2^n+100n^2+45n, \quad n2^n+n^22^n$$

22.3 示例：确定大 O

> **要点提示**：本节给出多个示例，为循环、顺序以及选择语句确定大 O。

示例 1

考虑下面循环的时间复杂度：

```
for (int i = 1; i <= n; i++) {
  k = k + 5;
}
```

执行下面的语句

```
k = k + 5;
```

需要一个常量时间 c，因为循环执行了 n 次，所以其时间复杂度是

$$T(n) = c \times n = O(n)$$

理论分析预测了算法的性能。为了观察这个算法的执行，运行程序清单 22-1 中的代码来获得 $n = 1\,000\,000$、$10\,000\,000$、$100\,000\,000$ 以及 $1\,000\,000\,000$ 的运行时间。

程序清单 22-1 PerformanceTest.java

```
1  public class PerformanceTest {
2    public static void main(String[] args) {
3      getTime(1000000);
4      getTime(10000000);
```

```java
5       getTime(100000000);
6       getTime(1000000000);
7   }
8
9   public static void getTime(long n) {
10      long startTime = System.currentTimeMillis();
11      long k = 0;
12      for (long i = 1; i <= n; i++) {
13        k = k + 5;
14      }
15      long endTime = System.currentTimeMillis();
16      System.out.println("Execution time for n = " + n
17        + " is " + (endTime - startTime) + " milliseconds");
18  }
19  }
```

```
Execution time for n = 1,000,000 is 6 milliseconds
Execution time for n = 10,000,000 is 61 milliseconds
Execution time for n = 100,000,000 is 610 milliseconds
Execution time for n = 1,000,000,000 is 6048 milliseconds
```

前面的分析预测了这个循环为线性时间复杂度。如示例运行所显示的，当输入问题的大小增加了 10 倍时，运行时间也增加了大约 10 倍。运行和预测是吻合的。

示例 2

下面循环的时间复杂度是多少？

```java
for (int i = 1; i <= n; i++) {
  for (int j = 1; j <= n; j++) {
    k = k + i + j;
  }
}
```

执行下面的语句

```java
k = k + i + j;
```

需要一个常量时间 c，外层循环执行了 n 次。对于外层循环的每次迭代，内层循环都会执行 n 次。因此，该循环的时间复杂度是

$$T(n) = c \times n \times n = O(n^2)$$

时间复杂度为 $O(n^2)$ 的算法称为二次方算法（quadratic algorithm），它体现了二次方的增长率。二次方算法随着问题规模的增加快速增长。如果输入规模加倍，算法时间就变成 4 倍。通常，两层嵌套循环的算法都是二次方的。

示例 3

考虑下面的循环：

```java
for (int i = 1; i <= n; i++) {
  for (int j = 1; j <= i; j++) {
    k = k + i + j;
  }
}
```

外层循环执行 n 次。对于 $i=1, 2, \cdots, n$，内层循环分别执行 1 次，2 次，$\cdots$，n 次。因此，该循环的时间复杂度是

$$\begin{aligned} T(n) &= c + 2c + 3c + 4c + \cdots + nc \\ &= cn(n+1)/2 \\ &= (c/2)\, n^2 + (c/2)n \\ &= O(n^2) \end{aligned}$$

示例 4

考虑下面的循环：

```
for (int i = 1; i <= n; i++) {
  for (int j = 1; j <= 20; j++) {
    k = k + i + j;
  }
}
```

内层循环执行 20 次，外层循环执行 n 次。因此，该循环的时间复杂度是

$$T(n) = 20 \times c \times n = O(n)$$

示例 5

考虑下面的语句：

```
for (int j = 1; j <= 10; j++) {
  k = k + 4;
}
for (int i = 1; i <= n; i++) {
  for (int j = 1; j <= 20; j++) {
    k = k + i + j;
  }
}
```

第一个循环执行 10 次，第二个循环执行 $20 \times n$ 次。因此，该循环的时间复杂度是

$$T(n) = 10 \times c + 20 \times c \times n = O(n)$$

示例 6

考虑下面的选择语句：

```
if (list.contains(e)) {
  System.out.println(e);
}
else
  for (Object t: list) {
    System.out.println(t);
  }
```

假设线性表中包含 n 个元素，那么 list.contains(e) 的执行时间是 $O(n)$。在 else 子句中的循环耗费的时间是 $O(n)$。因此，整个语句的时间复杂度是

$$T(n) = \text{if 测试时间} + \text{最差情况时间 (if 子句, else 子句)}$$
$$= O(n) + O(n) = O(n)$$

示例 7

考虑计算 a^n，n 为整数。一个简单的算法就是将 a 乘 n 次，如下所示：

```
result = 1;
for (int i = 1; i <= n; i++)
  result *= a;
```

这个算法耗费的时间是 $O(n)$。不失一般性，假设 $n=2^k$。可以使用下面的方法提高算法的效率：

```
result = a;
for (int i = 1; i <= k; i++)
  result = result * result;
```

这个算法耗费的时间是 $O(\log n)$。针对任意的 n，可以修改算法并证明其复杂度仍然是 $O(\log n)$（参见复习题 22.3.5）。

注意：具有 $O(\log n)$ 时间复杂度的算法称为对数算法（logarithmic algorithm），体现了呈对数增长的复杂度。log 的底为 2，但是底不会影响对数增长率，因此可以将其忽略。在

对数算法中，底通常为2。

✓ **复习题**

22.3.1 下列循环中的循环次数是多少？

```
int count = 1;
while (count < 30) {
  count = count * 2;
}
```
a)

```
int count = 15;
while (count < 30) {
  count = count * 3;
}
```
b)

```
int count = 1;
while (count < n) {
  count = count * 2;
}
```
c)

```
int count = 15;
while (count < n) {
  count = count * 3;
}
```
d)

22.3.2 如果 n 为 10，那么下面的代码将显示多少个星号？如果 n 为 20，将显示多少个星号？使用大 O 表示法估算时间复杂度。

```
for (int i = 0; i < n; i++) {
  System.out.print('*');
}
```
a)

```
for (int i = 0; i < n; i++) {
  for (int j = 0; j < n; j++) {
    System.out.print('*');
  }
}
```
b)

```
for (int k = 0; k < n; k++) {
  for (int i = 0; i < n; i++) {
    for (int j = 0; j < n; j++) {
      System.out.print('*');
    }
  }
}
```
c)

```
for (int k = 0; k < 10; k++) {
  for (int i = 0; i < n; i++) {
    for (int j = 0; j < n; j++) {
      System.out.print('*');
    }
  }
}
```
d)

22.3.3 使用大 O 表示法估算下列方法的时间复杂度。

```
public static void mA(int n) {
  for (int i = 0; i < n; i++) {
    System.out.print(Math.random());
  }
}
```
a)

```
public static void mB(int n) {
  for (int i = 0; i < n; i++) {
    for (int j = 0; j < i; j++)
      System.out.print(Math.random());
  }
}
```
b)

```
public static void mC(int[ ] m) {
  for (int i = 0; i < m.length; i++) {
    System.out.print(m[i]);
  }

  for (int i = m.length - 1; i >= 0; )
  {
    System.out.print(m[i]);
    i--;
  }
}
```
c)

```
public static void mD(int[] m) {
  for (int i = 0; i < m.length; i++) {
    for (int j = 0; j < i; j++)
      System.out.print(m[i] * m[j]);
  }
}
```
d)

22.3.4 设计一个 $O(n)$ 时间的算法，计算从 $n1$ 到 $n2$ 的数字和（$n1 < n2$）。你可以设计一个 $O(1)$ 复杂度的算法来执行同样的任务吗？

22.3.5 22.3 节的示例 7 假设 $n = 2^k$。针对任意的 n 修改算法，并证明其复杂度仍然为 $O(\log n)$。

22.4 分析算法的时间复杂度

☞ **要点提示**：本节将分析几种著名算法的复杂度，这些算法包括：二分查找法、选择排序法和汉诺塔法。

22.4.1 分析二分查找算法

程序清单 7-7 中给出的二分查找算法，是在一个排好序的数组中查找键。算法中的每次迭代都包含固定次数的操作，次数由 c 来表示。设 $T(n)$ 表示在包含 n 个元素的线性表中进行二分查找的时间复杂度。不失一般性，假定 n 是 2 的幂，且 $k=\log n$。在两次比较之后，二分查找排除了输入的一半，

$$T(n) = T\left(\frac{n}{2}\right) + c = T\left(\frac{n}{2^2}\right) + c + c = T\left(\frac{n}{2^k}\right) + kc$$
$$= T(1) + c\log n = 1 + (\log n)c$$
$$= O(\log n)$$

忽略常量和非主导项，二分查找算法的复杂度为 $O(\log n)$。这就是对数算法。随着问题规模的增长，对数算法的复杂度增长得比较缓慢。在二分查找的情形下，每次将数组的大小翻倍，最多需要增加一次比较。如果将任意对数算法的输入规模平方，那么算法的时间复杂度只会翻倍。因此，对数时间算法是很高效的。

22.4.2 分析选择排序算法

程序清单 7-8 中给出的选择排序算法，是在线性表中找到最小元素，并将其和第一个元素交换。然后在剩下的元素中找到最小元素，将其和剩余的列表中的第一个元素交换，这样一直做下去，直到线性表中仅剩一个元素为止。对于第一次迭代，比较次数为 $n-1$，第二次迭代的比较次数为 $n-2$，以此类推。设 $T(n)$ 表示选择排序的复杂度，c 表示每次迭代中其他操作如赋值和额外比较的总数。这样，

$$T(n) = (n-1) + c + (n-2) + c + \cdots + 2 + c + 1 + c$$
$$= \frac{(n-1)(n-1+1)}{2} + c(n-1) = \frac{n^2}{2} - \frac{n}{2} + cn - c$$
$$= O(n^2)$$

因此，选择排序算法的复杂度为 $O(n^2)$。

22.4.3 分析汉诺塔问题

程序清单 18-8 中给出的汉诺塔问题，是按如下方式借助塔 C 将 n 个盘子从塔 A 递归地移动到塔 B：

1）借助塔 B 将前 $n-1$ 个盘子从塔 A 移动到塔 C。
2）将盘子 n 从塔 A 移动到塔 B。
3）借助塔 A 将 $n-1$ 个盘子从塔 C 移动到塔 B。

该算法的复杂度由移动的次数来衡量。设 $T(n)$ 表示使用该算法从塔 A 到塔 B 移动 n 个盘子所需要的移动次数，$T(1)$ 为 1。因此，

$$T(n) = T(n-1) + 1 + T(n-1)$$
$$= 2T(n-1) + 1$$
$$= 2(2T(n-2) + 1) + 1$$
$$= 2(2(2T(n-3) + 1) + 1) + 1$$
$$= 2^{n-1}T(1) + 2^{n-2} + \cdots + 2 + 1$$
$$= 2^{n-1} + 2^{n-2} + \cdots + 2 + 1 = 2^n - 1 = O(2^n)$$

具有 $O(2^n)$ 时间复杂度的算法称为指数算法（exponential algorithm），体现为指数级的增长率。随着输入规模的增长，指数算法耗费的时间呈指数增长。指数算法在大的输入规模下并不实用。假设盘子一次移动耗时 1 秒，则需要耗费 $2^{32}/(365 \times 24 \times 60 \times 60) = 136$ 年来移动 32 个盘子，以及 $2^{64}/(365 \times 24 \times 60 \times 60) = 5850$ 亿年来移动 64 个盘子。

22.4.4 常用的递推关系

递推关系（recurrence relation）是分析算法复杂度的有用工具。如前面例子所示，二分查找、选择排序以及汉诺塔问题的复杂度分别为：$T(n) = T\left(\dfrac{n}{2}\right) + c$，$T(n) = T(n-1) + O(n)$，$T(n) = 2T(n-1) + O(1)$。表 22-2 总结了常用的递推关系。

表 22-2 常用的递推关系

递推关系	结果	示例
$T(n) = T(n/2) + O(1)$	$T(n) = O(\log n)$	二分查找，欧几里得法求最大公约数
$T(n) = T(n-1) + O(1)$	$T(n) = O(n)$	线性查找
$T(n) = 2T(n/2) + O(1)$	$T(n) = O(n)$	复习题 22.8.3
$T(n) = 2T(n/2) + O(n)$	$T(n) = O(n\log n)$	归并排序（第 23 章）
$T(n) = T(n-1) + O(n)$	$T(n) = O(n^2)$	选择排序
$T(n) = 2T(n-1) + O(1)$	$T(n) = O(2^n)$	汉诺塔
$T(n) = T(n-1) + T(n-2) + O(1)$	$T(n) = O(2^n)$	递归的斐波那契算法

22.4.5 比较常用的增长函数

前面分析了几个算法的复杂度。表 22-3 列出了一些常用的增长函数，然后显示当输入规模从 $n = 25$ 加倍到 $n = 50$ 时，增长率是如何变化的。

表 22-3 增长率的变化

函数	名称	$n = 25$	$n = 50$	$f(50)/f(25)$
$O(1)$	常量时间	1	1	1
$O(\log n)$	对数时间	4.64	5.64	1.21
$O(n)$	线性时间	25	50	2
$O(n\log n)$	对数-线性时间	116	282	2.43
$O(n^2)$	二次时间	625	2500	4
$O(n^3)$	三次时间	15 625	125 000	8
$O(2^n)$	指数时间	3.36×10^7	1.27×10^{15}	3.35×10^7

这些函数的排序（如图 22-1 所示）如下：
$$O(1) < O(\log n) < O(n) < O(n\log n) < O(n^2) < O(n^3) < O(2^n)$$

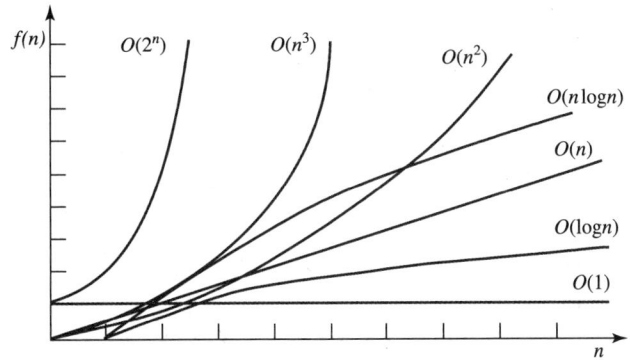

图 22-1 随着 n 的增大函数的增长趋势

复习题

22.4.1 对下面的增长函数排序：

$$\frac{5n^3}{4032}, 44\log n, 10n\log n, 500, 2n^2, \frac{2^n}{45}, 3n$$

22.4.2 估算将两个 $n \times m$ 矩阵相加的时间复杂度，以及将 $n \times m$ 矩阵与 $m \times k$ 矩阵相乘的时间复杂度。

22.4.3 描述寻找数组中最大元素出现次数的算法。分析该算法的复杂度。

22.4.4 描述从数组中删除重复元素的算法。分析该算法的复杂度。

22.4.5 分析下面的排序算法：

```
for (int i = 0; i < list.length - 1; i++) {
  if (list[i] > list[i + 1]) {
    swap list[i] with list[i + 1];
    i = -1;
  }
}
```

22.4.6 分析分别使用暴力法和 Horner 方法计算对于给定 x 值的 n 阶多项式 $f(x)$ 的复杂度。暴力法是计算多项式中的每项并将其相加。Horner 方法在 6.7 节介绍过。

$$f(x) = a_n x^n + a_{n-1} x^{n-1} + a_{n-2} x^{n-2} + \cdots + a_1 x^1 + a_0$$

22.5 使用动态编程求斐波那契数

要点提示：本节使用动态编程分析和设计一个高效算法来求斐波那契数。

18.3 节给出了一个求斐波那契数的递归方法，如下所示：

```
/** The method for finding the Fibonacci number */
public static long fib(long index) {
  if (index == 0) // Base case
    return 0;
  else if (index == 1) // Base case
    return 1;
  else // Reduction and recursive calls
    return fib(index - 1) + fib(index - 2);
}
```

现在，我们可以证明这个算法的复杂度是 $O(2^n)$。为方便起见，令下标为 n。假设 $T(n)$ 表示求解 fib(n) 的算法复杂度，而 c 表示将下标与 0 和 1 进行比较所耗费的常量时间。因此，

$$T(n) = T(n-1) + T(n-2) + c$$
$$\leq 2T(n-1) + c$$
$$\leq 2(2T(n-2) + c) + c$$
$$= 2^2 T(n-2) + 2c + c$$

和汉诺塔问题的分析类似，我们可以得出 $T(n)$ 是 $O(2^n)$。

这个算法的效率并不高。能找出求斐波那契数的高效算法吗？递归的 `fib` 方法中的问题在于重复地调用同样参数的方法。例如，为了计算 `fib(4)`，要调用 `fib(3)` 和 `fib(2)`。为了计算 `fib(3)`，要调用 `fib(2)` 和 `fib(1)`。注意，`fib(2)` 被重复调用。我们可以通过避免重复调用同样参数的 `fib` 方法来提高效率。注意到一个新的斐波那契数是通过对数列中的前两个数相加得到的。如果使用两个变量 `f0` 和 `f1` 来存储前面的两个数，那么可以通过将 `f0` 和 `f1` 相加立即获得新数 `f2`。现在，应该通过将 `f1` 赋给 `f0`，将 `f2` 赋给 `f1` 来更新 `f0` 和 `f1`，如图 22-2 所示。

```
         f0 f1 f2
斐波那契数列:  0  1  1  2  3  5  8 13 21 34 55 89 ...
下标:        0  1  2  3  4  5  6  7  8  9 10 11
            f0 f1 f2
斐波那契数列:  0  1  1  2  3  5  8 13 21 34 55 89 ...
下标:        0  1  2  3  4  5  6  7  8  9 10 11
                                    f0 f1 f2
斐波那契数列:  0  1  1  2  3  5  8 13 21 34 55 89 ...
下标:        0  1  2  3  4  5  6  7  8  9 10 11
```

图 22-2　变量 `f0`、`f1` 和 `f2` 存储数列中的三个连续斐波那契数

新的方法在程序清单 22-2 中实现。

程序清单 22-2　ImprovedFibonacci.java

```java
 1  import java.util.Scanner;
 2
 3  public class ImprovedFibonacci {
 4    /** Main method */
 5    public static void main(String args[]) {
 6      // Create a Scanner
 7      Scanner input = new Scanner(System.in);
 8      System.out.print("Enter an index for the Fibonacci number: ");
 9      int index = input.nextInt();
10
11      // Find and display the Fibonacci number
12      System.out.println(
13        "Fibonacci number at index " + index + " is " + fib(index));
14    }
15
16    /** The method for finding the Fibonacci number */
17    public static long fib(long n) {
18      long f0 = 0; // For fib(0)
19      long f1 = 1; // For fib(1)
20      long f2 = 1; // For fib(2)
21
22      if (n == 0)
23        return f0;
24      else if (n == 1)
25        return f1;
26      else if (n == 2)
27        return f2;
28
29      for (int i = 3; i <= n; i++) {
```

```
30        f0 = f1;
31        f1 = f2;
32        f2 = f0 + f1;
33      }
34
35      return f2;
36    }
37  }
```

```
Enter an index for the Fibonacci number: 6 ↵Enter
Fibonacci number at index 6 is 8
```

```
Enter an index for the Fibonacci number: 7 ↵Enter
Fibonacci number at index 7 is 13
```

很显然，新算法的复杂度是 $O(n)$，比递归的 $O(2^n)$ 算法提高了不少。

- 算法设计提示：这里给出的求斐波那契数的算法使用了一种称为动态编程（dynamic programming）的方法。动态编程是通过解决子问题，然后将子问题的结果合并来获得整个问题的解的过程。这自然地引向递归求解。然而，使用递归效率不高，因为子问题相互重叠了。动态编程的关键思想是只解决子问题一次，并将子问题的结果存储以备后用，从而避免了重复的子问题求解。

复习题

22.5.1 什么是动态编程？给出一个动态编程的示例。
22.5.2 为什么递归的斐波那契算法是低效的，而非递归的斐波那契算法是高效的？

22.6 使用欧几里得算法求最大公约数

- 要点提示：本节给出几个求两个整数的最大公约数的算法，以在其中找出一个高效的算法。

两个整数的最大公约数（Greatest Common Divisor, GCD）是能被这两个整数整除的最大数字。程序清单 5-9 给出了一个求两个整数 m 和 n 的最大公约数的暴力算法。

- 算法设计提示：暴力（brute force）法指使用最简单和最直接，或者显而易见的方式解决问题的算法。结果是，为了解决一个给定问题，这样的算法相比更聪明或者更复杂的算法而言，可能导致做更多的工作。另一方面，暴力算法相对于复杂的算法而言，通常更加易于实现，并且因为其简单性有时可以更加高效。

暴力算法检测 k（k=2,3,4,…）是否是 n1 和 n2 的公约数，直到 k 大于 n1 或 n2。该算法可以如下描述：

```
public static int gcd(int m, int n) {
  int gcd = 1;

  for (int k = 2; k <= m && k <= n; k++) {
    if (m % k == 0 && n % k == 0)
      gcd = k;
  }

  return gcd;
}
```

假设 $m \geq n$，那么，显然该算法的复杂度是 $O(n)$。

是否还有求最大公约数的更好算法？不从 1 向上开始查找可能的除数，而是从 n 开始向下查找，这样会更高效。一旦找到一个除数，该除数就是最大公约数。因此，可以使用下面的循环来改进算法：

```java
for (int k = n; k >= 1; k--) {
  if (m % k == 0 && n % k == 0) {
    gcd = k;
    break;
  }
}
```

这个算法比前一个效率高,但是其最坏情况的时间复杂度依旧是 $O(n)$。

数字 n 的除数不可能比 $n/2$ 大。因此,可以使用下面的循环进一步改进算法:

```java
for (int k = n / 2; k >= 1; k--) {
  if (m % k == 0 && n % k == 0) {
    gcd = k;
    break;
  }
}
```

但是,该算法是不正确的,因为 n 可能会是 m 的除数。这种情况必须考虑到。正确的算法如程序清单 22-3 所示。

程序清单 22-3 GCD.java

```java
 1  import java.util.Scanner;
 2
 3  public class GCD {
 4    /** Find GCD for integers m and n */
 5    public static int gcd(int m, int n) {
 6      int gcd = 1;
 7
 8      if (m % n == 0) return n;
 9
10      for (int k = n / 2; k >= 1; k--) {
11        if (m % k == 0 && n % k == 0) {
12          gcd = k;
13          break;
14        }
15      }
16
17      return gcd;
18    }
19
20    /** Main method */
21    public static void main(String[] args) {
22      // Create a Scanner
23      Scanner input = new Scanner(System.in);
24
25      // Prompt the user to enter two integers
26      System.out.print("Enter first integer: ");
27      int m = input.nextInt();
28      System.out.print("Enter second integer: ");
29      int n = input.nextInt();
30
31      System.out.println("The greatest common divisor for " + m +
32        " and " + n + " is " + gcd(m, n));
33    }
34  }
```

```
Enter first integer: 2525 ↵Enter
Enter second integer: 125 ↵Enter
The greatest common divisor for 2525 and 125 is 25
```

```
Enter first integer: 3 ↵Enter
Enter second integer: 3 ↵Enter
The greatest common divisor for 3 and 3 is 3
```

假设 $m \geq n$，那么这个 for 循环最多执行 $n/2$ 次，比前一个算法节省了一半的运行时间。该算法的时间复杂度仍然是 $O(n)$，但实际上，它比程序清单 5-9 中的算法快得多。

注意：大 O 表示法提供了对算法效率的一个很好的理论估算。但是，两个算法即使有相同的时间复杂度，它们的效率也不一定相同。如前面的例子所示，程序清单 5-9 和程序清单 22-3 中的两个算法具有相同的复杂度，但实际上，程序清单 22-3 中的算法显然更好些。

求最大公约数的一个更有效的算法是在公元前 300 年左右由欧几里得发现的，这是非常古老的著名算法之一。它可以递归地定义如下：

用 gcd(m,n) 表示整数 m 和 n 的最大公约数：

- 如果 m%n 为 0，那么 gcd(m,n) 为 n。
- 否则，gcd(m,n) 就是 gcd(n,m%n)。

不难证明这个算法的正确性。假设 m%n=r，那么，m=qn+r，这里的 q 是 m/n 的商。能整除 m 和 n 的任意数字必然也能整除 r。因此，gcd(m,n) 和 gcd(n,r) 相等，其中 r=m%n。该算法的实现如程序清单 22-4 所示。

程序清单 22-4 GCDEuclid.java

```java
 1  import java.util.Scanner;
 2
 3  public class GCDEuclid {
 4    /** Find GCD for integers m and n */
 5    public static int gcd(int m, int n) {
 6      if (m % n == 0)
 7        return n;
 8      else
 9        return gcd(n, m % n);
10    }
11
12    /** Main method */
13    public static void main(String[] args) {
14      // Create a Scanner
15      Scanner input = new Scanner(System.in);
16
17      // Prompt the user to enter two integers
18      System.out.print("Enter first integer: ");
19      int m = input.nextInt();
20      System.out.print("Enter second integer: ");
21      int n = input.nextInt();
22
23      System.out.println("The greatest common divisor for " + m +
24        " and " + n + " is " + gcd(m, n));
25    }
26  }
```

```
Enter first integer: 2525  ↵Enter
Enter second integer: 125  ↵Enter
The greatest common divisor for 2525 and 125 is 25
```

```
Enter first integer: 3  ↵Enter
Enter second integer: 3  ↵Enter
The greatest common divisor for 3 and 3 is 3
```

最好的情况是当 m % n 为 0 的时候，算法只用一步就能找出最大公约数。分析平均情况是很困难的。然而，我们可以证明最坏情况的时间复杂度是 $O(\log n)$。

假设 $m \geq n$，我们可以证明 m%n < m/2，如下所示：

- 如果 n<=m/2，那么 m%n < m/2，因为 m 除以 n 的余数总是小于 n。
- 如果 n>m/2，那么 m%n=m-n < m/2。因此，m%n < m/2。

欧几里得的算法递归地调用 gcd 方法。它首先调用 gcd(m,n)，接着调用 gcd(n,m%n)，然后是 gcd(m%n,n%(m%n))，以此类推，如下所示：

```
  gcd(m, n)
= gcd(n, m % n)
= gcd(m % n, n % (m % n))
= ...
```

因为 m%n<m/2 且 n%(m%n)<n/2，所以传递给 gcd 方法的参数在每两次迭代之后减少一半。在调用 gcd 两次之后，第二个参数小于 n/2。在调用 gcd 四次之后，第二个参数小于 n/4。在调用 gcd 六次之后，第二个参数小于 $n/2^3$。假设 k 是调用 gcd 方法的次数。在调用 gcd 方法 k 次之后，第二个参数小于 $n/2^{(k/2)}$，它是大于或等于 1 的。也就是

$$\frac{n}{2^{(k/2)}} \geq 1 \Rightarrow n \geq 2^{(k/2)} \Rightarrow \log n \geq k/2 \Rightarrow k \leq 2\log n$$

因此，$k \leq 2\log n$。所以该 gcd 方法的时间复杂度是 $O(\log n)$。

最坏情况发生在两个数导致了相除次数最多的时候。事实证明，两个连续的斐波那契数会造成相除次数最多的情况。回顾斐波那契数列是从 0 和 1 开始，然后每个数都是前两个数的和，例如：

0 1 1 2 3 5 8 13 21 34 55 89…

这个数列可以递归地定义为

```
fib(0) = 0;
fib(1) = 1;
fib(index) = fib(index - 2) + fib(index - 1); index >= 2
```

对于两个连续的斐波那契数 fib(index) 和 fib(index-1)，

```
  gcd(fib(index), fib(index - 1))
= gcd(fib(index - 1), fib(index - 2))
= gcd(fib(index - 2), fib(index - 3))
= gcd(fib(index - 3), fib(index - 4))
= ...
= gcd(fib(2), fib(1))
= 1
```

例如，

```
  gcd(21, 13)
= gcd(13, 8)
= gcd(8, 5)
= gcd(5, 3)
= gcd(3, 2)
= gcd(2, 1)
= 1
```

因此，gcd 方法被调用的次数和下标相等。我们可以证明 index ≤ 1.44 log$n$，其中 n=fib(index-1)。这是一个比 index ≤ 2log$n$ 更严格的限定。

表 22-4 总结了三个求最大公约数的算法的复杂度。

表 22-4　GCD 算法的比较

算法	复杂度	描述
程序清单 5-9	$O(n)$	暴力法，检查所有可能的除数
程序清单 22-3	$O(n)$	检查所有可能除数的一半
程序清单 22-4	$O(\log n)$	欧几里得算法

复习题

22.6.1 证明下面寻找两个整数 m 和 n 的 GCD 的算法是错误的。

```
int gcd = 1;
for (int k = Math.min(Math.sqrt(n), Math.sqrt(m)); k >= 1; k--) {
  if (m % k == 0 && n % k == 0) {
    gcd = k;
    break;
  }
}
```

22.7 求素数的高效算法

要点提示：本节给出了多个算法，以找到求素数的高效算法。

第一个发现至少具有 100 000 000 位十进制素数的个人或团体可以得到 150 000 美元的奖励（https://www.eff.org/awards/coop）。

你可以设计一个求素数的快速算法吗？

对于一个大于 1 的整数，如果其除数只有 1 和它本身，那么它就是一个素数（prime）。例如，2、3、5、7 都是素数，但是 4、6、8、9 都不是。

如何确定一个数字 n 是否是素数？程序清单 5-15 给出了一个求素数的暴力算法。算法检测 2,3,4,5,…,n-1 是否能整除 n。如果不能，那么 n 就是素数。这个算法耗费 $O(n)$ 时间来检测 n 是否是一个素数。注意，只需要检测 2,3,4,5,…,n/2 是否能整除 n。如果不能，那么 n 就是素数。算法的效率只稍微提高了一点，它的复杂度仍然是 $O(n)$。

实际上，我们可以证明，如果 n 不是素数，那么 n 必须有一个大于 1 且小于或等于 $\sqrt{n}$ 的因子。下面是其证明过程：因为 n 不是素数，所以会存在两个数 p 和 q，满足 n=pq 且 $1<p\leq q$。注意，$n=\sqrt{n}\sqrt{n}$。p 必须小于或等于 $\sqrt{n}$。因此，只需要检测 2,3,4,5,… 或者 $\sqrt{n}$ 是否能被 n 整除。如果不能，n 就是素数。这会显著地将算法的时间复杂度降低为 $O(\sqrt{n})$。

现在考虑找出不超过 n 的所有素数的算法。一个直观的实现方法就是检测 i 是否是素数，这里 i=2,3, 4,…,n。该程序在程序清单 22-5 中给出。

程序清单 22-5 PrimeNumbers.java

```
1  import java.util.Scanner;
2
3  public class PrimeNumbers {
4    public static void main(String[] args) {
5      Scanner input = new Scanner(System.in);
6      System.out.print("Find all prime numbers <= n, enter n: ");
7      int n = input.nextInt();
8
9      final int NUMBER_PER_LINE = 10; // Display 10 per line
10     int count = 0; // Count the number of prime numbers
11     int number = 2; // A number to be tested for primeness
12
13     System.out.println("The prime numbers are:");
14
15     // Repeatedly find prime numbers
16     while (number <= n) {
17       // Assume the number is prime
18       boolean isPrime = true; // Is the current number prime?
19
20       // Test if number is prime
21       for (int divisor = 2; divisor <= (int)(Math.sqrt(number));
22            divisor++) {
23         if (number % divisor == 0) { // If true, number is not prime
```

```
24            isPrime = false; // Set isPrime to false
25            break; // Exit the for loop
26          }
27        }
28
29        // Print the prime number and increase the count
30        if (isPrime) {
31          count++; // Increase the count
32
33          if (count % NUMBER_PER_LINE == 0) {
34            // Print the number and advance to the new line
35            System.out.printf("%7d\n", number);
36          }
37          else
38            System.out.printf("%7d", number);
39        }
40
41        // Check if the next number is prime
42        number++;
43      }
44
45      System.out.println("\n" + count +
46        " prime(s) less than or equal to " + n);
47    }
48  }
```

```
Find all prime numbers <= n, enter n: 1000 ⏎Enter
The prime numbers are:
      2      3      5      7     11     13     17     19     23     29
     31     37     41     43     47     53     59     61     67     71
...
...
168 prime(s) less than or equal to 1000
```

如果 for 循环的每次迭代都必须计算 Math.sqrt(number)（第 21 行），那么该程序的效率不高。一个好的编译器应该为整个 for 循环只计算一次 Math.sqrt(number)。为确保这一点，可以显式地用下面两行替换第 21 行：

```
int squareRoot = (int)(Math.sqrt(number));
for (int divisor = 2; divisor <= squareRoot; divisor++) {
```

实际上，没有必要对每个 number 来确切计算 Math.sqrt(number)。只需要找出完全平方数，例如，4、9、16、25、36、49，等等。注意到对于 36 和 48 之间并包括 36 和 48 的数，它们的 (int)(Math.sqrt(number)) 值为 6。看清楚这一点，就可以用下面的代码替换第 16 ~ 26 行：

```
...
int squareRoot = 1;
// Repeatedly find prime numbers
while (number <= n) {
  // Assume the number is prime
  boolean isPrime = true; // Is the current number prime?

  if (squareRoot * squareRoot < number) squareRoot++;

  // Test if number is prime
  for (int divisor = 2; divisor <= squareRoot; divisor++) {
    if (number % divisor == 0) { // If true, number is not prime
      isPrime = false; // Set isPrime to false
      break; // Exit the for loop
    }
  }
  ...
```

现在，我们专注于分析该程序的复杂度。因为它在 for 循环中耗费 $\sqrt{i}$ 步（第 21～27 行）来检查数字 i 是否是素数，所以算法耗费 $\sqrt{2}+\sqrt{3}+\sqrt{4}+\cdots+\sqrt{n}$ 步来找出所有小于或等于 n 的素数。观察到

$$\sqrt{2}+\sqrt{3}+\sqrt{4}+\cdots+\sqrt{n} \leqslant n\sqrt{n}$$

因此，该算法的复杂度为 $O(n\sqrt{n})$。

为了确定 i 是否是素数，算法需要检测 2,3,4,5,… 以及 $\sqrt{i}$ 是否能被 i 整除。可以进一步提高该算法的效率，因为只需要检测从 2 到 $\sqrt{i}$ 之间的素数能否被 i 整除。

我们可以证明，如果 i 不是素数，那就必须存在一个素数 p，满足 i=pq 且 p ⩽ q。下面是其证明过程。假设 i 不是素数，且 p 是 i 的最小因子。那么 p 肯定是素数，否则 p 就有一个因子 k，且 2 ⩽ k < p。k 也是 i 的一个因子，这和 p 是 i 的最小因子冲突。因此，如果 i 不是素数，那么可以得到从 2 到 $\sqrt{i}$ 之间被 i 整除的素数。这会得到一个求解不超过 n 的所有素数的更有效的算法，如程序清单 22-6 所示。

程序清单 22-6 EfficientPrimeNumbers.java

```java
 1  import java.util.Scanner;
 2
 3  public class EfficientPrimeNumbers {
 4    public static void main(String[] args) {
 5      Scanner input = new Scanner(System.in);
 6      System.out.print("Find all prime numbers <= n, enter n: ");
 7      int n = input.nextInt();
 8
 9      // A list to hold prime numbers
10      java.util.List<Integer> list =
11        new java.util.ArrayList<>();
12
13      final int NUMBER_PER_LINE = 10; // Display 10 per line
14      int count = 0; // Count the number of prime numbers
15      int number = 2; // A number to be tested for primeness
16      int squareRoot = 1; // Check whether number <= squareRoot
17
18      System.out.println("The prime numbers are \n");
19
20      // Repeatedly find prime numbers
21      while (number <= n) {
22        // Assume the number is prime
23        boolean isPrime = true; // Is the current number prime?
24
25        if (squareRoot * squareRoot < number) squareRoot++;
26
27        // Test whether number is prime
28        for (int k = 0; k < list.size()
29                        && list.get(k) <= squareRoot; k++) {
30          if (number % list.get(k) == 0) { // If true, not prime
31            isPrime = false; // Set isPrime to false
32            break; // Exit the for loop
33          }
34        }
35
36        // Print the prime number and increase the count
37        if (isPrime) {
38          count++; // Increase the count
39          list.add(number); // Add a new prime to the list
40          if (count % NUMBER_PER_LINE == 0) {
41            // Print the number and advance to the new line
42            System.out.println(number);
43          }
```

```
44          else
45            System.out.print(number + " ");
46        }
47
48        // Check whether the next number is prime
49        number++;
50      }
51
52      System.out.println("\n" + count +
53        " prime(s) less than or equal to " + n);
54    }
55  }
```

```
Find all prime numbers <= n, enter n: 1000 ↵Enter
The prime numbers are:
     2    3    5    7   11   13   17   19   23   29
    31   37   41   43   47   53   59   61   67   71
...
...
168 prime(s) less than or equal to 1000
```

假设 $\pi(i)$ 表示小于或等于 i 的素数的个数。20 以下的素数是 2、3、5、7、11、13、17 和 19。因此，$\pi(2)$ 是 1，$\pi(3)$ 是 2，$\pi(6)$ 是 3，而 $\pi(20)$ 是 8。已经证明 $\pi(i)$ 近似为 $\frac{i}{\log i}$（参见 primes.utm.edu/howmany.shtml）。

对每个数字 i，该算法检查小于或等于 $\sqrt{i}$ 的素数是否能被 i 整除。小于或等于 $\sqrt{i}$ 的素数的个数是

$$\frac{\sqrt{i}}{\log \sqrt{i}} = \frac{2\sqrt{i}}{\log i}$$

这样，找出不超过 n 的所有素数的复杂度为

$$\frac{2\sqrt{2}}{\log 2} + \frac{2\sqrt{3}}{\log 3} + \frac{2\sqrt{4}}{\log 4} + \frac{2\sqrt{5}}{\log 5} + \frac{2\sqrt{6}}{\log 6} + \frac{2\sqrt{7}}{\log 7} + \frac{2\sqrt{8}}{\log 8} + \cdots + \frac{2\sqrt{n}}{\log n}$$

对于 $i < n$ 且 $n \geq 16$，由于 $\frac{\sqrt{i}}{\log i} < \frac{\sqrt{n}}{\log n}$，所以

$$\frac{2\sqrt{2}}{\log 2} + \frac{2\sqrt{3}}{\log 3} + \frac{2\sqrt{4}}{\log 4} + \frac{2\sqrt{5}}{\log 5} + \frac{2\sqrt{6}}{\log 6} + \frac{2\sqrt{7}}{\log 7} + \frac{2\sqrt{8}}{\log 8} + \cdots + \frac{2\sqrt{n}}{\log n} < \frac{2n\sqrt{n}}{\log n}$$

因此，这个算法的复杂度为 $O\left(\frac{n\sqrt{n}}{\log n}\right)$。

这个算法是动态编程的另外一个示例。该算法在数组线性表中存储子问题的结果，之后使用它们来检查一个新的数字是否是素数。

还有比 $O\left(\frac{n\sqrt{n}}{\log n}\right)$ 更好的算法吗？让我们考察一下著名的找素数的埃拉托色尼算法。埃拉托色尼（Eratosthenes，公元前 276—公元前 194）是一位希腊数学家，他设计了一个称为埃拉托色尼筛选法（sieve of Eratosthenes）的聪明算法，该算法求出所有小于或等于 n 的素数。该算法使用一个包含 n 个布尔值的数组 primes。初始时，primes 的所有元素都设置为 true。

因为 2 的倍数都不是素数,所以对于所有的 $2 \leqslant i < n/2$,都将 primes[2*i] 设置为 false,如图 22-3 所示。因为我们不关注 primes[0] 和 primes[1],所以这些值在图中被标注上 ×。

素数数组																												
下标	0	1	2	3	4	5	6	7	8	9	10	11	12	13	14	15	16	17	18	19	20	21	22	23	24	25	26	27
初始值	×	×	T	T	T	T	T	T	T	T	T	T	T	T	T	T	T	T	T	T	T	T	T	T	T	T	T	T
k=2	×	×	T	T	F	T	F	T	F	T	F	T	F	T	F	T	F	T	F	T	F	T	F	T	F	T	F	T
k=3	×	×	T	T	F	T	F	T	F	F	F	T	F	T	F	F	F	T	F	T	F	F	F	T	F	T	F	F
k=5	×	×	ⓉⓉ	F	ⓉⓉ	F	Ⓣ	F	F	F	F	Ⓣ	F	Ⓣ	F	F	F	Ⓣ	F	Ⓣ	F	F	F	Ⓣ	F	F	F	F

图 22-3 primes 中的值随着每个素数 k 而改变

因为 3 的倍数不是素数,所以对于所有的 $3 \leqslant i \leqslant n/3$,都将 primes[3*i] 设置为 false。因为 5 的倍数不是素数,所以对于所有的 $5 \leqslant i \leqslant n/5$,都将 primes[5*i] 设置为 false。注意,无须考虑 4 的倍数,因为 4 的倍数也是 2 的倍数,这已经考虑过了。同样,6、8、9 的倍数也无须考虑。只需要考虑素数 k=2,3,5,7,11,… 的倍数,并且将 primes 中对应的元素设置为 false。之后,如果 primes[i] 仍然为 true,那么 i 就是素数。如图 22-3 所示,2、3、5、7、11、13、17、19、23 都是素数。程序清单 22-7 给出使用埃拉托色尼筛选法求素数的程序。

程序清单 22-7 SieveOfEratosthenes.java

```
1  import java.util.Scanner;
2
3  public class SieveOfEratosthenes {
4    public static void main(String[] args) {
5      Scanner input = new Scanner(System.in);
6      System.out.print("Find all prime numbers <= n, enter n: ");
7      int n = input.nextInt();
8
9      boolean[] primes = new boolean[n + 1]; // Prime number sieve
10
11     // Initialize primes[i] to true
12     for (int i = 0; i < primes.length; i++) {
13       primes[i] = true;
14     }
15
16     for (int k = 2; k <= n / k; k++) {
17       if (primes[k]) {
18         for (int i = k; i <= n / k; i++) {
19           primes[k * i] = false; // k * i is not prime
20         }
21       }
22     }
23
24     final int NUMBER_PER_LINE = 10; // Display 10 per line
25     int count = 0; // Count the number of prime numbers found so far
26     // Print prime numbers
27     for (int i = 2; i < primes.length; i++) {
28       if (primes[i]) {
29         count++;
30         if (count % NUMBER_PER_LINE == 0)
31           System.out.printf("%7d\n", i);
32         else
33           System.out.printf("%7d", i);
34       }
35     }
36
37     System.out.println("\n" + count +
```

```
38            " prime(s) less than or equal to " + n);
39       }
40  }
```

```
Find all prime numbers <= n, enter n: 1000  ↵Enter
The prime numbers are:
      2     3     5     7    11    13    17    19    23    29
     31    37    41    43    47    53    59    61    67    71
...
...
168 prime(s) less than or equal to 1000
```

注意，k<=n/k（第 16 行）。否则，k*i 可能会大于 n（第 19 行）。该算法的时间复杂度是多少？

对于每个素数 k（第 17 行），算法将 primes[k*i] 设置为 false（第 19 行）。这在 for 循环中执行了 n/k-k+1 次（第 18 行）。这样，找出不超过 n 的所有素数的复杂度就是

$$\frac{n}{2}-2+1+\frac{n}{3}-3+1+\frac{n}{5}-5+1+\frac{n}{7}-7+1+\frac{n}{11}-11+1\cdots$$

$$=O\left(\frac{n}{2}+\frac{n}{3}+\frac{n}{5}+\frac{n}{7}+\frac{n}{11}+\cdots\right)<O(n\pi(n))$$

$$=O\left(n\frac{\sqrt{n}}{\log n}\right)$$

该数列的项数为 $\pi(n)$

该上限 $O\left(\frac{n\sqrt{n}}{\log n}\right)$ 是非常宽松的。实际的时间复杂度比 $O\left(\frac{n\sqrt{n}}{\log n}\right)$ 好很多。埃拉托色尼筛选法对于小的 n 值而言是一个好算法，这样 primes 数组可以载入内存。

表 22-5 总结了找出不超过 n 的所有素数的三个算法的复杂度。

表 22-5 素数算法的比较

算法	复杂度	描述
程序清单 5-15	$O(n^2)$	暴力法，检测所有可能的因子
程序清单 22-5	$O(n\sqrt{n})$	检测直到 $\sqrt{n}$ 的因子
程序清单 22-6	$O\left(\frac{n\sqrt{n}}{\log n}\right)$	检测直到 $\sqrt{n}$ 的素数因子
程序清单 22-7	$O\left(\frac{n\sqrt{n}}{\log n}\right)$	埃拉托色尼筛选法

✔复习题

22.7.1 证明如果 n 不是素数，那么必然存在一个素数 p，使得 $p \leq \sqrt{n}$ 并且 p 是 n 的一个因子。
22.7.2 描述埃拉托色尼筛选算法是如何找到素数的。

教学注意：下面几节介绍有趣而富有挑战性的问题。是时候开始学习高级算法，成为一名熟练的程序员了。我们建议你学习算法，并在练习中实现它们。

22.8 使用分治法寻找最近点对

要点提示：本节给出使用分治法找到最近点对的高效算法。

给定一个点集，那么最近点对问题就是找出两个相距最近的点。如图 22-4 所示，在最近点对动画中画出一条直线连接最近的两个点。

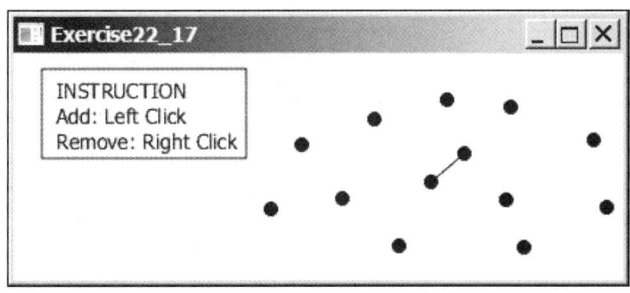

图 22-4 最近点对动画中，当交互式地增加和移除点时，画一条直线动态连接最近的点对（来源：Oracle 或其附属公司版权所有 ©1995 ~ 2016，经授权使用）

8.6 节给出了一个求最近点对的暴力算法。该算法计算所有点对之间的距离，并且求出最小距离的点对。显然，这个算法耗费 $O(n^2)$ 时间。可以设计一个更有效的算法吗？我们将使用一种称为分治（divide-and-conquer）的方法来解决这个问题。

算法设计提示：分治法将问题分解为子问题，解决子问题，然后将子问题的解合并从而得到整个问题的解。和动态编程方法不一样的是，分治法中的子问题不会重叠。子问题类似于初始问题，但是大小更小，因此可以应用递归来解决这样的问题。事实上，所有递归问题的解决方案都遵循分治法。

程序清单 22-8 描述了如何使用分治法来解决最近点对问题。

程序清单 22-8 寻找最近点对的算法

步骤 1：以 x 坐标的升序对点进行排序。对于 x 坐标相同的点，按它的 y 坐标排序。这样就能得到一个由排好序的点构成的线性表 S。

步骤 2：使用排好序的线性表的中点将 S 分为两个大小相等的子集 S_1 和 S_2。让中点位于 S_1 中。递归地找到 S_1 和 S_2 中的最近点对。设 d_1 和 d_2 分别表示两个子集中最近点对的距离。

步骤 3：找出 S_1 中的点和 S_2 中的点之间距离最近的点对，它们之间的距离用 d_3 表示。最近的点对是距离为 $\min(d_1, d_2, d_3)$ 的点对。

选择排序耗费 $O(n^2)$ 时间。在第 23 章中，我们将介绍归并排序和堆排序。这些排序算法耗费 $O(n\log n)$ 时间。所以，步骤 1 可以在 $O(n\log n)$ 时间内完成。

步骤 3 可以在 $O(n)$ 时间内完成。设 $d=\min(d_1, d_2)$。我们已经了解到最近点对的距离不可能大于 d。对于 S_1 中的点和 S_2 中的点，要形成一个最近点对集 S，左边的点必须在 stripL 中，而右边的点必须在 stripR 中，如图 22-5a 所示。

对于 stripL 中的点 p，只需要考虑在 $d \times 2d$ 矩形中的右边点，如图 22-5b 所示。矩形外的任何右边点都不能与 p 形成最近点对。因为在 S_2 最近点对的距离大于或等于 d，在矩形中最多有 6 个点。因此，对于 stripL 中的每个点，最多考虑 stripR 中的 6 个点。

对于 stripL 中的每个点 p，如何定位在 stripR 中对应的 $d \times 2d$ 矩形中的点？如果 stripL 和 stripR 中的点都以 y 坐标的升序排列，就可以高效地完成。设 pointsOrderedOnY 是一个以 y 坐标升序排列的点构成的线性表，它可以在算法中提前得出。stripL 和 stripR 都可以从步骤 3 的 pointsOrderedOnY 中获取，如程序清单 22-9 所示。

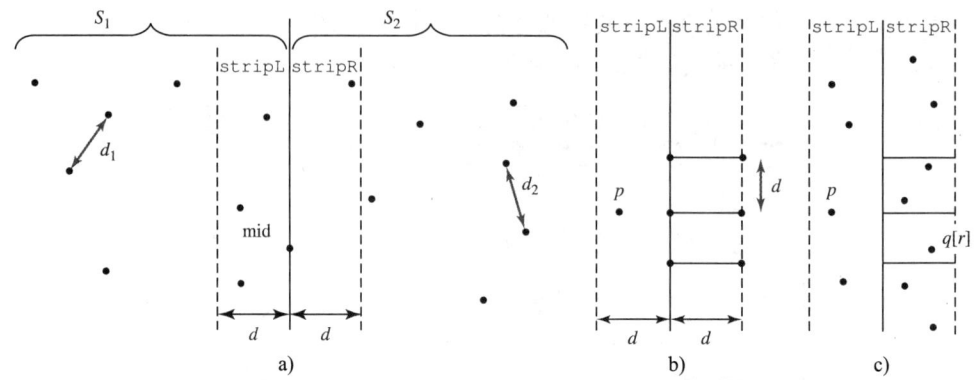

图 22-5 中间点将点集分为两个相同大小的集合

程序清单 22-9 获取 stripL 和 stripR 的算法

```
1  for each point p in pointsOrderedOnY
2    if (p is in S1 and mid.x - p.x <= d)
3      append p to stripL;
4    else if (p is in S2 and p.x - mid.x <= d)
5      append p to stripR;
```

假设 stripL 中的点和 stripR 中的点分别是 $\{p_0, p_1, \cdots, p_k\}$ 和 $\{q_0, q_1, \cdots, q_t\}$，如图 22-5c 所示。stripL 中的点和 stripR 中的点之间的最近点对可以使用程序清单 22-10 所描述的算法找到。

程序清单 22-10 在步骤 3 中找出最近点对的算法

```
1  d = min(d1, d2);
2  r = 0; // r is the index of a point in stripR
3  for (each point p in stripL) {
4    // Skip the points in stripR below p.y - d
5    while (r < stripR.length && q[r].y <= p.y - d)
6      r++;
7
8    let r1 = r;
9    while (r1 < stripR.length && |q[r1].y - p.y| <= d) {
10     // Check if (p, q[r1]) is a possible closest pair
11     if (distance(p, q[r1]) < d) {
12       d = distance(p, q[r1]);
13       (p, q[r1]) is now the current closest pair;
14     }
15
16     r1 = r1 + 1;
17   }
18 }
```

以 $p_0, p_1, \cdots, p_k$ 的顺序考虑 stripL 中的点。对于 stripL 中的点 p，跳过 stripR 中 p.y-d 下面的点（第 5～6 行）。一旦跳过某个点，就不再考虑这个点。while 循环（第 9～17 行）检查 (p, q[r1]) 是否是可能的最近点对。这里最多有 6 个这样的 q[r1] 点对，因为 stripR 中的两点距离不能小于 d。因此在步骤 3 中找出最近点对的复杂度是 $O(n)$。

注意程序清单 22-8 中的步骤 1 只被执行一次以预先对点进行排序。假设所有的点都预先排好序了。设 T(n) 表示算法的复杂度，那么

$$T(n) = 2T(n/2) + O(n) = O(n\log n)$$

（步骤 2 ↓，步骤 3 ↓）

因此，找出最近点对耗费的时间是 $O(n\log n)$。这个算法的完整实现留作练习题（参见编

程练习题 22.7）。

✓ 复习题

22.8.1 什么是分治法？给出一个示例。
22.8.2 分治以及动态编程之间的区别是什么？
22.8.3 如何使用分治法设计一个算法，以找到一个线性表中的最小元素？该算法的复杂度是多少？

22.9 使用回溯法解决八皇后问题

🔑 **要点提示**：本节使用回溯法解决八皇后问题。

八皇后问题是要找到一个解决方案，可以在一个棋盘的每行上放一个皇后，并且没有两个皇后可以相互攻击。这个问题可以用递归方法解决（参见编程练习题 18.34）。本节中，我们将介绍一个称为回溯法（backtracking）的通用算法设计技术来解决这个问题。

🔑 **算法设计提示**：有许多可能的备选算法吗？如何得出一个解决方案呢？回溯法渐进地寻找一个备选方案，一旦确定该备选方案不可能是一个有效方案就放弃，继而寻找一个新的备选方案。

可以使用二维数组来表示一个棋盘。然而，由于每行只能放一个皇后，因此使用一个一维数组足以表示每行皇后的位置了。可以如下定义一个 queens 数组：

```
int[] queens = new int[8];
```

将 queens[i] 赋值为 j 表示将一个皇后放置在第 i 行第 j 列。图 22-6 显示了棋盘的 queens 数组的内容。

搜索从 k = 0 的第一行开始，其中 k 为考虑的当前行的下标。算法为当前行检查一个皇后是否可能放在第 j 列，按照 j = 0,1,…,7 的次序检查。搜索按下列步骤进行：

- 如果成功了，则继续为下一行的皇后搜索一个位置。如果当前行是最后一行，则解决方案已经找到。
- 如果不成功，则回溯到前一行，继续在前一行的下一列上搜索一个新的放置位置。
- 如果算法回溯到第一行并且不能在该行找到一个新的位置放置皇后，则此问题无解。

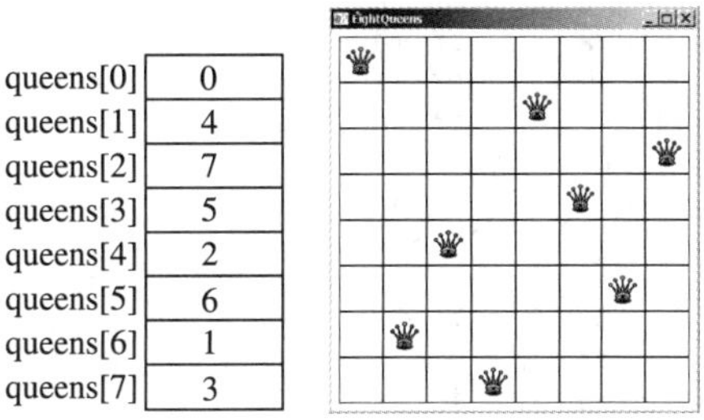

图 22-6 queens[i] 表示第 i 行的皇后的位置（来源：Oracle 或其附属公司版权所有©1995～2016，经授权使用）

算法的运行过程动画可以参见网址 http://liveexample.pearsoncmg.com/dsanimation/EightQueens.html。

程序清单 22-11 给出了显示八皇后问题解决方案的程序。

程序清单 22-11 EightQueens.java

```
1  import javafx.application.Application;
2  import javafx.geometry.Pos;
3  import javafx.stage.Stage;
4  import javafx.scene.Scene;
5  import javafx.scene.control.Label;
6  import javafx.scene.image.Image;
7  import javafx.scene.image.ImageView;
8  import javafx.scene.layout.GridPane;
```

```java
 9
10   public class EightQueens extends Application {
11     public static final int SIZE = 8; // The size of the chessboard
12     // queens are placed at (i, queens[i])
13     // -1 indicates that no queen is currently placed in the ith row
14     // Initially, place a queen at (0, 0) in the 0th row
15     private int[] queens = {-1, -1, -1, -1, -1, -1, -1, -1};
16
17     @Override // Override the start method in the Application class
18     public void start(Stage primaryStage) {
19       search(); // Search for a solution
20
21       // Display chessboard
22       GridPane chessBoard = new GridPane();
23       chessBoard.setAlignment(Pos.CENTER);
24       Label[][] labels = new Label[SIZE][SIZE];
25       for (int i = 0; i < SIZE; i++)
26         for (int j = 0; j < SIZE; j++) {
27           chessBoard.add(labels[i][j] = new Label(), j, i);
28           labels[i][j].setStyle("-fx-border-color: black");
29           labels[i][j].setPrefSize(55, 55);
30         }
31
32       // Display queens
33       Image image = new Image("image/queen.jpg");
34       for (int i = 0; i < SIZE; i++)
35         labels[i][queens[i]].setGraphic(new ImageView(image));
36
37       // Create a scene and place it in the stage
38       Scene scene = new Scene(chessBoard, 55 * SIZE, 55 * SIZE);
39       primaryStage.setTitle("EightQueens"); // Set the stage title
40       primaryStage.setScene(scene); // Place the scene in the stage
41       primaryStage.show(); // Display the stage
42     }
43
44     /** Search for a solution */
45     private boolean search() {
46       // k - 1 indicates the number of queens placed so far
47       // We are looking for a position in the kth row to place a queen
48       int k = 0;
49       while (k >= 0 && k < SIZE) {
50         // Find a position to place a queen in the kth row
51         int j = findPosition(k);
52         if (j < 0) {
53           queens[k] = -1;
54           k--; // Backtrack to the previous row
55         } else {
56           queens[k] = j;
57           k++;
58         }
59       }
60
61       if (k == -1)
62         return false; // No solution
63       else
64         return true; // A solution is found
65     }
66
67     public int findPosition(int k) {
68       int start = queens[k] + 1; // Search for a new placement
69
70       for (int j = start; j < SIZE; j++) {
71         if (isValid(k, j))
72           return j; // (k, j) is the place to put the queen now
73       }
74
75       return -1;
```

```
 76     }
 77
 78   /** Return true if a queen can be placed at (row, column) */
 79   public boolean isValid(int row, int column) {
 80     for (int i = 1; i <= row; i++)
 81       if (queens[row - i] == column // Check column
 82         || queens[row - i] == column - i // Check upleft diagonal
 83         || queens[row - i] == column + i)  // Check upright diagonal
 84         return false; // There is a conflict
 85     return true; // No conflict
 86   }
 87 }
```

该程序调用 search()（第 19 行）来搜索一个解决方案。开始时，任何一行上都没有皇后（第 15 行）。现在搜索从 k = 0 的第一行开始（第 48 行），找到一个位置放置皇后（第 51 行）。如果成功了，则将其放置在该行上（第 56 行），然后考虑下一行（第 57 行）。如果没有成功，则回溯到前一行（第 53 ～ 54 行）。

findPosition(k) 方法为在第 k 行放置一个皇后寻找一个可行的位置，从 queen[k] + 1 开始搜索（第 68 行）。它依次检测一个皇后是否可以放置在 start, start + 1, …, start + 7（第 70 ～ 73 行）。如果可以，则返回列的下标（第 72 行）；否则，返回 −1（第 75 行）。

isValid(row, column) 方法用于检查将一个皇后放置在指定位置是否会引起和之前所放置的皇后的冲突（第 71 行）。它确保皇后没有被放置在同一列上（第 81 行）、左上角对角线上（第 82 行）或者右上角对角线上（第 83 行），如图 22-7 所示。

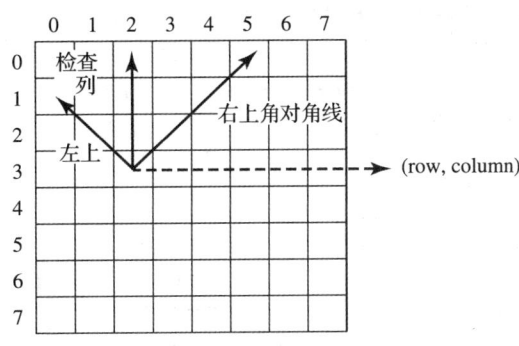

图 22-7 调用 isValid(row, column) 检查一个皇后是否可以放置在 (row, column)

复习题

22.9.1 什么是回溯？给出一个示例。

22.9.2 如果将八皇后问题推广到 $n \times n$ 棋盘上的 n 皇后问题，算法的复杂度将是多少？

22.10 计算几何：寻找凸包

要点提示：本节给出一个在点集中寻找凸包的高效算法。

计算几何为几何问题研究算法。它在计算机图形学、游戏、模式识别、图像处理、机器人、地理信息系统以及计算机辅助设计和制造方面有着广泛应用。22.8 节给出了一个寻找最近点对的几何算法。本节介绍一个寻找凸包的几何算法。

给定一个点集，凸包（convex hull）是指包围所有这些点的最小凸多边形，如图 22-8a 所示。如果连接两个顶点的任意直线都在多边形里面，则这个多边形是凸的。例如，图 22-8a 中的顶点 v0、v1、v2、v3、v4、v5 构成了一个凸多边形，但是图 22-8b 中的不是，因为连接 v3 和 v1 的点不在多边形里面。

凸包在游戏编程、模式识别和图像处理方面有许多应用。在介绍算法之前，使用位于网址 liveexample. pearsoncmg.com/dsanimation/ConvexHull.html 的交互式工具有助于熟悉概念，如图 22-8c 所示。通过这个工具可以添加和移除一些点，然后动态地显示相应的凸包。

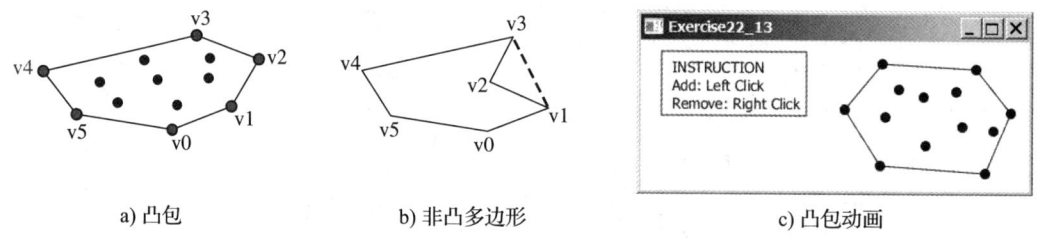

a) 凸包　　　　　　　b) 非凸多边形　　　　　　c) 凸包动画

图 22-8　凸包是包含一个点集的最小凸多边形（来源：Oracle 或其附属公司版权所有 ©1995 ~ 2016，经授权使用）

已经有许多用于寻找凸包的算法，本节介绍两个流行的算法：卷包裹算法和格雷厄姆算法。

22.10.1　卷包裹算法

卷包裹算法（gift-wrapping algorithm）是一种直观方法，工作机制如程序清单 22-12 所示。

程序清单 22-12　使用卷包裹算法寻找凸包

步骤 1：给定一个点的线性表 S，将 S 中的点标记为 s_0, s_1, …, s_k。选择 S 中最右下角的点，如图 22-9a 所示，h_0 即为这样的一个点。将 h_0 添加到线性表 H 中（H 初始时为空，当算法结束时，H 将包含凸包中的所有点），将 t_0 赋值为 h_0。

步骤 2：将 t_1 赋值为 s_0。
　　　　对于 S 中的每个点 p，
　　　　　　如果 p 位于 t_0 到 t_1 的直线的右侧，则将 t_1 赋值为 p。
　　　　（步骤 2 之后，没有点位于 t_0 到 t_1 的直线的右侧，如图 22-9b 所示。）

步骤 3：如果 t_1 为 h_0（参见图 22-9d），则 H 中的点构建了一个 S 点集的凸包。否则，将 t_1 添加到 H 中，将 t_0 赋值为 t_1，回到步骤 2（参见图 22-9c）。

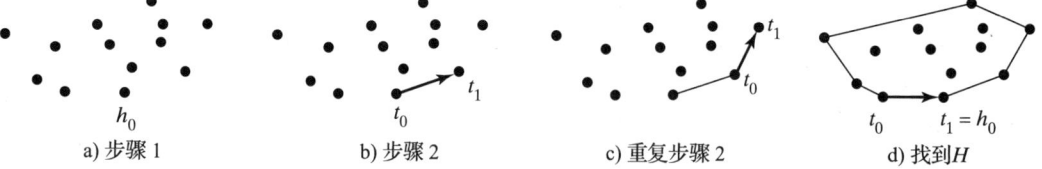

a) 步骤 1　　　　　b) 步骤 2　　　　　c) 重复步骤 2　　　　　d) 找到 H

图 22-9　a) h_0 为 S 中最右下角的点；b) 步骤 2 找到点 t_1；c) 凸包不断重复地扩张；d) 当 t_1 成为 h_0 时，一个凸包被找到

凸包渐进地扩张。正确性由以下事实确保：步骤 2 后，没有点位于 t_0 到 t_1 的直线的右侧。这保证了连接 S 中两个点的每条线段都位于多边形里面。

步骤 1 中寻找最右下角的点可以在 $O(n)$ 时间内完成。无论点在直线的左侧、右侧，还是在直线上，都可以在 $O(1)$ 时间内确定（参见编程练习题 3.32）。因此，步骤 2 将耗费 $O(n)$ 时间找到一个新的点 t_1。步骤 2 重复 h 次，其中 h 为凸包的边数。因此，算法耗费 $O(hn)$ 时间。最坏情况下，h 等于 n。

该算法的实现留作练习题（参见编程练习题 22.9）。

22.10.2 格雷厄姆算法

一种更为有效的算法是 1972 年由罗纳德·格雷厄姆（Ronald Graham）开发的，如程序清单 22-13 所示。

程序清单 22-13 使用格雷厄姆算法找到凸包

步骤 1：给定一个点的线性表 S，选择 S 中最右下角的点并命名为 p_0。如图 22-10a 所示，p_0 即为这样的一个点。

步骤 2：将 S 中的点按照以 p_0 为原点的 x 轴夹角进行排序，如图 22-10b 所示。如果出现同样的值，即两个点具有同样的角度，则去掉离 p_0 较近的那个点。现在 S 中的点被排序为 p_0，p_1，p_2，…，p_{n-1}。

步骤 3：将 p_0、p_1 和 p_2 加入栈 H（算法结束后，H 将包含凸包中的所有点）。

步骤 4：
```
i = 3;
while ( i < n ){
    将 t₁ 和 t₂ 作为栈 H 中顶部的第 1 和第 2 个元素;
    if (pᵢ 位于 t₂ 到 t₁ 的直线的左侧){
        将 pᵢ 压入栈 H;
        i ++;  // 考察 S 中的下一个点
    }
    else
        将栈 H 的顶部元素弹出
}
```

步骤 5：H 中的点形成了一个凸包。

凸包是逐步得到的。首先，p_0、p_1 以及 p_2 构成了一个凸包。p_3 在当前的凸包之外，因为点是根据它们的夹角按照增序进行排列的。如果 p_3 严格地位于 p_1 到 p_2 的直线的左侧（参见图 22-10c），则将 p_3 压入 H 中。现在 p_0、p_1、p_2 以及 p_3 构成了一个凸包。如果 p_3 位于 p_1 到 p_2 的直线的右侧（参见图 22-10d），则将 p_2 从 H 中弹出，并且将 p_3 压入 H 中。现在 p_0、p_1、p_3 构成了一个凸包，而 p_2 位于凸包内。可以通过推理证明，步骤 5 后 H 中的所有点针对输入线性表 S 中的所有点构成了一个凸包。

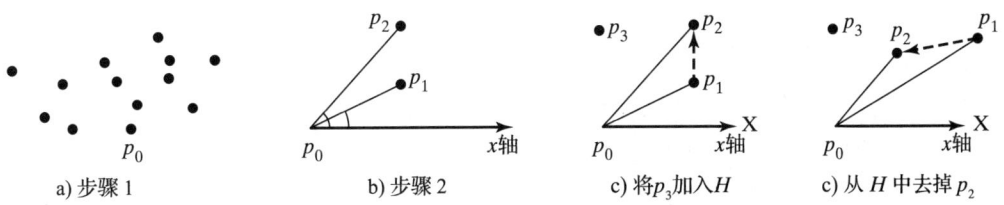

a) 步骤 1　　　　b) 步骤 2　　　　c) 将 p_3 加入 H　　　　c) 从 H 中去掉 p_2

图 22-10　a) p_0 是 S 中最右下角的点；b) 根据点的夹角排序；c～d) 逐步得到凸包

步骤 1 中寻找最右下角的点可以在 $O(n)$ 时间内完成。角度可以使用三角函数进行计算。然而，可以不计算角度而进行排序。观察到当且仅当 p_2 位于 p_0 到 p_1 的直线的左侧时，p_2 将比 p_1 形成一个更大的角度。一个点是否位于一条直线的左侧可以在 $O(1)$ 时间内确定，如编程练习题 3.32 所示。步骤 2 中的排序可以使用归并排序或者堆排序算法在 $O(n\log n)$ 时间内完成，这两种排序算法将在第 23 章中介绍。步骤 4 可以在 $O(n)$ 时间内完成。因此，算法需要 $O(n\log n)$ 时间。

该算法的实现留作练习题（参见编程练习题 22.11）。

✓ 复习题

22.10.1 什么是凸包？

22.10.2 描述如何使用卷包裹算法来得到凸包。线性表 H 应该使用 `ArrayList` 或者 `LinkedList` 来实现吗？

22.10.3 描述如何使用格雷厄姆算法来得到凸包。为什么算法使用栈来存储凸包中的点？

22.11 字符串匹配

⛀ 要点提示： 本节给出用于实现字符串匹配的暴力算法、Boyer-Moore 算法以及 Knuth-Morris-Pratt 算法。

字符串匹配是指在一个字符串中找到一个针对子字符串的匹配。字符串通常被作为文本，而子字符串通常称为模式。字符串匹配是计算机编程中常见的任务。`String` 类用 `text.contains(pattern)` 方法来测试一种模式是否存在于文本中，用 `text.indexOf(pattern)` 方法来返回文本中匹配该模式的第一个子串的下标。针对字符串匹配，开展了许多研究来寻找高效算法。本节给出三种算法：暴力算法，Boyer-Moore 算法，Knuth-Morris-Pratt 算法。

22.11.1 暴力算法

暴力算法简单地将模式与文本中的每个可能的子字符串进行比较。假设文本和模式的长度分别为 n 和 m。算法可以描述如下：

```
for i from 0 to n − m {
  test if pattern matches text[i .. i + m − 1]
}
```

这里，`text[i..j]` 表示文本中从下标 i 到下标 j 的子字符串。该算法的动画参见 https://liveexample.pearsoncmg.com/dsanimation/StringMatch.html。

程序清单 22-14 给出了暴力算法的实现。

程序清单 22-14 `StringMatch.java`

```java
 1  public class StringMatch {
 2    public static void main(String[] args) {
 3      java.util.Scanner input = new java.util.Scanner(System.in);
 4      System.out.print("Enter a string text: ");
 5      String text = input.nextLine();
 6      System.out.print("Enter a string pattern: ");
 7      String pattern = input.nextLine();
 8
 9      int index = match(text, pattern);
10      if (index >= 0)
11        System.out.println("matched at index "+ index);
12      else
13        System.out.println("unmatched");
14    }
15
16    // Return the index of the first match. −1 otherwise.
17    public static int match(String text, String pattern) {
18      for (int i = 0; i < text.length() − pattern.length() + 1; i++) {
19        if (isMatched(i, text, pattern))
20          return i;
21      }
22
23      return −1;
24    }
25
```

```
26    // Test if pattern matches text starting at index i
27    private static boolean isMatched(int i, String text,
28        String pattern) {
29      for (int k = 0; k < pattern.length(); k++) {
30        if (pattern.charAt(k) != text.charAt(i + k)) {
31          return false;
32        }
33      }
34
35      return true;
36    }
37  }
```

match(text, pattern) 方法（第 17 ～ 24 行）测试 pattern 是否匹配 text 中的一个子字符串。isMatched(i, text, pattern) 方法（第 27 ～ 36 行）测试 pattern 是否匹配从下标 i 开始的 text[i..i+m-1]。

显然，该算法需要 $O(nm)$ 时间，因为测试 pattern 是否匹配 text[i..i+m-1] 需要 $O(m)$ 时间。

22.11.2 Boyer-Moore 算法

暴力算法通过检测所有的对齐来搜索文本中模式的匹配。这并不是必要的。Boyer-Moore 算法通过从右到左将模式和文本中的子字符串进行比较来得到匹配。如果文本中的一个字符与模式中的不匹配，并且该字符串也没有在模式的剩余部分中，可以将模式整个滑过该字符。网址 https://liveexample.pearsoncmg.com/dsanimation/StringMatchBoyerMoore.html 中的动画展示了这一点。

Boyer-Moore 算法可以如下描述：

```
i = m − 1;
while i <= n − 1
```
从右到左逐个字符地将模式与 text[i − (m − 1) .. i] 进行比较，如图 22-11 所示。
如果它们全部都匹配，完成任务。否则，设 text[k] 为第一个不匹配模式中相应字符的字符。
考虑两种情形：
情形 1：如果 text[k] 不在模式的剩余部分中，将模式滑过 text[k]，如图 22-12 所示。设置 i = k + m。
情形 2：如果 text[k] 在模式中，找到模式中匹配 text[k] 的最后一个字符，假设为 pattern[j]，并且将模式向右滑动以将 pattern[j] 与 text[k] 对齐，如图 22-13 所示。设置 i = k + m − j − 1。

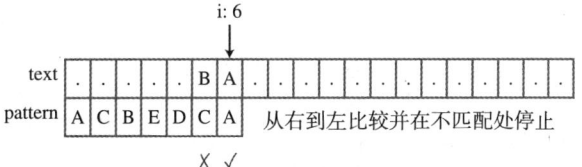

图 22-11 通过从右到左比较字符来测试模式是否匹配子字符串，并在不匹配处停止

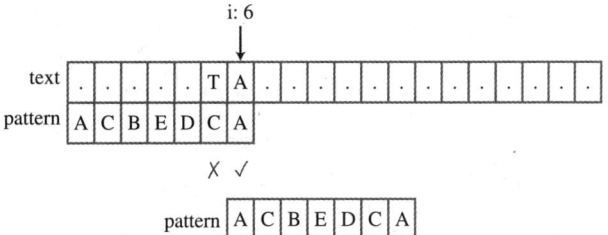

图 22-12 因为 T 没有在模式的剩余部分中，将模式滑过 T 并开始下一个测试

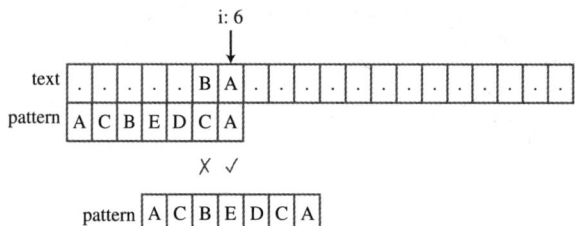

图 22-13 B 不匹配 C。B 按照从右到左的顺序在模式的剩余部分中第一次匹配到 pattern[2]，则将模式滑动以将文本中的 B 与 pattern[2] 对齐

程序清单 22-15 实现了 Boyer-Moore 算法。

程序清单 22-15 StringMatchBoyerMoore.java

```java
 1  public class StringMatchBoyerMoore {
 2    public static void main(String[] args) {
 3      java.util.Scanner input = new java.util.Scanner(System.in);
 4      System.out.print("Enter a string text: ");
 5      String text = input.nextLine();
 6      System.out.print("Enter a string pattern: ");
 7      String pattern = input.nextLine();
 8
 9      int index = match(text, pattern);
10      if (index >= 0)
11        System.out.println("matched at index "+ index);
12      else
13        System.out.println("unmatched");
14    }
15
16    // Return the index of the first match. -1 otherwise.
17    public static int match(String text, String pattern) {
18      int i = pattern.length() - 1;
19      while (i < text.length()) {
20        int k = i;
21        int j = pattern.length() - 1;
22        while (j >= 0) {
23          if (text.charAt(k) == pattern.charAt(j)) {
24            k--; j--;
25          }
26          else {
27            break;
28          }
29        }
30
31        if (j < 0)
32          return i = pattern.length() + 1; // A match found
33
34        int u = findLastIndex(text.charAt(k), j - 1, pattern);
35        if (u >= 0) { // text[k] is in the remaining part of the pattern
36          i = k + pattern.length() - 1 - u;
37        }
38        else { // text[k] is in the remaining part of the pattern
39          i = k + pattern.length();
40        }
41      }
42
43      return -1;
44    }
45
46    // Return the index of the last element in pattern[0 .. j]
47    // that matches ch. -1 otherwise.
48    private static int findLastIndex(char ch, int j, String pattern) {
49      for (int k = j; k >= 0; k--) {
```

```
50            if (ch == pattern.charAt(k)) {
51              return k;
52            }
53          }
54
55          return - 1;
56        }
57      }
```

match(text, pattern)方法（第17～44行）测试pattern是否匹配text中的子字符串。i表示子字符串的最后字符下标。它开始于i=pattern.length()-1（第21行），并将text[i]与pattern[j]比较，text[i-1]与pattern[j-1]比较，以此类推（第22～29行）。如果j<0，则找到一个匹配（第31和32行）。否则，使用findLastIndex方法（第34行）找到pattern[0..j-1]中text[k]的最后一个匹配元素的下标。如果index ≥ 0，则设为k+m-1-index（第36行），这里m为pattern.length()。否则，设置i为k+m（第39行）。

最坏情况下，Boyer-Moore算法需要$O(nm)$时间。达到最坏情况的示例如图22-14所示。

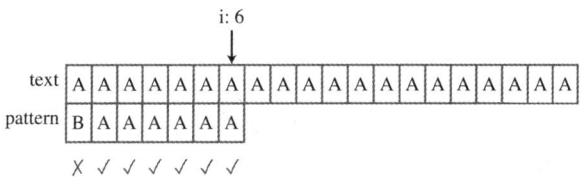

图22-14 文本全部为A，而模式为BAAAAAA。针对每个可能的对齐，模式中的每个字符都与文本中的每个字符进行比较

然而，平均时间上，Boyer-Moore算法是高效的，因为该算法经常可以跳过很大一部分文本。有许多Boyer-Moore算法的变种，我们在本节中只提供了一个简化版本。

22.11.3 Knuth-Morris-Pratt算法

Knuth-Morris-Pratt（KMP）算法是一种高效算法，在最坏情况下可以达到$O(m+n)$。这是最佳选择，因为在最坏情况下，必须至少检查一次文本和模式中的每个字符。在暴力算法或者Boyer-Moore算法中，一旦找到一个不匹配，算法就重新开始搜索下一个可能的匹配，对暴力算法来说是将模式向右移动一个位置，而对于Boyer-Moore算法则可能是右移多个位置。这个过程中，不匹配之前的成功字符匹配被略去了。KMP算法在继续下一次搜索之前考虑了成功的匹配，从而得到模式中要移动的位置数目。

要得到模式中要移动的位置的最大数目，我们首先定义一个失败函数fail(k)作为pattern的最大前缀的长度，该长度是pattern[0..k]的后缀。对于一个给定的模式来说，失败函数可以提前计算好。失败函数实质上是一个具有m个元素的数组。假设模式是ABCABCDABC。该模式的失败函数如图22-15所示。

```
       0 1 2 3 4 5 6 7 8 9
k
pattern A B C A B C D A B C
fail    0 0 0 1 2 3 0 1 2 3
```

图22-15 失败函数fail(k)是与pattern[0..k]中的后缀匹配的最长前缀的长度

例如，fail(5)为3，因为ABC是针对ABCABC后缀的最长前缀。fail(7)为1，因为A是针对ABCABCDA后缀的最长前缀。当比较文本和模式时，一旦一个不匹配在模式的下标k处找到，就可以将模式移动到模式的下标 fail(k-1) 与文本对齐的地方，如图 22-16 所示。

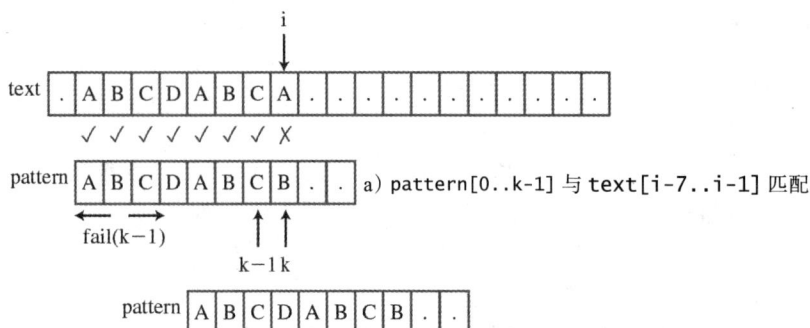

图 22-16 当在 text[i] 处不匹配时，将模式右移以将前缀中的第一个 fail[k-1] 元素与 text[i-1] 对齐

KMP 算法可以如下描述：

步骤 1：首先，我们提前计算失败函数。现在，从 i=0 和 k=0 开始。

步骤 2：将 text[i] 与 pattern[k] 比较。考虑两种情形。

情形 1（text[i] 等于 pattern[k]）：如果 k 为 m-1，则找到了一个匹配并返回 i-m+1。否则，将 i 和 k 加 1。

情形 2（text[i] 不等于 pattern[k]）：如果 k > 0，通过设置 k=fail[k-1]，将匹配 pattern[k-1] 后缀的最长前缀移动，使得该前缀中的最后一个字符与 text[i-1] 对齐；否则将 i 加 1。

步骤 3：如果 i＜n，重复步骤 2。

现在，我们专注于计算失败函数。这可以通过将模式与自身比较实现，如下所述：

步骤 1：失败函数是一个具有 m 个元素的数组。开始时，设置所有元素为 0。我们从 i=1 以及 k=0 开始。

步骤 2：将 pattern[i] 与 pattern[k] 比较。考虑两种情形：

情形 1（pattern[i]==pattern[k]）：fail[i]=k+1。将 i 和 k 加 1。

情形 2（pattern[i]!=pattern[k]）：如果 k > 0，则设置 k=fail[k-1]，否则将 i 加 1。

步骤 3：如果 i＜m，重复步骤 2。注意，如果 pattern[i]==pattern[k] 的话，k 代表了 pattern[0..i-1] 中的最长前缀的长度。

这是一个复杂的算法。以下动画有助于你理解该算法的高效：https://liveexample.pearsoncmg.com/dsanimation/StringMatchKMP.html。

以下动画演示了如何获得失败函数：https://liveexample.pearsoncmg.com/dsanimation/StringMatchKMPFail.html。

程序清单 22-16 给出了 KMP 算法的实现。

程序清单 22-16 StringMatchKMP.java

```
1  public class StringMatchKMP {
2    public static void main(String[] args) {
3      java.util.Scanner input = new java.util.Scanner(System.in);
4      System.out.print("Enter a string text: ");
5      String text = input.nextLine();
6      System.out.print("Enter a string pattern: ");
7      String pattern = input.nextLine();
8
9      int index = match(text, pattern);
```

```
10      if (index >= 0)
11        System.out.println("matched at index "+ index);
12      else
13        System.out.println("unmatched");
14    }
15
16    // Return the index of the first match. -1 otherwise.
17    public static int match(String text, String pattern) {
18      int[] fail = getFailure(pattern);
19      int i = 0; // Index on text
20      int k = 0; // Index on pattern
21      while (i < text.length()) {
22        if (text.charAt(i) == pattern.charAt(k)) {
23          if (k == pattern.length() - 1) {
24            return i- pattern.length() + 1; // pattern matched
25          }
26          i++; // Compare the next pair of characters
27          k++;
28        }
29        else {
30          if (k > 0) {
31            k = fail[k -1]; // Matching prefix position
32          }
33          else {
34            i++; // No prefix
35          }
36        }
37      }
38
39      return -1;
40    }
41
42    // Compute failure function
43    private static int[] getFailure(String pattern) {
44      int[] fail = new int [pattern.length()];
45      int i = 1;
46      int k = 0;
47      while (i < pattern.length()) {
48        if (pattern.charAt(i) == pattern.charAt(k)) {
49          fail[i] = k + 1;
50          i++;
51          k++;
52        }
53        else if (k > 0) {
54          k = fail[k -1];
55        }
56        else {
57          i++;
58        }
59      }
60
61      return fail;
62    }
63  }
```

match(text, pattern) 方法（第 17 ～ 40 行）测试 pattern 是否匹配 text 中的一个子字符串。i 代表了文本中的当前位置。该位置只会向前移动。k 代表了模式中的当前位置。如果 text[i]==pattern[k]（第 22 行），则 i 和 k 都加 1（第 26 和 27 行）。否则，如果 k > 0，则设置 fail(k-1) 为 k，从而滑动模式以将 pattern[k] 和 text[i] 对齐（第 31 行），否则将 i 加 1（第 34 行）。

getFailure(pattern) 方法（第 43 ～ 62 行）将 pattern 与 pattern 比较以获得 fail[k]，这是与 pattern[0..k] 中后缀相同的最长前缀的长度。该方法将数组 fail 初始化为 0（第

44 行），并将 i 和 k 分别设置为 1 和 0（第 45 和 46 行）。i 表明了第一个模式中的当前位置，它一直是向前移动的。k 表明了 pattern[0..i] 中当前既是前缀又是后缀的子串的最大长度。如果 pattern[i]==pattern[k]，则设置 fail[i] 为 k+1（第 49 行）并将 i 和 k 都加 1（第 50 和 51 行）。否则，如果 k > 0，则设置 k 为 fail[k-1]，以滑动第二个模式从而将第一个模式中的 pattern[i] 和第二个模式中的 pattern[k] 对齐（第 54 行），否则将 i 加 1（第 57 行）。

要分析运行时间，考虑三种情形：

情形 1：text[i] 等于 pattern[k]。i 向前移动一个位置。

情形 2：text[i] 不等于 pattern[k] 并且 k 为 0。i 向前移动一个位置。

情形 3：text[i] 不等于 pattern[k] 并且 k > 0。模式至少向前移动一个位置。

在任意情形下，要么是 i 在文本中向前移动一个位置，要么是模式至少向右移动一个位置。因此，match 方法中的 while 循环的遍历次数最多为 $2n$。类似地，getFailure 方法中的遍历次数最多为 $2m$。因此，KMP 算法的运行时间为 $O(n+m)$。

关键术语

average-case analysis（平均情况分析）
backtracking approach（回溯法）
best-case input（最佳情况输入）
big O notation（大 O 表示法）
brute force（暴力法）
constant time（常量时间）
convex hull（凸包）
divide-and-conquer approach（分治法）

dynamic programming approach（动态编程法）
exponential time（指数时间）
growth rate（增长率）
logarithmic time（对数时间）
quadratic time（二次方时间）
space complexity（空间复杂度）
time complexity（时间复杂度）
worst-case input（最差情况输入）

本章小结

1. 大 O 表示法是分析算法性能的理论方法。它估计算法的执行时间随着输入规模的增加会有多快的增长。因此，可以通过考察两个算法的增长率来比较它们。
2. 导致最短执行时间的输入称为最佳情况输入，而导致最长执行时间的输入称为最差情况输入。最佳情况和最差情况都不具有代表性，但是最差情况分析非常有用。可以确保算法永远不会比最差情况还慢。
3. 平均情况分析试图在所有可能的相同规模的输入中确定平均时间。平均情况分析是比较理想的，但是实现很困难，因为对于许多问题，要确定不同输入实例的相对概率和分布是相当困难的。
4. 如果执行时间与输入规模无关，我们就认为该算法需要常量时间，以符号 $O(1)$ 表示。
5. 线性查找需要 $O(n)$ 时间。具有 $O(n)$ 时间复杂度的算法称为线性算法，它表现为线性增长率。二分查找需要 $O(\log n)$ 时间。具有 $O(\log n)$ 时间复杂度的算法称为对数算法，它表现为对数增长率。
6. 选择排序的最差情况时间复杂度为 $O(n^2)$。具有 $O(n^2)$ 时间复杂度的算法称为二次方算法，它表现为平方级增长率。
7. 汉诺塔问题的时间复杂度是 $O(2^n)$。具有 $O(2^n)$ 时间复杂度的算法称为指数算法，它表现为指数增长率。
8. 求出给定下标处的斐波那契数可以使用动态编程法在 $O(n)$ 时间内完成。
9. 动态编程是通过解决子问题，然后合并子问题的结果来求解整个问题的过程。动态编程的关键思想是只解决子问题一次，并将子问题的结果存储以备后用，从而避免了重复的子问题求解。
10. 欧几里得的 GCD 算法需要 $O(\log n)$ 时间。
11. 所有小于等于 n 的素数可以在 $O\left(\dfrac{n\sqrt{n}}{\log n}\right)$ 时间内找到。

12. 使用分治法可以在 $O(n\log n)$ 时间内找到最近点对。
13. 分治法将问题分解为子问题，解决子问题，然后将子问题的解合并从而获得整个问题的解。和动态编程方法不一样的是，分治法中的子问题不会重叠。子问题类似于初始问题，但是大小更小，因此可以应用递归来解决这样的问题。
14. 可以使用回溯法解决八皇后问题。
15. 回溯法渐进地寻找一个备选方案，一旦确定该备选方案不可能是一个有效方案，则放弃，继而寻找一个新的备选方案。
16. 使用卷包裹法可以在 $O(n^2)$ 时间内找到一个点集的凸包，使用格雷厄姆算法则需要 $O(n\log n)$ 时间。
17. 暴力法和 Boyer-Moore 字符串匹配算法需要 $O(nm)$ 时间，而 KMP 字符串匹配算法需要 $O(n+m)$ 时间。

测试题

回答位于本书配套网站上的本章测试题。

编程练习题

*22.1 （最大连续递增的有序子字符串）编写一个程序，提示用户输入一个字符串，然后显示最大连续递增的有序子字符串。分析你的程序的时间复杂度。下面是一个运行示例：

```
Enter a string: abcabcdgabxy ↵Enter
Maximum consecutive substring is abcdg
```

```
Enter a string: abcabcdgabmnsxy ↵Enter
Maximum consecutive substring is abmnsxy
```

**22.2 （最大递增子序列）编写一个程序，提示用户输入一个字符串，然后显示最大递增的有序字符子序列。分析你的程序的时间复杂度。下面是一个运行示例：

```
Enter a string: Welcome ↵Enter
Maximum increasingly ordered subsequence is Welo
```

*22.3 （模式匹配）编写一个时间复杂度为 $O(n)$ 的程序，提示用户输入两个字符串，然后检测第二个字符串是否是第一个字符串的子串。假定在字符串中相邻的字符是不同的。（不要使用 String 类中的 indexOf 方法。）下面是该程序的一个运行示例：

```
Enter a string s1: Welcome to Java ↵Enter
Enter a string s2: come ↵Enter
matched at index 3
```

*22.4 （改进 Boyer-Moore 算法）改进程序清单 22-15 中的 Boyer-Moore 算法实现，在 $O(1)$ 的时间内测试不匹配字符位于模式中的哪里，使用由模式中所有字符组成的规则集。如果测试为假，算法可以将模式滑过不匹配字符。

```
Enter a string s1: Mississippi ↵Enter
Enter a string s2: sip ↵Enter
matched at index 6
```

*22.5 （同样数字的子序列）编写一个时间复杂度为 $O(n)$ 的程序，提示用户输入一个以 0 结束的整数序列，找出有同样数字的最长子序列。下面是该程序的一个运行示例：

```
Enter a series of numbers ending with 0:
2 4 4 8 8 8 8 2 4 4 0 ↵Enter
The longest same number sequence starts at index 3 with 4 values of 8
```

*22.6 （GCD 的执行时间）编写一个程序，使用程序清单 22-3 和程序清单 22-4 中的算法，求下标从 40 到 45 的每两个连续的斐波那契数的 GCD，并求其执行时间。你的程序应该打印如下所示的表格：

	40	41	42	43	44	45
程序清单 22-3 GCD						
程序清单 22-4 GCDEuclid						

（提示：可以使用下面的代码模板来获取执行时间。）

```
long startTime = System.currentTimeMillis();
perform the task;
long endTime = System.currentTimeMillis();
long executionTime = endTime - startTime;
```

**22.7 （最近点对）22.8 节介绍了一个使用分治法求最近点对的算法。实现这个算法，使其满足下面的要求：
- 定义一个名为 `Pair` 的类，其数据域 p1 和 p2 表示两个点，名为 `getDistance()` 的方法返回这两个点之间的距离。
- 实现下面的方法：

```
/** Return the distance of the closest pair of points */
public static Pair getClosestPair(double[][] points)

/** Return the distance of the closest pair of points */
public static Pair getClosestPair(Point2D[] points)

/** Return the distance of the closest pair of points
  * in pointsOrderedOnX[low..high]. This is a recursive
  * method. pointsOrderedOnX and pointsOrderedOnY are
  * not changed in the subsequent recursive calls.
  */
public static Pair distance(Point2D[] pointsOrderedOnX,
    int low, int high, Point2D[] pointsOrderedOnY)

/** Compute the distance between two points p1 and p2 */
public static double distance(Point2D p1, Point2D p2)

/** Compute the distance between points (x1, y1) and (x2, y2) */
public static double distance(double x1, double y1,
    double x2, double y2)
```

**22.8 （不大于 10 000 000 000 的所有素数）编写一个程序，找出不大于 10 000 000 000 的所有素数。大概有 455 052 511 个这样的素数。你的程序应该满足下面的要求：
- 应该将这些素数都存储在一个名为 PrimeNumber.dat 的二进制数据文件中。当找到一个新素数时，将该数字追加到这个文件中。
- 为了判断一个新的数是否为素数，程序应该从数据文件加载这些素数到一个大小为 10000 的 `long` 型数组中。如果数组中没有任何数是这个新数的除数，继续从数据文件中读取接下来的 10000 个素数，直到找到除数或者读取完文件中的所有数字。如果没找到除数，这个新的数字就是素数。
- 因为执行该程序要花很长时间，所以应该把它作为 UNIX 机器上的一个批处理任务来运行。如果机器被关闭或重启，程序应该使用二进制数据文件中存储的素数来继续，而不是从零开始启动。

**22.9 （几何：找到凸包的卷包裹算法）22.10.1 节介绍了一个为点集找到凸包的卷包裹算法。使用下面的方法实现该算法：

```
/** Return the points that form a convex hull */
public static ArrayList<Point2D> getConvexHull(double[][] s)
```

`Point2D` 在 9.6.3 节中定义。

编写一个测试程序，提示用户输入点集的大小以及点，然后显示构成一个凸包的点的信息。注意，当你调试代码时，会发现算法忽视了两种情形：（1）t1=t0；（2）有点位于 t0 到 t1 的同一条线上。当其中一种情形发生时，如果从 t0 到 p 的距离大于从 t0 到 t1 的距离，则将 t1 替换为点 p。下面是一个运行示例：

```
How many points are in the set? 6 ↵Enter
Enter 6 points: 1 2.4 2.5 2 1.5 34.5 5.5 6 6 2.4 5.9 ↵Enter
The convex hull is
    (2.5, 2.0) (6.0, 2.4) (5.5, 9.0) (1.5, 34.5) (1.0, 2.4)
```

22.10 （素数的个数）编程练习题 22.8 将素数存储在一个名为 PrimeNumbers.dat 的文件中。编写一个程序，找出小于或等于 10、100、1000、10000、100000、1000000、10000000、100000000、1000000000、10000000000 的素数个数。你的程序应该从 PrimeNumbers.dat 文件中读取数据。

**22.11 （几何：寻找凸包的格雷厄姆算法）22.10.2 节介绍了为一个点集寻找凸包的格雷厄姆算法。假定使用 Java 的坐标系统表示点。使用下面的方法实现该算法：

```java
/** Return the points that form a convex hull */
public static ArrayList<MyPoint> getConvexHull(double[][] s)
MyPoint is a static inner class defined as follows:
  private static class MyPoint implements Comparable<MyPoint> {
    double x, y;

    MyPoint rightMostLowestPoint;

    MyPoint(double x, double y) {
      this.x = x; this.y = y;
    }

    public void setRightMostLowestPoint(MyPoint p) {
      rightMostLowestPoint = p;
    }

    @Override
    public int compareTo(MyPoint o) {
      // Implement it to compare this point with point o
      // angularly along the x-axis with rightMostLowestPoint
      // as the center, as shown in Figure 22.10b. By implementing
      // the Comparable interface, you can use the Array.sort
      // method to sort the points to simplify coding.
    }
  }
```

编写一个测试程序，提示用户输入点集的大小和点，然后显示构成一个凸包的点。下面是一个运行示例：

```
How many points are in the set? 6 ↵Enter
Enter six points: 1 2.4 2.5 2 1.5 34.5 5.5 6 6 2.4 5.9 ↵Enter
The convex hull is
    (2.5, 2.0) (6.0, 2.4) (5.5, 9.0) (1.5, 34.5) (1.0, 2.4)
```

*22.12 （最后的 100 个素数）编程练习题 22.8 将素数存储在一个名为 PrimeNumbers.dat 的文件中。编写一个高效程序，从该文件中读取最后 100 个素数。
（提示：不要从文件中读取所有的数字，跳过文件中最后 100 个数之前的所有数。）

**22.13 （几何：凸包动画）编程练习题 22.11 为从控制台输入的点集中找到凸包。编写一个程序，可以让用户通过单击鼠标左 / 右键来添加 / 移除点，然后显示凸包，如图 22-8c 所示。

*22.14 （素数的执行时间）编写一个程序，使用程序清单 22-5 到程序清单 22-7 中的算法，找出小于 8000000、10000000、12000000、14000000、16000000 和 18000000 的所有素数，获取其执行时间。你的程序应该打印如下所示的一个表格：

	8 000 000	10 000 000	12 000 000	14 000 000	16 000 000	18 000 000
程序清单 22-5						
程序清单 22-6						
程序清单 22-7						

**22.15 （几何：无交叉多边形）编写一个程序，可以让用户通过单击鼠标左 / 右键来添加 / 移除点，然后显示一个连接所有点的无交叉多边形，如图 22-17a 所示。如果一个多边形有两条或者更多条边是相交的，则认为是交叉多边形，如图 22-17b 所示。使用如下算法来从一个点集中构建一个多边形：

步骤 1：给定一个点集 S，选择 S 中的最右下角点 p_0。

步骤 2：将 S 中的点按照以 p_0 为原点的 x 轴夹角进行排序。如果出现同样的值，即两个点具有同样的角度，则认为离 p_0 较近的那个点具有更大的角度。S 中的点现在排序为 $p_0, p_1, p_2, \cdots, p_{n-1}$。

步骤 3：排好序的点形成了一个无交叉多边形。

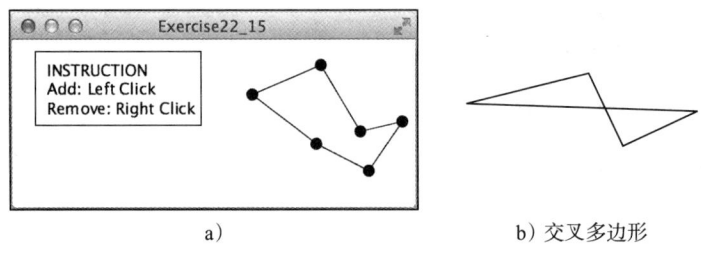

a)   b) 交叉多边形

图 22-17 a）编程练习题 22.15 为一个点集显示一个无交叉多边形（来源：Oracle 或其附属公司版权所有 ©1995 ～ 2016，经授权使用）；b) 在一个交叉多边形中，两条或者更多条边是相交的

**22.16 （线性查找动画）编写一个程序，显示线性查找的动画。创建一个包含从 1 到 20 的 20 个不同数字并且顺序随机的数组。数组元素以直方图显示，如图 22-18 所示。你需要在文本域中输入一个查找键值。单击 step 按钮将引发程序执行算法中的一次比较，重绘直方图并且其中一个条块显示查找的位置。这个按钮同时冻结文本域以防止其中的值被改变。当算法结束时，在 border 面板的顶部标签中显示状态，从而给用户以提示信息。单击 Reset 按钮创建一个新的随机数组，开始一次新的查找。这个按钮也使得文本域可以编辑。

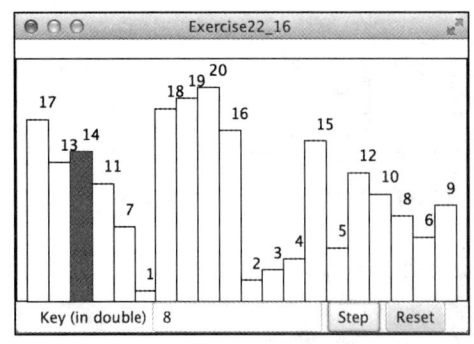 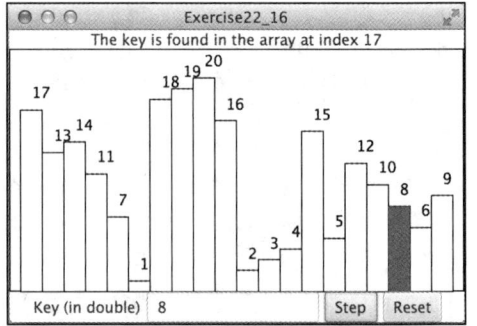

图 22-18 该程序显示线性查找的动画（来源：Oracle 或其附属公司版权所有 ©1995 ～ 2016，经授权使用）

**22.17 （最近点对的动画）编写一个程序，可以让用户通过单击鼠标左 / 右键来添加 / 移除点，然后显

示一条连接最近点对的直线，如图22-4所示。

**22.18 （二分查找动画）编写一个程序，显示二分查找的动画。创建一个包含从1到20的按顺序排列的数字数组。数组元素以直方图显示，如图22-19所示。你需要在文本域中输入一个查找键值。单击Step按钮将引发程序执行算法中的一次比较。使用灰色来绘制代表目前查找范围内数字的条块，使用黑色绘制表示查找范围的中间数的条块。按下Step按钮同时会冻结文本域以防止其中的值被改变。当算法结束时，在border面板的顶部标签中显示状态信息。单击Reset按钮创建一个新的随机数组，从而开始一次新的查找。该按钮也使得文本域可编辑。

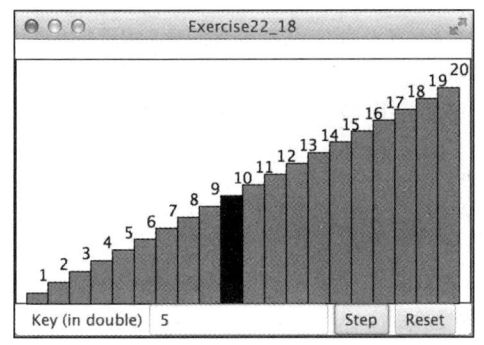

图22-19 该程序显示二分查找的动画（来源：Oracle或其附属公司版权所有©1995～2016，经授权使用）

*22.19 （最大块）编程练习题8.35描述了寻找最大块的问题。设计一个动态编程的算法，在$O(n^2)$时间内求解这个问题。编写一个测试程序，显示一个10×10的方矩阵，如图22-20a所示。矩阵中的每个元素为0或者1，单击Refresh按钮可以随机生成。在一个文本域的中央显示每个数字。对每个条目使用一个文本域。允许用户改变条目的值。单击Find Largest Block按钮找到包含1值的最大子块。高亮显示块中的数字，如图22-20b所示。

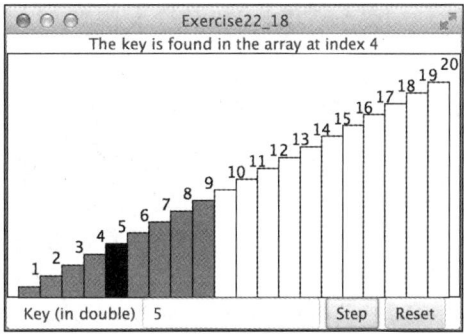

图22-20 该程序找到包含1的最大块（来源：Oracle或其附属公司版权所有©1995～2016，经授权使用）

***22.20 （游戏：数独的多个解）补充材料VI.A给出了数独问题的完整解。数独问题可能有多个解。修改补充材料VI.A中的Sudoku.java显示解的总数。如果存在多个解，则显示两个解。

***22.21 （游戏：数独）补充材料VI.C给出了数独问题的完整解。编写一个程序，提示用户从文本域输入数字，如图22-21a所示。单击Solve按钮显示结果，如图22-21b～图22-21c所示。

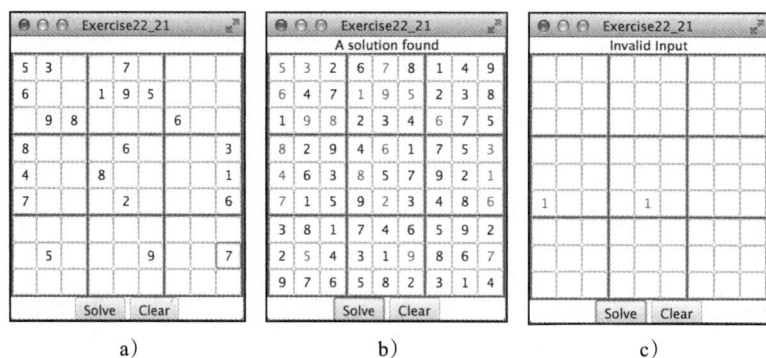

图 22-21 求解数独问题的程序（来源：Oracle 或其附属公司版权所有 ©1995 ～ 2016，经授权使用）

***22.22 （游戏：递归数独）为数独问题编写一个递归的解法。
***22.23 （游戏：八皇后问题的多个解）编写一个程序，在一个滚动面板中显示八皇后问题的所有可能解，如图 22-22 所示。对于每个解，使用标签标记解的数字。（提示：将所有解的面板放在一个 HBox 中，然后将其放入一个 ScrollPane 中。如果程序遇到了 StackOverflowError，在命令窗口中使用 java -Xss200m Exercise22_23 运行该程序。）

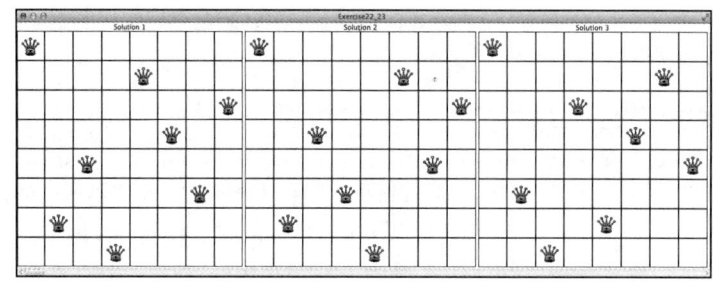

图 22-22 所有的解都放置在一个滚动面板中（来源：Oracle 或其附属公司版权所有 ©1995 ～ 2016，经授权使用）

**22.24 （找到最小数字）编写一个方法，使用分治法找到线性表中的最小数字。
***22.25 （游戏：数独）修改编程练习题 22.21，显示数独的所有解，如图 22-23a 所示。当单击 Solve 按钮时，程序将所有的解保存在一个 ArrayList 中。表中的每个元素都是一个二维的 9×9 网格。如果程序有多个解，则如图 22-23b 显示 Next 按钮。可以单击 Next 按钮显示下一个解，同样添加一个标签显示解的数字。单击 Clear 按钮时，则清除单元格，隐藏 Next 按钮，如图 22-23c 所示。

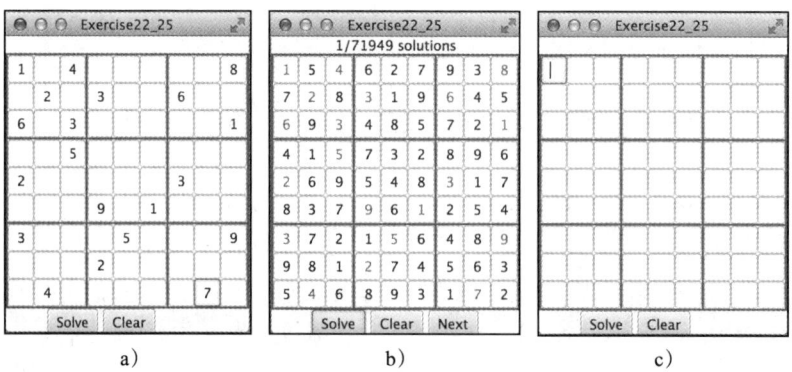

图 22-23 该程序显示数独的多个解（来源：Oracle 或其附属公司版权所有 ©1995 ～ 2016，经授权使用）

*22.26 （首先用最小的物品装箱）装箱问题是将各种重量的物品装入集装箱中。假设每个箱子最多可容纳 10 磅。该程序使用的算法是，将具有最小重量的物品放到它可放入的第一个箱子中。你的程序应该提示用户输入物品的总数和每个物品的重量。该程序显示了装箱物品所需的箱子总数，以及每个箱子所装的物品。下面是该程序的一个运行示例：

```
Enter the number of objects: 6
Enter the weights of the objects: 7 5 2 3 5 8 ↵Enter
Container 1 contains objects with weight 2 3 5
Container 2 contains objects with weight 5
Container 3 contains objects with weight 7
Container 4 contains objects with weight 8
```

这个程序是否产生了一个最优的解决方案，即是否找到最小数量的箱子来装下这些物品？

**22.27 （装箱问题最优解）重写之前的程序，使之可以找到一种最优的方案，用最少数量的箱子装下所有的物品。下面是该程序的一个运行示例：

```
Enter the number of objects: 6 ↵Enter
Enter the weights of the objects: 7 5 2 3 5 8 ↵Enter
Container 1 contains objects with weight 7 3
Container 2 contains objects with weight 5 5
Container 3 contains objects with weight 2 8
The optimal number of bins is 3
```

你的程序的时间复杂度是多少呢？

第 23 章

Introduction to Java Programming and Data Structures, Comprehensive Version, Twelfth Edition

排　　序

教学目标
- 研究和分析各种排序算法的时间复杂度（23.2 ～ 23.7 节）。
- 设计、实现和分析插入排序（23.2 节）。
- 设计、实现和分析冒泡排序（23.3 节）。
- 设计、实现和分析归并排序（23.4 节）。
- 设计、实现和分析快速排序（23.5 节）。
- 设计和实现一个二叉堆（23.6 节）。
- 设计、实现和分析堆排序（23.6 节）。
- 设计、实现和分析桶排序和基数排序（23.7 节）。
- 设计、实现和分析针对具有大量数据的文件的外部排序（23.8 节）。

23.1 引言

要点提示：排序算法是学习算法设计和分析的极好例子。

当总统 Barack Obama 2007 年访问 Google 公司时，Google 的 CEO 候选人 Eric Schmidt 问了 Obama 一个问题，对 100 万个 32 位整数排序的最有效方式是什么（www.youtube.com/watch?v=k4RRi_ntQc8）？ Obama 回答冒泡算法是错误的选择。他的回答正确吗？我们将在本章中考察各种排序算法，然后看看他是否正确。

在计算机科学中，排序是一个经典的主题。学习排序算法的原因有三个。
- 首先，排序算法展示了许多问题求解的创造性的方法，并且这些方法还可用于解决其他问题。
- 其次，排序算法有助于使用选择语句、循环、方法和数组来练习基本的程序设计技术。
- 最后，排序算法是演示算法性能的优秀示例。

要排序的数据可能是整数、双精度浮点数、字符或者对象。7.11 节给出了选择排序。在 19.5 节中将选择排序算法扩展到针对对象数组的排序。Java API 在 java.util.Arrays 和 java.util.Collections 类中包含了对基本类型值和对象进行排序的多个重载方法。为简单起见，本章假定：

1）要排序的数据是整数。
2）数据存储在数组中。
3）数据以升序排列。

这些程序可以很容易地修改为对其他类型数据的排序，以降序排列或者对 ArrayList 或

LinkedList 中的数据排序。

目前已有许多排序算法。你已经学习了选择排序。本章将介绍插入排序、冒泡排序、归并排序、快速排序、桶排序、基数排序和外部排序。

23.2 插入排序

要点提示：插入排序重复地将新的元素插入一个排好序的子线性表中，直到整个线性表排好序。

图 23-1 描述如何用插入排序法对线性表 {2,9,5,4,8,1,6} 进行排序。可以通过网址 liveexample.pearsoncmg.com/dsanimation/InsertionSortNeweBook.html 看到插入排序的工作方式的交互式演示。

这个算法可以描述如下：

```
for(int i=1; i<list.length; i++){
    将 list[i] 插入已排好序的子线性表中，这样 list[0..i] 也是排好序的
}
```

为了将 list[i] 插入 list[0..i-1]，需要将 list[i] 存储在一个名为 currentElement 的临时变量中。如果 list[i-1]>currentElement，就将 list[i-1] 移到 list[i] 的位置；如果 list[i-2]>currentElement，就将 list[i-2] 移到 list[i-1] 的位置，以此类推，直到 list[i-k]<=currentElement 或者 k>i（传递的是排好序的线性表的第一个元素）。将 currentElement 赋值给 list[i-k+1]。例如，为了在图 23-2 的步骤 4 中将 4 插入 {2,5,9} 中，由于 9>4，所以把 list[2](9) 移到 list[3]，又因为 5>4，所以把 list[1](5) 移到 list[2]。最后，把 currentElement(4) 移到 list[1]。

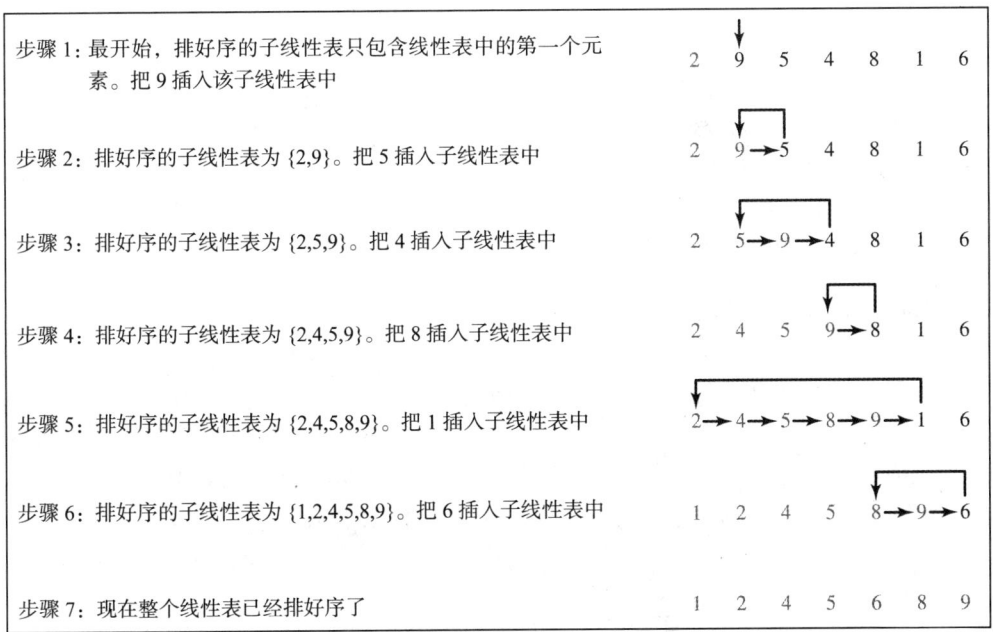

图 23-1 插入排序重复地将新元素插入已排好序的子线性表中

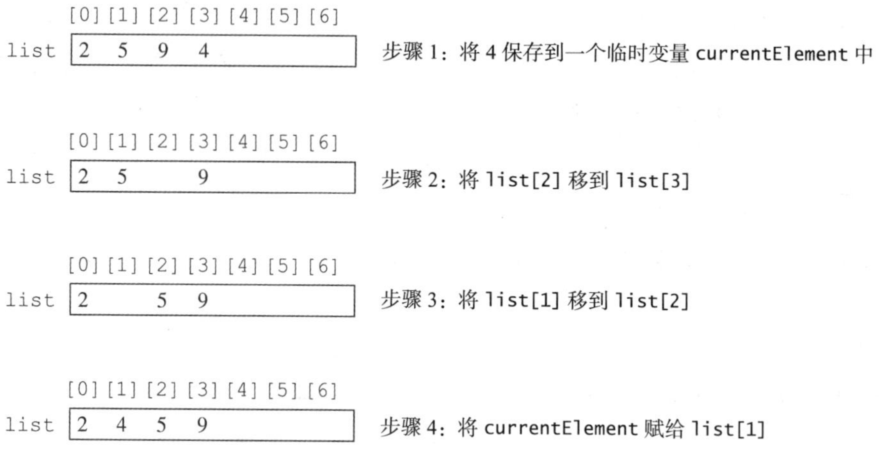

```
        [0][1][2][3][4][5][6]
    list │2  5  9  4          │   步骤1：将4保存到一个临时变量 currentElement 中

        [0][1][2][3][4][5][6]
    list │2  5     9          │   步骤2：将 list[2] 移到 list[3]

        [0][1][2][3][4][5][6]
    list │2     5  9          │   步骤3：将 list[1] 移到 list[2]

        [0][1][2][3][4][5][6]
    list │2  4  5  9          │   步骤4：将 currentElement 赋给 list[1]
```

图 23-2　一个新的元素插入排好序的子线性表中

算法可以扩展和实现，如程序清单 23-1 所示。

程序清单 23-1 InsertionSort.java

```java
1   public class InsertionSort {
2     /** The method for sorting the numbers */
3     public static void insertionSort(int[] list) {
4       for (int i = 1; i < list.length; i++) {
5         /** Insert list[i] into a sorted sublist list[0..i-1] so that
6             list[0..i] is sorted. */
7         int currentElement = list[i];
8         int k;
9         for (k = i - 1; k >= 0 && list[k] > currentElement; k--) {
10          list[k + 1] = list[k];
11        }
12
13        // Insert the current element into list[k + 1]
14        list[k + 1] = currentElement;
15      }
16    }
17  }
```

insertionSort(int[] list) 方法对一个 int 元素的数组进行排序。该方法是用嵌套的 for 循环实现的。外层循环（循环控制变量 i）（第 4 行）的迭代是为了获取已排好序的子线性表，其范围从 list[0] 到 list[i]。内层循环（循环控制变量 k）将 list[i] 插入从 list[0] 到 list[i-1] 的子线性表中。

为了更好地理解这个方法，使用下面的语句跟踪这个方法：

```
int[] list = {1, 9, 4, 6, 5, -4};
InsertionSort.insertionSort(list);
```

这里给出的插入排序算法重复地将一个新的元素插入一个排好序的部分数组中，直到整个数组排好序。在第 k 次迭代中，为了将一个元素插入一个大小为 k 的数组中，将进行 k 次比较来找到插入的位置，还要进行 k 次的移动来插入元素。使用 $T(n)$ 表示插入排序的复杂度，c 表示诸如每次迭代中的赋值和额外的比较这样的其他操作的总数，则

$$T(n) = (2+c) + (2 \times 2 + c) + \cdots + (2 \times (n-1) + c)$$
$$= 2(1 + 2 + \cdots + n - 1) + c(n-1)$$
$$= 2\frac{(n-1)n}{2} + cn - c = n^2 - n + cn - c$$
$$= O(n^2)$$

那么，插入排序算法的复杂度为 $O(n^2)$。因此，选择排序和插入排序具有同样的时间复杂度。

✓ 复习题

23.2.1 描述插入排序是如何工作的。插入排序的时间复杂度为多少？

23.2.2 使用图 23-1 作为一个例子来演示如何在 {45,11,50,59,60,2,4,7,10} 上应用插入排序。

23.2.3 如果一个线性表已经排好序了，insertionSort 方法将执行多少次比较？

23.3 冒泡排序

要点提示：冒泡排序算法多次遍历数组，在每次遍历中，如果元素没有按照顺序排列，则连续互换相邻的元素。

冒泡排序算法多次遍历数组。在每次遍历中，连续比较相邻的元素。如果某一对元素是降序，则互换其值；否则，保持不变。由于较小的值像"气泡"一样逐渐浮向顶部，而较大的值沉向底部，所以称这种技术为冒泡排序（bubble sort）或下沉排序（sinking sort）。第一次遍历之后，最后一个元素成为数组中的最大数。第二次遍历之后，倒数第二个元素成为数组中的第二大数。整个过程持续到所有元素都已排好序。

图 23-3a 给出 6 个元素（2 9 5 4 8 1）的数组经过第一次冒泡排序的遍历情况。首先比较第一对元素（2 和 9），因为这两个数已经是顺序排列的，所以不需要交换。接着比较第二对元素（9 和 5），因为 9 大于 5，所以交换 9 和 5。然后比较第三对元素（9 和 4）并交换 9 和 4。再比较第四对元素（9 和 8）并交换 9 和 8。最后比较第五对元素（9 和 1）并交换 9 和 1。在图 23-3 中，突出显示比较的数对，用斜体表示已经排好序的数字。可以通过网址 liveexample.pearsoncmg.com/dsanimation/BubbleSortNeweBook.html 观看冒泡排序如何工作的交互式演示。

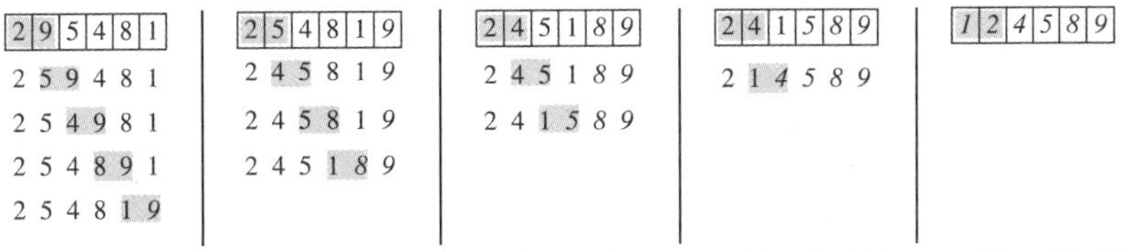

```
2 9 5 4 8 1      2 5 4 8 1 9      2 4 5 1 8 9      2 4 1 5 8 9      1 2 4 5 8 9
2 5 9 4 8 1      2 4 5 8 1 9      2 4 5 1 8 9      2 1 4 5 8 9
2 5 4 9 8 1      2 4 5 8 1 9      2 4 1 5 8 9
2 5 4 8 9 1      2 4 5 1 8 9
2 5 4 8 1 9

a) 第 1 次遍历    b) 第 2 次遍历    c) 第 3 次遍历    d) 第 4 次遍历    e) 第 5 次遍历
```

图 23-3 每次遍历都依次对元素对进行比较和排序

经过第 1 次遍历后，最大数（9）被放置在数组的末尾。在如图 23-3b 所示的第 2 次遍历中，依次对元素进行比较和排序。因为数组中的最后一个元素已经是最大的，所以不必考虑最后一对元素。在如图 23-3c 所示的第 3 次遍历中，因为最后两个元素已排好序，所以对除了它们之外的元素对进行依次比较和排序。因此，在第 k 次遍历时，不需要考虑后面的 k-1 个元素，因为它们已经排好序了。

冒泡排序算法在程序清单 23-2 中进行了描述。

程序清单 23-2 冒泡排序算法

```
1  for (int k = 1; k < list.length; k++) {
2    // Perform the kth pass
3    for (int i = 0; i < list.length - k; i++) {
4      if (list[i] > list[i + 1])
```

```
5     swap list[i] with list[i + 1];
6   }
7 }
```

注意，如果在某次遍历中没有发生交换，那么就不必进行下一次遍历，因为所有的元素都已经排好序了。可以使用该特性改进上面程序清单 23-2 中的算法，如程序清单 23-3 所示。

程序清单 23-3 改进的冒泡排序算法

```
1  boolean needNextPass = true;
2  for (int k = 1; k < list.length && needNextPass; k++) {
3    // Array may be sorted and next pass not needed
4    needNextPass = false;
5    // Perform the kth pass
6    for (int i = 0; i < list.length - k; i++) {
7      if (list[i] > list[i + 1]) {
8        swap list[i] with list[i + 1];
9        needNextPass = true; // Next pass still needed
10     }
11   }
12 }
```

算法可以如程序清单 23-4 所示实现。

程序清单 23-4 BubbleSort.java

```
1  public class BubbleSort {
2    /** Bubble sort method */
3    public static void bubbleSort(int[] list) {
4      boolean needNextPass = true;
5
6      for (int k = 1; k < list.length && needNextPass; k++) {
7        // Array may be sorted and next pass not needed
8        needNextPass = false;
9        for (int i = 0; i < list.length - k; i++) {
10         if (list[i] > list[i + 1]) {
11           // Swap list[i] with list[i + 1]
12           int temp = list[i];
13           list[i] = list[i + 1];
14           list[i + 1] = temp;
15
16           needNextPass = true; // Next pass still needed
17         }
18       }
19     }
20   }
21
22   /** A test method */
23   public static void main(String[] args) {
24     int[] list = {2, 3, 2, 5, 6, 1, -2, 3, 14, 12};
25     bubbleSort(list);
26     for (int i = 0; i < list.length; i++)
27       System.out.print(list[i] + " ");
28   }
29 }
```

```
-2 1 2 2 3 3 5 6 12 14
```

在最佳情况下，冒泡排序算法只需要一次遍历就能确定数组已排好序，不需要进行下一次遍历。由于第一次遍历的比较次数为 $n-1$，因此在最佳情况下，冒泡排序的时间为 $O(n)$。

在最差情况下，冒泡排序算法需要进行 $n-1$ 次遍历。第 1 次遍历需要 $n-1$ 次比较；第 2 次遍历需要 $n-2$ 次比较；依此进行，最后一次遍历需要 1 次比较。因此，比较的总数为：

$$(n-1)+(n-2)+\cdots+2+1$$
$$=\frac{(n-1)n}{2}=\frac{n^2}{2}-\frac{n}{2}=O(n^2)$$

因此，在最差情况下，冒泡排序的时间为 $O(n^2)$。

✓ 复习题

23.3.1 描述冒泡排序法是如何工作的。冒泡排序的时间复杂度是多少？
23.3.2 使用图 23-3 作为一个例子，演示如何将冒泡排序应用在 {45,11,50,59,60,2,4,7,10} 上。
23.3.3 如果一个线性表已经排好序了，bubbleSort 方法还需要进行多少次比较？

23.4 归并排序

요점 요점提示：归并排序可以如下递归地描述：归并排序算法将数组分为两半，对每部分递归地应用归并排序。在两部分都排好序后，对它们进行归并。

归并排序的算法在程序清单 23-5 中给出。

程序清单 23-5 归并排序算法

```
1  public static void mergeSort(int[] list) {
2    if (list.length > 1) {
3      mergeSort(list[0 ... list.length / 2);
4      mergeSort(list[list.length / 2 + 1 ... list.length]);
5      merge list[0 ... list.length / 2] with
6        list[list.length / 2 + 1 ... list.length];
7    }
8  }
```

图 23-4 演示了对 8 个元素（2 9 5 4 8 1 6 7）的数组进行归并排序。初始数组拆分为（2 9 5 4）和（8 1 6 7）。对这两个子数组递归地应用归并排序，将（2 9 5 4）拆分为（2 9）和（5 4），并将（8 1 6 7）拆分为（8 1）和（6 7）。继续进行这个过程直到子数组只包含一个元素为止。例如，将数组（2 9）分为（2）和（9）。由于（2）包含的是单一元素，所以它不能再细分了。现在，将（2）和（9）归并为一个新的有序数组（2 9），将（5）和（4）归并为一个新的有序数组（4 5）。然后将（2 9）和（4 5）归并为一个新的有序数组（2 4 5 9），最后将（2 4 5 9）和（1 6 7 8）归并为一个新的有序数组（1 2 4 5 6 7 8 9）。

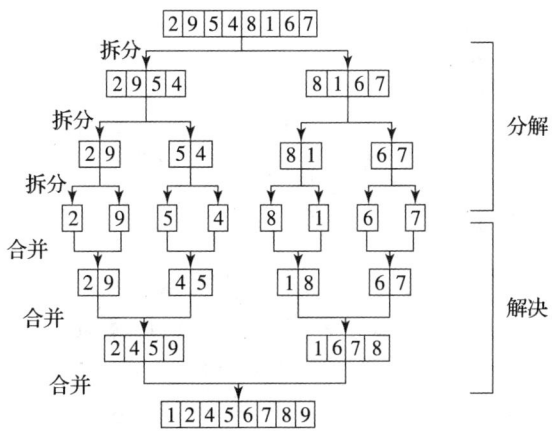

图 23-4 归并排序使用分治法对数组排序

递归调用持续将数组划分为子数组，直到每个子数组只包含一个元素。然后，该算法将这些小的子数组归并为稍大的有序子数组，直到最后形成一个有序的数组。

在程序清单 23-6 中实现了归并排序算法。

程序清单 23-6 MergeSort.java

```java
 1  public class MergeSort {
 2    /** The method for sorting the numbers */
 3    public static void mergeSort(int[] list) {
 4      if (list.length > 1) {
 5        // Merge sort the first half
 6        int[] firstHalf = new int[list.length / 2];
 7        System.arraycopy(list, 0, firstHalf, 0, list.length / 2);
 8        mergeSort(firstHalf);
 9
10        // Merge sort the second half
11        int secondHalfLength = list.length - list.length / 2;
12        int[] secondHalf = new int[secondHalfLength];
13        System.arraycopy(list, list.length / 2,
14          secondHalf, 0, secondHalfLength);
15        mergeSort(secondHalf);
16
17        // Merge firstHalf with secondHalf into list
18        merge(firstHalf, secondHalf, list);
19      }
20    }
21
22    /** Merge two sorted lists */
23    public static void merge(int[] list1, int[] list2, int[] temp) {
24      int current1 = 0; // Current index in list1
25      int current2 = 0; // Current index in list2
26      int current3 = 0; // Current index in temp
27
28      while (current1 < list1.length && current2 < list2.length) {
29        if (list1[current1] < list2[current2])
30          temp[current3++] = list1[current1++];
31        else
32          temp[current3++] = list2[current2++];
33      }
34
35      while (current1 < list1.length)
36        temp[current3++] = list1[current1++];
37
38      while (current2 < list2.length)
39        temp[current3++] = list2[current2++];
40    }
41
42    /** A test method */
43    public static void main(String[] args) {
44      int[] list = {2, 3, 2, 5, 6, 1, -2, 3, 14, 12};
45      mergeSort(list);
46      for (int i = 0; i < list.length; i++)
47        System.out.print(list[i] + " ");
48    }
49  }
```

mergeSort 方法（第 3 ～ 20 行）创建一个新数组 firstHalf，该数组是 list 前半部分的一个拷贝（第 7 行）。该算法在 firstHalf 上递归地调用 mergeSort（第 8 行）。firstHalf 的长度为 list.length/2，而 secondHalf 的长度为 list.length-list.length/2。创建的新数组 secondHalf 包含初始数组 list 的后半部分。算法在 secondHalf 上递归地调用 mergeSort（第 15 行）。在 firstHalf 和 secondHalf 都排好序之后，将它们归并成一个新的有序数组 list（第 18 行）。这样，数组 list 就排好序了。

merge 方法（第 23 ～ 40 行）归并两个有序数组 list1 和 list2 为数组 temp。current1 和 current2 指向 list1 和 list2 中要考虑的当前元素（第 24 ～ 26 行）。该方法重复比较 list1 和 list2 中的当前元素，并将较小的一个元素移动到 temp 中。如果较小元素在 list1

中,则 current1 加 1(第 30 行);如果较小元素在 list2 中,则 current2 加 1(第 32 行)。最后,其中一个数组中的所有元素都被移动到 temp 中。如果 list1 中仍有未移动的元素,则将它们复制到 temp 中(第 35 ~ 36 行)。如果 list2 中仍有未移动的元素,则将它们复制到 temp 中(第 38 ~ 39 行)。

图 23-5 演示了如何将两个数组 list1(2 4 5 9)和 list2(1 6 7 8)进行归并。初始时,要考虑的数组中的两个当前元素是 2 和 1。比较这两个数,并将较小元素 1 移到 temp 中,如图 23-5a 所示。current2 和 current3 加 1。继续比较这两个数组中的当前元素,并将较小的数移动到 temp 中,直到其中一个数组移动完毕。如图 23-5b 所示,list2 中的所有元素都被移动到 temp 中,而 current1 指向 list1 中的元素 9。将 9 复制到 temp 中,如图 23-5c 所示。可以通过网址 liveexample.pearsoncmg.com/dsanimation/MergeSortNeweBook.html 看到归并排序的工作方式的交互式演示。

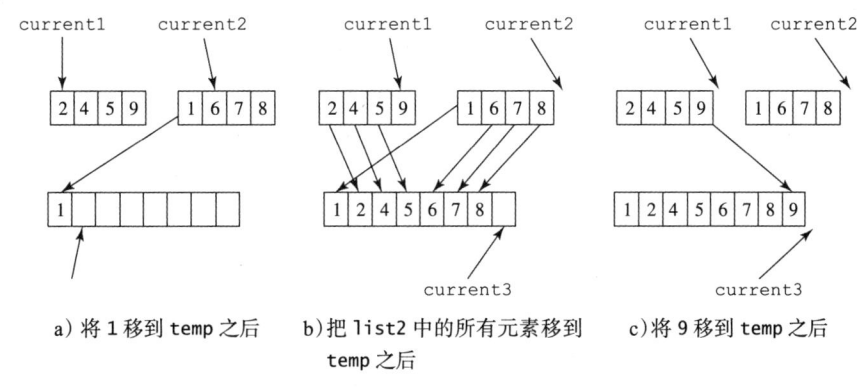

图 23-5 将两个有序数组归并为一个排好序的数组

MergeSort 方法在分解过程中创建了两个临时数组(第 6 和 12 行),将数组的前半部分和后半部分复制到临时数组中(第 7 和 13 行),对临时数组排序(第 8 和 15 行),然后将它们归并到原始数组中(第 18 行),如图 23-6a 所示。可以改写该代码,递归地对数组的前半部分和后半部分进行排序,而不创建新的临时数组,然后把两个数组归并到一个临时数组中并将它的内容复制到初始数组中,如图 23-6b 所示。这个留作编程练习题 23.20。

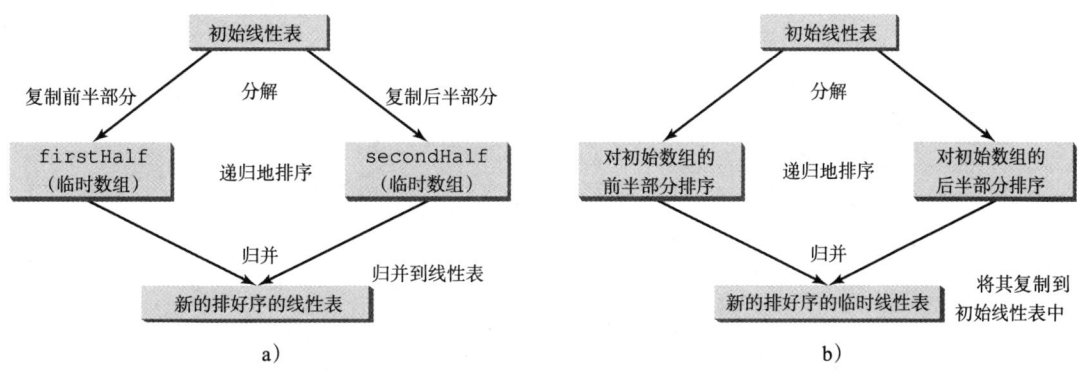

图 23-6 创建临时数组以支持归并排序

注意:归并排序可以使用并行处理高效执行。参见 32.16 节中归并排序的并行实现。

现在我们分析归并排序的运行时间。假设 $T(n)$ 表示使用归并排序对 n 个元素的数组进行排序所需的时间。不失一般性,假设 n 是 2 的幂。归并排序算法将数组分为两个子数组,

使用同样的算法对子数组进行递归排序，然后将子数组进行归并。因此

$$T(n) = T\left(\frac{n}{2}\right) + T\left(\frac{n}{2}\right) + 归并用时$$

第一项是对数组的前半部分排序所需的时间，而第二项是对数组的后半部分排序所需的时间。要归并两个子数组，最多需要 $n-1$ 次比较来比较两个子数组中的元素，以及 n 次移动将元素移到临时数组中。于是，总时间为 $2n-1$。因此

$$T(n) = T\left(\frac{n}{2}\right) + T\left(\frac{n}{2}\right) + 2n - 1 = O(n\log n)$$

归并排序的复杂度为 $O(n\log n)$。该算法优于选择排序、插入排序和冒泡排序，因为这些排序算法的复杂度为 $O(n^2)$。java.util.Arrays 类中的 sort 方法是使用归并排序算法的一种变体来实现的。

✓ 复习题

23.4.1 描述归并排序是如何工作的。归并排序的时间复杂度为多少？

23.4.2 以图 23-4 为例，演示如何在 {45, 11, 50, 59, 60, 2, 4, 7, 10} 上使用归并排序。

23.4.3 如果程序清单 23-6 中的第 6 ～ 15 行被下面代码替代，会有什么错误？

```
// Merge sort the first half
int[] firstHalf = new int[list.length / 2 + 1];
System.arraycopy(list, 0, firstHalf, 0, list.length / 2 + 1);
mergeSort(firstHalf);

// Merge sort the second half
int secondHalfLength = list.length - list.length / 2 - 1;
int[] secondHalf = new int[secondHalfLength];
System.arraycopy(list, list.length / 2 + 1,
  secondHalf, 0, secondHalfLength);
mergeSort(secondHalf);
```

23.5 快速排序

要点提示：快速排序工作机制是，算法在数组中选择一个称为基准（pivot）的元素，将数组分为两部分，使得第一部分中的所有元素都小于或等于基准元素，而第二部分中的所有元素都大于基准元素。对第一部分递归地应用快速排序算法，然后同样应用于第二部分。

快速排序是由 C. A. R. Hoare 于 1962 年开发的，该算法在程序清单 23-7 中描述。

程序清单 23-7 快速排序算法

```
1  public static void quickSort(int[] list) {
2    if (list.length > 1) {
3      选择一个基准;
4      将列表划分为 list1 和 list2, 从而
5        list1 中的所有元素 <=pivot 并且
6        list2 中的所有元素 >pivot;
7      quickSort(list1)
8      quickSort(list2)
9    }
10 }
```

该算法的每次划分都将基准元素放在恰当的位置。它将列表划分为两个子列表，如下图所示。

基准元素的选择会影响算法的性能。理想情况下,应该选择能平均划分两部分的基准元素。为了简单起见,假定将数组的第一个元素选为基准元素。(编程练习题 23.4 提出了选择基准元素的一个替代策略。)

图 23-7 演示了如何使用快速排序算法对数组(5 2 9 3 8 4 0 1 6 7)排序。选择第一个元素 5 作为基准元素将该数组划分为两部分,如图 23-7b 所示。突出显示的基准元素放在数组的恰当位置。分别对两个子数组(4 2 1 3 0)和(8 9 6 7)应用快速排序。基准元素 4 将(4 2 1 3 0)仅划分为一个数组(0 2 1 3),如图 23-7c 所示。然后对(0 2 1 3)应用快速排序。基准元素 0 将(0 2 1 3)也仅分为一个数组(2 1 3),如图 23-7d 所示。再对(2 1 3)应用快速排序。基准元素 2 将(2 1 3)分为(1)和(3),如图 23-7e 所示。再对(1)应用快速排序。由于该数组只包含一个元素,所以无须进一步划分。

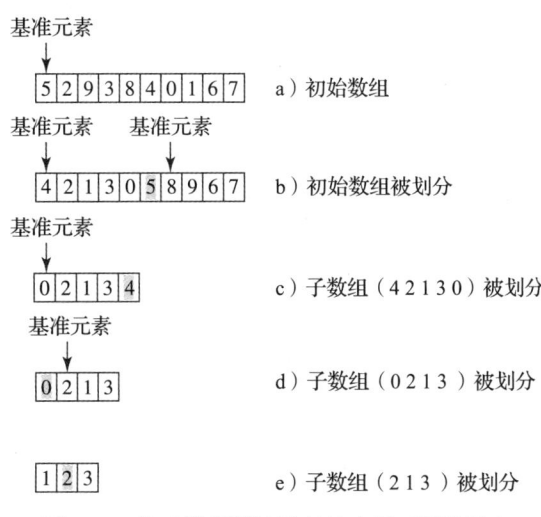

图 23-7 快速排序算法递归地应用于子数组上

程序清单 23-8 实现了快速排序算法。该类中有两个重载的 quickSort 方法。第一个方法(第 2 行)用来对数组进行排序。第二个是一个辅助方法(第 6 行),用于对指定范围内的子数组进行排序。

程序清单 23-8 QuickSort.java

```
1  public class QuickSort {
2    public static void quickSort(int[] list) {
3      quickSort(list, 0, list.length - 1);
4    }
5
6    public static void quickSort(int[] list, int first, int last) {
7      if (last > first) {
8        int pivotIndex = partition(list, first, last);
9        quickSort(list, first, pivotIndex - 1);
10       quickSort(list, pivotIndex + 1, last);
11     }
12   }
13
14   /** Partition the array list[first..last] */
15   public static int partition(int[] list, int first, int last) {
16     int pivot = list[first]; // Choose the first element as the pivot
```

```java
17      int low = first + 1; // Index for forward search
18      int high = last; // Index for backward search
19
20      while (high > low) {
21        // Search forward from left
22        while (low <= high && list[low] <= pivot)
23          low++;
24
25        // Search backward from right
26        while (low <= high && list[high] > pivot)
27          high--;
28
29        // Swap two elements in the list
30        if (high > low) {
31          int temp = list[high];
32          list[high] = list[low];
33          list[low] = temp;
34        }
35      }
36
37      while (high > first && list[high] >= pivot)
38        high--;
39
40      // Swap pivot with list[high]
41      if (pivot > list[high]) {
42        list[first] = list[high];
43        list[high] = pivot;
44        return high;
45      }
46      else {
47        return first;
48      }
49    }
50
51    /** A test method */
52    public static void main(String[] args) {
53      int[] list = {2, 3, 2, 5, 6, 1, -2, 3, 14, 12};
54      quickSort(list);
55      for (int i = 0; i < list.length; i++)
56        System.out.print(list[i] + " ");
57    }
58  }
```

```
-2 1 2 2 3 3 5 6 12 14
```

partition 方法（第 15～49 行）使用基准元素划分数组 list[first..last]。将子数组的第一个元素选为基准元素（第 16 行）。在初始情况下，low 指向子数组中的第二个元素（第 17 行），而 high 指向子数组中的最后一个元素（第 18 行）。

该方法在数组中从左侧开始查找第一个大于基准元素的元素（第 22～23 行），然后从数组右侧开始查找第一个小于或等于基准元素的元素（第 26～27 行）。然后交换这两个元素，并在 while 循环中重复相同的查找和交换操作，直到所有元素都查找完为止（第 20～35 行）。

如果基准元素被移动了，方法返回将子数组分为两部分的基准元素的新下标（第 44 行）；否则，返回基准元素的初始下标（第 47 行）。

图 23-8 演示了如何划分数组（5 2 9 3 8 4 0 1 6 7）。选择第一个元素 5 作为基准元素。初始时，low 是指向元素 2 的下标，而 high 是指向元素 7 的下标，如图 23-8a 所示。向前推进下标 low 以查找第一个大于基准元素的元素（9），然后从下标 high 往回推查找第一个小于或等于基准元素的元素（1），如图 23-8b 所示。交换 9 和 1，如图 23-8c 所示。继续查找并移动 low 使其指向元素 8，移动 high 使其指向元素 0，如图 23-8d 所示。交换 8 和

0，如图23-8e所示。继续移动low直到它超过high，如图23-8f所示。现在所有的元素都检测过了。交换基准元素与下标high处的元素4。最终的划分情况如图23-8g所示。当方法结束时，返回基准元素的下标。可以通过网址liveexample.pearsoncmg.com/dsanimation/QuickSortNeweBook.html查看快速排序如何工作的交互式演示。

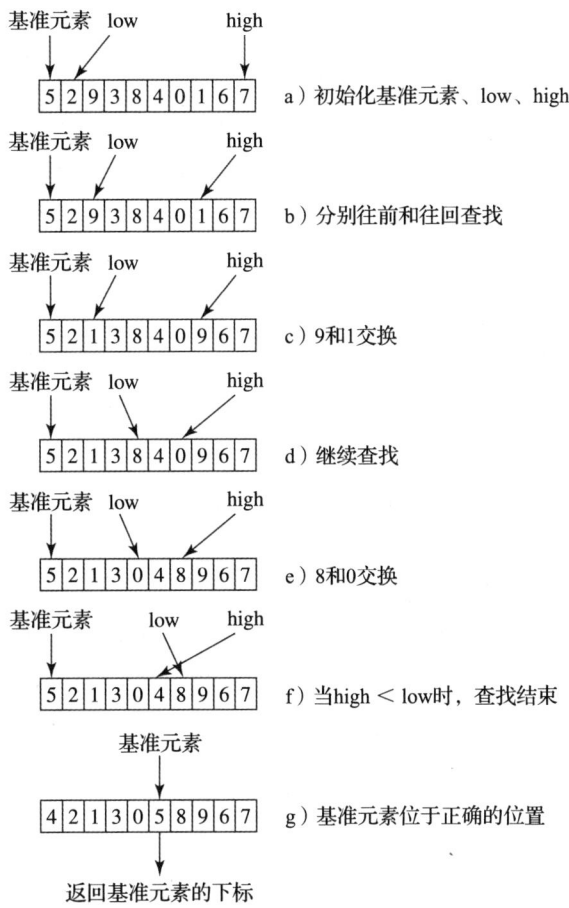

图23-8 partition方法在基准元素放置在正确的位置后返回基准元素的下标

在最差情况下，划分 n 个元素的数组需要进行 n 次比较和 n 次移动。因此，划分所需时间为 $O(n)$。

在最差情况下，基准元素每次会将数组划分为一个大的子数组和另外一个空数组。这个大的子数组的大小是划分前的子数组的大小减1。该算法需要 $(n-1) + (n-2) + \cdots + 2 + 1 = O(n^2)$ 时间。

在最佳情况下，基准元素每次将数组划分为规模大致相等的两部分。设 $T(n)$ 表示使用快速排序算法对包含 n 个元素的数组排序所需的时间，因此

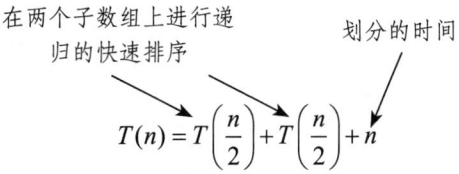

和归并排序的分析相似，快速排序的 $T(n) = O(n\log n)$。

在平均情况下，基准元素不会每次将数组分为规模相等的两部分或是一个空的部分。从统计上说，两部分的规模会非常接近。因此，平均时间为 $O(n\log n)$。精确的平均情况分析已经超出了本书的范围。

归并排序和快速排序都使用了分治法。对于归并排序，大量的工作是将两个子线性表进行归并，归并是在子线性表都排好序后进行的。对于快速排序，大量的工作是将线性表划分为两个子线性表，划分是在子线性表排好序前进行的。在最差情况下，归并排序的效率高于快速排序，但在平均情况下，两者的效率相同。归并排序在归并两个子数组时需要一个临时数组，而快速排序不需要额外的数组空间。因此，快速排序的空间效率高于归并排序。

复习题

23.5.1 描述快速排序是如何工作的。快速排序的时间复杂度是多少？

23.5.2 为什么快速排序比归并排序的空间效率更高？

23.5.3 以图 23-7 为例，演示如何在 {45, 11, 50, 59, 60, 2, 4, 7, 10} 上面应用快速排序。

23.5.4 如果删除 QuickSort 程序中的第 37～38 行，它还能工作吗？举一个反例来说明它不能工作了。

23.6 堆排序

要点提示：堆排序使用的是二叉堆。它首先将所有的元素添加到一个堆上，然后不断移除最大的元素以获得一个排好序的线性表。

堆排序（heap sort）使用一个二叉堆，该二叉堆是一棵完全二叉树。二叉树是一种分层体系结构。它可能是空的，也可能包含一个称为根（root）的元素以及称为左子树（left subtree）和右子树（right subtree）的两棵不同的二叉树。一条路径的长度（length）是指这条路径上的边数。一个结点的深度（depth）是指从根结点到该结点的路径的长度。一个结点如果没有子树，那么称为叶子结点。

二叉堆（binary heap）是一棵具有以下属性的二叉树：

- 形状属性：它是一棵完全二叉树。
- 堆属性：每个结点大于或等于它的任意一个子结点。

如果一棵二叉树的每一层都是满的，或者最后一层虽然没满并且最后一层的叶子结点都是靠最左放置的，那么这棵二叉树就是完全的（complete）。例如，在图 23-9 中，a）和 b）中的二叉树都是完全的，但是 c）和 d）中的二叉树都不是完全的。而且，a 中的二叉树是一个堆，但是 b 中的二叉树不是堆，因为根（39）小于它的右孩子（42）。

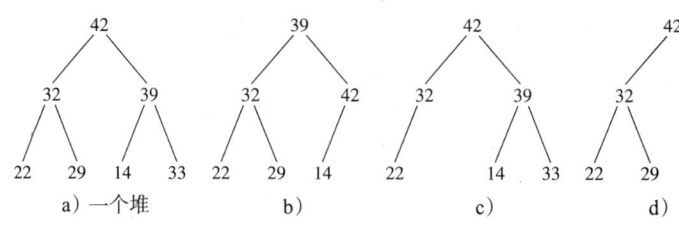

图 23-9 二叉堆是一种特殊的完全二叉树

注意：堆是一个在计算机科学中具有许多含义的词汇。本章中，堆表示一个二叉堆。

教学注意：堆在插入键和删除根结点时，执行效率很高。在网址 liveexample.pearsoncmg.

com/dsanimation/HeapeBook.html 上可以通过一个交互式的演示看到堆是如何工作的，如图 23-10 所示。

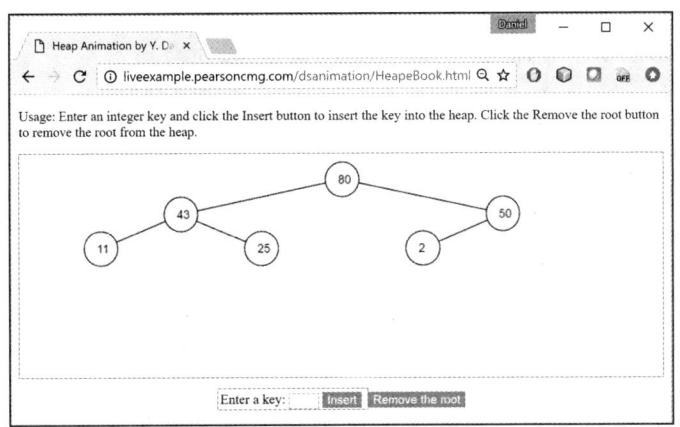

图 23-10　堆的动画工具允许可视化地插入键以及删除根结点（来源：Oracle 或其附属公司版权所有 ©1995 ～ 2016，经授权使用）

23.6.1　堆的存储

如果堆的大小是事先已知的，那么可以将堆存储在一个 `ArrayList` 或一个数组中。图 23-11a 中的堆可以使用图 23-11b 中的数组来存储。根在位置 0 处，它的两个子结点在位置 1 和位置 2 处。对于位置 i 处的结点，它的左子结点在位置 $2i+1$ 处，它的右子结点在位置 $2i+2$ 处，而它的父结点在位置 $(i-1)/2$ 处。例如，元素 39 的结点在位置 4 处，因此，它的左子结点（元素 14）在位置 9 处（$2 \times 4+1$），它的右子结点（元素 33）在位置 10 处（$2 \times 4+2$），而它的父结点（元素 42）在位置 1 处（$(4-1)/2$）。

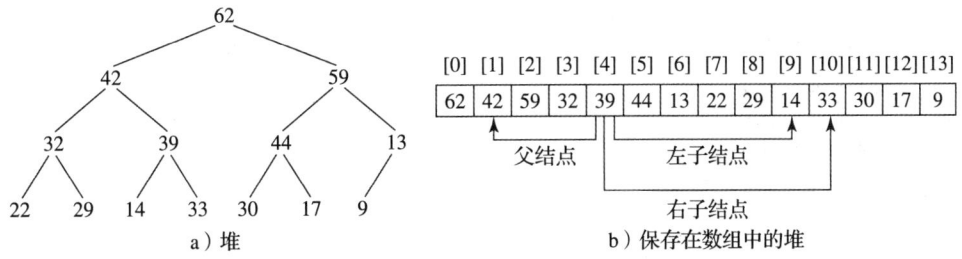

图 23-11　可以使用数组实现二叉堆

23.6.2　添加一个新结点

为了给堆添加一个新结点，首先将它添加到堆的末尾，然后按如下方式重建这棵树：

```
将最后一个结点作为当前结点；
while（当前结点大于它的父结点）{
    将当前结点和它的父结点交换；
    现在当前结点往上面进了一个层次；
}
```

假设该堆初始化为空。在顺序添加数字 3、5、1、19、11 和 22 之后，这个堆如图 23-12 所示。

现在考虑向堆中添加数字 88。将新结点 88 放在树的末尾，如图 23-13a 所示。互换 88 和 19，如图 23-13b 所示。互换 88 和 22，如图 23-13c 所示。

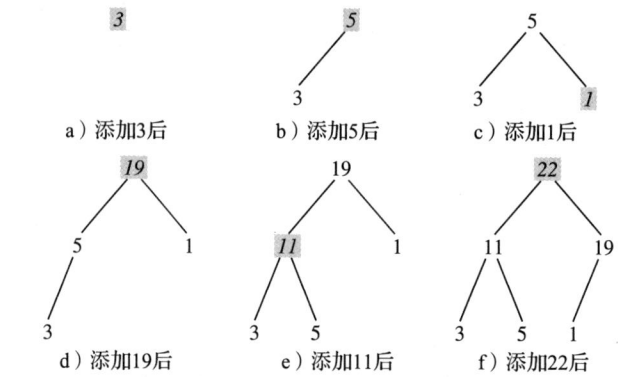

图 23-12　将元素 3、5、1、19、11 和 22 插入堆中

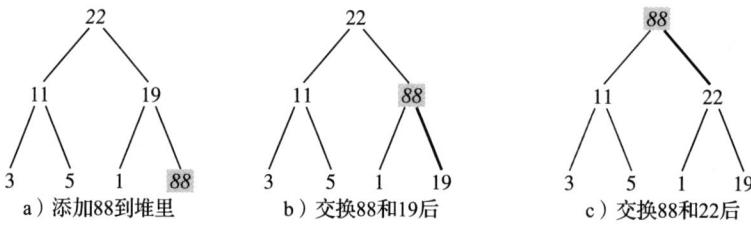

图 23-13　在添加一个元素之后重建这个堆

23.6.3　删除根结点

经常需要从堆中删除最大的元素，也就是这个堆的根结点。在删除根结点之后，必须重建这棵树以保持堆的属性。重建该树的算法如下所示：

```
用最后一个结点替换根结点；
让根结点成为当前结点；
while（当前结点具有子结点并且当前结点小于它的子结点）{
    将当前结点和它的较大子结点交换；
    现在当前结点往下面退了一个层次；
}
```

图 23-14 给出了从图 23-11a 中删除根结点 62 之后重建堆的过程。将最后的结点 9 移到根结点处，如图 23-14a 所示。互换 9 和 59，如图 23-14b 所示。互换 9 和 44，如图 23-14c 所示。互换 9 和 30，如图 23-14d 所示。

图 23-15 给出了从图 23-14d 中删除根结点 59 之后重建堆的过程。将最后的结点 17 移到根结点处，如图 23-15a 所示。互换 17 和 44，如图 23-15b 所示。互换 17 和 30，如图 23-15c 所示。

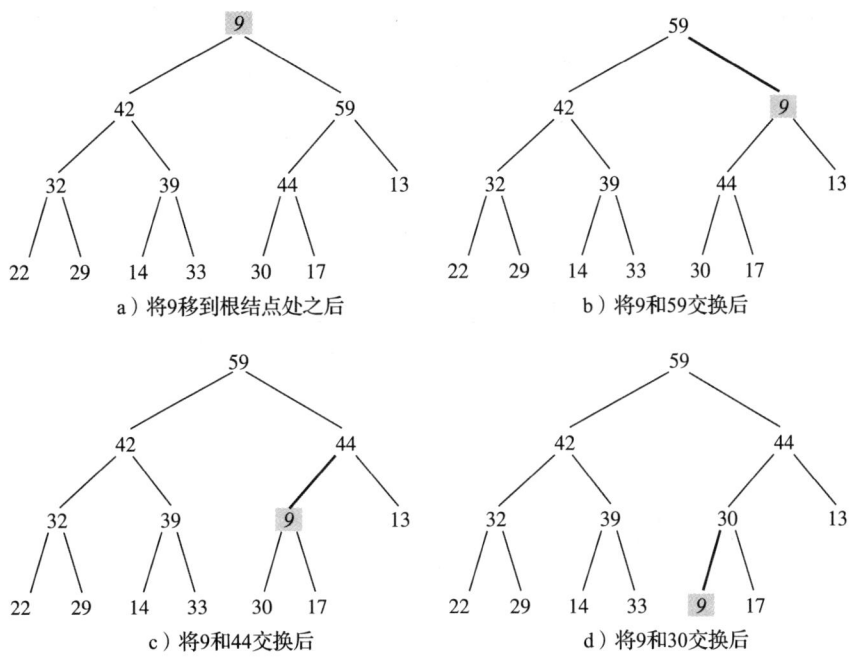

图 23-14 在删除根结点 62 之后重建堆

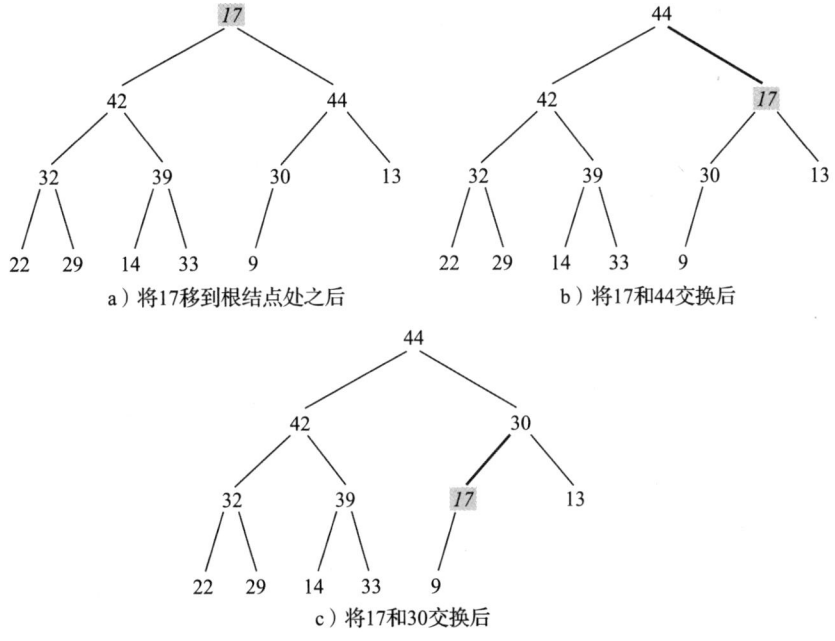

图 23-15 在删除根结点 59 之后重建堆

23.6.4 Heap 类

现在，可以设计和实现 Heap 类了。其类图如图 23-16 所示。其实现在程序清单 23-9 中给出。

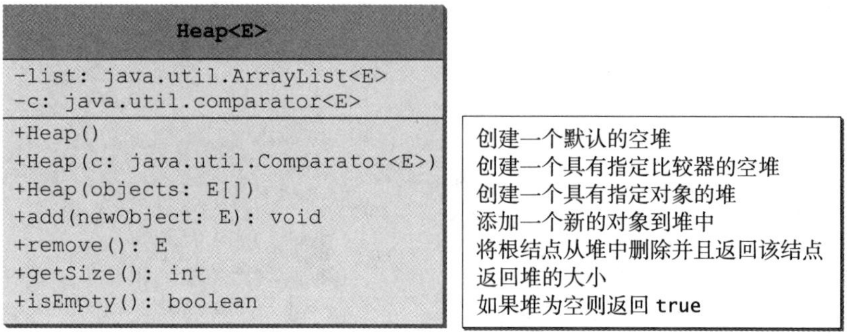

```
          Heap<E>
-list: java.util.ArrayList<E>
-c: java.util.comparator<E>
+Heap()                                    创建一个默认的空堆
+Heap(c: java.util.Comparator<E>)          创建一个具有指定比较器的空堆
+Heap(objects: E[])                        创建一个具有指定对象的堆
+add(newObject: E): void                   添加一个新的对象到堆中
+remove(): E                               将根结点从堆中删除并且返回该结点
+getSize(): int                            返回堆的大小
+isEmpty(): boolean                        如果堆为空则返回 true
```

图 23-16 Heap 类提供了处理堆的操作

程序清单 23-9 Heap.java

```java
 1  public class Heap<E> {
 2    private java.util.ArrayList<E> list = new java.util.ArrayList<>();
 3    private java.util.Comparator<? super E> c;
 4
 5    /** Create a default heap */
 6    public Heap() {
 7      this.c = (e1, e2) -> ((Comparable<E>)e1).compareTo(e2);
 8    }
 9
10    /** Create a heap with a specified comparator */
11    public Heap(java.util.Comparator<E> c) {
12      this.c = c;
13    }
14
15    /** Create a heap from an array of objects */
16    public Heap(E[] objects) {
17      this.c = (e1, e2) -> ((Comparable<E>)e1).compareTo(e2);
18      for (int i = 0; i < objects.length; i++)
19        add(objects[i]);
20    }
21
22    /** Add a new object into the heap */
23    public void add(E newObject) {
24      list.add(newObject); // Append to the heap
25      int currentIndex = list.size() - 1; // The index of the last node
26
27      while (currentIndex > 0) {
28        int parentIndex = (currentIndex - 1) / 2;
29        // Swap if the current object is greater than its parent
30        if (c.compare(list.get(currentIndex),
31            list.get(parentIndex)) > 0) {
32          E temp = list.get(currentIndex);
33          list.set(currentIndex, list.get(parentIndex));
34          list.set(parentIndex, temp);
35        }
36        else
37          break; // The tree is a heap now
38
39        currentIndex = parentIndex;
40      }
41    }
42
43    /** Remove the root from the heap */
44    public E remove() {
45      if (list.size() == 0) return null;
46
47      E removedObject = list.get(0);
48      list.set(0, list.get(list.size() - 1));
```

```
49       list.remove(list.size() - 1);
50
51       int currentIndex = 0;
52       while (currentIndex < list.size()) {
53         int leftChildIndex = 2 * currentIndex + 1;
54         int rightChildIndex = 2 * currentIndex + 2;
55
56         // Find the maximum between two children
57         if (leftChildIndex >= list.size()) break; // The tree is a heap
58         int maxIndex = leftChildIndex;
59         if (rightChildIndex < list.size()) {
60           if (c.compare(list.get(maxIndex),
61               list.get(rightChildIndex)) < 0) {
62             maxIndex = rightChildIndex;
63           }
64         }
65
66         // Swap if the current node is less than the maximum
67         if (c.compare(list.get(currentIndex),
68                 list.get(maxIndex)) < 0) {
69           E temp = list.get(maxIndex);
70           list.set(maxIndex, list.get(currentIndex));
71           list.set(currentIndex, temp);
72           currentIndex = maxIndex;
73         }
74         else
75           break; // The tree is a heap
76       }
77
78       return removedObject;
79     }
80
81     /** Get the number of nodes in the tree */
82     public int getSize() {
83       return list.size();
84     }
85
86     /** Return true if heap is empty */
87     public boolean isEmpty() {
88       return list.size() == 0;
89     }
90   }
```

堆在内部是使用数组线性表来表示的（第 2 行）。可以将它改为其他的数据结构，但是 Heap 类的合约保持不变。

堆中的元素可以使用一个比较器进行比较，如果没有指定比较器，则根据自然顺序比较。

无参构造方法（第 6 行）创建一个空堆，并使用元素的自然顺序作为比较器。比较器使用 lambda 表达式创建（第 7 行）。

add(E newObject) 方法（第 23～41 行）将一个对象添加到树中，如果该对象大于它的父结点，就互换它们。此过程持续到该新对象成为根结点，或者新对象不大于它的父结点。

remove() 方法（第 44～79 行）删除并返回根结点。为保持堆的特征，该方法将最后的对象移到根结点处，如果该对象小于它的较大的子结点，就互换它们。此过程持续到最后一个对象成为叶子结点，或者该对象不小于它的子结点。

23.6.5 使用 Heap 类进行排序

要使用堆对数组排序，首先使用 Heap 类创建一个对象，使用 add 方法将所有元素添加到堆中，然后使用 remove 方法从堆中删除所有元素。元素以降序被删除。程序清单 23-10 给出了使用堆对数组排序的算法。

程序清单 23-10 HeapSort.java

```java
 1  import java.util.Comparator;
 2
 3  public class HeapSort {
 4    /** Heap sort method */
 5    public static <E> void heapSort(E[] list) {
 6      // Create a Heap of integers
 7      heapSort(list, (e1, e2) -> ((Comparable<E>)e1).compareTo(e2));
 8    }
 9
10    /** Heap sort method */
11    public static <E> void heapSort(E[] list, Comparator<E> c) {
12      // Create a Heap of integers
13      Heap<E> heap = new Heap<>(c);
14
15      // Add elements to the heap
16      for (int i = 0; i < list.length; i++)
17        heap.add(list[i]);
18
19      // Remove elements from the heap
20      for (int i = list.length - 1; i >= 0; i--)
21        list[i] = heap.remove();
22    }
23
24    /** A test method */
25    public static void main(String[] args) {
26      Integer[] list = {-44, -5, -3, 3, 3, 1, -4, 0, 1, 2, 4, 5, 53};
27      heapSort(list);
28      for (int i = 0; i < list.length; i++)
29        System.out.print(list[i] + " ");
30    }
31  }
```

```
-44 -5 -4 -3 0 1 1 2 3 3 4 5 53
```

这里给出了两个 heapSort 方法。heapSort(E[] list) 方法（第 5 行）使用 Comparable 接口以自然顺序对一个线性表进行排序。heapSort(E[] list, Comparator<E>c) 方法（第 11 行）使用一个指定的比较器对线性表进行排序。

23.6.6 堆排序的时间复杂度

下面将注意力转到分析堆排序的时间复杂度上。设 h 表示包含 n 个元素的堆的高度。一棵非空树的高度是从根结点到最远的叶结点的一条最长路径。只有单个结点的树的高度为 0。按照惯例认为空树的高度是 -1。由于堆是完全二叉树，所以，第一层有 1（2^0）个结点，第二层有 2（2^1）个结点，第 k 层有 2^{k-1} 个结点，第 h 层有 2^{h-1} 个结点，而最后第 $h+1$ 层最少有一个结点且最多有 2^h 个结点。因此

$$1 + 2 + \cdots + 2^{h-1} < n \leq 1 + 2 + \cdots + 2^{h-1} + 2^h$$

也就是

$$2^h - 1 < n \leq 2^{h+1} - 1$$
$$2^h < n+1 \leq 2^{h+1}$$
$$h < \log(n+1) \leq h+1$$

这样，$h<\log(n+1)$ 且 $h \geq \log(n+1)-1$。因此，$\log(n+1)-1 \leq h<\log(n+1)$。所以，堆的高度为 $O(\log n)$。更准确地说，对于一个非空树，你可以证明 $h = \lfloor \log n \rfloor$。

由于 add 方法追踪从叶子结点到根结点的路径，因此向堆中添加一个新元素最多需要 h 步。所以，对于一个包含 n 个元素的数组构建一个初始堆，共需要 $O(n\log n)$ 时间。因为 remove 方法要追踪从根结点到叶子结点的路径，因此从堆中删除根结点后，重建堆最多需要 h 步。由于要调用 n 次 remove 方法，所以从堆产生一个有序数组需要的总时间为 $O(n\log n)$。

归并排序和堆排序需要的时间都为 $O(n\log n)$。归并排序需要一个临时数组来归并两个子数组，而堆排序不需要额外的数组空间。因此，堆排序的空间效率高于归并排序。

✔ 复习题

23.6.1 什么是完全二叉树？什么是堆？描述如何从堆中删除根结点，以及如何向堆中添加一个新对象。

23.6.2 顺序添加元素 4，5，1，2，9 和 3 到一个堆中，画一个图来演示添加每个元素后堆的情况。

23.6.3 显示使用 {45，11，50，59，60，2，4，7，10} 创建一个堆的步骤。

23.6.4 显示图 23-15c 中堆的根结点被删除后的堆。

23.6.5 给定下面的堆，描述从堆中删除所有结点的步骤。

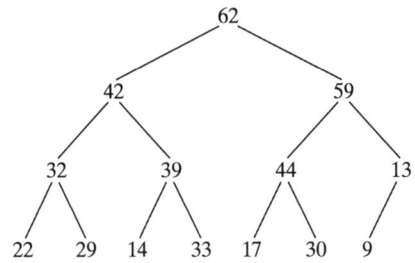

23.6.6 下面的语句哪个是错误的？

```
1  Heap<Object> heap1 = new Heap<>();
2  Heap<Number> heap2 = new Heap<>();
3  Heap<BigInteger> heap3 = new Heap<>();
4  Heap<Calendar> heap4 = new Heap<>();
5  Heap<String> heap5 = new Heap<>();
```

23.6.7 如果堆为空的话，调用 remove 方法的返回值是什么？

23.6.8 插入一个新元素到一个堆中的时间复杂度是多少？从一个堆中删除一个元素的时间复杂度是多少？

23.6.9 一个非空堆的高度是多少？一个具有 16、17 和 512 个元素的堆的高度是多少？如果堆的高度是 5，那么堆中结点的最大数目是多少？

23.7 桶排序和基数排序

🔑 要点提示：桶排序和基数排序是对整数进行排序的高效算法。

目前所讨论的所有排序算法都是可以用在任何键值类型（例如，整数、字符串以及任何可比较的对象）上的通用排序算法。这些算法都是通过比较它们的键值来对元素排序的。已经证明，基于比较的排序算法的复杂度不会优于 $O(n\log n)$。但是，如果键值是整数，那么可以使用桶排序，而无须比较这些键值。

桶排序算法的工作方式如下。假设键值的范围是从 0 到 t。我们需要 t+1 个标记为 0，1，…，t 的桶。如果元素的键值是 i，那么就将该元素放入桶 i 中。每个桶中都放着具有相同键值的元素。

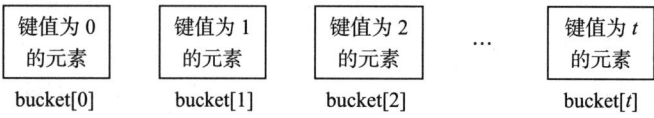

可以使用 ArrayList 来实现一个桶。应用桶排序算法对一个元素线性表进行排序的过程可以描述如下：

```java
public static void bucketSort(E[] list) {
  E[] bucket = (E[])new java.util.ArrayList[t+1];

  // Distribute the elements from list to buckets
  for (int i = 0; i < list.length; i++) {
    int key = list[i].getKey(); // Assume element has the getKey() method

    if (bucket[key] == null)
      bucket[key] = new java.util.ArrayList<>();

    bucket[key].add(list[i]);
  }

  // Now move the elements from the buckets back to list
  int k = 0; // k is an index for list
  for (int i = 0; i < bucket.length; i++) {
    if (bucket[i] != null) {
      for (int j = 0; j < bucket[i].size(); j++)
        list[k++] = bucket[i].get(j);
    }
  }
}
```

很明显，它需要 $O(n+t)$ 时间来对线性表排序，使用的空间是 $O(n+t)$，其中 n 是线性表的大小。

注意，如果 t 太大，那么桶排序不可取。作为替代，可以使用基数排序。基数排序是基于桶排序的，但是它只使用 10 个桶。

值得注意的是，桶排序是稳定的（stable），这意味着，如果原始线性表中的两个元素有相同的键值，那么它们在有序线性表中的顺序是不变的。也就是说，如果元素 e_1 和元素 e_2 有相同的键值，并且在原始线性表中，e_1 在 e_2 之前，那么在排好序的线性表中，e_1 还是在 e_2 之前。

假定键值是正整数。基数排序（radix sort）的思路就是将这些键值基于它们的基数位置分为子组。然后反复地从最小的基数位置开始，对其上的键值应用桶排序。

考虑对具有以下键值的元素进行排序：

331, 454, 230, 34, 343, 45, 59, 453, 345, 231, 9

在最后一位基数位置上应用桶排序。这些元素按如下方式放在桶中：

230	331 231		343 453	454 34	45 345				59 9
bucket[0]	bucket[1]	bucket[2]	bucket[3]	bucket[4]	bucket[5]	bucket[6]	bucket[7]	bucket[8]	bucket[9]

从桶中收集元素之后，将它们以下面的顺序排列：

230, 331, 231, 343, 453, 454, 34, 45, 345, 59, 9

在倒数第二位基数位置上应用桶排序。这些元素按如下方式放在桶中：

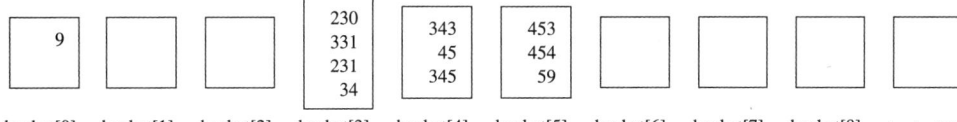

从桶中收集元素之后，将它们以下面的顺序排列：

9, 230, 331, 231, 34, 343, 45, 345, 453, 454, 59

（注意，9 是 009。）

在倒数第三位基数位置上应用桶排序。这些元素按如下方式放在桶中：

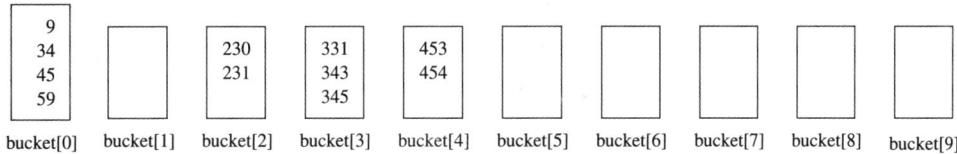

从桶中收集元素之后，将它们以下面的顺序排列：

9, 34, 45, 59, 230, 231, 331, 343, 345, 453, 454

现在这些元素是有序的了。

基数排序需要 $O(dn)$ 时间对带整数键值的 n 个元素排序，其中 d 是所有键值中基数位数的最大值。

✓ 复习题

23.7.1 可以使用桶排序来对一个字符串线性表进行排序吗？

23.7.2 使用数字 454、34、23、43、74、86 以及 76 来演示基数排序是如何进行排序的。

23.8 外部排序

要点提示：可以使用外部排序来对大量数据进行排序。

前面几节讨论的所有排序算法，都假定要排序的所有数据某个时刻在内存中都可用，如位于一个数组中。要对存储在外部文件中的数据排序，首先要将数据放入内存，然后内部地对它们进行排序。然而，如果文件太大，那么文件中的所有数据不能同时放入内存中。本节将讨论如何在大的外部文件中对数据排序。这称为外部排序（external sort）。

为简单起见，假定将 200 万个 int 值存储在一个名为 largedata.dat 的二进制文件中。该文件是使用程序清单 23-11 中的程序创建的。

程序清单 23-11 CreateLargeFile.java

```
1  import java.io.*;
2
3  public class CreateLargeFile {
4    public static void main(String[] args) throws Exception {
5      DataOutputStream output = new DataOutputStream(
6        new BufferedOutputStream(
7        new FileOutputStream("largedata.dat")));
8
```

```
 9      for (int i = 0; i < 2_000_000; i++)
10        output.writeInt((int)(Math.random() * 1000000));
11
12      output.close();
13
14      // Display first 100 numbers
15      DataInputStream input = new DataInputStream(
16        new BufferedInputStream(new FileInputStream("largedata.dat")));
17      for (int i = 0; i < 100; i++)
18        System.out.print(input.readInt() + " ");
19
20      input.close();
21    }
22  }
```

```
569193 131317 608695 776266 767910 624915 458599 5010 ... (omitted)
```

可以使用归并排序的一种变体对这个文件进行两阶段排序。

阶段 I：重复地将数据从文件读入数组中，并使用内部排序算法对数组排序，然后将数据从数组输出到一个临时文件中。该过程如图 23-17 所示。理想情况下，你会希望创建一个大型数组，但是数组的最大尺寸依赖于操作系统分配给 JVM 的内存大小。假定数组的最大尺寸为 100 000 个 int 值，那么在临时文件中就是对每 100 000 个 int 值进行排序。将它们标记为 $S_1, S_2, \cdots, S_k$，其中最后一段 S_k 包含的数值可能会少于 100 000 个。

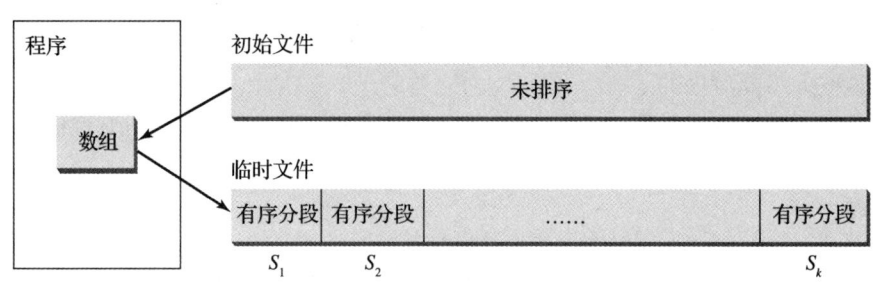

图 23-17　对原始文件分段排序

阶段 II：将每对有序分段（比如 S_1 和 S_2，S_3 和 S_4，…）归并到一个更大的有序分段中，并将新分段存储到新的临时文件中。继续同样的过程，直到只有一个有序分段结果。图 23-18 演示了如何对 8 个分段进行归并。

注意：不一定要归并两个相邻分段。例如，在第一步归并中，可以归并 S_1 和 S_5，S_2 和 S_6，S_3 和 S_7，S_4 和 S_8。这在高效实现阶段 II 时很有用。

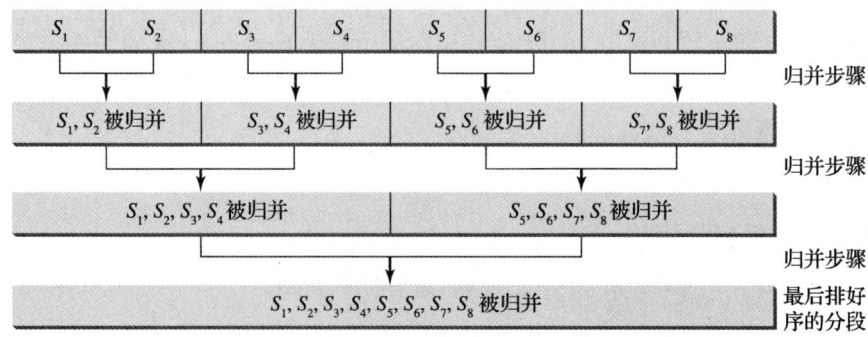

图 23-18　对排好序的分段迭代地进行归并

23.8.1 实现阶段 I

程序清单 23-12 给出一个方法,它从文件中读取每个数据段,并对该段进行排序,然后将排好序的分段存在一个新文件中。该方法返回分段的个数。

程序清单 23-12 创建初始的排好序的分段

```
1   /** Sort original file into sorted segments */
2   private static int initializeSegments
3      (int segmentSize, String originalFile, String f1)
4      throws Exception {
5     int[] list = new int[segmentSize];
6     DataInputStream input = new DataInputStream(
7       new BufferedInputStream(new FileInputStream(originalFile)));
8     DataOutputStream output = new DataOutputStream(
9       new BufferedOutputStream(new FileOutputStream(f1)));
10
11    int numberOfSegments = 0;
12    while (input.available() > 0) {
13      numberOfSegments++;
14      int i = 0;
15      for ( ; input.available() > 0 && i < segmentSize; i++) {
16        list[i] = input.readInt();
17      }
18
19      // Sort an array list[0..i-1]
20      java.util.Arrays.sort(list, 0, i);
21
22      // Write the array to f1.dat
23      for (int j = 0; j < i; j++) {
24        output.writeInt(list[j]);
25      }
26    }
27
28    input.close();
29    output.close();
30
31    return numberOfSegments;
32  }
```

该方法在第 5 行创建一个具有最大尺寸的数组,在第 6 行为原始文件创建一个数据输入流,在第 8 行为临时文件创建一个数据输出流。缓冲流用于提高程序性能。

第 14 ~ 17 行从文件中读取一段数据到数组中。第 20 行对数组进行排序。第 23 ~ 25 行将数组中的数据写入临时文件中。

第 31 行返回分段的个数。注意,除了最后一个分段的元素数可能较少外,其他分段都有 MAX_ARRAY_SIZE 个元素。

23.8.2 实现阶段 II

在每步归并中,都将两个有序分段归并成一个新的分段。新段的大小是原来的两倍,因此,每次归并后分段的个数减少一半。一个分段太大,不能放入内存的数组中。为了实现归并步骤,要将文件 f1.dat 中一半数目的分段复制到临时文件 f2.dat 中。然后,将 f1.dat 剩下部分的第一个分段与 f2.dat 中的第一个分段归并到名为 f3.dat 的临时文件中,如图 23-19 所示。

> **注意**:f1.dat 可能会比 f2.dat 多一个分段。如果是这样,在归并后将最后一个分段移到 f3.dat 中。

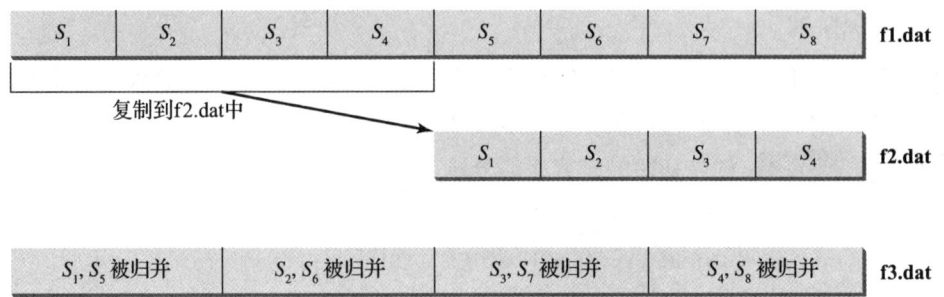

图 23-19 迭代地归并排好序的分段

程序清单 23-13 给出一个方法，将 f1.dat 中的前半部分复制到 f2.dat 中。程序清单 23-14 给出一个方法，将 f1.dat 和 f2.dat 中的一对分段进行归并。程序清单 23-15 给出一个方法，对两个分段进行归并。

程序清单 23-13 复制前半部分的分段

```
1  private static void copyHalfToF2(int numberOfSegments,
2      int segmentSize, DataInputStream f1, DataOutputStream f2)
3      throws Exception {
4    for (int i = 0; i < (numberOfSegments / 2) * segmentSize; i++) {
5      f2.writeInt(f1.readInt());
6    }
7  }
```

程序清单 23-14 归并所有分段

```
1  private static void mergeSegments(int numberOfSegments,
2      int segmentSize, DataInputStream f1, DataInputStream f2,
3      DataOutputStream f3) throws Exception {
4    for (int i = 0; i < numberOfSegments; i++) {
5      mergeTwoSegments(segmentSize, f1, f2, f3);
6    }
7
8    // If f1 has one extra segment, copy it to f3
9    while (f1.available() > 0) {
10     f3.writeInt(f1.readInt());
11   }
12 }
```

程序清单 23-15 归并两个分段

```
1  private static void mergeTwoSegments(int segmentSize,
2      DataInputStream f1, DataInputStream f2,
3      DataOutputStream f3) throws Exception {
4    int intFromF1 = f1.readInt();
5    int intFromF2 = f2.readInt();
6    int f1Count = 1;
7    int f2Count = 1;
8
9    while (true) {
10     if (intFromF1 < intFromF2) {
11       f3.writeInt(intFromF1);
12       if (f1.available() == 0 || f1Count++ >= segmentSize) {
13         f3.writeInt(intFromF2);
14         break;
15       }
16       else {
17         intFromF1 = f1.readInt();
```

```
18      }
19    }
20    else {
21      f3.writeInt(intFromF2);
22      if (f2.available() == 0 || f2Count++ >= segmentSize) {
23        f3.writeInt(intFromF1);
24        break;
25      }
26      else {
27        intFromF2 = f2.readInt();
28      }
29    }
30  }
31
32  while (f1.available() > 0 && f1Count++ < segmentSize) {
33    f3.writeInt(f1.readInt());
34  }
35
36  while (f2.available() > 0 && f2Count++ < segmentSize) {
37    f3.writeInt(f2.readInt());
38  }
39 }
```

23.8.3 结合两个阶段

程序清单 23-16 给出一个完整的程序，对 largedata.dat 中的 int 值进行排序，并将已排好序的数据存储在 sortedfile.dat 中。

程序清单 23-16 SortLargeFile.java

```
1  import java.io.*;
2
3  public class SortLargeFile {
4    public static final int MAX_ARRAY_SIZE = 100000;
5    public static final int BUFFER_SIZE = 100000;
6
7    public static void main(String[] args) throws Exception {
8      // Sort largedata.dat to sortedfile.dat
9      sort("largedata.dat", "sortedfile.dat");
10
11     // Display the first 100 numbers in the sorted file
12     displayFile("sortedfile.dat");
13   }
14
15   /** Sort data in source file and into target file */
16   public static void sort(String sourcefile, String targetfile)
17       throws Exception {
18     // Implement Phase 1: Create initial segments
19     int numberOfSegments =
20       initializeSegments(MAX_ARRAY_SIZE, sourcefile, "f1.dat");
21
22     // Implement Phase 2: Merge segments recursively
23     merge(numberOfSegments, MAX_ARRAY_SIZE,
24       "f1.dat", "f2.dat", "f3.dat", targetfile);
25   }
26
27   /** Sort original file into sorted segments */
28   private static int initializeSegments
29       (int segmentSize, String originalFile, String f1)
30       throws Exception {
31     // Same as Listing 23.12, so omitted
32   }
33
```

```java
 34    private static void merge(int numberOfSegments, int segmentSize,
 35        String f1, String f2, String f3, String targetfile)
 36        throws Exception {
 37      if (numberOfSegments > 1) {
 38        mergeOneStep(numberOfSegments, segmentSize, f1, f2, f3);
 39        merge((numberOfSegments + 1) / 2, segmentSize * 2,
 40          f3, f1, f2, targetfile);
 41      }
 42      else { // Rename f1 as the final sorted file
 43        File sortedFile = new File(targetfile);
 44        if (sortedFile.exists()) sortedFile.delete();
 45        new File(f1).renameTo(sortedFile);
 46      }
 47    }
 48
 49    private static void mergeOneStep(int numberOfSegments,
 50        int segmentSize, String f1, String f2, String f3)
 51        throws Exception {
 52      DataInputStream f1Input = new DataInputStream(
 53        new BufferedInputStream(new FileInputStream(f1), BUFFER_SIZE));
 54      DataOutputStream f2Output = new DataOutputStream(
 55        new BufferedOutputStream(new FileOutputStream(f2), BUFFER_SIZE));
 56
 57      // Copy half number of segments from f1.dat to f2.dat
 58      copyHalfToF2(numberOfSegments, segmentSize, f1Input, f2Output);
 59      f2Output.close();
 60
 61      // Merge remaining segments in f1 with segments in f2 into f3
 62      DataInputStream f2Input = new DataInputStream(
 63        new BufferedInputStream(new FileInputStream(f2), BUFFER_SIZE));
 64      DataOutputStream f3Output = new DataOutputStream(
 65        new BufferedOutputStream(new FileOutputStream(f3), BUFFER_SIZE));
 66
 67      mergeSegments(numberOfSegments / 2,
 68        segmentSize, f1Input, f2Input, f3Output);
 69
 70      f1Input.close();
 71      f2Input.close();
 72      f3Output.close();
 73    }
 74
 75    /** Copy first half number of segments from f1.dat to f2.dat */
 76    private static void copyHalfToF2(int numberOfSegments,
 77        int segmentSize, DataInputStream f1, DataOutputStream f2)
 78        throws Exception {
 79      // Same as Listing 23.13, so omitted
 80    }
 81
 82    /** Merge all segments */
 83    private static void mergeSegments(int numberOfSegments,
 84        int segmentSize, DataInputStream f1, DataInputStream f2,
 85        DataOutputStream f3) throws Exception {
 86      // Same as Listing 23.14, so omitted
 87    }
 88
 89    /** Merges two segments */
 90    private static void mergeTwoSegments(int segmentSize,
 91      DataInputStream f1, DataInputStream f2,
 92      DataOutputStream f3) throws Exception {
 93      // Same as Listing 23.15, so omitted
 94    }
 95
 96    /** Display the first 100 numbers in the specified file */
 97    public static void displayFile(String filename) {
 98      try {
 99        DataInputStream input =
```

```
100            new DataInputStream(new FileInputStream(filename));
101        for (int i = 0; i < 100; i++)
102          System.out.print(input.readInt() + " ");
103        input.close();
104      }
105      catch (IOException ex) {
106        ex.printStackTrace();
107      }
108    }
109  }
```

```
0 1 1 1 2 2 2 3 3 4 5 6 8 8 9 9 9 10 10 11...(omitted)
```

在运行该程序之前，应首先运行程序清单 23-11 来创建 largedata.dat。调用 sort("large-data.dat","sortedfile.dat")（第 9 行）从 largedata.dat 中读取数据并向 sortedfile.dat 写入排好序的数据。调用 displayFile("sortedfile.dat")（第 12 行）显示指定文件中的前 100 个数字。注意，这个文件是用二进制 I/O 创建的，因而不能使用记事本这样的文本编辑器来查看它。

sort 方法首先从原始数组中创建初始分段，并且将排好序的分段存入新文件 f1.dat 中（第 19 ~ 20 行），然后在 targetfile 中就产生了一个排好序的文件（第 23 ~ 24 行）。

merge 方法

```
merge(int numberOfSegments, int segmentSize,
    String f1, String f2, String f3, String targetfile)
```

使用 f2 作为辅助文件将 f1 中的分段归并到 f3 中。merge 方法在很多归并步骤中都会被递归调用。每步归并都会使分段数 numberOfSegments 减少一半，同时使有序分段的大小翻倍。在完成一个归并步骤后，下一个归并步骤使用 f1 作为辅助文件将 f3 中的新分段归并到 f2 中。调用新归并方法的语句为：

```
merge((numberOfSegments + 1) / 2, segmentSize * 2,
    f3, f1, f2, targetfile);
```

下一个归并步骤中的 numberOfSegments 为 (numberOfSegments+1)/2。例如，如果 numberOfSegments 为 5，那么，下一个归并步骤的 numberOfSegments 为 3，因为每两个分段进行归并时会留下一个未归并的分段。

当 numberOfSegments 为 1 时，结束递归的 merge 方法。在这种情况下，f1 中包含已排好序的数据。文件 f1 被重命名为 targetfile（第 45 行）。

23.8.4 外部排序复杂度

在外部排序中，主要开销是在 I/O 上。假设 n 是文件中要排序的元素个数。在阶段 I，从原始文件中读取 n 个元素，然后将其输出到一个临时文件。因此，阶段 I 的 I/O 复杂度为 $O(n)$。

对于阶段 II，在第一个合并步骤之前，排好序的分段的个数为 $\frac{n}{c}$，其中 c 是 MAX_ARRAY_SIZE。每一个合并步骤都会使分段的个数减半。因此，在第一次合并步骤之后，分段个数为 $\frac{n}{2c}$。在第二次合并步骤之后，分段个数为 $\frac{n}{2^2c}$。在第三次合并步骤之后，分段个数为 $\frac{n}{2^3c}$。在第 $\log\left(\frac{n}{c}\right)$ 次合并步骤之后，分段个数减到 1。因此，合并步骤的总数为 $\log\left(\frac{n}{c}\right)$。

在每次合并步骤中，从文件 f1 读取一半数量的分段，然后将它们写入一个临时文件 f2。合并 f1 中剩余的分段和 f2 中的分段。每一个合并步骤中 I/O 的次数为 $O(n)$。因为合并步骤的总数是 $\log\left(\dfrac{n}{c}\right)$，I/O 的总数是

$$O(n) \times \log\left(\dfrac{n}{c}\right) = O(n \log n)$$

因此，外部排序的复杂度是 $O(n\log n)$。

复习题

23.8.1 描述外部排序是如何工作的。外部排序算法的复杂度是多少？

23.8.2 10 个数字 {2,3,4,0,5,6,7,9,8,1} 保存在外部文件 largedata.dat 中。设 MAX_ARRAY_SIZE 为 2，手工跟踪 SortLargeFile 程序。

关键术语

bubble sort（冒泡排序）
bucket sort（桶排序）
complete binary tree（完全二叉树）
external sort（外部排序）
heap（堆）

heap sort（堆排序）
height of a heap（堆的高度）
merge sort（归并排序）
quick sort（快速排序）
radix sort（基数排序）

本章小结

1. 选择排序、插入排序、冒泡排序和快速排序的最差时间复杂度为 $O(n^2)$。
2. 归并排序的平均情况和最差情况的复杂度为 $O(n\log n)$。快速排序的平均时间也是 $O(n\log n)$。
3. 对于设计排序这样的高效算法，堆是一个很有用的数据结构。本章学习了如何定义和实现一个堆类，以及如何向 / 从堆中插入和删除元素。
4. 堆排序的时间复杂度为 $O(n\log n)$。
5. 桶排序和基数排序都是针对整数键的特定排序算法。这些算法不是通过比较键而是使用桶来对键排序的，它们会比一般的排序算法效率更高。
6. 可以使用归并排序的一种变体——称为外部排序，对外部文件中的大量数据进行排序。

测试题

回答位于本书配套网站上的本章测试题。

编程练习题

23.3 ~ 23.5 节

23.1 （泛型冒泡排序）使用冒泡排序编写下面两个泛型方法。第一个方法使用 Comparable 接口对元素排序，第二个方法使用 Comparator 接口对元素排序。

```
public static <E extends Comparable<E>>
  void bubbleSort(E[] list)
public static <E> void bubbleSort(E[] list,
  Comparator<? super E> comparator)
```

23.2 （泛型归并排序）使用归并排序编写下面两个泛型方法。第一个方法使用 Comparable 接口对元素排序，第二个方法使用 Comparator 接口对元素排序。

```
public static <E extends Comparable<E>>
  void mergeSort(E[] list)
public static <E> void mergeSort(E[] list,
  Comparator<? super E> comparator)
```

23.3 （泛型快速排序）使用快速排序编写下面两个泛型方法。第一个方法使用 Comparable 接口对元素排序，第二个方法使用 Comparator 接口对元素排序。

```
public static <E extends Comparable<E>>
  void quickSort(E[] list)
public static <E> void quickSort(E[] list,
  Comparator<? super E> comparator)
```

23.4 （改进快速排序）本书提供的快速排序算法选择线性表中的第一个元素作为基准元素。修改该算法，在线性表中的第一个元素、中间元素和最后一个元素中选择一个中位数作为基准元素。

*23.5 （修改归并排序）重写 mergeSort 方法，递归地对数组的前半部分和后半部分进行排序，而不创建新的临时数组。然后将两个部分归并到一个临时数组中并拷贝其内容到初始数组中，如图 23-6b 所示。

23.6 （检查顺序）编写下面的重载方法，用于检查数组是按升序还是降序排列的。默认情况下，该方法是检查升序的。为检查降序，则将 false 传递给方法中的 ascending 参数。

```
public static boolean ordered(int[] list)
public static boolean ordered(int[] list, boolean ascending)
public static boolean ordered(double[] list)
public static boolean ordered
  (double[] list, boolean ascending)
public static <E extends Comparable<E>>
  boolean ordered(E[] list)
public static <E extends Comparable<E>> boolean ordered
  (E[] list, boolean ascending)
public static <E> boolean ordered(E[] list,
  Comparator<? super E> comparator)
public static <E> boolean ordered(E[] list,
  Comparator<? super E> comparator, boolean ascending)
```

23.6 节

23.7 （最小堆）本书中介绍的堆也称为最大堆（max-heap），其中的每个结点都大于或等于它的任何一个子结点。最小堆（min-heap）是指每个结点都小于或等于它的任何一个子结点的堆。最小堆常用于实现优先队列。修改程序清单 23-9 中的 Heap 类以实现最小堆。

*23.8 （泛型插入排序）使用插入排序编写下面两个泛型方法。

```
public static <E extends Comparable<E>>
  void insertionSort(E[] list)
public static <E> void insertionSort(E[] list,
  Comparator<? super E> comparator)
```

*23.9 （泛型堆排序）使用堆排序编写下面两个泛型方法。第一个方法使用 Comparable 接口对元素排序，第二个方法使用 Comparator 接口对元素排序。（提示：使用编程练习题 23.5 中的 Heap 类。）

```
public static <E extends Comparable<E>>
  void heapSort(E[] list)
public static <E> void heapSort(E[] list,
  Comparator<? super E> comparator)
```

**23.10 （堆的可视化）编写一个程序，图形化显示一个堆，如图 23-10 所示。该程序允许用户向堆中插入元素和从堆中删除元素。

23.11 （堆的 clone 和 equals 方法）实现 Heap 类中的 clone 和 equals 方法。
对你的代码使用位于 liveexample.pearsoncmg.com/test/Exercise23_11.txt 的模板。

23.7 节

*23.12 （基数排序）编写一个程序，随机创建 1 000 000 个整数，然后使用基数排序对它们排序。

*23.13 （排序的执行时间）编写一个程序，获取输入规模为 50 000、100 000、150 000、200 000、250 000 和 300 000 时的选择排序、冒泡排序、归并排序、快速排序、堆排序以及基数排序的执行时间。该程序应随机地创建数据，然后打印如下所示的一个表格：

数组大小	选择排序	冒泡排序	归并排序	快速排序	堆排序	基数排序
50 000						
100 000						
150 000						
200 000						
250 000						
300 000						

（提示：可以使用下面的代码模板来获取执行时间。）

```
long startTime = System.nanoTime();
perform the task;
long endTime = System.nanoTime();
long executionTime = endTime - startTime;
```

23.8 节

*23.14 （外部排序的执行时间）编写程序，获取输入规模为 5 000 000、10 000 000、15 00 0000、20 000 000、25 000 000 和 30 000 000 时外部排序的执行时间。该程序应该打印出如下所示的一个表格：

文件大小	5 000 000	10 000 000	15 000 000	20 000 000	25 000 000	30 000 000
时间						

综合

*23.15 （选择排序动画）编写一个程序，实现选择排序算法的动画。创建一个数组，以随机顺序包含从 1 到 20 的 20 个不同数字。数组元素在一个直方图中显示，如图 23-20a 所示。单击 Step 按钮使程序执行算法中外部循环的一次迭代，然后为新的数组重画直方图。将排好序的子数组的最后一个条块标上颜色。当算法结束时，显示一条信息通知用户。单击 Reset 按钮为一次新的开始创建一个新的随机数组。（可以很容易地修改程序，来制作插入排序算法的动画。）

*23.16 （冒泡排序动画）编写一个程序，实现冒泡排序算法的动画。创建一个数组，以随机顺序包含从 1 到 20 的 20 个不同数字。数组元素在一个直方图中显示，如图 23-20b 所示。单击 Step 按钮使程序执行算法中的一次比较，然后为新的数组重画直方图。将表示考虑交换的数字的条块标上颜色。当算法结束时，显示一条信息通知用户。单击 Reset 按钮为一次新的开始创建一个新的随机数组。

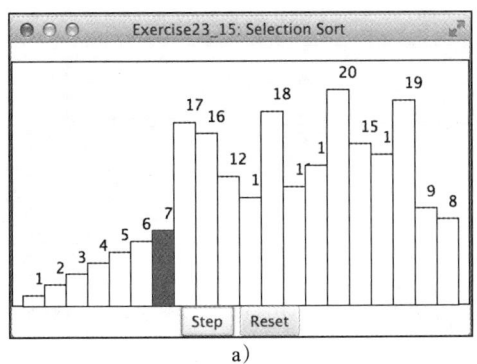
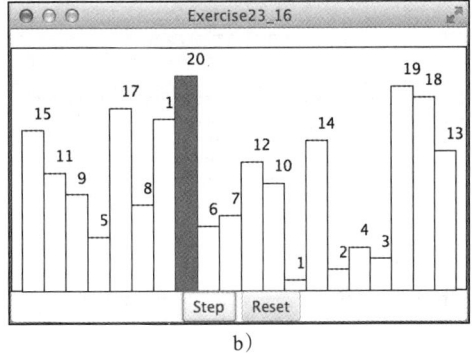

图 23-20 a）该程序实现选择排序的动画（来源：Oracle 或其附属公司版权所有©1995～2016，经授权使用）；b）该程序实现冒泡排序的动画

*23.17 （基数排序动画）编写一个程序，实现基数排序算法的动画。创建一个数组，以随机顺序包含从 1 到 1000 的 20 个不同数字。数组元素在一个直方图中显示，如图 23-21 所示。单击 Step 按钮使程序放置一个数字在一个桶中。刚放入的数字以红色显示。一旦所有的数字都放在桶中后，单击 Step 按钮从桶中收集所有的数字，将它们移回到数组中。当算法结束时，单击 Step 按钮显示一条信息通知用户。单击 Reset 按钮为一次新的开始创建一个新的随机数组。

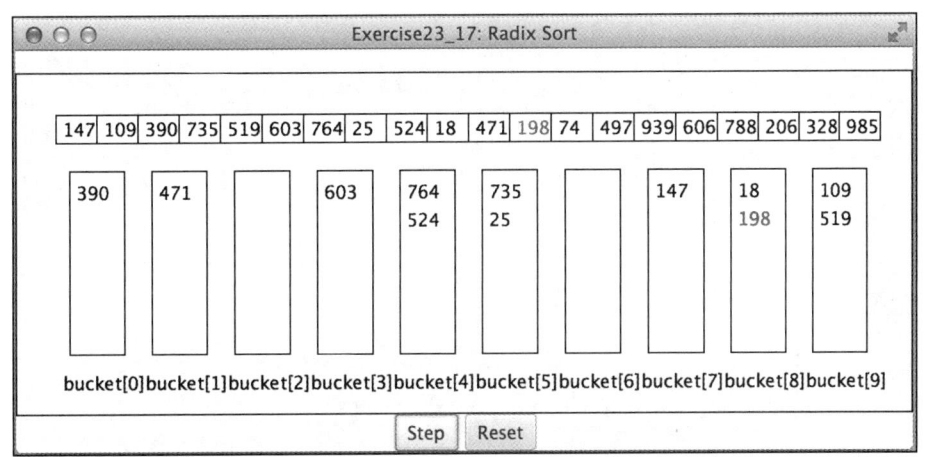

图 23-21 该程序实现基数排序的动画（来源：Oracle 或其附属公司版权所有©1995～2016，经授权使用）

*23.18 （归并排序动画）编写一个程序，实现归并两个排好序的线性表的动画。创建两个数组 list1 和 list2，每个包含从 1 到 999 的 8 个随机数字。数组元素如图 23-22a 所示。单击 Step 按钮使程序将 list1 或者 list2 中的一个元素移到 temp 中。单击 Reset 按钮为一个新的开始创建两个新的随机数组。当算法结束时，单击 Step 按钮显示一条信息通知用户。

*23.19 （快速排序分区动画）编写一个程序，实现快速排序的分区动画。程序创建一个包含从 1 到 999 的 20 个随机数字的线性表。线性表如图 23-22b 所示。单击 Step 按钮使程序将 low 移动到右边，或者 high 移动到左边，或者交换 low 和 high 位置的元素。单击 Reset 按钮为一个新的开始创建两个新的随机数组。当算法结束时，单击 Step 按钮显示一条信息通知用户。

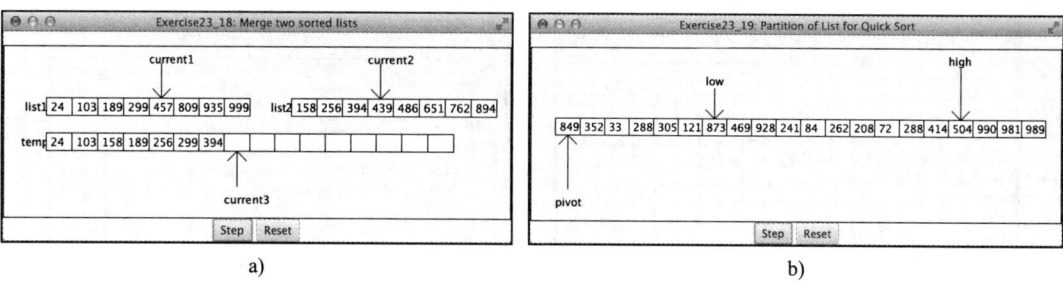

图 23-22　a）该程序实现归并两个排好序的线性表的动画（来源：Oracle 或其附属公司版权所有 ©1995 ～ 2016，经授权使用）；b）该程序实现快速排序的分区动画

第 24 章

实现线性表、栈、队列和优先队列

教学目标

- 在接口中设计线性表的通用操作，并且将该接口作为 Collection 的子类型（24.2 节）。
- 使用数组设计并实现数组线性表（24.3 节）。
- 使用链式结构设计并实现链表（24.4 节）。
- 使用数组线性表设计并实现一个栈类，使用链表设计并实现一个队列类（24.5 节）。
- 使用堆设计并实现优先队列（24.6 节）。

24.1 引言

要点提示：本章专注于实现数据结构。

线性表、栈、队列和优先队列都是典型的数据结构，在数据结构课程中会涵盖这些内容。Java API 中对它们提供了支持，关于它们的使用方法已在第 20 章中给出。本章将剖析这些数据结构是如何实现的。规则集和映射的实现将在第 27 章中讲述。通过这些例子，你将深刻认识数据结构，并学会设计和实现自定义的数据结构。

24.2 线性表的通用操作

要点提示：线性表的通用操作在 List 接口中定义。

线性表是一个顺序存储数据的常见数据结构——例如，学生线性表、可用房间线性表、城市线性表以及书籍线性表。可以在线性表上执行下面的操作：

- 从线性表中获取一个元素。
- 向线性表中插入一个新元素。
- 从线性表中删除一个元素。
- 得到线性表中元素的个数。
- 确定线性表中是否包含某个元素。
- 确定线性表是否为空。

实现线性表的方式有两种。一种是使用数组（array）存储线性表的元素。数组大小是固定的。如果元素个数超过了数组的容量，就需要创建一个新的更大数组，并将当前数组中的元素复制到新数组中。另一种是使用链式结构（linked structure）。链式结构由结点组成，每个结点都是动态创建的，用来存储一个元素。所有的结点链接成一个线性表。这样，就可以给线性表定义两个类。为了方便起见，分别称这两个类为 MyArrayList 和 MyLinkedList。这两个类具有相同的操作，但是具有不同的实现。

要查看数组线性表和链表如何工作的交互式演示，参见链接 liveexample.pearsoncmg.com/dsanimation/ArrayListeBook.html 和 liveexample.pearsoncmg.com/dsanimation/LinkedListeBook.html，如图 24-1 所示。

设计指南：在 Java 8 之前，一种常见的 Java 数据结构设计策略是定义通用的操作接口并提供便利抽象类。因此，具体的类可以简单地扩展便利抽象类，而无须实现完整的接口。Java 8 允许定义默认方法。你可以为接口中的某些方法提供默认实现，而不是放在便利抽象类中。使用默认方法便不需要使用便利抽象类了。

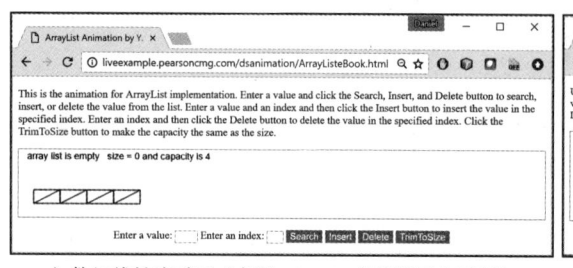
a) 数组线性表动画（来源：Oracle 或其附属公司版权所有 © 1995~2016，经授权使用）

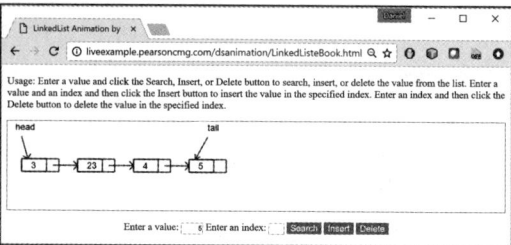
b) 链表动画

图 24-1 动画工具有助于了解数组线性表和链表是如何工作的

我们把这样的接口命名为 MyList，并将它定义为 Collection 的子类型，这样 Collection 接口中的通用操作同样可以在 MyList 中使用。图 24-2 展示了 Collection、MyList、MyArrayList 以及 MyLinkedList 之间的关系。图 24-3 列出了 MyList 中实现的方法。程序清单 24-1 给出了 MyList 的源代码。

图 24-2 MyList 定义了 MyArrayList 和 MyLinkedList 的通用接口

程序清单 24-1 MyList.java

```
1  import java.util.Collection;
2
3  public interface MyList<E> extends Collection<E> {
4    /** Add a new element at the specified index in this list */
5    public void add(int index, E e);
6
7    /** Return the element from this list at the specified index */
8    public E get(int index);
9
10   /** Return the index of the first matching element in this list.
11    *  Return -1 if no match. */
12   public int indexOf(Object e);
13
14   /** Return the index of the last matching element in this list
15    *  Return -1 if no match. */
16   public int lastIndexOf(E e);
17
18   /** Remove the element at the specified position in this list
19    *  Shift any subsequent elements to the left.
20    *  Return the element that was removed from the list. */
21   public E remove(int index);
22
```

```java
23    /** Replace the element at the specified position in this list
24     *  with the specified element and returns the new set. */
25    public E set(int index, E e);
26
27    @Override /** Add a new element at the end of this list */
28    public default boolean add(E e) {
29      add(size(), e);
30      return true;
31    }
32
33    @Override /** Return true if this list contains no elements */
34    public default boolean isEmpty() {
35      return size() == 0;
36    }
37
38    @Override /** Remove the first occurrence of the element e
39     * from this list. Shift any subsequent elements to the left.
40     * Return true if the element is removed. */
41    public default boolean remove(Object e) {
42      if (indexOf(e) >= 0) {
43        remove(indexOf(e));
44        return true;
45      }
46      else
47        return false;
48    }
49
50    @Override
51    public default boolean containsAll(Collection<?> c) {
52      // Left as an exercise
53      return true;
54    }
55
56    @Override
57    public default boolean addAll(Collection<? extends E> c) {
58      // Left as an exercise
59      return true;
60    }
61
62    @Override
63    public default boolean removeAll(Collection<?> c) {
64      // Left as an exercise
65      return true;
66    }
67
68    @Override
69    public default boolean retainAll(Collection<?> c) {
70      // Left as an exercise
71      return true;
72    }
73
74    @Override
75    public default Object[] toArray() {
76      // Left as an exercise
77      return null;
78    }
79
80    @Override
81    public default <T> T[] toArray(T[] array) {
82      // Left as an exercise
83      return null;
84    }
85  }
```

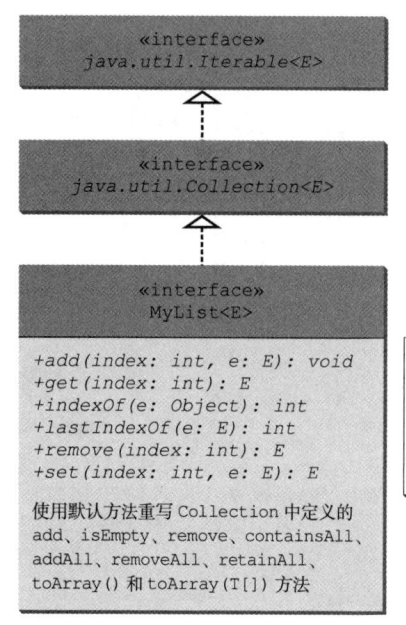

图 24-3 MyList 定义了操作线性表的方法，并且部分地实现了 Collection 接口中定义的一些方法

isEmpty()、add(E)、remove(E)、containsAll、addAll、removeAll、retainAll、toArray() 和 toArray(T[]) 方法都在 Collection 接口中定义。由于这些方法在 MyList 中实现，它们在 MyList 接口中被重写为默认方法。isEmpty()、add(E) 和 remove(E) 已经提供了实现，其余默认方法的实现留作编程练习题 24.1。

下面的两节分别给出 MyArrayList 和 MyLinkedList 的实现。

✓ 复习题

24.2.1 假设 list 是 MyList 的一个实例，可以使用 list.iterator() 得到 list 的一个迭代器吗？
24.2.2 可以使用 new MyList() 创建一个线性表吗？
24.2.3 Collection 中的什么方法在 MyList 中作为默认方法被重写？
24.2.4 将 Collection 中的方法在 MyList 中作为默认方法重写有什么好处？

24.3 数组线性表

O╾ 要点提示：数组线性表采用数组来实现。

数组是一种大小固定的数据结构。一旦数组被创建，它的大小就无法改变。尽管如此，仍然可以使用数组来实现动态的数据结构。处理的方法是，当数组不能再存储线性表中的新元素时，创建一个更大的新数组来替换当前数组。

初始时，用默认大小创建一个类型为 E[] 的数组 data。向数组中插入一个新元素时，首先确认数组是否有足够的空间。若数组的空间不够，则创建大小为当前数组两倍的新数组，然后将当前数组中的元素复制到新的数组中。现在，新数组就变成了当前数组。在指定下标处插入一个新元素之前，必须将指定下标后面的所有元素都向右移动一个位置并且将该线性表的大小增加 1，如图 24-4 所示。

O╾ 注意：该数据数组的类型是 E[]，所以数组中每个元素实际存储的是对象的引用。

删除指定下标处的一个元素时，应该将该下标后面的元素都向左移动一个位置，并将线性表的大小减 1，如图 24-5 所示。

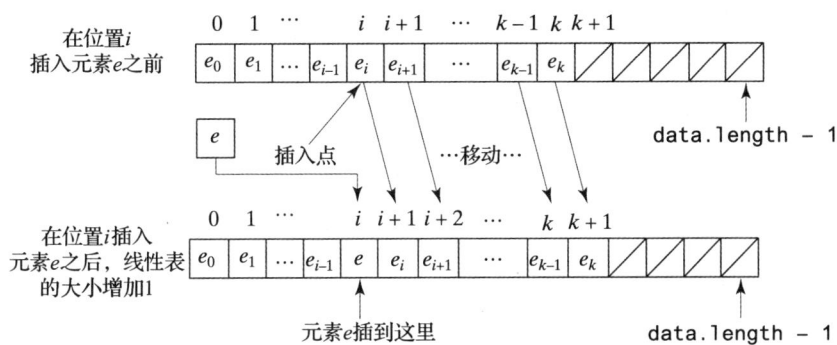

图 24-4 向数组中插入一个新元素,要求插入点之后的所有元素都向右移动一个位置,以便在插入点插入新元素

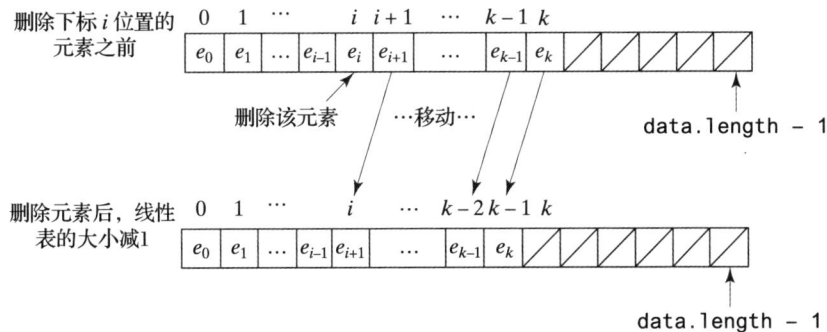

图 24-5 从数组中删除一个元素,需要删除点之后的元素都向左移动一个位置

MyArrayList 使用数组来实现 MyList,如图 24-6 所示。它的实现在程序清单 24-2 中给出。

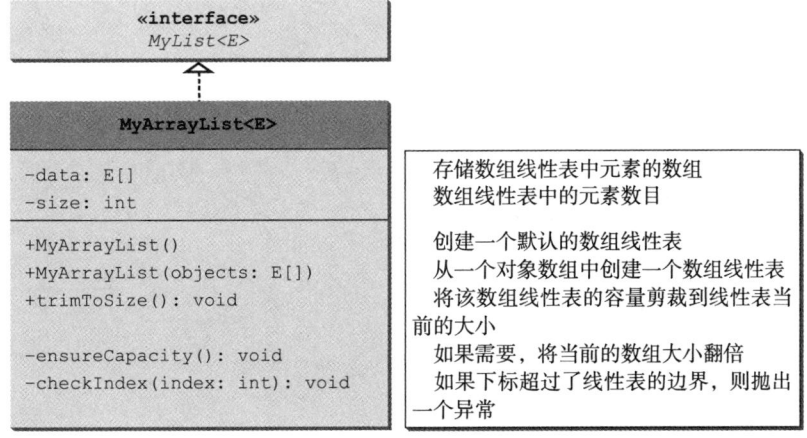

图 24-6 MyArrayList 使用数组实现线性表

程序清单 24-2 MyArrayList.java

```
1  public class MyArrayList<E> implements MyList<E> {
2    public static final int INITIAL_CAPACITY = 16;
3    private E[] data = (E[])new Object[INITIAL_CAPACITY];
4    private int size = 0; // Number of elements in the list
5
6    /** Create an empty list */
```

```java
 7    public MyArrayList() {
 8    }
 9
10    /** Create a list from an array of objects */
11    public MyArrayList(E[] objects) {
12      for (int i = 0; i < objects.length; i++)
13        add(objects[i]); // Warning: don't use super(objects)!
14    }
15
16    @Override /** Add a new element at the specified index */
17    public void add(int index, E e) {
18      // Ensure the index is in the right range
19      if (index < 0 || index > size)
20        throw new IndexOutOfBoundsException
21          ("Index: " + index + ", Size: " + size);
22
23      ensureCapacity();
24
25      // Move the elements to the right after the specified index
26      for (int i = size - 1; i >= index; i--)
27        data[i + 1] = data[i];
28
29      // Insert new element to data[index]
30      data[index] = e;
31
32      // Increase size by 1
33      size++;
34    }
35
36    /** Create a new larger array, double the current size + 1 */
37    private void ensureCapacity() {
38      if (size >= data.length) {
39        E[] newData = (E[])(new Object[size * 2 + 1]);
40        System.arraycopy(data, 0, newData, 0, size);
41        data = newData;
42      }
43    }
44
45    @Override /** Clear the list */
46    public void clear() {
47      data = (E[])new Object[INITIAL_CAPACITY];
48      size = 0;
49    }
50
51    @Override /** Return true if this list contains the element */
52    public boolean contains(Object e) {
53      for (int i = 0; i < size; i++)
54        if (e.equals(data[i])) return true;
55
56      return false;
57    }
58
59    @Override /** Return the element at the specified index */
60    public E get(int index) {
61      checkIndex(index);
62      return data[index];
63    }
64
65    private void checkIndex(int index) {
66      if (index < 0 || index >= size)
67        throw new IndexOutOfBoundsException
68          ("Index: " + index + ", Size: " + size);
69    }
70
71    @Override /** Return the index of the first matching element
72     *  in this list. Return -1 if no match. */
```

```java
73    public int indexOf(Object e) {
74      for (int i = 0; i < size; i++)
75        if (e.equals(data[i])) return i;
76
77      return -1;
78    }
79
80    @Override /** Return the index of the last matching element
81     * in this list. Return -1 if no match. */
82    public int lastIndexOf(E e) {
83      for (int i = size - 1; i >= 0; i--)
84        if (e.equals(data[i])) return i;
85
86      return -1;
87    }
88
89    @Override /** Remove the element at the specified position
90     * in this list. Shift any subsequent elements to the left.
91     * Return the element that was removed from the list. */
92    public E remove(int index) {
93      checkIndex(index);
94
95      E e = data[index];
96
97      // Shift data to the left
98      for (int j = index; j < size - 1; j++)
99        data[j] = data[j + 1];
100
101     data[size - 1] = null; // This element is now null
102
103     // Decrement size
104     size--;
105
106     return e;
107   }
108
109   @Override /** Replace the element at the specified position
110    * in this list with the specified element. */
111   public E set(int index, E e) {
112     checkIndex(index);
113     E old = data[index];
114     data[index] = e;
115     return old;
116   }
117
118   @Override
119   public String toString() {
120     StringBuilder result = new StringBuilder("[");
121
122     for (int i = 0; i < size; i++) {
123       result.append(data[i]);
124       if (i < size - 1) result.append(", ");
125     }
126
127     return result.toString() + "]";
128   }
129
130   /** Trims the capacity to current size */
131   public void trimToSize() {
132     if (size != data.length) {
133       E[] newData = (E[])(new Object[size]);
134       System.arraycopy(data, 0, newData, 0, size);
135       data = newData;
136     } // If size == capacity, no need to trim
137   }
138
```

```java
139    @Override /** Override iterator() defined in Iterable */
140    public java.util.Iterator<E> iterator() {
141      return new ArrayListIterator();
142    }
143
144    private class ArrayListIterator
145        implements java.util.Iterator<E> {
146      private int current = 0; // Current index
147
148      @Override
149      public boolean hasNext() {
150        return current < size;
151      }
152
153      @Override
154      public E next() {
155        return data[current++];
156      }
157
158      @Override // Remove the element returned by the last next()
159      public void remove() {
160        if (current == 0)  // next() has not been called yet
161          throw new IllegalStateException();
162        MyArrayList.this.remove(--current);
163      }
164    }
165
166    @Override /** Return the number of elements in this list */
167    public int size() {
168      return size;
169    }
170  }
```

常量 INITIAL_CAPACITY（第 2 行）用于创建一个初始数组 data（第 3 行）。由于泛型类型擦除（参见 19.8 节的限制 2），所以不能使用语法 new e[INITIAL_CAPACITY] 创建泛型数组。为了规避这个限制，第 3 行创建了一个 Object 类型的数组，并将它转换为 E[] 类型。数据域 size 跟踪线性表中元素的个数（第 4 行）。

add(int index,E e) 方法（第 17～34 行）将元素 e 插入数组的指定下标 index 处。该方法首先调用 ensureCapacity() 方法（第 23 行），以确保数组中还有存储新元素的空间。在插入新元素之前，将指定下标后面的所有元素都向右移动一个位置（第 26～27 行）。添加新元素之后，将 size 加 1（第 33 行）。

ensureCapacity() 方法（第 37～43 行）用来检验数组是否已满。如果数组已满，则创建一个容量为当前数组大小两倍 +1 的新数组，并使用 System.arraycopy 方法将当前数组的所有元素复制到新数组中，再把新数组设为当前数组。注意在调用 trimToSize() 方法之后，当前大小可能为 0，而 new Object[2*size+1]（第 39 行）确保了新的大小不为 0。

clear() 方法（第 46～49 行）创建一个大小为 INITIAL_CAPACITY 的新数组，并设置变量 size 为 0。如果删除第 47 行，类仍然可以工作，但是将会产生内存泄漏，因为尽管元素已经不再被需要，但是它们依然在数组中。通过创建一个新数组并且将其赋值给 data，老的数组和保存在老数组中的元素变成了垃圾，将被 JVM 自动回收。

contains(Object e) 方法（第 52～57 行）使用 equals 方法将元素 e 与数组中的所有元素逐一比较，以判断数组中是否包含元素 e。

get(int index) 方法（第 60～63 行）检查 index 是否在范围内，如果 index 在范围内，则返回 data[index]。

checkIndex(int index) 方法（第 65 ～ 69 行）检查 index 是否在范围内，如果不在，则方法抛出一个 IndexOutOfBoundsException（第 67 行）。

indexOf(Object e) 方法（第 73 ～ 78 行）从第一个元素开始，将元素 e 与数组中的每一个元素逐一比较。如果匹配，则返回匹配元素的下标；否则，返回 -1。

lastIndexOf(Object e) 方法（第 82 ～ 87 行）从最后一个元素开始，将元素 e 与数组中的每一个元素逐一比较。如果匹配，则返回匹配元素的下标；否则，返回 -1。

remove(int index) 方法（第 92 ～ 107 行）将指定下标之后的所有元素向左移动一个位置（第 98 ～ 99 行），并将数组大小 size 减 1（第 104 行）。最后一个元素不再使用，设置为 null（第 101 行）。

set(int index,E e) 方法（第 111 ～ 116 行）只是简单地将 e 赋给 data[index]，将数组中指定下标处的元素用 e 替换。

toString() 方法（第 119 ～ 128 行）重写 Object 类中的 toString 方法，返回一个表示线性表中所有元素的字符串。

trimToSize() 方法（第 131 ～ 137 行）创建一个新数组，它的大小与当前数组线性表的大小匹配（第 133 行），使用 System.arraycopy 方法将当前数组复制到新的数组中（第 134 行），然后将新数组设置为当前数组（第 135 行）。注意，如果 size == capacity，则无须裁剪数组的大小。

java.lang.Iterable 接口中定义的 iterator() 方法被实现为返回一个 java.util.Iterator 的实例（第 140 ～ 142 行）。ArrayListIterator 类通过 hasNext、next 以及 remove 的具体方法实现了 Iterator（第 144 ～ 164 行）。它使用 current 来标识被遍历的元素的当前位置（第 146 行）。

size() 方法简单地返回数组线性表的元素数目（第 167 ～ 169 行）。

程序清单 24-3 给出一个使用 MyArrayList 创建线性表的例子。它使用 add 方法来给线性表添加字符串，并使用 remove 方法来删除字符串。由于 MyArrayList 实现了 Iterable，元素可以使用一个 foreach 循环来进行遍历（第 35 ～ 36 行）。

程序清单 24-3 TestMyArrayList.java

```
1   public class TestMyArrayList {
2     public static void main(String[] args) {
3       // Create a list
4       MyList<String> list = new MyArrayList<>();
5
6       // Add elements to the list
7       list.add("America"); // Add it to the list
8       System.out.println("(1) " + list);
9
10      list.add(0, "Canada"); // Add it to the beginning of the list
11      System.out.println("(2) " + list);
12
13      list.add("Russia"); // Add it to the end of the list
14      System.out.println("(3) " + list);
15
16      list.add("France"); // Add it to the end of the list
17      System.out.println("(4) " + list);
18
19      list.add(2, "Germany"); // Add it to the list at index 2
20      System.out.println("(5) " + list);
21
22      list.add(5, "Norway"); // Add it to the list at index 5
```

```
23         System.out.println("(6) " + list);
24
25         // Remove elements from the list
26         list.remove("Canada"); // Same as list.remove(0) in this case
27         System.out.println("(7) " + list);
28
29         list.remove(2); // Remove the element at index 2
30         System.out.println("(8) " + list);
31
32         list.remove(list.size() - 1); // Remove the last element
33         System.out.print("(9) " + list + "\n(10) ");
34
35         for (String s: list)
36             System.out.print(s.toUpperCase() + " ");
37     }
38 }
```

```
(1)  [America]
(2)  [Canada, America]
(3)  [Canada, America, Russia]
(4)  [Canada, America, Russia, France]
(5)  [Canada, America, Germany, Russia, France]
(6)  [Canada, America, Germany, Russia, France, Norway]
(7)  [America, Germany, Russia, France, Norway]
(8)  [America, Germany, France, Norway]
(9)  [America, Germany, France]
(10) AMERICA GERMANY FRANCE
```

✔ 复习题

24.3.1 数组数据类型的局限性是什么？

24.3.2 MyArrayList 是使用数组来实现的，而数组是一种大小固定的数据结构。那么为什么认为 MyArrayList 是动态的数据结构呢？

24.3.3 执行下面的语句后，给出 MyArrayList 中数组的长度。

```
1  MyArrayList<Double> list = new MyArrayList<>();
2  list.add(1.5);
3  list.trimToSize();
4  list.add(3.4);
5  list.add(7.4);
6  list.add(17.4);
```

24.3.4 如果程序清单 24-2 中的第 11～12 行

```
for (int i = 0; i < objects.length; i++)
  add(objects[i]);
```

被下面的语句

```
data = objects;
size = objects.length;
```

代替，会出现什么错误？

24.3.5 如果将程序清单 24-2 中第 33 行的代码从

```
E[] newData = (E[])(new Object[size * 2 + 1]);
```

改为

```
E[] newData = (E[])(new Object[size * 2]);
```

程序就是错误的。你能找出原因吗？

24.3.6 如果第 41 行的以下代码被删除，MyArrayList 类会有内存泄漏吗？

```
      data = (E[])new Object[INITIAL_CAPACITY];
```

24.3.7 如果下标越界，get(index) 方法调用 checkIndex(index) 方法（程序清单 24-2 的第 59～63 行）会抛出 IndexOutOfBoundsException。假设 add(index,e) 如下实现：

```
public void add(int index, E e) {
  checkIndex(index);

   // Same as lines 23-33 in Listing 24.2 MyArrayList.java
}
```

那么运行下面的代码会发生什么情况？

```
MyArrayList<String> list = new MyArrayList<>();
list.add("New York");
```

24.4 链表

☞ **要点提示**：链表采用链式结构实现。

由于 MyArrayList 是用数组实现的，所以 get(int index) 和 set(int index, E e) 方法可以通过下标访问和修改元素，也可以用 add(E e) 方法在线性表末尾添加元素，它们是高效的。但是，add(int index, E e) 和 remove(int index) 方法的效率很低，因为这两个方法实际上需要移动大量元素。为提高在表的开始位置添加和删除元素的效率，可以采用链式结构来实现线性表。

24.4.1 结点

链表中的每个元素都包含一个称为结点（node）的结构。当向链表中加入一个新的元素时，就会产生一个结点包含它。每个结点都和它的相邻结点相链接，如图 24-7 所示。

结点可以按如下方式定义为一个类：

```
class Node<E> {
  E element;
  Node<E> next;

  public Node(E e) {
    element = e;
  }
}
```

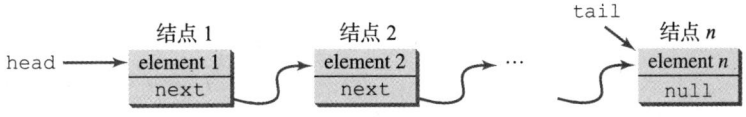

图 24-7 链表由链接在一起的任意多个结点构成

我们使用变量 head 指向链表的第一个结点，而变量 tail 指向最后一个结点。如果链表为空，head 和 tail 这两个变量均为 null。下面就是一个创建存储三个结点的链表的例子，其中每个结点存储一个字符串元素。

步骤 1：声明 head 和 tail。

```
Node<String> head = null;    线性表现在为空
Node<String> tail = null;
```

head 和 tail 都为 null，该线性表为空。

步骤 2：创建第一个结点并将它追加到线性表中，如图 24-8 所示。在将第一个结点插入线性表之后，head 和 tail 都指向这个结点。

```
head = new Node<>("Chicago");      第一个结点插入后
tail = head;
```

图 24-8　向线性表追加第一个结点

步骤 3：创建第二个结点并将它追加到线性表中，如图 24-9a 所示。为了将第二个结点追加到线性表中，需要将新结点和第一个结点链接起来，现在，新结点就是尾结点。所以，应该移动 tail，使它指向该新结点，如图 24-9b 所示。

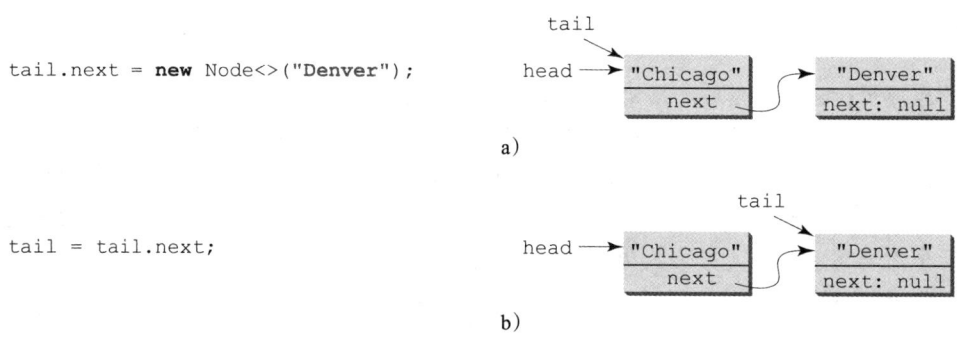

```
tail.next = new Node<>("Denver");
```

a)

```
tail = tail.next;
```

b)

图 24-9　向线性表追加第二个结点

步骤 4：创建第三个结点并将它追加到线性表中，如图 24-10a 所示。为了向线性表追加新的结点，链接新结点和线性表中的最后一个结点。现在，新结点就是尾结点。所以，应该移动 tail，使它指向该新结点，如图 24-10b 所示。

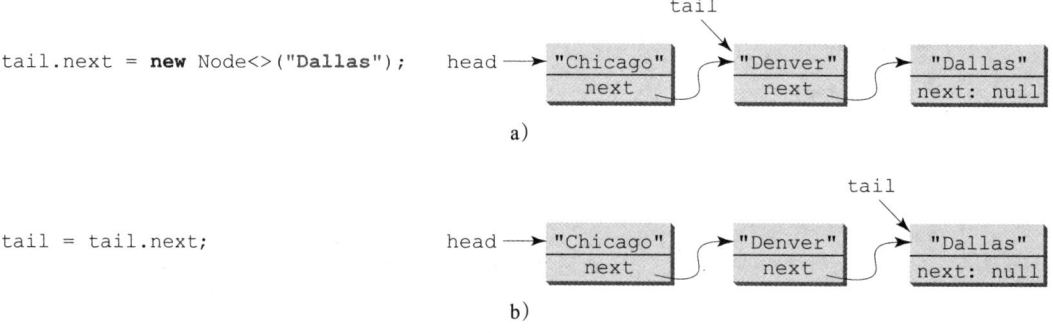

```
tail.next = new Node<>("Dallas");
```

a)

```
tail = tail.next;
```

b)

图 24-10　向线性表追加第三个结点

每个结点都包含元素和一个名为 next 的数据域，next 指向下一个元素。如果结点是线性表中的最后一个，那么它的指针数据域 next 所包含的值是 null。可以使用这个特性来检测某结点是否是最后的结点。例如，可以编写下面的循环来遍历线性表中的所有结点：

```
1  Node<E> current = head;
2  while (current != null) {
3    System.out.println(current.element);
4    current = current.next;
5  }
```

初始状态时,变量 current 指向线性表的第一个结点(第1行)。在循环中,获取当前结点的元素(第3行),然后 current 指向下一个结点(第4行)。循环持续到当前结点为 null 时为止。

24.4.2 MyLinkedList 类

MyLinkedList 类使用链式结构实现动态线性表,它实现了 MyList。此外,它还提供 addFirst、addLast、removeFirst、removeLast、getFirst 和 getLast 等方法,如图 24-11 所示。

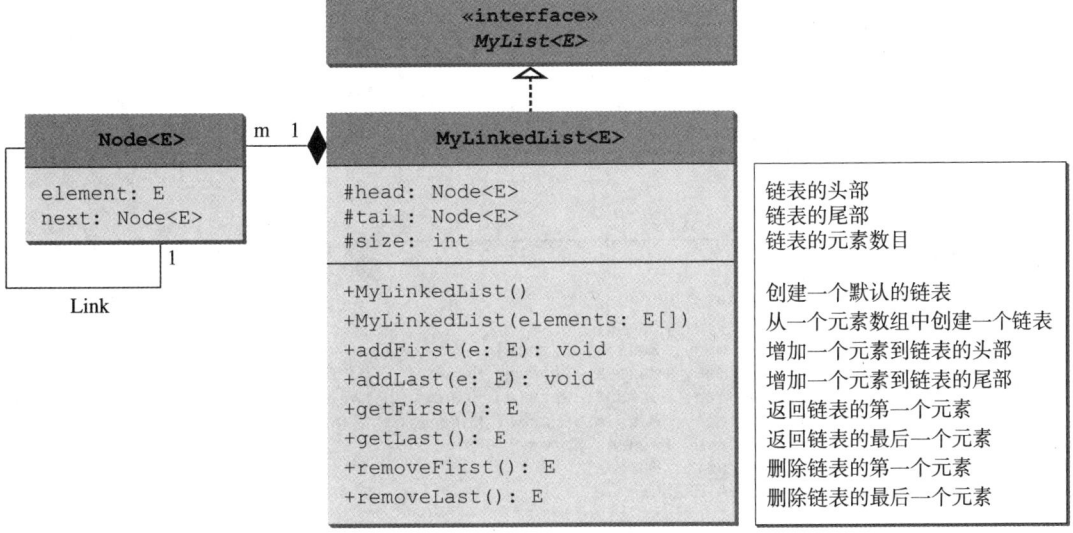

图 24-11 MyLinkedList 使用链接在一起的结点实现线性表

假设已经实现了这个类,程序清单 24-4 给出使用该类的测试程序。

程序清单 24-4 TestMyLinkedList.java

```
1  public class TestMyLinkedList {
2    /** Main method */
3    public static void main(String[] args) {
4      // Create a list for strings
5      MyLinkedList<String> list = new MyLinkedList<>();
6
7      // Add elements to the list
8      list.add("America"); // Add it to the list
9      System.out.println("(1) " + list);
10
11     list.add(0, "Canada"); // Add it to the beginning of the list
12     System.out.println("(2) " + list);
13
14     list.add("Russia"); // Add it to the end of the list
15     System.out.println("(3) " + list);
16
17     list.addLast("France"); // Add it to the end of the list
18     System.out.println("(4) " + list);
19
20     list.add(2, "Germany"); // Add it to the list at index 2
21     System.out.println("(5) " + list);
22
23     list.add(5, "Norway"); // Add it to the list at index 5
24     System.out.println("(6) " + list);
25
```

```
26    list.add(0, "Poland"); // Same as list.addFirst("Poland")
27    System.out.println("(7) " + list);
28
29    // Remove elements from the list
30    list.remove(0); // Same as list.remove("Poland") in this case
31    System.out.println("(8) " + list);
32
33    list.remove(2); // Remove the element at index 2
34    System.out.println("(9) " + list);
35
36    list.remove(list.size() - 1); // Remove the last element
37    System.out.print("(10) " + list + "\n(11) ");
38
39    for (String s: list)
40      System.out.print(s.toUpperCase() + " ");
41
42    list.clear();
43    System.out.println("\nAfter clearing the list, the list size is "
44      + list.size());
45  }
46 }
```

```
(1) [America]
(2) [Canada, America]
(3) [Canada, America, Russia]
(4) [Canada, America, Russia, France]
(5) [Canada, America, Germany, Russia, France]
(6) [Canada, America, Germany, Russia, France, Norway]
(7) [Poland, Canada, America, Germany, Russia, France, Norway]
(8) [Canada, America, Germany, Russia, France, Norway]
(9) [Canada, America, Russia, France, Norway]
(10) [Canada, America, Russia, France]
(11) CANADA AMERICA RUSSIA FRANCE
After clearing the list, the list size is 0
```

24.4.3 实现 MyLinkedList

现在，我们将注意力转移到 MyLinkedList 类的实现上。下面将讨论如何实现 addFirst、addLast、add(index,e)、removeFirst、removeLast 和 remove(index) 方法，而将 MyLinkedList 类中的其他方法留作练习题。

实现 addFirst(e) 方法

addFirst(e) 方法创建一个包含元素 e 的新结点。该新结点成为链表的第一个结点。该方法可以如下实现：

```
1  public void addFirst(E e) {
2    Node<E> newNode = new Node<>(e); // Create a new node
3    newNode.next = head; // link the new node with the head
4    head = newNode; // head points to the new node
5    size++; // Increase list size
6
7    if (tail == null) // The new node is the only node in list
8      tail = head;
9  }
```

addFirst(e) 方法创建一个新结点来存储元素（第 2 行），并将该结点插入链表的起始位置（第 3 行），如图 24-12a 所示。插入后，head 应该指向该新元素结点（第 4 行），如图 24-12b 所示。

如果链表为空（第 7 行），那么 head 和 tail 都将指向该新结点（第 8 行）。在创建完该结点之后，size 应该加 1（第 5 行）。

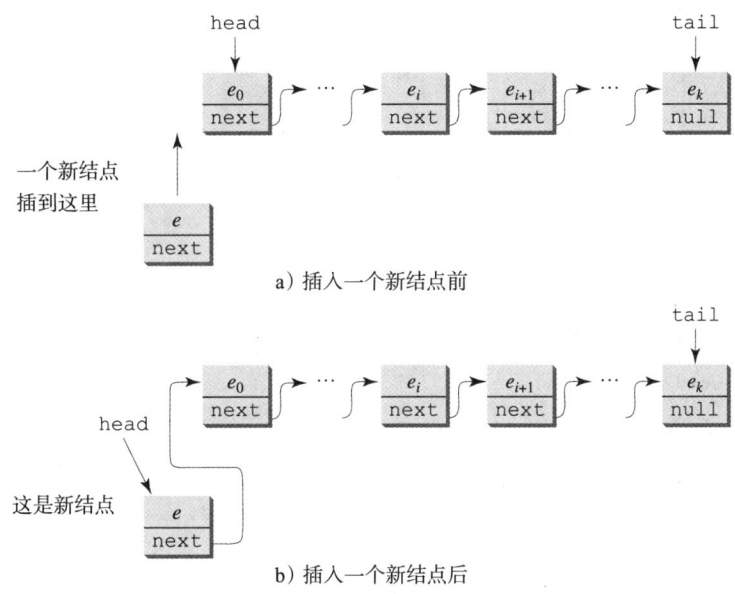

a) 插入一个新结点前

b) 插入一个新结点后

图 24-12 将一个新元素添加到链表的起始位置

实现 addLast(e) 方法

addLast(e) 方法创建一个包含元素的新结点,并将它追加到链表的末尾。该方法可以如下实现:

```
1  public void addLast(E e) {
2    Node<E> newNode = new Node<>(e); // Create a new node for e
3
4    if (tail == null) {
5      head = tail = newNode; // The only node in list
6    }
7    else {
8      tail.next = newNode; // Link the new node with the last node
9      tail = newNode; // tail now points to the last node
10   }
11
12   size++; // Increase size
13 }
```

addLast(e) 方法创建一个新结点来存储元素(第 2 行),并且将它追加到链表的末尾。考虑以下两种情况:

1) 如果链表为空(第 4 行),那么 head 和 tail 都将指向该新结点(第 5 行)。

2) 否则,将该结点和该链表的最后一个结点相链接(第 8 行)。现在,tail 应该指向该新结点(第 9 行)。图 24-13a 和图 23-13b 展示了插入前后包含元素 e 的新结点。

不论哪种情况,在创建一个结点后,size 都增加 1(第 12 行)。

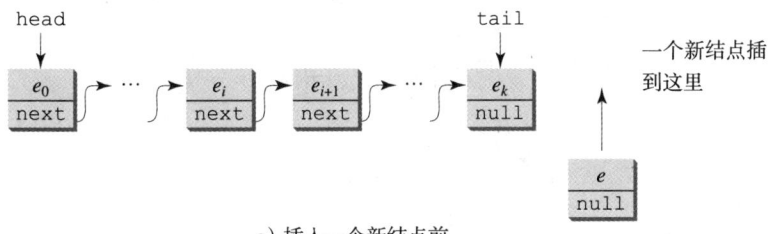

a) 插入一个新结点前

图 24-13 将一个新元素添加到链表的末尾

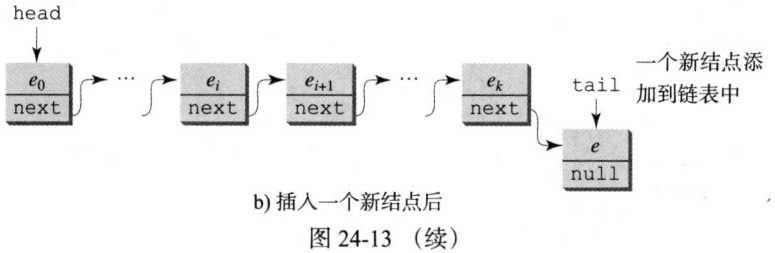

b) 插入一个新结点后

图 24-13 （续）

实现 add(index,e) 方法

add(index,e) 方法将一个元素插入到链表的指定下标处。该方法可以如下实现：

```
1  public void add(int index, E e) {
2    if (index == 0) addFirst(e); // Insert first
3    else if (index >= size) addLast(e); // Insert last
4    else { // Insert in the middle
5      Node<E> current = head;
6      for (int i = 1; i < index; i++)
7        current = current.next;
8      Node<E> temp = current.next;
9      current.next = new Node<>(e);
10     (current.next).next = temp;
11     size++;
12   }
13 }
```

将一个元素插入链表中时，会出现以下三种情况：

1）当指定下标 index 为 0 时，调用 addFirst(e) 方法（第 2 行）将该元素插入到链表的起始位置。

2）当 index 大于或等于链表的大小 size 时，调用 addLast(e) 方法（第 3 行）将元素 e 添加到链表的末尾。

3）否则，创建一个新结点来存储新元素，并确定它的插入位置。新的结点应该插到结点 current 和 temp 之间，如图 24-14a 所示。该方法将新结点赋给 current.next，并将 temp 赋给新结点的 next，如图 24-14b 所示。链表的大小现在加 1（第 11 行）。

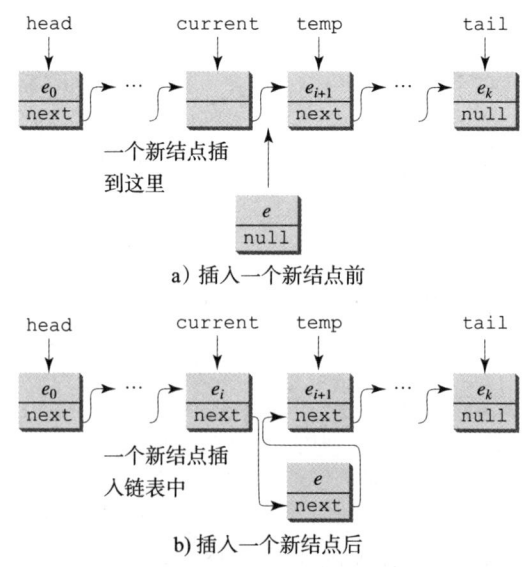

a) 插入一个新结点前

b) 插入一个新结点后

图 24-14 将新元素插到链表的中间位置

实现 removeFirst() 方法

removeFirst() 方法从链表中删除第一个元素。该方法可以如下实现：

```
1  public E removeFirst() {
2    if (size == 0) return null; // Nothing to delete
3    else {
4      Node<E> temp = head; // Keep the first node temporarily
5      head = head.next; // Move head to point to next node
6      size--; // Reduce size by 1
7      if (head == null) tail = null; // List becomes empty
8      return temp.element; // Return the deleted element
9    }
10 }
```

考虑以下两种情况：

1）如果该链表为空，它就没有什么可删除的，因此返回 null（第 2 行）。

2）否则，通过将 head 指向第二个结点以从链表中删除第一个结点。图 24-15a 和图 24-15b 展示了删除之前和删除之后的链表。在删除之后，链表的大小减 1（第 6 行）。如果链表为空，那么删除该元素之后，tail 应该设置为 null（第 7 行）。

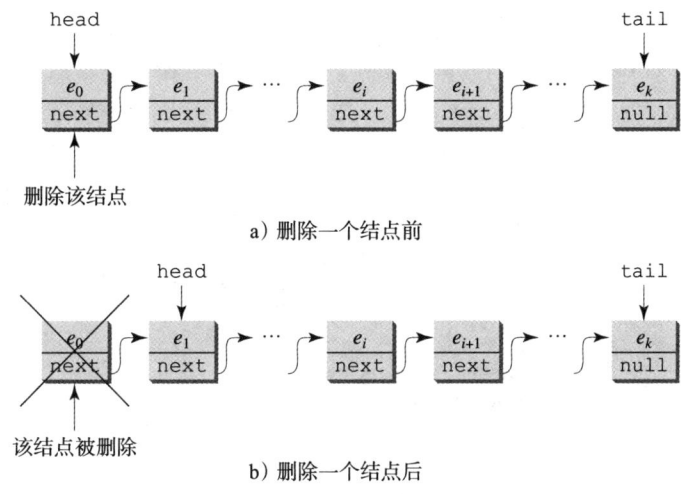

图 24-15 从链表中删除第一个结点

实现 removeLast() 方法

removeLast() 方法从链表中删除最后一个元素，如图 24-16 所示。该方法可以如下实现：

```
1  public E removeLast() {
2    if (size == 0 || size == 1) {
3      return removeFirst();
4    }
5    else {
6      Node<E> current = head;
7      for (int i = 0; i < size - 2; i++) {
8        current = current.next;
9      }
10
11     E temp = tail.element;
12     tail = current;
13     tail.next = null;
14     size--;
15     return temp;
16   }
17 }
```

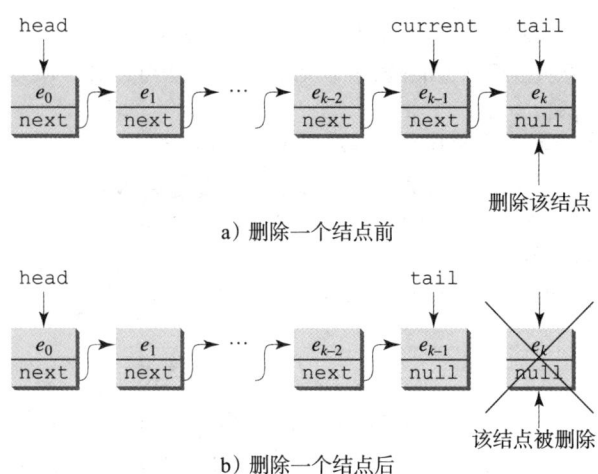

a) 删除一个结点前

b) 删除一个结点后

图 24-16　从链表中删除最后一个结点

考虑以下两种情况：

1）如果链表为空或者有单个元素，调用 removeFist() 将处理这种情况（第 2 ～ 4 行）。

2）否则，将 current 指向倒数第二个结点（第 6 ～ 9 行）。将最后一个结点的值保存到 temp（第 11 行）。将 tail 设为 current（第 12 行）。tail 现在被重新定位，指向倒数第二个结点并销毁最后一个结点（第 13 行）。删除后大小减 1（第 14 行），并返回删除结点的元素值（第 15 行）。

实现 remove(index) 方法

remove(index) 方法找到指定下标处的结点，然后将它删除。该方法可以如下实现：

```
 1  public E remove(int index) {
 2    if (index < 0 || index >= size) return null; // Out of range
 3    else if (index == 0) return removeFirst(); // Remove first
 4    else if (index == size - 1) return removeLast(); // Remove last
 5    else {
 6      Node<E> previous = head;
 7
 8      for (int i = 1; i < index; i++) {
 9        previous = previous.next;
10      }
11
12      Node<E> current = previous.next;
13      previous.next = current.next;
14      size--;
15      return current.element;
16    }
17  }
```

考虑以下四种情况：

1）如果 index 超出链表的范围（即 index<0||index>=size），则返回 null（第 2 行）。

2）如果 index 为 0，则调用 removeFirst() 方法删除链表的第一个结点（第 3 行）。

3）当 index 为 size-1 时，调用 removeLast() 方法删除链表的最后一个结点（第 4 行）。

4）否则，定位指定 index 位置的结点。让 current 指向该结点，并让 previous 指向该结点的前一个结点，如图 24-17a 所示。将 current.next 赋给 previous.next 以消除当前结点，如图 24-17b 所示。

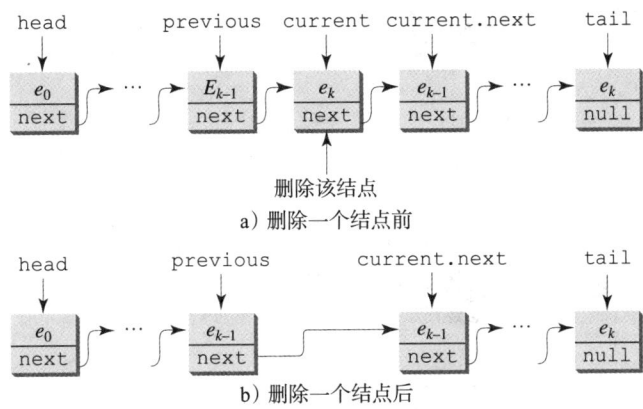

图 24-17 从链表中删除一个结点

程序清单 24-5 给出了 MyLinkedList 的实现。这里省略了 get(index)、indexOf(e)、lastIndexOf(e)、contains(e) 和 set(index,e) 方法的实现，将它们留作练习题。该程序实现了 java.lang.Iterable 接口中定义的方法 iterator()，以返回一个 java.util.Iterator 的实例（第 128～130 行）。LinkedListIterator 类针对 hasNext、next 以及 remove 等方法给出具体实现，从而实现了 Iterator 接口（第 132～152 行）。该实现使用 current 来指向被遍历的元素的当前位置（第 134 行）。开始时，current 指向链表的头部。

程序清单 24-5 MyLinkedList.java

```
1   public class MyLinkedList<E> implements MyList<E> {
2     private Node<E> head, tail;
3     private int size = 0; // Number of elements in the list
4
5     /** Create an empty list */
6     public MyLinkedList() {
7     }
8
9     /** Create a list from an array of objects */
10    public MyLinkedList(E[] objects) {
11      for (int i = 0; i < objects.length; i++)
12        add(objects[i]);
13    }
14
15    /** Return the head element in the list */
16    public E getFirst() {
17      if (size == 0) {
18        return null;
19      }
20      else {
21        return head.element;
22      }
23    }
24
25    /** Return the last element in the list */
26    public E getLast() {
27      if (size == 0) {
28        return null;
29      }
30      else {
31        return tail.element;
32      }
33    }
34
35    /** Add an element to the beginning of the list */
```

```java
36    public void addFirst(E e) {
37      // Implemented in Section 24.4.3.1, so omitted here
38    }
39
40    /** Add an element to the end of the list */
41    public void addLast(E e) {
42      // Implemented in Section 24.4.3.2, so omitted here
43    }
44
45    @Override /** Add a new element at the specified index
46      * in this list. The index of the head element is 0 */
47    public void add(int index, E e) {
48      // Implemented in Section 24.4.3.3, so omitted here
49    }
50
51    /** Remove the head node and
52      * return the object that is contained in the removed node. */
53    public E removeFirst() {
54      // Implemented in Section 24.4.3.4, so omitted here
55    }
56
57    /** Remove the last node and
58      * return the object that is contained in the removed node. */
59    public E removeLast() {
60      // Implemented in Section 24.4.3.5, so omitted here
61    }
62
63    @Override /** Remove the element at the specified position in this
64      * list. Return the element that was removed from the list. */
65    public E remove(int index) {
66      // Implemented earlier in Section 24.4.3.6, so omitted
67    }
68
69    @Override /** Override toString() to return elements in the list */
70    public String toString() {
71      StringBuilder result = new StringBuilder("[");
72
73      Node<E> current = head;
74      for (int i = 0; i < size; i++) {
75        result.append(current.element);
76        current = current.next;
77        if (current != null) {
78          result.append(", "); // Separate two elements with a comma
79        }
80        else {
81          result.append("]"); // Insert the closing ] in the string
82        }
83      }
84
85      return result.toString();
86    }
87
88    @Override /** Clear the list */
89    public void clear() {
90      size = 0;
91      head = tail = null;
92    }
93
94    @Override /** Return true if this list contains the element e */
95    public boolean contains(Object e) {
96      // Left as an exercise
97      return true;
98    }
99
100   @Override /** Return the element at the specified index */
101   public E get(int index) {
```

```java
102      // Left as an exercise
103      return null;
104    }
105
106    @Override /** Return the index of the first matching element in
107     * this list. Return -1 if no match. */
108    public int indexOf(Object e) {
109      // Left as an exercise
110      return 0;
111    }
112
113    @Override /** Return the index of the last matching element in
114     * this list. Return -1 if no match. */
115    public int lastIndexOf(E e) {
116      // Left as an exercise
117      return 0;
118    }
119
120    @Override /** Replace the element at the specified position
121     * in this list with the specified element. */
122    public E set(int index, E e) {
123      // Left as an exercise
124      return null;
125    }
126
127    @Override /** Override iterator() defined in Iterable */
128    public java.util.Iterator<E> iterator() {
129      return new LinkedListIterator();
130    }
131
132    private class LinkedListIterator
133        implements java.util.Iterator<E> {
134      private Node<E> current = head; // Current index
135
136      @Override
137      public boolean hasNext() {
138        return (current != null);
139      }
140
141      @Override
142      public E next() {
143        E e = current.element;
144        current = current.next;
145        return e;
146      }
147
148      @Override
149      public void remove() {
150        // Left as an exercise
151      }
152    }
153
154    private static class Node<E> {
155      E element;
156      Node<E> next;
157
158      public Node(E element) {
159        this.element = element;
160      }
161    }
162
163    @Override /** Return the number of elements in this list */
164    public int size() {
165      return size;
166    }
167  }
```

24.4.4 MyArrayList 和 MyLinkedList

MyArrayList 和 MyLinkedList 都可以用来存储线性表。MyArrayList 使用数组实现，MyLinkedList 使用链表实现。MyArrayList 的开销比 MyLinkedList 的小。但是，如果需要在线性表的开始位置插入和删除元素，那么 MyLinkedList 的效率更高。表 24-1 总结了 MyArrayList 和 MyLinkedList 中方法的时间复杂度。注意，MyArrayList 和 java.util.ArrayList 一样，而 MyLinkedList 和 java.util.LinkedList 一样，除 MyLinkedList 使用单链表实现而 LinkedList 使用双向链表实现。我们将在 24.4.5 节中介绍双向链表。

表 24-1 MyArrayList 和 MyLinkedList 中方法的时间复杂度

方法	MyArrayList/ArrayList	MyLinkedList/LinkedList
add(e: E)	$O(1)$	$O(1)$
add(index: int, e: E)	$O(n)$	$O(n)$
clear()	$O(1)$	$O(1)$
contains(e: E)	$O(n)$	$O(n)$
get(index: int)	$O(1)$	$O(n)$
indexOf(e: E)	$O(n)$	$O(n)$
isEmpty()	$O(1)$	$O(1)$
lastIndexOf(e: E)	$O(n)$	$O(n)$
remove(e: E)	$O(n)$	$O(n)$
size()	$O(1)$	$O(1)$
remove(index: int)	$O(n)$	$O(n)$
set(index: int, e: E)	$O(1)$	$O(n)$
addFirst(e: E)	$O(n)$	$O(1)$
removeFirst()	$O(n)$	$O(1)$

注意，可以不使用 size 数据域实现 MyLinkedList。但是这样 size() 方法将需要 $O(n)$ 时间。

注意：MyArrayList 使用数组实现。在 MyLinkedList 中，每个元素被包裹在一个对象中。MyArrayList 的开销比 MyLinkedList 的开销小。应该仅在程序涉及经常在列表的头部插入 / 删除元素的时候使用 MyLinkedList。

24.4.5 链表的变体

前一节介绍的链表称为单链表（singly linked list）。它包含一个指向线性表第一个结点的指针。每个结点都包含一个指针指向下一个结点。在某些应用中，链表的几种变体是很有用的。

循环单链表（circular, singly linked list）除了链表中的最后一个结点的指针指回到第一个结点以外，其他都很像单链表，如图 24-18a 所示。注意，在循环单链表中不需要 tail。head 指向链表中的当前结点。插入和删除操作都在当前结点处。循环单链表的一个很适合的应用是在以分时方式服务多个用户的操作系统中，系统会从循环链表中选择一个用户，确保分给他一小部分 CPU 时间，然后继续移动到链表中的下一个用户。

双向链表（doubly linked list）包含带两个指针的结点，一个指针指向下一个结点，而另一个指针指向前一个结点，如图 24-18b 所示。为方便起见，这两个指针分别称为前向指针

（forward pointer）和后向指针（backward pointer）。因此，双向链表既可以向前遍历，也可以向后遍历。java.util.LinkedList 类使用双向链表实现，支持使用 ListIterator 向前或者向后遍历链表。

循环双向链表（circular, doubly linked list）除了链表中最后一个结点的前向指针指向第一个结点，且第一个结点的后向指针指向最后一个结点以外，其他都和双向链表一样，如图 24-18c 所示。

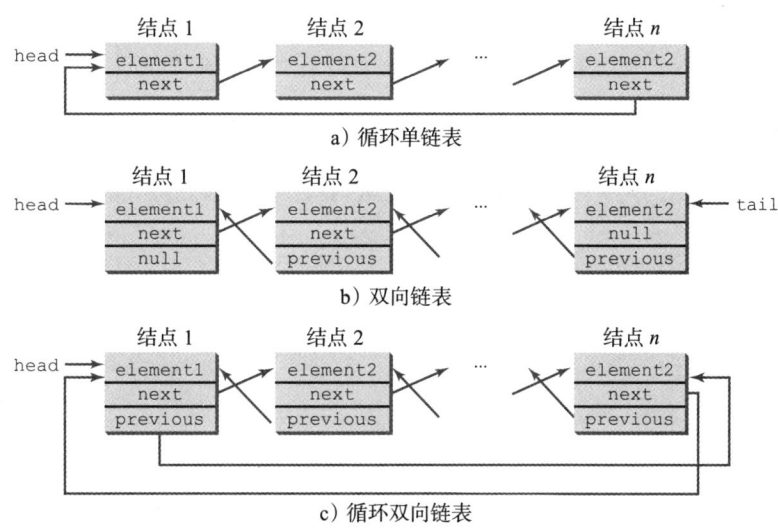

图 24-18 链表可表现为不同形式

这些链表的实现都留作练习题。

> 注意：在一个单链表中，removeLast() 需要 $O(n)$ 的时间。在一个双向链表中，removeLast() 可以以 $O(1)$ 时间实现。Java API 中的 removeLast 使用双向链表实现。参见复习题 24.4.11。

✓ 复习题

24.4.1 如果链表不包含任何结点，那么 head 和 tail 中的值是多少？

24.4.2 如果链表只包含一个结点，那么 head==tail 吗？列出所有 head==tail 为真的情况。

24.4.3 绘图展示以下语句执行后的链表。

```
MyLinkedList<Double> list = new MyLinkedList<>();
list.add(1.5);
list.add(6.2);
list.add(3.4);
list.add(7.4);
list.remove(1.5);
list.remove(2);
```

24.4.4 当一个新的结点被插入链表的头部时，head 和 tail 会改变吗？

24.4.5 当一个新的结点被添加到链表的尾部时，head 和 tail 会改变吗？

24.4.6 使用条件表达式简化程序清单 24-5 中第 77～82 行的代码。

24.4.7 MyLinkedList 中的 addFirst(e) 和 removeFirst() 的时间复杂度是多少？

24.4.8 假设你需要存储一个元素线性表。如果程序中的元素个数是固定的，应该使用什么数据结构？如果程序中的元素个数是变化的，应该使用什么数据结构？

24.4.9 如果需要在线性表的开始位置添加或删除元素，应该选择 MyArrayList 还是 MyLinkedList？如果线性表上的大量操作都涉及在一个给定下标处获取元素，应该选择 MyArrayList 还是

MyLinkedList?

24.4.10 MyArrayList 和 MyLinkedList 都用于存储一个对象线性表。为什么两种线性表我们都需要？

24.4.11 当 size 小于等于 1 的时候，通过调用 removeFirst() 方法来简化 removeLast() 方法的代码。新代码在执行时是否更高效？

24.5 栈和队列

要点提示：可以使用数组线性表实现栈，使用链表实现队列。

栈可以看作一种特殊类型的线性表，访问、插入和删除其中的元素只能在栈尾（栈顶）进行，如图 10-11 所示。队列表示一个等待列表，它也可以看作一种特殊类型的列表，元素只能从队列的末端（队列尾）插入，从开始端（队列头）访问和删除，如图 24-19 所示。

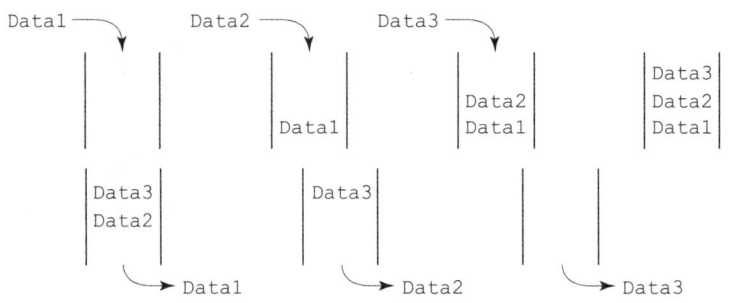

图 24-19 队列以先进先出的方式保存对象

教学注意：参见网址 liveexample.pearsoncmg .com/dsanimation/StackeBook.html 和 liveexample.pearsoncmg.com/dsanimation/QueueeBook.html，通过交互性演示查看栈和队列是如何工作的，如图 24-20 所示。

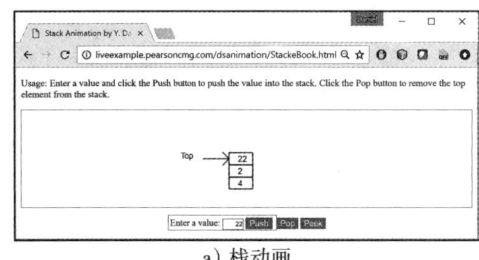

a) 栈动画　　　　　　　　　　　　b) 队列动画

图 24-20 动画工具有助于了解栈和队列是如何工作的（来源：Oracle 或其附属公司版权所有 © 1995~2016，经授权使用）

由于栈只在末尾处进行插入与删除操作，所以用数组线性表来实现栈比用链表来实现效率更高。由于队列删除是在线性表的起始位置进行的，所以用链表实现队列比用数组线性表实现效率更高。本节将用数组线性表来实现栈，用链表来实现队列。

有两种方法可用来设计栈和队列的类。

- 使用继承：可以通过继承数组线性表类 ArrayList 来定义栈类，以及通过继承链表类 LinkedList 来定义队列类，如图 24-21a 所示。
- 使用组合：可以将数组线性表定义为栈类中的数据域，将链表定义为队列类中的数据域,如图 24-21b 所示。

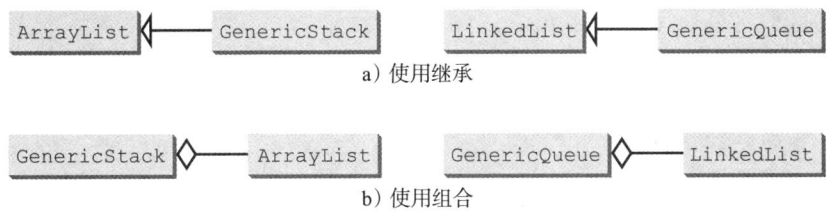

图 24-21 可以使用继承或组合实现 GenericStack 和 GenericQueue

这两种设计方法都可以,但是相比之下,使用组合更好一些,因为它使你可以定义一个新的栈类和队列类,而不需要继承数组线性表类与链表类中不必要和不合适的方法。使用组合方式实现的栈类见程序清单 19-1。程序清单 24-6 使用组合方式实现队列类 GenericQueue。图 24-22 给出该类的 UML 图。

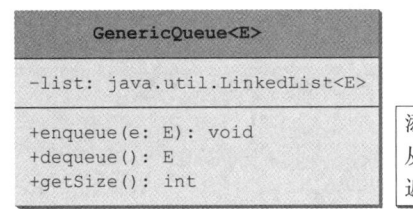

图 24-22 GenericQueue 使用链表来提供先进先出的数据结构

程序清单 24-6 GenericQueue.java

```
1  public class GenericQueue<E> {
2    private java.util.LinkedList<E> list
3      = new java.util.LinkedList<>();
4
5    public void enqueue(E e) {
6      list.addLast(e);
7    }
8
9    public E dequeue() {
10     return list.removeFirst();
11   }
12
13   public int getSize() {
14     return list.size();
15   }
16
17   @Override
18   public String toString() {
19     return "Queue: " + list.toString();
20   }
21 }
```

该程序创建一个链表来存储队列中的元素(第 2~3 行)。enqueue(e) 方法(第 5~7 行)将元素 e 添加到队列末尾。dequeue() 方法(第 9~11 行)从队列头部删除一个元素,并返回该被删除的元素。getSize() 方法(第 13~15 行)返回队列中元素的个数。

程序清单 24-7 给出一个使用 GenericStack 创建栈和使用 GenericQueue 创建队列的例子。它使用 push(enqueue) 方法向栈(或队列)中添加字符串,使用 pop(dequeue) 方法从栈(或队列)中删除字符串。

程序清单24-7 TestStackQueue.java

```java
1  public class TestStackQueue {
2    public static void main(String[] args) {
3      // Create a stack
4      GenericStack<String> stack = new GenericStack<>();
5
6      // Add elements to the stack
7      stack.push("Tom"); // Push Tom to the stack
8      System.out.println("(1) " + stack);
9
10     stack.push("Susan"); // Push Susan to the the stack
11     System.out.println("(2) " + stack);
12
13     stack.push("Kim"); // Push Kim to the stack
14     stack.push("Michael"); // Push Michael to the stack
15     System.out.println("(3) " + stack);
16
17     // Remove elements from the stack
18     System.out.println("(4) " + stack.pop());
19     System.out.println("(5) " + stack.pop());
20     System.out.println("(6) " + stack);
21
22     // Create a queue
23     GenericQueue<String> queue = new GenericQueue<>();
24
25     // Add elements to the queue
26     queue.enqueue("Tom"); // Add Tom to the queue
27     System.out.println("(7) " + queue);
28
29     queue.enqueue("Susan"); // Add Susan to the queue
30     System.out.println("(8) " + queue);
31
32     queue.enqueue("Kim"); // Add Kim to the queue
33     queue.enqueue("Michael"); // Add Michael to the queue
34     System.out.println("(9) " + queue);
35
36     // Remove elements from the queue
37     System.out.println("(10) " + queue.dequeue());
38     System.out.println("(11) " + queue.dequeue());
39     System.out.println("(12) " + queue);
40   }
41 }
```

```
(1) stack: [Tom]
(2) stack: [Tom, Susan]
(3) stack: [Tom, Susan, Kim, Michael]
(4) Michael
(5) Kim
(6) stack: [Tom, Susan]
(7) Queue: [Tom]
(8) Queue: [Tom, Susan]
(9) Queue: [Tom, Susan, Kim, Michael]
(10) Tom
(11) Susan
(12) Queue: [Kim, Michael]
```

对栈来说，push(e)方法将一个元素添加到栈顶，而pop()方法将栈顶元素从栈中删除并返回该元素。很容易得出，push和pop方法的时间复杂度为$O(1)$。

对队列来说，enqueue(e)方法将一个元素添加到队列尾，而dequeue()方法从队列头删除元素。很容易得出，enqueue和dequeue方法的时间复杂度为$O(1)$。

✓ 复习题

24.5.1 可以采用继承或组合来为栈和队列设计数据结构，试讨论这两种方法的优缺点。

24.5.2 如果程序清单 24-6 中第 2～3 行的 LinkedList 被替换为 ArrayList，enqueue 和 dequeue 方法的时间复杂度为多少？

24.5.3 下面代码的哪些行有错误？

```
1  List<String> list = new ArrayList<>();
2  list.add("Tom");
3  list = new LinkedList<>();
4  list.add("Tom");
5  list = new GenericStack<>();
6  list.add("Tom");
```

24.6 优先队列

要点提示：可以用堆实现优先队列。

普通的队列是一种先进先出的数据结构，元素在队列的末尾追加，并从队列的头部删除。在优先队列（priority queue）中，元素被赋予优先级。当访问元素时，具有最高优先级的元素最先删除。例如，医院的急救室为病人赋予优先级，具有最高优先级的病人最先得到治疗。

可以使用堆实现优先队列，其中根结点是队列中具有最高优先级的对象。本书在 23.6 节中介绍过堆。优先队列的类图如图 24-23 所示，它的实现在程序清单 24-8 中给出。

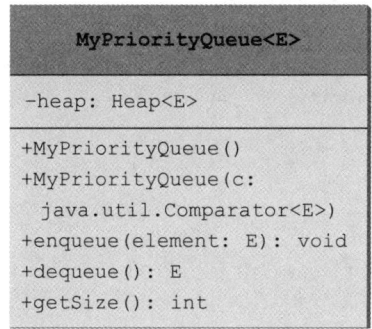

图 24-23 MyPriorityQueue 使用堆存储元素

程序清单 24-8 MyPriorityQueue.java

```
1  public class MyPriorityQueue<E> {
2    private Heap<E> heap;
3  
4    public void MyPriorityQueue<E> {
5      heap.add(new Heap<E>());
6    }
7  
8    public MyPriorityQueue(java.util.Comparator<E> c) {
9      heap = new Heap<E>(c);
10   }
11  
12   public void enqueue(E newObject) {
13     heap.add(newObject);
14   }
15  
16   public E dequeue() {
17     return heap.remove();
18   }
19  
20   public int getSize() {
21     return heap.getSize();
22   }
23  }
```

程序清单 24-9 给出了一个用于病人的优先队列的例子。Patient 类在第 19～37 行定义。第 3～6 行创建带优先级数值的 4 个病人实例。第 8 行创建一个优先队列。在第 10～13 行将病人实例加入队列。第 16 行从队列移除一个病人实例。

程序清单 24-9 TestPriorityQueue.java

```java
 1  public class TestPriorityQueue {
 2    public static void main(String[] args) {
 3      Patient patient1 = new Patient("John", 2);
 4      Patient patient2 = new Patient("Jim", 1);
 5      Patient patient3 = new Patient("Tim", 5);
 6      Patient patient4 = new Patient("Cindy", 7);
 7
 8      MyPriorityQueue<Patient> priorityQueue
 9        = new MyPriorityQueue<>();
10      priorityQueue.enqueue(patient1);
11      priorityQueue.enqueue(patient2);
12      priorityQueue.enqueue(patient3);
13      priorityQueue.enqueue(patient4);
14
15      while (priorityQueue.getSize() > 0)
16        System.out.print(priorityQueue.dequeue() + " ");
17    }
18
19    static class Patient implements Comparable<Patient> {
20      private String name;
21      private int priority;
22
23      public Patient(String name, int priority) {
24        this.name = name;
25        this.priority = priority;
26      }
27
28      @Override
29      public String toString() {
30        return name + "(priority:" + priority + ")";
31      }
32
33      @Override
34      public int compareTo(Patient patient) {
35        return this.priority - patient.priority;
36      }
37    }
38  }
```

```
Cindy(priority:7) Tim(priority:5) John(priority:2) Jim(priority:1)
```

复习题

24.6.1 什么是优先队列?

24.6.2 MyPriorityQueue 中的 enqueue、dequeue 以及 getSize 方法的时间复杂度为多少?

24.6.3 下面语句哪些有错误?

```java
1  MyPriorityQueue<Object> q1 = new MyPriorityQueue<>();
2  MyPriorityQueue<Number> q2 = new MyPriorityQueue<>();
3  MyPriorityQueue<Integer> q3 = new MyPriorityQueue<>();
4  MyPriorityQueue<Date> q4 = new MyPriorityQueue<>();
5  MyPriorityQueue<String> q5 = new MyPriorityQueue<>();
```

本章小结

1. 本章学习了如何实现数组线性表、链表、栈以及队列。

2. 定义一个数据结构本质上是定义一个类。为数据结构定义的类应该使用数据域来存储数据,并提供方法来支持诸如插入和删除等操作。
3. 创建一个数据结构是从该类创建一个实例。这样就可以将方法应用在实例上来处理数据结构,比如插入一个元素到数据结构中,或者从数据结构中删除一个元素。
4. 本章学习了如何采用堆来实现优先队列。

测试题

回答位于本书配套网站上的本章测试题。

编程练习题

24.1 (在 MyList 中实现操作)在 MyList 接口中省略了 addAll、removeAll、retainAll、toArray() 和 toArray(T[]) 方法的实现,请实现它们。使用网址 liveexample.pearsoncmg.com/test/ Exercise-24_01Test.txt 中的代码检测新 MyList 类。

*24.2 (实现 MyLinkedList) MyLinkedList 类中省略了 contains(E e)、get(int index)、indexOf(E e)、lastIndexOf(E e) 和 set(int index, E e) 方法的实现,请实现它们。定义一个继承自 MyLinkedList 的名为 MyLinkedListExtra 的新类以重写这些方法。使用位于 https://liveexample.pearsoncmg.com/test/Exercise24_02.txt 的代码测试你的新 MyList 类。

*24.3 (实现双向链表)程序清单 24-5 中使用的 MyLinkedList 类是一个单向链表,它只能单向遍历线性表。修改 Node 类,添加一个名为 previous 的数据域指向链表中的前一个结点,如下所示:

```
public class Node<E> {
  E element;
  Node<E> next;
  Node<E> previous;

  public Node(E e) {
    element = e;
  }
}
```

实现一个名为 TwoWayLinkedList 的新类,使用双向链表来存储元素。定义 TwoWayLinkedList 实现 MyList 接口。不仅要实现 listIterator() 和 listIterator(int index) 方法,还要实现定义在 MyLinkedList 中的所有方法。listIterator() 和 listIterator(int index) 方法都返回一个 java.util.ListIterator<E> 类型的实例(参见图 20-4)。前者设置光标指向线性表的头部,后者指向指定下标的元素。使用 https://liveexample.pearsoncmg.com/test/Exercise24_03.txt 处的代码测试你的新类。

24.4 (使用 GenericStack 类)编写一个程序,以降序显示前 50 个素数。使用栈存储素数。

24.5 (使用继承实现 GenericQueue) 24.5 节使用组合实现了 GenericQueue。通过继承 java.util.LinkedList 定义一个新的队列类。

*24.6 (修改 MyPriorityQueue)程序清单 24-8 使用堆实现了优先队列。修改实现方式,采用一个排好序的数组线性表来存储元素并将新类命名为 PriorityQueueUsingSortedArrayList。数组线性表中的元素按照它们的优先级升序排列,最后一个元素具有最高优先级。编写一个测试程序,产生 500 万个整数并将它们放入优先队列以及从队列中删除。对 MyPriorityQueue 和 PriorityQueueUsingSortedArrayList 使用相同的数字并比较它们的执行时间。

**24.7 (动画:链表)编写一个程序,实现链表的查找、插入和删除的动画,如图 24-1b 所示。按钮 Search 用来在链表中查找一个指定的值;按钮 Delete 用来从链表中删除一个特定值;按钮 Insert 用来在链表的指定下标处插入一个值,如果没有指定下标,则添加到链表的末尾。

*24.8 （动画：数组线性表）编写一个程序，实现数组线性表的查找、插入和删除的动画，如图 24-1a 所示。按钮 Search 用来查找一个指定的值是否在线性表中；按钮 Delete 用来从线性表中删除一个指定的值；按钮 Insert 用来在线性表的指定下标处插入一个值，如果没有指定下标，则添加到线性表的末尾。

*24.9 （动画：慢动作显示数组线性表）改进编程练习题 24.8 的动画效果，通过慢动作显示插入和删除操作。

*24.10 （动画：栈）编写一个程序，用动画实现栈的压入和弹出，如图 24-20a 所示。

*24.11 （动画：双向链表）编写一个程序，实现双向链表的查找、插入和删除的动画，如图 24-24 所示。按钮 Search 用来查找一个指定的值是否在链表中；按钮 Delete 用来从链表中删除一个指定值；按钮 Insert 用来在链表的指定下标处插入一个值，如果没有指定下标，则添加到链表的末尾。同时，添加两个名为 Forward Traversal 和 Backward Traversal 的按钮，用于采用迭代器分别以向前和向后的顺序来显示元素，如图 24-24 所示。这些元素都显示在标签中。

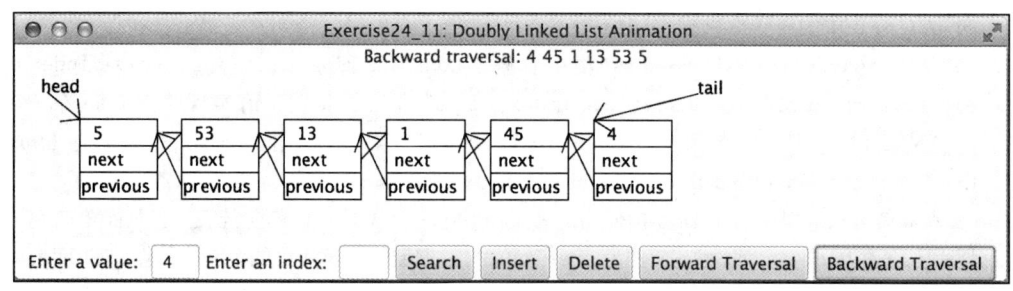

图 24-24 该程序实现双向链表的运行动画（来源：Oracle 或其附属公司版权所有 © 1995~2016，经授权使用）

*24.12 （动画：队列）编写一个程序，实现队列的 enqueue 和 dequeue 操作的动画，如图 24-20b 所示。

*24.13 （斐波那契数迭代器）定义一个名为 FibonacciIterator 的迭代器，用于遍历斐波那契数。构造方法带有一个参数，用于指定斐波那契数的上限。比如，new FibonacciIterator (23302) 创建一个迭代器，可以用于遍历小于或者等于 23302 的斐波那契数。编写一个测试程序，使用该迭代器显示所有小于或者等于 100000 的斐波那契数。

*24.14 （素数迭代器）定义一个名为 PrimeIterator 的迭代器类，用于遍历素数。构造方法带有一个参数，用于指定素数的上限。比如，new PrimeIterator(23302) 创建一个迭代器，可以用于遍历小于或者等于 23302 的素数。编写一个测试程序，使用该迭代器显示所有小于或者等于 100000 的素数。

**24.15 （测试 MyArrayList）设计和编写一个完整的测试程序，用于测试程序清单 24-2 中的 MyArrayList 类是否符合所有的要求。

**24.16 （测试 MyLinkedList）设计和编写一个完整的测试程序，用于测试程序清单 24-5 中的 MyLinkedList 类是否符合所有的要求。

第 25 章

二叉搜索树

教学目标

- 设计并实现二叉搜索树（25.2 节）。
- 使用链式数据结构表示二叉树（25.3 节）。
- 在二叉搜索树中查找元素（25.4 节）。
- 在二叉搜索树中插入元素（25.5 节）。
- 遍历二叉树中的元素（25.6 节）。
- 设计和实现 Tree 接口以及 BST 类（25.7 节）。
- 从二叉搜索树中删除元素（25.8 节）。
- 图形化地显示二叉树（25.9 节）。
- 创建迭代器来遍历二叉树（25.10 节）。
- 使用二叉树实现用于压缩数据的霍夫曼编码（25.11 节）。

25.1 引言

☞ 要点提示：二叉搜索树可以比线性表更高效地进行搜索、插入和删除操作。

前一章给出了数组线性表和链表的实现，在这些数据结构中，查找、插入和删除操作的时间复杂度是 $O(n)$。这一章给出了一种称为二叉搜索树的新数据结构。它花费 $O(\log n)$ 的平均时间来进行查找、插入和删除元素。

25.2 二叉搜索树基础

☞ 要点提示：对于二叉搜索树中的每个结点，其左子结点的值小于该结点的值，而右子结点的值大于该结点的值。

回顾一下，线性表、栈和队列都是由一系列元素组成的线性结构。二叉树（binary tree）是一种层次结构，它要么是空集，要么是由一个称为根（root）的元素和两棵不同的二叉树组成的，这两棵二叉树分别称为左子树（left subtree）和右子树（right subtree），其中的一棵或者两棵可能为空，如图 25-1a 所示。二叉树的示例如图 25-1a 和图 25-1b 所示。

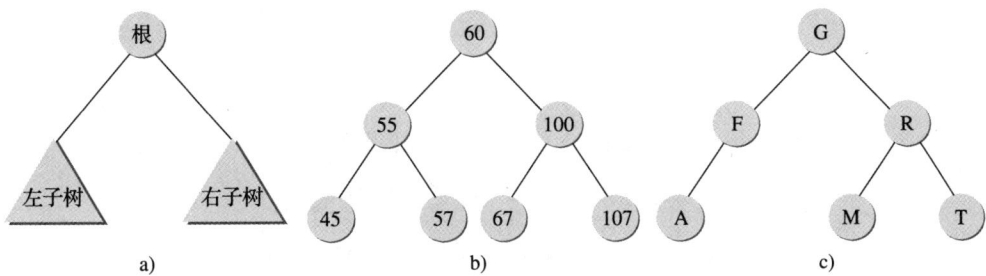

图 25-1　二叉树的每个结点有 0 棵、1 棵或 2 棵子树

一条路径的长度（length）是指在该条路径上的边的数量。一个结点的深度（depth）是指从根结点到该结点的路径长度。具有某个给定深度的所有结点的集合有时称为该树的层（level）。兄弟结点（sibling）是共享同一父结点的结点。一个结点的左（右）子树的根称为这个结点的左（右）子结点（left（right）child）。没有子结点的结点称为叶子结点（leaf）。非空树的高度为从根结点到最远的叶子结点的路径长度。只包含一个结点的树的高度为 0。习惯上，将空树的高度定为 -1。考虑图 25-1b 中的树。从结点 60 到 45 的路径长度为 2。结点 60 的深度为 0，结点 55 的深度为 1，而结点 45 的深度为 2。这棵树的高度为 2。结点 45 和 57 是兄弟结点。结点 45、57、67 和 107 位于同一层。

一种称为二叉搜索树（binary search tree，BST）的特殊类型的二叉树非常有用。一个 BST（没有重复元素）的特征是：对于树中的每一个结点，它的左子结点的值都小于该结点的值，而它的右子结点的值都大于该结点的值。图 25-1 中的二叉树都是 BST。

教学注意：参见链接 liveexample.pearsoncmg.com/dsanimation/BSTeBook.html 查看 BST 如何工作的交互式 GUI 演示，如图 25-2 所示。

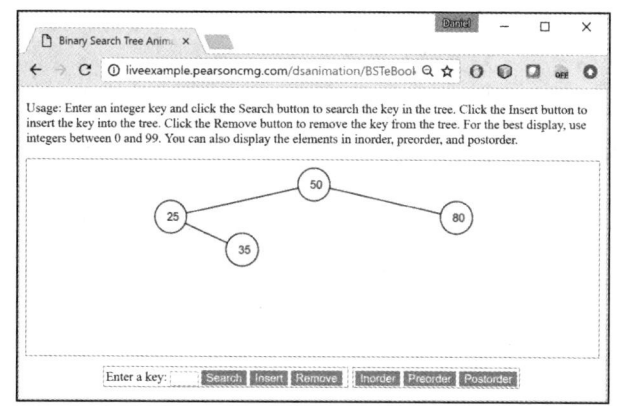

图 25-2　动画工具可以让你插入、删除和查找元素（来源：Oracle 或其附属公司版权所有 © 1995 ~ 2016，经授权使用）

25.3　表示二叉搜索树

要点提示：可以使用链接结构表示一棵二叉搜索树。

可以使用一个相互链接的结点集合来表示一棵二叉树。每个结点包含一个值以及两个称为 left 和 right 的链接，分别引用左子结点和右子结点，如图 25-3 所示。

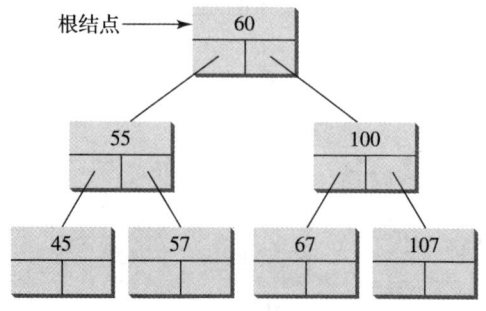

图 25-3　二叉树可以使用链接结点的集合表示

结点可以定义为一个类，如下所示：

```java
class TreeNode<E> {
  protected E element;
  protected TreeNode<E> left;
  protected TreeNode<E> right;

  public TreeNode(E e) {
    element = e;
  }
}
```

我们使用变量 root 指向树的根结点。如果树为空，那么 root 的值为 null。下面的代码创建了如图 25-3 所示的树的前三个结点：

```java
// Create the root node
TreeNode<Integer> root = new TreeNode<>(60);

// Create the left child node
root.left = new TreeNode<>(55);

// Create the right child node
root.right = new TreeNode<>(100);
```

25.4 查找一个元素

要点提示：BST 支持类似二分查找的高效搜索。

要在 BST 中查找一个元素，可从根结点开始向下扫描，直到找到一个匹配元素，或者直到达到一棵空子树为止。该算法在程序清单 25-1 中描述。让 current 指向根结点（第 2 行），重复下面的步骤直到 current 为 null（第 4 行）或者元素匹配到 current.element（第 12 行）。

- 如果 e 小于 current.element，则将 current.left 赋给 current（第 6 行）。
- 如果 e 大于 current.element，则将 current.right 赋给 current（第 9 行）。
- 如果 e 等于 current.element，则返回 true（第 12 行）。

如果 current 为 null，那么子树为空且该元素不在这棵树中（第 14 行）。

程序清单 25-1 在 BST 中查找一个元素

```java
 1  public boolean search(E e) {
 2    TreeNode<E> current = root; // Start from the root
 3
 4    while (current != null)
 5      if (e < current.element) {
 6        current = current.left; // Go left
 7      }
 8      else if (e > current.element) {
 9        current = current.right; // Go right
10      }
11      else // Element e matches current.element
12        return true; // Element e is found
13
14    return false; // Element e is not in the tree
15  }
```

✓ **复习题**

25.4.1 使用递归实现 search(element) 方法。

25.5 在 BST 中插入一个元素

要点提示：新元素插在叶子结点处。

为了在 BST 中插入一个元素，需要定位在树中插入元素的位置。关键思路是确定新结点的父结点所在的位置。程序清单 25-2 给出该算法。

程序清单 25-2 在 BST 中插入一个元素

```
1   boolean insert(E e) {
2     if (tree is empty)
3       // Create the node for e as the root;
4     else {
5       // Locate the parent node
6       parent = current = root;
7       while (current != null)
8         if (e < the value in current.element) {
9           parent = current; // Keep the parent
10          current = current.left; // Go left
11        }
12        else if (e > the value in current.element) {
13          parent = current; // Keep the parent
14          current = current.right; // Go right
15        }
16        else
17          return false; // Duplicate node not inserted
18
19      // Create a new node for e and attach it to parent
20
21      return true; // Element inserted
22    }
23  }
```

如果该树为空，则使用新元素创建一个根结点（第 2～3 行）；否则，寻找新元素结点的父结点的位置（第 6～17 行）。为该元素创建一个新结点，然后将该结点链接到它的父结点上。如果新元素小于父元素，则将新元素的结点设置为父结点的左子结点；如果新元素的值大于父元素的值，则将新元素的结点设置为父结点的右子结点。

例如，要将数据 101 插入图 25-3 所示的树中，在算法中的 while 循环结束之后，parent 指向存储数据 107 的结点，如图 25-4a 所示。存储数据 101 的新结点将成为父结点的左子结点。要将数据 59 插入树中，在算法中的 while 循环结束之后，父结点指向存储数据 57 的结点，如图 25-4b 所示。存储数据 59 的新结点成为父结点的右子结点。

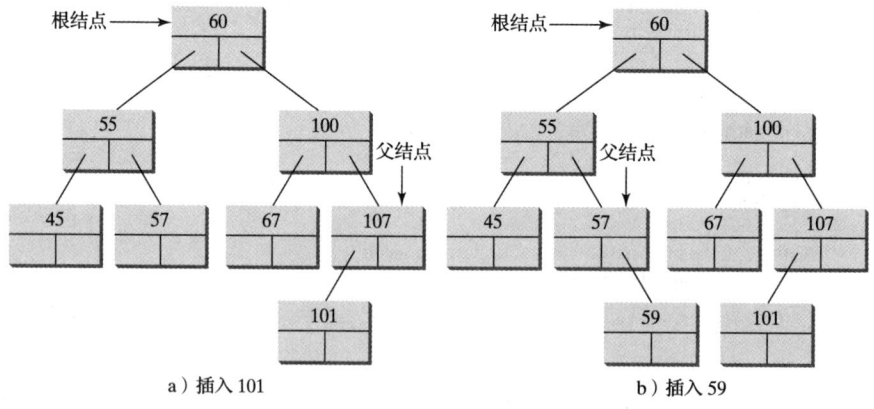

图 25-4 在树中插入两个新元素

复习题

25.5.1 显示将 44 插入图 25-4b 后的结果。

25.5.2 在 BST 中插入一个元素的时间复杂度是多少？

25.6 树的遍历

要点提示：中序、前序、后序、深度优先和广度优先是遍历二叉树中元素的常见方式。

树的遍历（tree traversal）是只访问树中每个结点一次的过程。遍历树的方法有多种。本节将介绍中序（inorder）、前序（preorder）、后序（postorder）、深度优先（depth-first）和广度优先（breadth-first）等遍历方法。

中序遍历（inorder traversal）首先递归地访问当前结点的左子树，然后访问当前结点，最后递归地访问该结点的右子树。中序遍历法以递增顺序显示 BST 中的所有结点，如图 25-5 所示。

中序：45 55 57 59 60 67 100 101 107

后序：45 59 57 55 67 101 107 100 60

前序：60 55 45 57 59 100 67 107 101

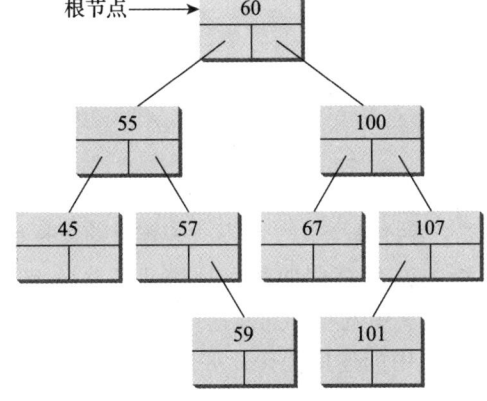

图 25-5 树的遍历是以某种顺序访问树中的每个结点

后序遍历（postorder traversal）首先递归地访问当前结点的左子树，然后递归地访问该结点的右子树，最后访问该结点本身。

前序遍历（preorder traversal）首先访问当前结点，然后递归地访问当前结点的左子树，最后递归地访问当前结点的右子树。

注意：可以采用前序插入元素的方法重构一棵二叉搜索树。重构的树保留了原始的二叉搜索树中父子结点的关系。

对于深度优先遍历而言，首先访问根结点，接着以任意顺序递归地访问它的左子树和右子树。前序遍历可以看作是深度优先遍历的一个特例，递归地访问它的左子树，然后是右子树。

广度优先遍历逐层访问树中的结点。首先访问根结点，然后从左往右访问根结点的所有子结点，再从左往右访问根结点的所有孙子结点，以此类推。

例如，对于图 25-5 中的树，中序遍历为

45 55 57 59 60 67 100 101 107

后序遍历为

45 59 57 55 67 101 107 100 60

前序遍历为

60 55 45 57 59 100 67 107 101

深度优先遍历为

60 55 45 57 59 100 67 107 101

广度优先遍历为

60 55 100 45 57 67 107 59 101

可以使用下面的简单树来帮助记忆中序、后序以及前序：

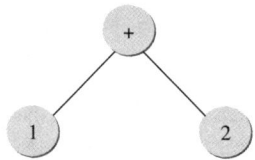

中序是 1 + 2，后序是 1 2 +，前序是 + 1 2。

✓ 复习题

25.6.1 显示对图 25-1c 中二叉树中元素的中序、前序、后序遍历。

25.6.2 如果将由相同元素构成的集合以两个不同的次序插入 BST 中，这两棵对应的 BST 是否一样？中序遍历是否一样？后序遍历是否一样？前序遍历是否一样？

25.7 BST 类

🔑 **要点提示**：BST 类定义了一个用于存储和操控二叉搜索树中元素的数据结构。

我们遵循 Java 集合框架的设计模式，并且利用 Java 8 中的默认方法，使用一个名为 Tree 的接口来定义树的所有共同操作，并定义 Tree 为 Collection 的子类型，从而可以为树使用 Collection 中的通用操作，如图 25-6 所示。一个具体的 BST 类可以定义为实现 Tree，如图 25-7 所示。

程序清单 25-3 给出了 Tree 的实现。它提供了如下的默认实现方法：add，isEmpty，remove，containsAll，addAll，removeAll，retainAll，toArray()，toArray(T[])。这些方法继承自 Collection 接口，还有定义在 Tree 接口中的 inorder()、preorder() 和 postorder() 方法。

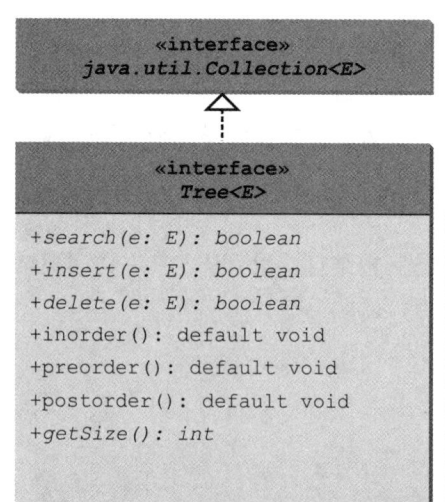

图 25-6 Tree 接口定义了树的通用操作，并且部分地实现了 Collection

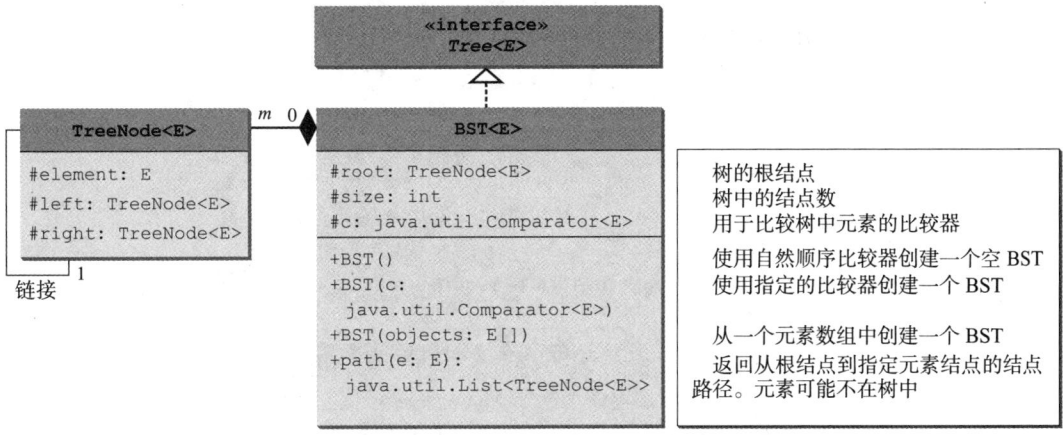

图 25-7 BST 类定义了一个具体的 BST

程序清单 25-3 Tree.java

```java
import java.util.Collection;

public interface Tree<E> extends Collection<E> {
  /** Return true if the element is in the tree */
  public boolean search(E e);

  /** Insert element e into the binary tree
   * Return true if the element is inserted successfully */
  public boolean insert(E e);

  /** Delete the specified element from the tree
   * Return true if the element is deleted successfully */
  public boolean delete(E e);

  /** Get the number of elements in the tree */
  public int getSize();

  /** Inorder traversal from the root*/
  public default void inorder() {
  }

  /** Postorder traversal from the root */
  public default void postorder() {
  }

  /** Preorder traversal from the root */
  public default void preorder() {
  }

  @Override /** Return true if the tree is empty */
  public default boolean isEmpty() {
    return size() == 0;
  }

  @Override
  public default boolean contains(Object e) {
    return search((E)e);
  }

  @Override
  public default boolean add(E e) {
    return insert(e);
  }
```

```java
45    @Override
46    public default boolean remove(Object e) {
47      return delete((E)e);
48    }
49
50    @Override
51    public default int size() {
52      return getSize();
53    }
54
55    @Override
56    public default boolean containsAll(Collection<?> c) {
57      // Left as an exercise
58      return false;
59    }
60
61    @Override
62    public default boolean addAll(Collection<? extends E> c) {
63      // Left as an exercise
64      return false;
65    }
66
67    @Override
68    public default boolean removeAll(Collection<?> c) {
69      // Left as an exercise
70      return false;
71    }
72
73    @Override
74    public default boolean retainAll(Collection<?> c) {
75      // Left as an exercise
76      return false;
77    }
78
79    @Override
80    public default Object[] toArray() {
81      // Left as an exercise
82      return null;
83    }
84
85    @Override
86    public default <T> T[] toArray(T[] array) {
87      // Left as an exercise
88      return null;
89    }
90  }
```

程序清单 25-4 给出了 BST 类的实现。

程序清单 25-4 BST.java

```java
1  public class BST<E> implements Tree<E> {
2    protected TreeNode<E> root;
3    protected int size = 0;
4    protected java.util.Comparator<E> c;
5
6    /** Create a default BST with a natural order comparator */
7    public BST() {
8      this.c = (e1, e2) -> ((Comparable<E>)e1).compareTo(e2);
9    }
10
11   /** Create a BST with a specified comparator */
12   public BST(java.util.Comparator<E> c) {
13     this.c = c;
```

```java
14    }
15
16    /** Create a binary tree from an array of objects */
17    public BST(E[] objects) {
18      this.c = (e1, e2) -> ((Comparable<E>)e1).compareTo(e2);
19      for (int i = 0; i < objects.length; i++)
20        add(objects[i]);
21    }
22
23    @Override /** Return true if the element is in the tree */
24    public boolean search(E e) {
25      TreeNode<E> current = root; // Start from the root
26
27      while (current != null) {
28        if (c.compare(e, current.element) < 0) {
29          current = current.left;
30        }
31        else if (c.compare(e, current.element) > 0) {
32          current = current.right;
33        }
34        else // element matches current.element
35          return true; // Element is found
36      }
37
38      return false;
39    }
40
41    @Override /** Insert element e into the binary tree
42     * Return true if the element is inserted successfully */
43    public boolean insert(E e) {
44      if (root == null)
45        root = createNewNode(e); // Create a new root
46      else {
47        // Locate the parent node
48        TreeNode<E> parent = null;
49        TreeNode<E> current = root;
50        while (current != null)
51          if (c.compare(e, current.element) < 0) {
52            parent = current;
53            current = current.left;
54          }
55          else if (c.compare(e, current.element) > 0) {
56            parent = current;
57            current = current.right;
58          }
59          else
60            return false; // Duplicate node not inserted
61
62        // Create the new node and attach it to the parent node
63        if (c.compare(e, parent.element) < 0)
64          parent.left = createNewNode(e);
65        else
66          parent.right = createNewNode(e);
67      }
68
69      size++;
70      return true; // Element inserted successfully
71    }
72
73    protected TreeNode<E> createNewNode(E e) {
74      return new TreeNode<>(e);
75    }
76
77    @Override /** Inorder traversal from the root */
78    public void inorder() {
79      inorder(root);
```

```java
 80    }
 81
 82    /** Inorder traversal from a subtree */
 83    protected void inorder(TreeNode<E> root) {
 84      if (root == null) return;
 85      inorder(root.left);
 86      System.out.print(root.element + " ");
 87      inorder(root.right);
 88    }
 89
 90    @Override /** Postorder traversal from the root */
 91    public void postorder() {
 92      postorder(root);
 93    }
 94
 95    /** Postorder traversal from a subtree */
 96    protected void postorder(TreeNode<E> root) {
 97      if (root == null) return;
 98      postorder(root.left);
 99      postorder(root.right);
100      System.out.print(root.element + " ");
101    }
102
103    @Override /** Preorder traversal from the root */
104    public void preorder() {
105      preorder(root);
106    }
107
108    /** Preorder traversal from a subtree */
109    protected void preorder(TreeNode<E> root) {
110      if (root == null) return;
111      System.out.print(root.element + " ");
112      preorder(root.left);
113      preorder(root.right);
114    }
115
116    /** This inner class is static, because it does not access
117        any instance members defined in its outer class */
118    public static class TreeNode<E> {
119      protected E element;
120      protected TreeNode<E> left;
121      protected TreeNode<E> right;
122
123      public TreeNode(E e) {
124        element = e;
125      }
126    }
127
128    @Override /** Get the number of nodes in the tree */
129    public int getSize() {
130      return size;
131    }
132
133    /** Returns the root of the tree */
134    public TreeNode<E> getRoot() {
135      return root;
136    }
137
138    /** Returns a path from the root leading to the specified element */
139    public java.util.ArrayList<TreeNode<E>> path(E e) {
140      java.util.ArrayList<TreeNode<E>> list =
141        new java.util.ArrayList<>();
142      TreeNode<E> current = root; // Start from the root
143
144      while (current != null) {
145        list.add(current); // Add the node to the list
```

```java
146        if (c.compare(e, current.element) < 0) {
147          current = current.left;
148        }
149        else if (c.compare(e, current.element) > 0) {
150          current = current.right;
151        }
152        else
153          break;
154      }
155
156      return list; // Return an array list of nodes
157    }
158
159    @Override /** Delete an element from the binary tree.
160     * Return true if the element is deleted successfully
161     * Return false if the element is not in the tree */
162    public boolean delete(E e) {
163      // Locate the node to be deleted and also locate its parent node
164      TreeNode<E> parent = null;
165      TreeNode<E> current = root;
166      while (current != null) {
167        if (c.compare(e, current.element) < 0) {
168          parent = current;
169          current = current.left;
170        }
171        else if (c.compare(e, current.element) > 0) {
172          parent = current;
173          current = current.right;
174        }
175        else
176          break; // Element is in the tree pointed at by current
177      }
178
179      if (current == null)
180        return false; // Element is not in the tree
181
182      // Case 1: current has no left child
183      if (current.left == null) {
184        // Connect the parent with the right child of the current node
185        if (parent == null) {
186          root = current.right;
187        }
188        else {
189          if (c.compare(e, parent.element) < 0)
190            parent.left = current.right;
191          else
192            parent.right = current.right;
193        }
194      }
195      else {
196        // Case 2: The current node has a left child
197        // Locate the rightmost node in the left subtree of
198        // the current node and also its parent
199        TreeNode<E> parentOfRightMost = current;
200        TreeNode<E> rightMost = current.left;
201
202        while (rightMost.right != null) {
203          parentOfRightMost = rightMost;
204          rightMost = rightMost.right; // Keep going to the right
205        }
206
207        // Replace the element in current by the element in rightMost
208        current.element = rightMost.element;
209
210        // Eliminate rightmost node
211        if (parentOfRightMost.right == rightMost)
```

```
212            parentOfRightMost.right = rightMost.left;
213        else
214          // Special case: parentOfRightMost == current
215          parentOfRightMost.left = rightMost.left;
216      }
217
218      size--;
219      return true; // Element deleted successfully
220    }
221
222    @Override /** Obtain an iterator. Use inorder. */
223    public java.util.Iterator<E> iterator() {
224      return new InorderIterator();
225    }
226
227    // Inner class InorderIterator
228    private class InorderIterator implements java.util.Iterator<E> {
229      // Store the elements in a list
230      private java.util.ArrayList<E> list =
231        new java.util.ArrayList<>();
232      private int current = 0; // Point to the current element in list
233
234      public InorderIterator() {
235        inorder(); // Traverse binary tree and store elements in list
236      }
237
238      /** Inorder traversal from the root*/
239      private void inorder() {
240        inorder(root);
241      }
242
243      /** Inorder traversal from a subtree */
244      private void inorder(TreeNode<E> root) {
245        if (root == null) return;
246        inorder(root.left);
247        list.add(root.element);
248        inorder(root.right);
249      }
250
251      @Override /** More elements for traversing? */
252      public boolean hasNext() {
253        if (current < list.size())
254          return true;
255
256        return false;
257      }
258
259      @Override /** Get the current element and move to the next */
260      public E next() {
261        return list.get(current++);
262      }
263
264      @Override // Remove the element returned by the last next()
265      public void remove() {
266        if (current == 0) // next() has not been called yet
267          throw new IllegalStateException();
268
269        delete(list.get(--current));
270        list.clear(); // Clear the list
271        inorder(); // Rebuild the list
272      }
273    }
274
275    @Override /** Remove all elements from the tree */
276    public void clear() {
277      root = null;
```

```
278        size = 0;
279      }
280    }
```

insert(E e) 方法 (第 43 ～ 71 行) 为元素 e 创建一个结点，并将其插入树中。如果树为空，则该结点就成为根结点；否则，该方法为这个结点寻找一个能够保持树的顺序的父结点。如果此元素已经在树中，则该方法返回 false；否则，返回 true。

inorder() 方法 (第 78 ～ 88 行) 调用 inorder(root) 遍历整棵树。inorder(TreeNode root) 方法从指定的根结点遍历树。它是一个递归方法，先递归地遍历左子树，然后遍历根结点，最后遍历右子树。当树为空时，遍历结束。

preorder() 方法 (第 91 ～ 101 行) 和 postorder() 方法 (第 104 ～ 114 行) 使用递归进行了类似的实现。

path(E e) 方法 (第 139 ～ 157 行) 以数组线性表返回结点的路径。路径从根结点开始直到该元素所在的结点。元素可能不在树中。例如，在图 25-4a 中，path(45) 包含元素 60、55 和 45 的结点，而 path(58) 包含元素 60、55 和 57 的结点。

delete() 和 iterator() 的实现 (第 61 ～ 273 行) 将在 25.8 节和 25.10 节中讨论。

> **设计模式提示**：GreateNewNode() 方法的设计应用了工厂方法模式。通过方法值返回来创建一个对象，而不是在代码中应用构造方法来创建一个对象。假设工厂方法返回一个 A 类型的对象。这样的设计使你可以重写方法来创建 A 的子类型的对象。BST 类中的 GreateNewNode() 方法返回一个 TreeNode 对象。在后续章节中，我们将重写该方法，以返回一个 TreeNode 的子类型的对象。

程序清单 25-5 给出了一个例子，使用 BST (第 4 行) 创建一棵二叉搜索树。程序向树中添加一些字符串 (第 5 ～ 11 行)，然后对该树进行中序、后序和前序遍历 (第 14 ～ 20 行)，查找一个元素 (第 24 行)，以及获取一个从包含 Peter 的结点到根结点的路径 (第 28 ～ 31 行)。

程序清单 25-5 TestBST.java

```
1  public class TestBST {
2    public static void main(String[] args) {
3      // Create a BST
4      BST<String> tree = new BST<>();
5      tree.insert("George");
6      tree.insert("Michael");
7      tree.insert("Tom");
8      tree.insert("Adam");
9      tree.insert("Jones");
10     tree.insert("Peter");
11     tree.insert("Daniel");
12
13     // Traverse tree
14     System.out.print("Inorder (sorted): ");
15     tree.inorder();
16     System.out.print("\nPostorder: ");
17     tree.postorder();
18     System.out.print("\nPreorder: ");
19     tree.preorder();
20     System.out.print("\nThe number of nodes is " + tree.getSize());
21
22     // Search for an element
23     System.out.print("\nIs Peter in the tree? " +
24       tree.search("Peter"));
25
```

```
26      // Get a path from the root to Peter
27      System.out.print("\nA path from the root to Peter is: ");
28      java.util.ArrayList<BST.TreeNode<String>> path
29        = tree.path("Peter");
30      for (int i = 0; path != null && i < path.size(); i++)
31        System.out.print(path.get(i).element + " ");
32
33      Integer[] numbers = {2, 4, 3, 1, 8, 5, 6, 7};
34      BST<Integer> intTree = new BST<>(numbers);
35      System.out.print("\nInorder (sorted): ");
36      intTree.inorder();
37    }
38  }
```

```
Inorder (sorted): Adam Daniel George Jones Michael Peter Tom
Postorder: Daniel Adam Jones Peter Tom Michael George
Preorder: George Adam Daniel Michael Jones Tom Peter
The number of nodes is 7
Is Peter in the tree? true
A path from the root to Peter is: George Michael Tom Peter
Inorder (sorted): 1 2 3 4 5 6 7 8
```

该程序在第 30 行检查 path!=null，以确保在调用 path.get(i) 之前路径不为 null。这是一个避免潜在运行时错误的防御性编程示例。

程序创建另一棵树来存储 int 值（第 34 行）。在树中插入所有的元素后，该树应该如图 25-8 所示。

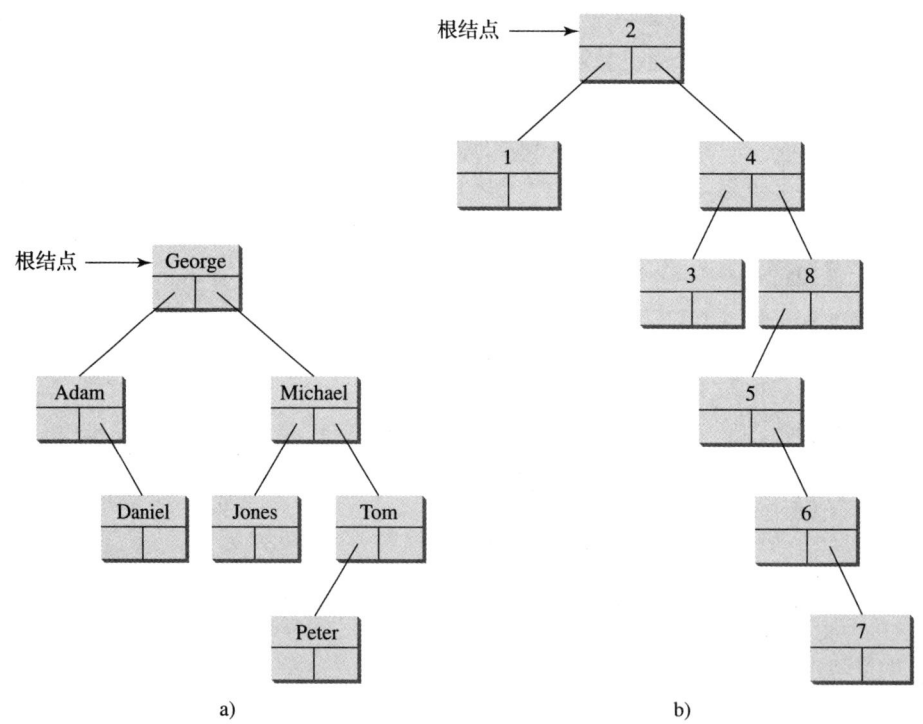

图 25-8 这里画程序清单 25-5 中创建的几个 BST

如果元素的插入顺序不同（例如，Daniel、Adam、Jones、Peter、Tom、Michael 和 George），那么树看起来可能不一样。但只要元素集合相同，中序遍历将以同样的顺序打印元素。中序遍历显示一个排好序的线性表。

复习题

25.7.1 编写代码,创建用于整数的 BST,插入数字 9,3,12,30 到树中,然后删除 9,以中序、前序和后序显示该树,查找 4,以及展示树中的元素个数。

25.7.2 在 BST 类中添加一个名为 getSmallest() 的新方法,用于返回树中的最小元素。

25.7.3 在 BST 类中添加一个名为 getSmallest() 的新方法,用于返回树中的最大元素。

25.8 删除 BST 中的一个元素

☞ **要点提示**:为了从 BST 中删除一个元素,首先需要定位该元素位置,然后在删除该元素以及重新连接树前,考虑两种情况——该结点有或者没有左子结点。

25.5 节给出了 insert(element) 方法。我们经常需要从二叉搜索树中删除一个元素,这比向二叉搜索树中添加一个元素复杂得多。

为了从二叉搜索树中删除一个元素,首先需要定位包含该元素的结点,以及它的父结点。假设 current 指向二叉搜索树中包含该元素的结点,而 parent 指向 current 结点的父结点。current 结点可能是 parent 结点的左子结点,也可能是右子结点。这里需要考虑以下两种情况:

情况 1:当前结点没有左子结点,如图 25-9a 所示。这时只需要将该结点的父结点和该结点的右子结点相连,如图 25-9b 所示。

例如,为了在图 25-10a 中删除结点 10,需要连接结点 10 的父结点和结点 10 的右子结点,如图 25-10b 所示。

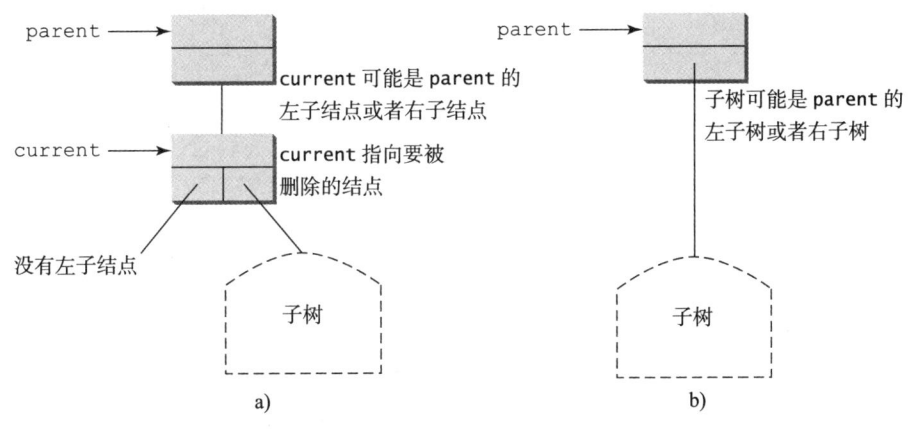

图 25-9 情况 1:当前结点没有左子结点

☞ **注意**:如果当前结点是叶子结点,这属于情况 1。例如,为了删除图 25-10a 中的元素 16,将结点 16 的右子结点(在这种情况下为 null)和它的父结点相连。

情况 2:current 结点有左子结点。让 rightMost 指向 current 结点的左子树中包含最大元素的结点,而 parentOfRightMost 指向 rightMost 结点的父结点,如图 25-11a 所示。注意,rightMost 结点不能有右子结点,但可能会有左子结点。使用 rightMost 结点中的元素值替换 current 结点中的元素值,将 parentOfRightMost 结点和 rightMost 结点的左子结点相连,然后删除 rightMost 结点,如图 25-11b 所示。

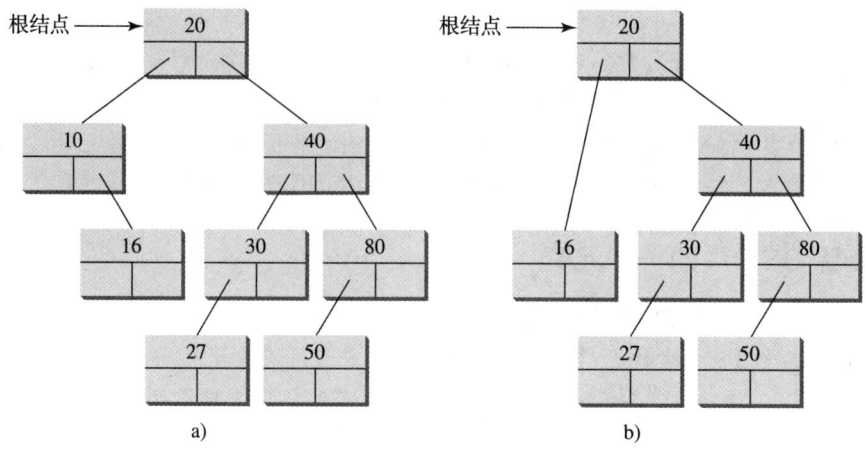

图 25-10 情况 1：从图 a 中删除结点 10 得到图 b

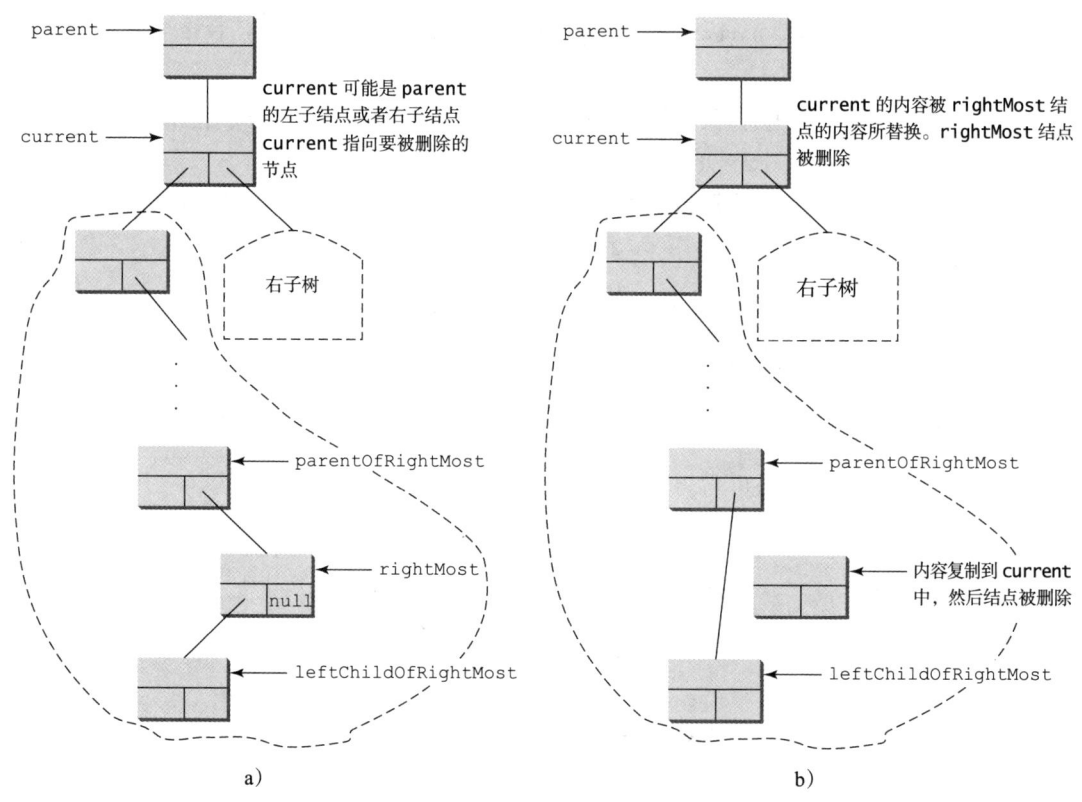

图 25-11 情况 2：当前结点有左子结点

例如，考虑删除图 25-12a 中的结点 20。rightMost 结点有一个值为 16 的元素。使用 current 结点中的 16 替换元素值 20，并将结点 10 作为结点 14 的父结点，如图 25-12b 所示。

注意：如果 current 的左子结点没有右子结点，那么 current.left 指向 current 左子树的最大元素。在这种情况下，rightMost 是 current.left，而 parentOfRightMost 是 current。必须考虑这种特殊情况，以重新连接 rightMost 的左子结点和 parentOfRightMost。

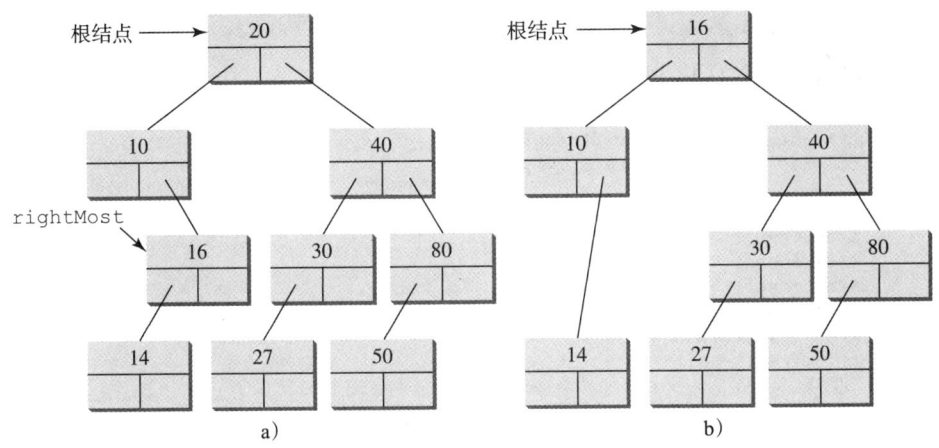

图 25-12 情况 2：从图 a 中删除结点 20 得到图 b

程序清单 25-6 描述了从二叉搜索树中删除一个元素的算法。

程序清单 25-6 从 BST 中删除一个元素

```
1   boolean delete(E e) {
2     Locate element e in the tree;
3     if element e is not found
4       return false;
5
6     Let current be the node that contains e and parent be
7       the parent of current;
8
9     if (current has no left child)  // Case 1
10      Connect the right child of current with parent;
11        Now current is not referenced, so it is eliminated;
12    else // Case 2
13      Locate the rightmost node in the left subtree of current.
14      Copy the element value in the rightmost node to current.
15      Connect the parent of the rightmost node to the left child
16        of rightmost node;
17
18    return true; // Element deleted
19  }
```

程序清单 25-4 中的第 162～220 行给出了方法 delete 的完整实现。该方法在第 164～177 行定位了要删除的结点（命名为 current），同时还定位了该结点的父结点（命名为 parent）。如果 current 为 null，那么该元素不在树内。所以，该方法返回 false（第 180 行）。注意，如果 current 是 root，则 parent 为 null。如果树为空，那么 current 和 parent 都为 null。

算法的情况 1 出现在第 183～194 行。在这种情况下，current 结点没有左子结点（即 current.left==null）。如果 parent 为 null，则将 current.right 赋给 root（第 185～187 行）；否则，根据 current 是 parent 的左子结点还是右子结点，将 current.right 赋给 parent.left 或者 parent.right（第 189～192 行）。

算法的情况 2 在第 195～216 行处理。在这种情况下，current 结点有左子结点。算法定位当前结点的左子树最右端的结点（命名为 rightMost），并且定位它的父结点（命名为 parentOfRightMost）（第 199～205 行）。用 rightMost 中的元素替换 current 中的元素（第 208 行）。根据 rightMost 是 parentOfRightMost 的右子结点还是左子结点，将 rightMost.left 赋给 parentOfRightMost.right 或者 parentOfRightMost.left（第 211～215 行）。

程序清单25-7给出从二叉搜索树中删除一个元素的测试程序。

程序清单 25-7 TestBSTDelete.java

```java
 1  public class TestBSTDelete {
 2    public static void main(String[] args) {
 3      BST<String> tree = new BST<>();
 4      tree.insert("George");
 5      tree.insert("Michael");
 6      tree.insert("Tom");
 7      tree.insert("Adam");
 8      tree.insert("Jones");
 9      tree.insert("Peter");
10      tree.insert("Daniel");
11      printTree(tree);
12
13      System.out.println("\nAfter delete George:");
14      tree.delete("George");
15      printTree(tree);
16
17      System.out.println("\nAfter delete Adam:");
18      tree.delete("Adam");
19      printTree(tree);
20
21      System.out.println("\nAfter delete Michael:");
22      tree.delete("Michael");
23      printTree(tree);
24    }
25
26    public static void printTree(BST tree) {
27      // Traverse tree
28      System.out.print("Inorder (sorted): ");
29      tree.inorder();
30      System.out.print("\nPostorder: ");
31      tree.postorder();
32      System.out.print("\nPreorder: ");
33      tree.preorder();
34      System.out.print("\nThe number of nodes is " + tree.getSize());
35      System.out.println();
36    }
37  }
```

```
Inorder (sorted): Adam Daniel George Jones Michael Peter Tom
Postorder: Daniel Adam Jones Peter Tom Michael George
Preorder: George Adam Daniel Michael Jones Tom Peter
The number of nodes is 7

After delete George:
Inorder (sorted): Adam Daniel Jones Michael Peter Tom
Postorder: Adam Jones Peter Tom Michael Daniel
Preorder: Daniel Adam Michael Jones Tom Peter
The number of nodes is 6

After delete Adam:
Inorder (sorted): Daniel Jones Michael Peter Tom
Postorder: Jones Peter Tom Michael Daniel
Preorder: Daniel Michael Jones Tom Peter
The number of nodes is 5

After delete Michael:
Inorder (sorted): Daniel Jones Peter Tom
Postorder: Peter Tom Jones Daniel
Preorder: Daniel Jones Tom Peter
The number of nodes is 4
```

图 25-13 ～图 25-15 显示了从树中删除元素时树的演变过程。

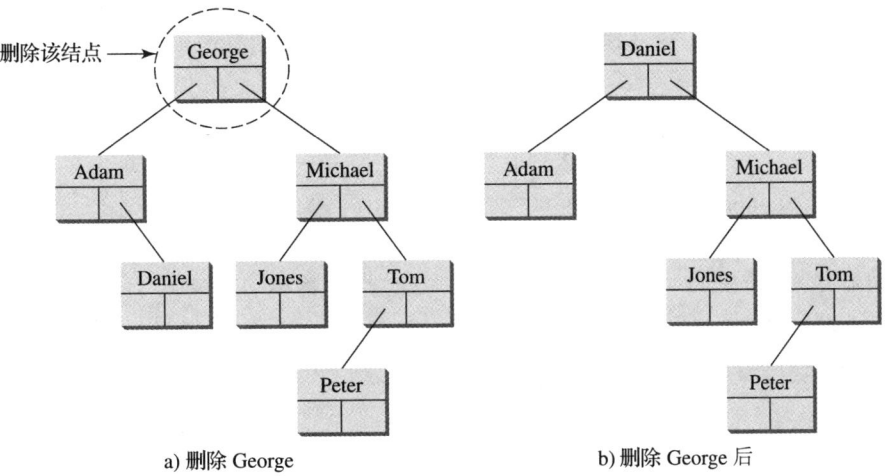

图 25-13　删除 George 属于情况 2

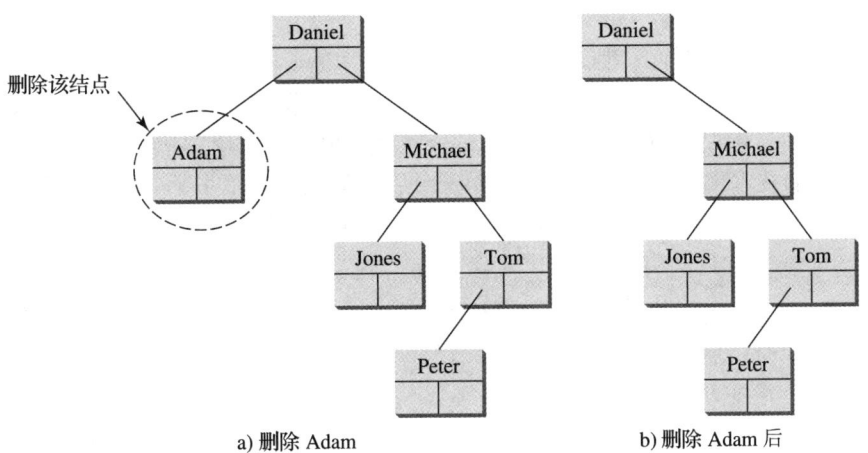

图 25-14　删除 Adam 属于情况 1

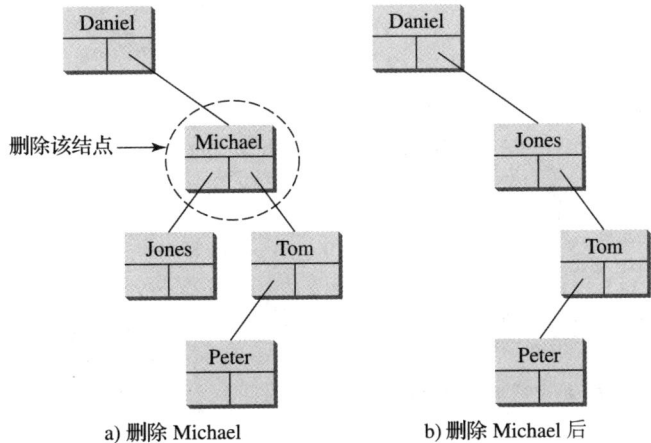

图 25-15　删除 Michael 属于情况 2

> **注意**：中序遍历、前序遍历和后序遍历的时间复杂度很明显都是 $O(n)$，因为每个结点都只遍历一次。查找、插入和删除的时间复杂度为树的高度。在最差的情况下，树的高度为 $O(n)$。平均而言，树的高度是 $O(\log n)$。因此，在一个 BST 中查找、插入、删除操作的平均时间为 $O(\log n)$。

✓ **复习题**

25.8.1 显示从图 25-4b 所示的树中删除 55 之后的结果。

25.8.2 显示从图 25-4b 所示的树中删除 60 之后的结果。

25.8.3 从 BST 中删除一个元素的时间复杂度是多少？

25.8.4 如果将程序清单 25-4 的 `delete()` 方法中属于情况 2 的第 211 ~ 215 行用下面的代码替换，算法还正确吗？

```
parentOfRightMost.right = rightMost.left;
```

25.9 树的可视化和 MVC

> **要点提示**：可以应用递归来显示一棵二叉树。

> **教学注意**：数据结构课程面临的一个挑战是激发学生的兴趣。用图形显示二叉树不仅有助于学生理解二叉树的工作机制，而且还会激发学生对程序设计的兴趣。本节介绍可视化二叉树的技术。学生也可以在其他项目中应用可视化技术。

如何显示一棵二叉树呢？它是一种递归的结构，因此你可以应用递归来显示一棵二叉树。可以简单地显示根结点，然后递归地显示两棵子树。可以应用程序清单 18-9 显示思瑞平斯基三角形的技术来显示二叉树。为简单起见，我们假设键是小于 100 的正整数。程序清单 25-8 和程序清单 25-9 给出该程序，图 25-16 显示了程序的一些运行示例。

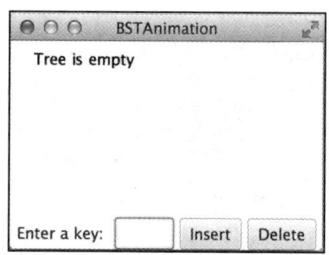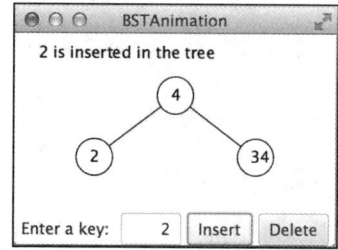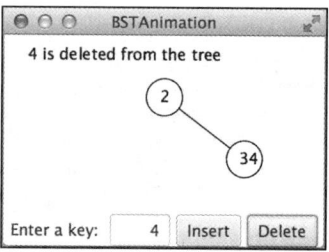

图 25-16　图形化显示一棵二叉树（来源：Oracle 或其附属公司版权所有 © 1995 ~ 2016，经授权使用）

程序清单 25-8　BSTAnimation.java

```java
 1  import javafx.application.Application;
 2  import javafx.geometry.Pos;
 3  import javafx.stage.Stage;
 4  import javafx.scene.Scene;
 5  import javafx.scene.control.Button;
 6  import javafx.scene.control.Label;
 7  import javafx.scene.control.TextField;
 8  import javafx.scene.layout.BorderPane;
 9  import javafx.scene.layout.HBox;
10
11  public class BSTAnimation extends Application {
12    @Override // Override the start method in the Application class
13    public void start(Stage primaryStage) {
14      BST<Integer> tree = new BST<>(); // Create a tree
```

```java
15
16      BorderPane pane = new BorderPane();
17      BTView view = new BTView(tree); // Create a View
18      pane.setCenter(view);
19
20      TextField tfKey = new TextField();
21      tfKey.setPrefColumnCount(3);
22      tfKey.setAlignment(Pos.BASELINE_RIGHT);
23      Button btInsert = new Button("Insert");
24      Button btDelete = new Button("Delete");
25      HBox hBox = new HBox(5);
26      hBox.getChildren().addAll(new Label("Enter a key: "),
27        tfKey, btInsert, btDelete);
28      hBox.setAlignment(Pos.CENTER);
29      pane.setBottom(hBox);
30
31      btInsert.setOnAction(e -> {
32        int key = Integer.parseInt(tfKey.getText());
33        if (tree.search(key)) { // key is in the tree already
34          view.displayTree();
35          view.setStatus(key + " is already in the tree");
36        }
37        else {
38          tree.insert(key); // Insert a new key
39          view.displayTree();
40          view.setStatus(key + " is inserted in the tree");
41        }
42      });
43
44      btDelete.setOnAction(e -> {
45        int key = Integer.parseInt(tfKey.getText());
46        if (!tree.search(key)) { // key is not in the tree
47          view.displayTree();
48          view.setStatus(key + " is not in the tree");
49        }
50        else {
51          tree.delete(key); // Delete a key
52          view.displayTree();
53          view.setStatus(key + " is deleted from the tree");
54        }
55      });
56
57      // Create a scene and place the pane in the stage
58      Scene scene = new Scene(pane, 450, 250);
59      primaryStage.setTitle("BSTAnimation"); // Set the stage title
60      primaryStage.setScene(scene); // Place the scene in the stage
61      primaryStage.show(); // Display the stage
62    }
63  }
```

程序清单 25-9 BTView.java

```java
1   import javafx.scene.layout.Pane;
2   import javafx.scene.paint.Color;
3   import javafx.scene.shape.Circle;
4   import javafx.scene.shape.Line;
5   import javafx.scene.text.Text;
6
7   public class BTView extends Pane {
8     private BST<Integer> tree = new BST<>();
9     private double radius = 15; // Tree node radius
10    private double vGap = 50; // Gap between two levels in a tree
11
12    BTView(BST<Integer> tree) {
13      this.tree = tree;
14      setStatus("Tree is empty");
```

```
15     }
16
17     public void setStatus(String msg) {
18       getChildren().add(new Text(20, 20, msg));
19     }
20
21     public void displayTree() {
22       this.getChildren().clear(); // Clear the pane
23       if (tree.getRoot() != null) {
24         // Display tree recursively
25         displayTree(tree.getRoot(), getWidth() / 2, vGap,
26           getWidth() / 4);
27       }
28     }
29
30     /** Display a subtree rooted at position (x, y) */
31     private void displayTree(BST.TreeNode<Integer> root,
32         double x, double y, double hGap) {
33       if (root.left != null) {
34         // Draw a line to the left node
35         getChildren().add(new Line(x - hGap, y + vGap, x, y));
36         // Draw the left subtree recursively
37         displayTree(root.left, x - hGap, y + vGap, hGap / 2);
38       }
39
40       if (root.right != null) {
41         // Draw a line to the right node
42         getChildren().add(new Line(x + hGap, y + vGap, x, y));
43         // Draw the right subtree recursively
44         displayTree(root.right, x + hGap, y + vGap, hGap / 2);
45       }
46
47       // Display a node
48       Circle circle = new Circle(x, y, radius);
49       circle.setFill(Color.WHITE);
50       circle.setStroke(Color.BLACK);
51       getChildren().addAll(circle,
52         new Text(x - 4, y + 4, root.element + ""));
53     }
54   }
```

在程序清单 25-8 中，创建了一棵树（第 14 行），并将一个树视图放置在面板中（第 18 行）。在将一个新的键插入树中之后（第 38 行），重新绘制这棵树（第 39 行）来反映该变化。在删除一个键之后（第 51 行），重新绘制这棵树（第 52 行）来反映该变化。

在程序清单 25-9 中，将结点显示为一个半径 radius 为 15 的圆（第 48 行）。在树中，将两层之间的距离定义为 vGap，取值 50（第 25 行）。hGap（第 32 行）定义两个结点之间的水平距离。当递归调用 displayTree 方法时，该值在下一层中减半（hGap/2）（第 37 和 44 行）。注意，在树中没有改变 vGap。

如果子树不为空，那么递归调用 displayTree 方法来显示一棵左子树（第 33～38 行）和一棵右子树（第 40～45 行）。一条直线添加到面板中来连接两个结点（第 35 和 42 行），注意该方法先将直线添加到面板中，然后添加两个圆到面板中（第 52 行），这样圆会在直线之上绘制，从而获得较好的视觉效果。

该程序假定键都是整数。可以很容易地修改该程序，使其采用泛型类型显示字符或者短字符串的键。

树的可视化是一个模型－视图－控制器（MVC）软件架构的例子。这是一个用于软件开发的重要架构。模型用于存储和处理数据，视图用于可视化地表达数据，控制器处理用户和模型的交互，并且控制视图，如图 25-17 所示。

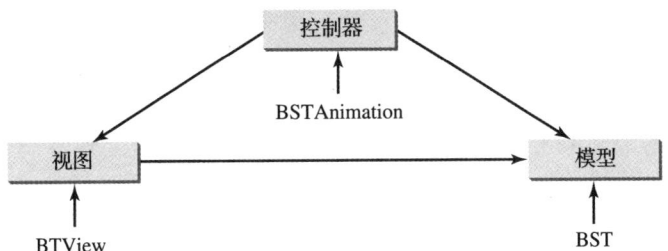

图 25-17 控制器获得数据并且将其存储在模型中。视图显示存储在模型中的数据

MVC 架构将数据的存储和处理与数据的可视化表示分离。它具有两个主要的好处：
- 使得多个视图成为可能，这样数据可以通过同样一个模型来分享。例如，你可以创建一个新的视图，将树显示为根结点在左边，而树水平向右生长（参见编程练习题 25.11）。
- 简化了编写复杂程序的任务，使得组件可扩展，并且易于维护。可以改变视图而不影响模型，反之亦然。

✓ 复习题

25.9.1 如果树为空，那么 displayTree 方法将被调用多少次？如果树有 100 个结点，那么 displayTree 方法将被调用多少次？

25.9.2 displayTree 方法以哪种顺序来访问树中的结点——中序、前序还是后序？

25.9.3 如果程序清单 25-9 中第 47 ~ 52 行的代码移到第 33 行，将会发生什么情况？

25.9.4 什么是 MVC？MVC 的好处是什么？

25.10 迭代器

⌐ 要点提示：BST 是可遍历的，因为它被定义为 java.lang.Iterable 接口的子类型。

inorder()、preorder() 和 postorder() 方法分别以中序、前序和后序方式显示二叉树中的元素。这些方法都限于显示树中的元素。如果要处理二叉树中的元素，而不是显示它们，就不能使用这些方法。回顾一下，遍历一个规则集或者线性表的元素时提供了一个迭代器。可以以同样的方式将迭代器应用到一棵二叉树上，从而提供一种统一的方式来遍历二叉树中的元素。

java.util.Iterator 接口定义了 iterator 方法，该方法返回一个 java.util.Iterator 的实例。java.util.Iterator 接口（如图 25-18 所示）定义了迭代器的通用特性。

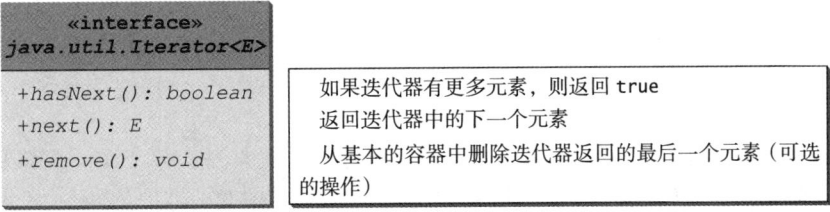

图 25-18 Iterator 接口定义了遍历一个容器中元素的统一方式

Tree 接口继承自 java.util.Collection。由于 Collection 继承自 java.lang.Iterable，所以 BST 是 Iterable 的子类。Iterable 接口包含 iterator() 方法，该方法返回 java.util.Iterator 的一个实例。

可以使用中序、前序或后序遍历二叉树。由于中序很常用，我们将使用中序来遍历一个二叉树中的元素。我们定义一个名为 InorderIterator 的迭代器类来实现 java.util.Iterator 接口，见程序清单 25-4（第 228 ～ 273 行）。iterator 方法简单地返回一个 InorderIterator 的实例（第 224 行）。

InorderIterator 的构造方法调用 inorder 方法（第 240 行）。inorder(root) 方法（第 239 ～ 249 行）在 list 中存储树中的所有元素。这些元素以中序方式遍历。

一旦创建一个 Iterator 对象，它的 current 值初始化为 0（第 232 行），它指向线性表中的第一个元素。调用 next() 方法返回当前元素，并将 current 移到指向线性表的下一个元素（第 261 行）。

hasNext() 方法检查 current 是否仍然在 list 的范围之内（第 253 行）。

remove() 方法从树中删除最后一次 next() 返回的元素（第 269 行）。此后，创建一个新的线性表（第 270 ～ 271 行）。注意，不需要改变 current。

程序清单 25-10 给出一个在 BST 中存储字符串的测试程序，并且以大写形式显示所有字符串。

程序清单 25-10 TestBSTWithIterator.java

```
1  public class TestBSTWithIterator {
2    public static void main(String[] args) {
3      BST<String> tree = new BST<>();
4      tree.insert("George");
5      tree.insert("Michael");
6      tree.insert("Tom");
7      tree.insert("Adam");
8      tree.insert("Jones");
9      tree.insert("Peter");
10     tree.insert("Daniel");
11
12     for (String s: tree)
13       System.out.print(s.toUpperCase() + " ");
14   }
15 }
```

```
ADAM DANIEL GEORGE JONES MICHAEL PETER TOM
```

foreach 循环（第 12 ～ 13 行）使用了一个迭代器来遍历树中的所有元素。

☛ **设计指南**：迭代器是一个重要的软件设计模式。它提供遍历容器内元素的统一方法，同时隐藏该容器的结构细节。通过实现相同的接口 java.util.Iterator，可以编写一个程序以统一的方式遍历所有容器的元素。

☛ **注意**：java.util.Iterator 定义了一个前向迭代器，它以前向的方向遍历迭代器中的元素，每个元素只能遍历一次。Java API 还提供 java.util.ListIterator，它支持前向遍历和后向遍历。如果你的数据结构要保证遍历的灵活性，可以将迭代器类定义为 java.util.ListIterator 的一个子类。

迭代器的实现不是很高效。每次通过迭代器删除一个元素时，整个线性表都要重新构造（程序清单 25-4 中第 270 ～ 271 行）。客户程序应该总是采用 BST 类中的 delete 方法来删除一个元素。为了防止用户使用迭代器中的 remove 方法，如下实现迭代器：

```
public void remove() {
  throw new UnsupportedOperationException
    ("Removing an element from the iterator is not supported");
}
```

在使得 remove 方法不被迭代器类支持后，可以使迭代器更加高效，因为无须为树中的元素维护一个线性表。可以使用栈来存储结点，栈顶的结点包含从 next() 方法返回的元素。如果树是平衡的，最大的栈尺寸将为 $O(\log n)$。

✓ 复习题

25.10.1 什么是迭代器？

25.10.2 java.lang.Iterable<E> 接口中定义了什么方法？

25.10.3 假设你从程序清单 25-3 的第 3 行删除了 implements Collection<E>，程序清单 25-10 还能编译吗？

25.10.4 作为 Iterable<E> 的子类型的好处是什么？

25.10.5 编写一条语句，显示名为 tree 的 BST 对象中的最大和最小元素（提示：使用 java.util.Collections 类中的 min 和 max 方法）。

25.11 示例学习：数据压缩

要点提示：霍夫曼编码通过使用较少的比特对较常出现的字符编码，从而压缩数据。字符的编码是基于字符在文本中出现的次数使用二叉树来构建的，该树称为霍夫曼编码树。

压缩数据是一个常见的任务。压缩文件的应用很多，本节介绍 David Huffman 在 1952 年发明的霍夫曼编码。

在 ASCII 码中，每个字符都被编码为 8 比特。如果一个文本中包含 100 个字符，则需要 800 比特来表示该文本。霍夫曼编码通过使用较少的比特对文本中常用的字符编码，以及较多的比特对不常用的字符编码来减少文件的整体大小。霍夫曼编码中，字符的编码基于字符在文本中出现的次数使用二叉树来构建，该树称为霍夫曼编码树（Huffman coding tree）。假设该文本是 Mississippi，它的霍夫曼树就如图 25-19a 所示。结点左侧的边和右侧的边分别被赋值 0 和 1。每个字符都是树中的一个叶结点。字符的编码由从根结点到叶结点的路径上的边的值所组成，如图 25-19b 所示。因为文本中 i 和 s 出现得比 M 和 p 多，所以它们被赋予更短的编码。

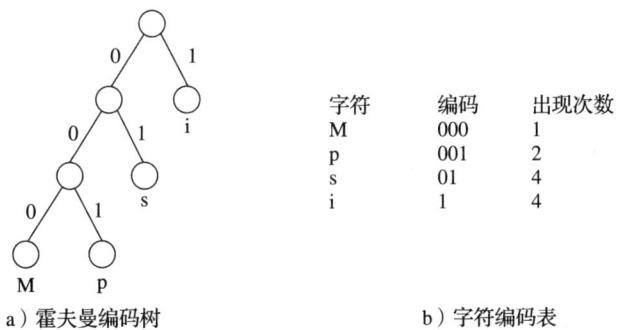

a) 霍夫曼编码树　　　　　　　　b) 字符编码表

图 25-19　使用编码树基于字符在文本中出现的次数来构建字符的编码

编码树也用于将一个比特序列解码为字符。为了做到这点，从序列中的第一个比特开始，基于比特值决定是走向树的根结点的左分支还是右分支。考虑下一个比特，然后继续基于比特值决定是走向左分支还是右分支。当到达一个叶结点时，就找到了一个字符。流中的下一个比特就是下一个字符的第一个比特。例如，数据流 011001 被解码为 sip，其中 01 匹配 s，1 匹配 i，001 匹配 p。

基于图 25-19 所示的编码方案，

Mississippi =========>编码为 000101011010110010011 ==========>解码为 Mississippi

为了构建一棵霍夫曼编码树，使用如下算法：

1）从由树构成的森林开始。每棵树都包含一个表示字符的结点。每个结点的权重为该字符在文本中出现的次数。

2）重复以下步骤来合并树，直到只有一棵树为止：选择两棵有最小权重的树，创建一个新结点作为它们的父结点。这棵新树的权重是子树的权重和。

3）对于每个内部结点，给它的左边赋值 0，右边赋值 1。所有的叶子结点都表示文本中的字符。

下面是一个为文本 Mississippi 构建编码树的例子。字符的出现次数表如图 25-19b 所示。初始情况下，森林包含单结点树，如图 25-20a 所示。重复组合树以形成更大的树，直到只留下一棵树，如图 25-20b ~ 图 25-20d 所示。

值得注意的是，任何编码不会是另外一个编码的前缀。这个属性保证了流可以无二义性地解码。

算法设计提示：这里使用的算法是贪婪算法（greedy algorithm）的一个示例。贪婪算法经常用于解决优化问题。算法做出局部最优的选择，并希望这样的选择会导致全局最优。这个示例中，算法总是选择具有最小权重的两棵树，并且创建一个新的结点作为它们的父结点。这种直观的最优局部解的确引向了最后构造霍夫曼树的最优解。

作为另外一个示例，考虑将钱兑换为可能的最少硬币。一种贪婪算法将优先使用最大的可能硬币。例如，对于 98 美分，将使用 3 个 quarter（25 美分）凑成 75 美分，然后加上两个 dime（10 美分）凑成 95 美分，最后再加上 3 个 penny（1 美分）来凑成 98 美分。贪婪算法找到了该问题的一个最优解。然而，贪婪算法并不是总能找到最优的结果。参见编程练习题 11.19 的装箱问题。

程序清单 25-11 给出了一个程序，提示用户输入一个字符串，然后显示文本中字符出现次数的表格，并且显示每个字符的霍夫曼编码。

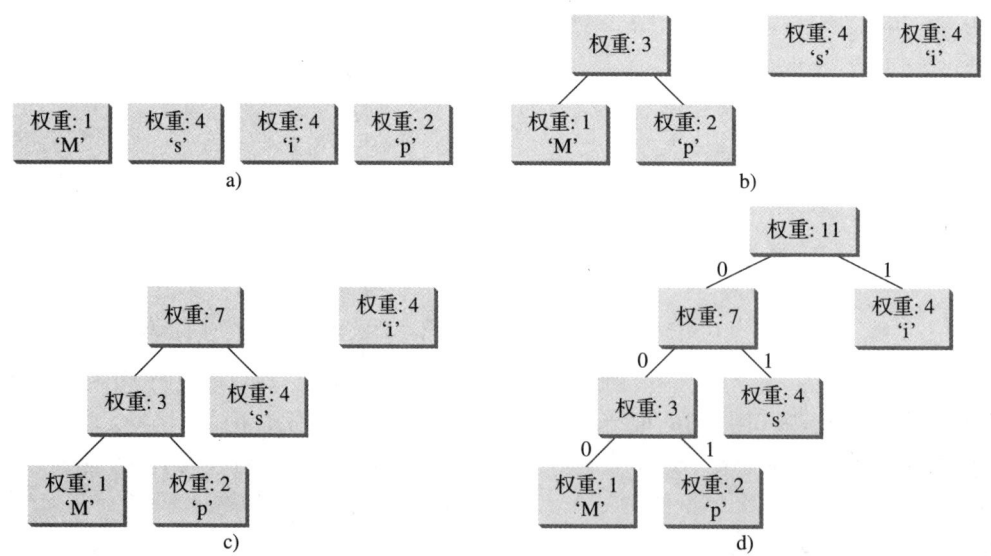

图 25-20 通过重复地组合两棵最小权重的树来构建编码树

程序清单 25-11 HuffmanCode.java

```java
 1  import java.util.Scanner;
 2
 3  public class HuffmanCode {
 4    public static void main(String[] args) {
 5      Scanner input = new Scanner(System.in);
 6      System.out.print("Enter text: ");
 7      String text = input.nextLine();
 8
 9      int[] counts = getCharacterFrequency(text); // Count frequency
10
11      System.out.printf("%-15s%-15s%-15s%-15s\n",
12        "ASCII Code", "Character", "Frequency", "Code");
13
14      Tree tree = getHuffmanTree(counts); // Create a Huffman tree
15      String[] codes = getCode(tree.root); // Get codes
16
17      for (int i = 0; i < codes.length; i++)
18        if (counts[i] != 0) // (char)i is not in text if counts[i] is 0
19          System.out.printf("%-15d%-15s%-15d%-15s\n",
20            i, (char)i + "", counts[i], codes[i]);
21    }
22
23    /** Get Huffman codes for the characters
24     * This method is called once after a Huffman tree is built
25     */
26    public static String[] getCode(Tree.Node root) {
27      if (root == null) return null;
28      String[] codes = new String[128];
29      assignCode(root, codes);
30      return codes;
31    }
32
33    /* Recursively get codes to the leaf node */
34    private static void assignCode(Tree.Node root, String[] codes) {
35      if (root.left != null) {
36        root.left.code = root.code + "0";
37        assignCode(root.left, codes);
38
39        root.right.code = root.code + "1";
40        assignCode(root.right, codes);
41      }
42      else {
43        codes[(int)root.element] = root.code;
44      }
45    }
46
47    /** Get a Huffman tree from the codes */
48    public static Tree getHuffmanTree(int[] counts) {
49      // Create a heap to hold trees
50      Heap<Tree> heap = new Heap<>(); // Defined in Listing 23.9
51      for (int i = 0; i < counts.length; i++) {
52        if (counts[i] > 0)
53          heap.add(new Tree(counts[i], (char)i)); // A leaf node tree
54      }
55
56      while (heap.getSize() > 1) {
57        Tree t1 = heap.remove(); // Remove the smallest-weight tree
58        Tree t2 = heap.remove(); // Remove the next smallest
59        heap.add(new Tree(t1, t2)); // Combine two trees
60      }
61
62      return heap.remove(); // The final tree
63    }
64
65    /** Get the frequency of the characters */
```

```java
 66   public static int[] getCharacterFrequency(String text) {
 67     int[] counts = new int[128]; // 128 ASCII characters
 68
 69     for (int i = 0; i < text.length(); i++)
 70       counts[(int)text.charAt(i)]++; // Count the characters in text
 71
 72     return counts;
 73   }
 74
 75   /** Define a Huffman coding tree */
 76   public static class Tree implements Comparable<Tree> {
 77     Node root; // The root of the tree
 78
 79     /** Create a tree with two subtrees */
 80     public Tree(Tree t1, Tree t2) {
 81       root = new Node();
 82       root.left = t1.root;
 83       root.right = t2.root;
 84       root.weight = t1.root.weight + t2.root.weight;
 85     }
 86
 87     /** Create a tree containing a leaf node */
 88     public Tree(int weight, char element) {
 89       root = new Node(weight, element);
 90     }
 91
 92     @Override /** Compare trees based on their weights */
 93     public int compareTo(Tree t) {
 94       if (root.weight < t.root.weight) // Purposely reverse the order
 95         return 1;
 96       else if (root.weight == t.root.weight)
 97         return 0;
 98       else
 99         return -1;
100     }
101
102     public class Node {
103       char element; // Stores the character for a leaf node
104       int weight; // weight of the subtree rooted at this node
105       Node left; // Reference to the left subtree
106       Node right; // Reference to the right subtree
107       String code = ""; // The code of this node from the root
108
109       /** Create an empty node */
110       public Node() {
111       }
112
113       /** Create a node with the specified weight and character */
114       public Node(int weight, char element) {
115         this.weight = weight;
116         this.element = element;
117       }
118     }
119   }
120 }
```

Enter text: Welcome ⏎Enter

ASCII Code	Character	Frequency	Code
87	W	1	110
99	c	1	111
101	e	2	10
108	l	1	011
109	m	1	010
111	o	1	00

该程序提示用户输入一个文本字符串（第5～7行），然后计算文本中字符的出现次数（第9行）。getCharacterFrequency方法（第66～73行）创建一个数组counts来统计文本中对应128个ASCII字符的每个字符的出现次数。如果文本中出现一个字符，它对应的计数器就加1（第70行）。

程序基于counts得到霍夫曼编码树（第14行）。该树由链接的结点构成。Node类在第102～118行定义。每个结点都包含属性element（存储字符）、weight（存储该结点下的子树的权重）、left（到左子树的链接）、right（到右子树的链接）和code（存储该字符的霍夫曼编码）。Tree类（第76～119行）包含根结点作为属性。可以从该根结点访问树中的所有结点。Tree类实现了Comparable。这些树基于它们的权重来进行比较。比较顺序被故意颠倒（第93～100行），从而最小权重的树首先从树的堆中删除。

getHuffmanTree方法返回一棵霍夫曼编码树。初始情况下，创建单结点树并将其添加到堆中（第50～54行）。在while循环的每次迭代中（第56～60行），将两棵最小权重的树从堆中删除，然后将它们组合成一棵大树，接着将新树添加到堆中。这个过程持续到堆中只包含一棵树为止，这就是我们给出的文本最终的霍夫曼树。

assignCode方法给树中的每个结点赋予编码（第34～45行）。getCode方法获取每个叶子结点中字符的编码（第26～31行）。codes[i]元素包含字符(char)i的编码，其中i从0到127。注意，如果(char)i不在文本中，那么codes[i]为null。

✓ 复习题

25.11.1 霍夫曼树中的每个内部结点具有两个子结点，对吗？

25.11.2 什么是贪婪算法？举一个例子。

25.11.3 如果程序清单25-9中第50行的Heap类替换为java.util.PriorityQueue，程序还能工作吗？

25.11.4 如何用一行代码替换程序清单25-11中的第94～99行？

关键术语

binary search tree（二叉搜索树）
binary tree（二叉树）
breadth-first traversal（广度优先遍历）
depth（深度）
depth-first traversal（深度优先遍历）
greedy algorithm（贪婪算法）
height（高度）
Huffman coding（霍夫曼编码）

inorder traversal（中序遍历）
leaf（叶子结点）
length（长度）
level（层）
postorder traversal（后序遍历）
preorder traversal（前序遍历）
sibling（兄弟结点）
tree traversal（树的遍历）

本章小结

1. 二叉搜索树（BST）是一种层次型数据结构。本章学习了如何定义和实现BST类，如何向/从BST插入和删除元素，以及如何使用中序、后序、前序、深度优先以及广度优先搜索来遍历BST。

2. 迭代器是一个对象，它提供了遍历像集合、线性表或二叉树这样的容器中的元素的统一方法。本章学习了如何定义和实现遍历二叉树中元素的迭代器类。

3. 霍夫曼编码是一种压缩数据的方案，它使用较少的比特来编码更常出现的字符。字符的编码是使用二叉树并基于它在文本中出现的次数来构建的，该二叉树称为霍夫曼编码树。

测试题

回答位于本书配套网站上的本章测试题。

编程练习题

25.2 ~ 25.6 节

*25.1 （树的高度）定义一个名为 BSTWithHeight 的新类，继承自 BST，并具有以下方法：

```
/** Return the height of this binary tree */
public int height()
```

使用 https://liveexample.pearsoncmg.com/test/Exercise25_01.txt 测试你的代码。

**25.2 （不使用递归实现中序遍历）使用栈替代递归，实现 BST 中的 inorder 方法。编写一个测试程序，提示用户输入 10 个整数，将它们保存在一个 BST 中，然后调用 inorder 方法来显示这些元素。

**25.3 （测试完美二叉树）完美二叉树是指所有层都被填满的完全二叉树。定义一个名为 BSTWithTestPerfect 的新类，继承自 BST 类，并具有以下方法（提示：非空的完美二叉树中的结点个数是 $2^{height+1}-1$）。

```
/** Returns true if the tree is a perfect binary tree */
public boolean isPerfectBST()
```

使用 https://liveexample.pearsoncmg.com/test/Exercise25_03.txt 测试你的代码。

**25.4 （不使用递归实现前序遍历）使用栈替代递归，实现 BST 中的 preorder 方法。编写一个测试程序，提示用户输入 10 个整数，将它们保存在一个 BST 中，然后调用 preorder 方法来显示这些元素。

**25.5 （不使用递归实现后序遍历）使用栈替代递归，实现 BST 中的 postorder 方法。编写一个测试程序，提示用户输入 10 个整数，将它们保存在一个 BST 中，然后调用 postorder 方法来显示这些元素。

**25.6 （找出叶子结点）定义一个名为 BSTWithNumberOfLeaves 的新类，继承自 BST 并具有以下方法：

```
/** Return the number of leaf nodes */
public int getNumberOfLeaves()
```

使用 https://liveexample.pearsoncmg.com/test/Exercise25_06.txt 测试你的代码。

**25.7 （找出非叶子结点）定义一个名为 BSTWithNumberOfLeaves 的新类，继承自 BST 并具有以下方法：

```
/** Return the number of nonleaf nodes */
public int getNumberofNonLeaves()
```

使用 https://liveexample.pearsoncmg.com/test/Exercise25_07.txt 测试你的代码。

***25.8 （实现双向迭代器）java.util.Iterator 接口定义了一个前向迭代器。Java API 也提供了 java.util.ListIterator 接口以定义一个双向迭代器。研究 ListIterator 并定义一个 BST 类的双向迭代器。

**25.9 （树的 clone 和 equals 方法）实现 BST 类中的 clone 和 equals 方法。两棵 BST 树如果包含相同的元素，则它们是相等的。clone 方法返回一棵 BST 树的一个相同的副本。

25.10 （前序迭代器）添加以下方法到 BST 类中，返回一个迭代器，用于前序遍历 BST 中的元素。

```
/** Return an iterator for traversing the elements in preorder */
java.util.Iterator<E> preorderIterator()
```

25.11 （显示树）编写一个新的类，用于水平地显示树，根结点在左边，如图 25-21 所示。

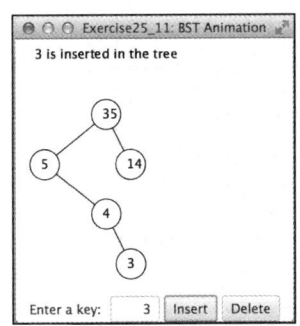

图 25-21 水平地显示一棵二叉树

**25.12 （测试 BST）设计和编写一个完整的测试程序，测试程序清单 25-4 中的 BST 类是否符合所有要求。

**25.13 （在 BSTAnimation 中添加新按钮）修改程序清单 25-8，添加三个新按钮——Show Inorder、Show Preorder 和 Show Postorder，用于在标签中显示结果，如图 25-22 所示。还需要修改程序清单 25-4 来实现 inorderList()、preorderList() 和 postorderList() 方法，这样，这些方法就能以中序、前序和后序返回一个由结点元素构成的 List，如下所示：

```
public java.util.List<E> inorderList();
public java.util.List<E> preorderList();
public java.util.List<E> postorderList();
```

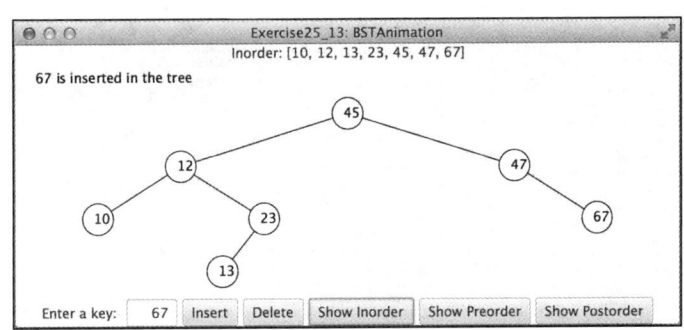

图 25-22 单击图中的 Show Inorder、Show Preorder 或者 Show Postorder 按钮，就会在标签中分别以中序、前序和后序显示元素（来源：Oracle 或其附属公司版权所有 © 1995 ~ 2016，经授权使用）

*25.14 （广度优先遍历）定义一个名为 BSTWithBFT 的新类，继承自 BST 并具有以下方法：

```
/** Display the nodes in a breadth-first traversal */
public void breadthFirstTraversal()
```

使用 https://liveexample.pearsoncmg.com/test/Exercise25_14.txt 测试你的代码。

***25.15 （BST 的父引用）重新定义 TreeNode，添加一个对结点的父结点的引用，如下所示：

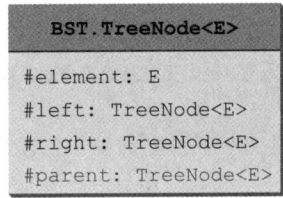

重新实现 BST 类中的 insert 和 delete 方法，为树中的每个结点更新父结点。在 BST 中

添加以下新方法：

```
/** Return the node for the specified element.
  * Return null if the element is not in the tree. */
private TreeNode<E> getNode(E element)

/** Return true if the node for the element is a leaf */
private boolean isLeaf(E element)

/** Return the path of elements from the specified element
  * to the root in an array list. */
public ArrayList<E> getPath(E e)
```

编写一个测试程序，提示用户输入 10 个整数，将它们添加到树中，从树中删除第一个整数，然后显示到所有叶子结点的路径。下面是一个运行示例：

```
Enter 10 integers: 45 54 67 56 50 45 23 59 23 67  ↵Enter
[50, 54, 23]
[59, 56, 67, 54, 23]
```

***25.16 （数据压缩：霍夫曼编码）编写一个程序，提示用户输入一个文件名，显示文件中字符出现次数的表格以及每个字符的霍夫曼编码。

***25.17 （数据压缩：霍夫曼编码的动画）编写一个程序，允许用户输入一个文本，然后显示基于该文本的霍夫曼编码树，如图 25-23a 所示。在一棵子树的代表根结点的圆中显示该子树的权重，显示每个叶子结点的字符，在标签中显示对文本编码的比特。当用户单击 Decode Text 按钮时，一个比特字符串被解码为一个文本显示在标签中，如图 25-23b 所示。

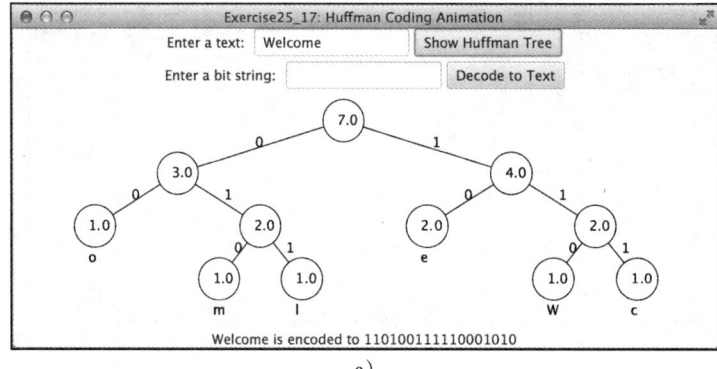

a)

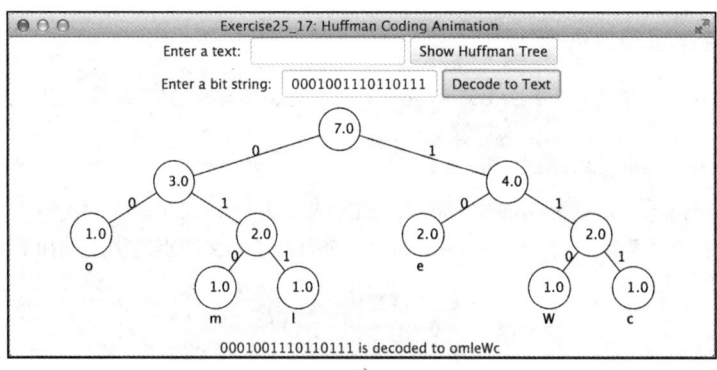

b)

图 25-23　a）动画显示给定文本字符串的编码树，并在标签中显示文本编码成的比特；b）输入一个比特串，在标签中显示对应的文本（来源：Oracle 或其附属公司版权所有 © 1995～2016，经授权使用）

***25.18 （压缩一个文件）编写一个程序，使用霍夫曼编码将源文件压缩为目标文件。首先使用 `ObjectOutputStream` 将霍夫曼编码输出到目标文件中，然后使用编程练习题 17.17 的 `BitOutputStream` 输出编码后的二进制内容到目标文件中。通过命令行使用如下命令传递文件：

```
java Exercise25_18 sourcefile targetfile
```

***25.19 （解压缩一个文件）前一个练习题压缩一个文件。压缩的文件包含了霍夫曼编码以及压缩的内容。编写一个程序，使用以下命令将一个源文件解压缩为目标文件：

```
java Exercise25_19 sourcefile targetfile
```

第 26 章

Introduction to Java Programming and Data Structures, Comprehensive Version, Twelfth Edition

AVL 树

教学目标
- 了解 AVL 树是什么（26.1 节）。
- 理解如何使用 LL 旋转、LR 旋转、RR 旋转以及 RL 旋转来重新平衡一棵树（26.2 节）。
- 通过继承 BST 类设计 AVLTree 类（26.3 节）。
- 在 AVL 树中插入元素（26.4 节）。
- 实现树的重新平衡（26.5 节）。
- 从 AVL 树中删除元素（26.6 节）。
- 实现 AVLTree 类（26.7 节）。
- 测试 AVLTree 类（26.8 节）。
- 分析在 AVL 树中查找、插入和删除操作的复杂度（26.9 节）。

26.1 引言

要点提示：AVL 树是平衡二叉搜索树。

第 25 章介绍了二叉搜索树。二叉树的查找、插入和删除操作的时间依赖于树的高度。最坏情形下，高度为 $O(n)$。如果一棵树是完全平衡的（perfectly balanced），即一棵完全二叉树，它的高度是 $\log n$。我们可以维护一棵完全平衡的树吗？可以的，但是这样做的代价比较大。一个权衡后的做法是维护一棵良好平衡的树，即每个结点的两个子树的高度基本一样。本章介绍 AVL 树。奖励章节第 40 和 41 章分别介绍 2-4 树和红黑树。

AVL 树是良好平衡的。AVL 树由两个俄罗斯计算机学家 G. M. Adelson-Velsky 和 E. M. Landis（因此命名为 AVL）于 1962 年发明。在一棵 AVL 树中，每个结点的子树的高度差距为 0 或者 1。可以得出一个 AVL 树的最大高度为 $O(\log n)$。

在一棵 AVL 树中插入或者删除一个元素的过程与在一棵普通二叉搜索树中一样，不同的是可能需要在插入或者删除操作之后重新进行平衡。一个结点的平衡因子（balance factor）是其右子树的高度减去左子树的高度。例如，图 26-1a 中结点 87 的平衡因子是 0，结点 67 的

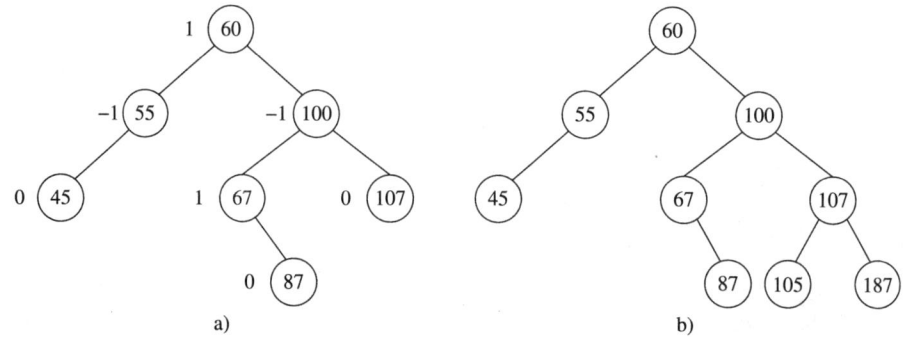

图 26-1 平衡因子决定一个结点是否平衡

是 1，结点 55 的是 -1。如果一个结点的平衡因子为 -1、0 或者 1，那么称该结点是平衡的（balanced）。如果结点的平衡因子为 -1 或更小，则该结点被认为是左偏重（left-heavy）的，如果平衡因子为 +1 或更大，则认为是右偏重（right-heavy）的。

教学注意：可以参见网址 liveexample.pearsoncmg.com/dsanimation/AVLTreeeBook.html，通过交互式 GUI 来观察 AVL 树是如何工作的，如图 26-2 所示。

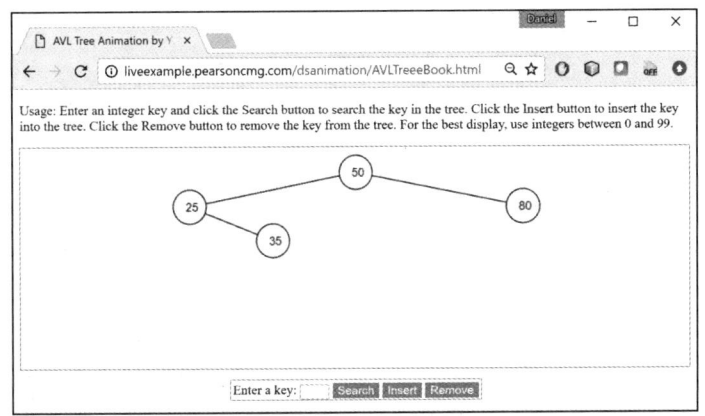

图 26-2　动画工具使你可以插入、删除和查找元素（来源：Oracle 或其附属公司版权所有 © 1995～2016，经授权使用）

26.2　重新平衡树

要点提示：从 AVL 树中插入或者删除一个元素后，如果树变得不平衡了，执行一个旋转操作来重新平衡该树。

如果一个结点在插入或者删除操作后不平衡了，需要重新进行平衡。一个结点重新获得平衡的过程称为旋转（rotation）。有 4 种可能的旋转：LL、RR、LR 以及 RL。

LL 旋转：LL 类型的不平衡（LL imbalance）发生在结点 A 的如下情况：A 有一个 -2 的平衡因子，而左子结点 B 有一个 -1 或者 0 的平衡因子，如图 26-3a 所示。这种类型的不平衡可以通过执行 A 上的一次右旋转来修复，如图 26-3b 所示。

RR 旋转：RR 类型的不平衡（RR imbalance）发生在结点 A 的如下情况：A 有一个 +2 的平衡因子，而右子结点 B 有一个 +1 或者 0 的平衡因子，如图 26-4a 所示。这种类型的不平衡可以通过执行 A 上的一次左旋转来修复，如图 26-4b 所示。

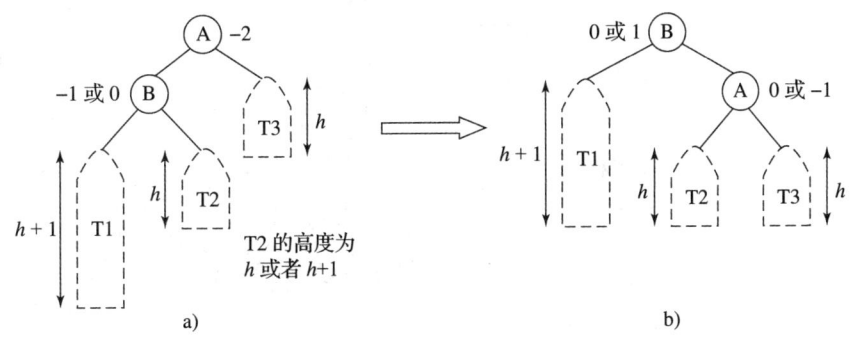

图 26-3　一个 LL 旋转修复 LL 类型不平衡

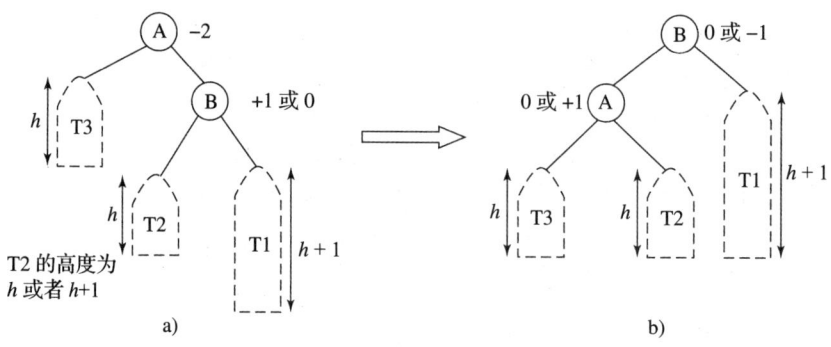

图 26-4　一个 RR 旋转修复 RR 类型不平衡

LR 旋转：LR 类型的不平衡（LR imbalance）发生在结点 A 的如下情况：A 有一个 -2 的平衡因子，而左子结点 B 有一个 +1 的平衡因子，如图 26-5a 所示。假设 B 的右子结点为 C。这种类型的不平衡可以通过执行两次旋转（首先在 B 上的一次左旋转，然后在 A 上的一次右旋转）来修复，如图 26-5b 所示。

RL 旋转：RL 类型的不平衡（RL imbalance）发生在结点 A 的如下情况：A 有一个 +2 的平衡因子，而右子结点 B 有一个 -1 的平衡因子，如图 26-6a 所示。假设 B 的左子结点为 C。这种类型的不平衡可以通过执行两次旋转（首先在 B 上的一次右旋转，然后在 A 上的一次左旋转）来修复，如图 26-6b 所示。

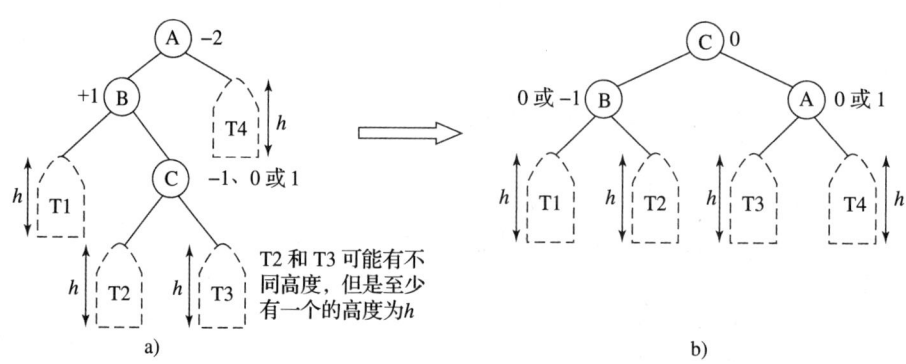

图 26-5　一个 LR 旋转修复 LR 类型不平衡

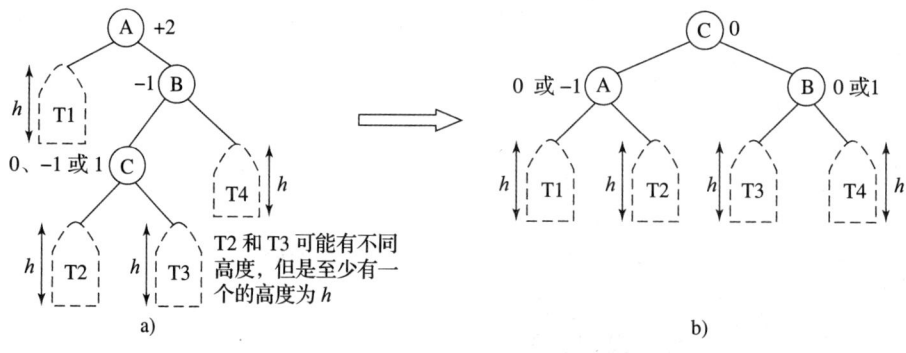

图 26-6　一个 RL 旋转修复 RL 类型不平衡

复习题

26.2.1　什么是 AVL 树？描述下面的词汇：平衡因子、左偏重、右偏重。

26.2.2 给出如图 26-1 中所示树的每个结点的平衡因子。

26.2.3 描述一个 AVL 树的 LL 旋转、RR 旋转、LR 旋转以及 RL 旋转。

26.3 为 AVL 树设计类

☛ 要点提示：由于 AVL 树是二叉搜索树，所以将 AVLTree 设计为 BST 的子类。

由于 AVL 树是二叉搜索树，所以可以通过继承 BST 类来定义 AVLTree 类，如图 26-7 所示。BST 和 TreeNode 类在 25.2.5 节中定义。

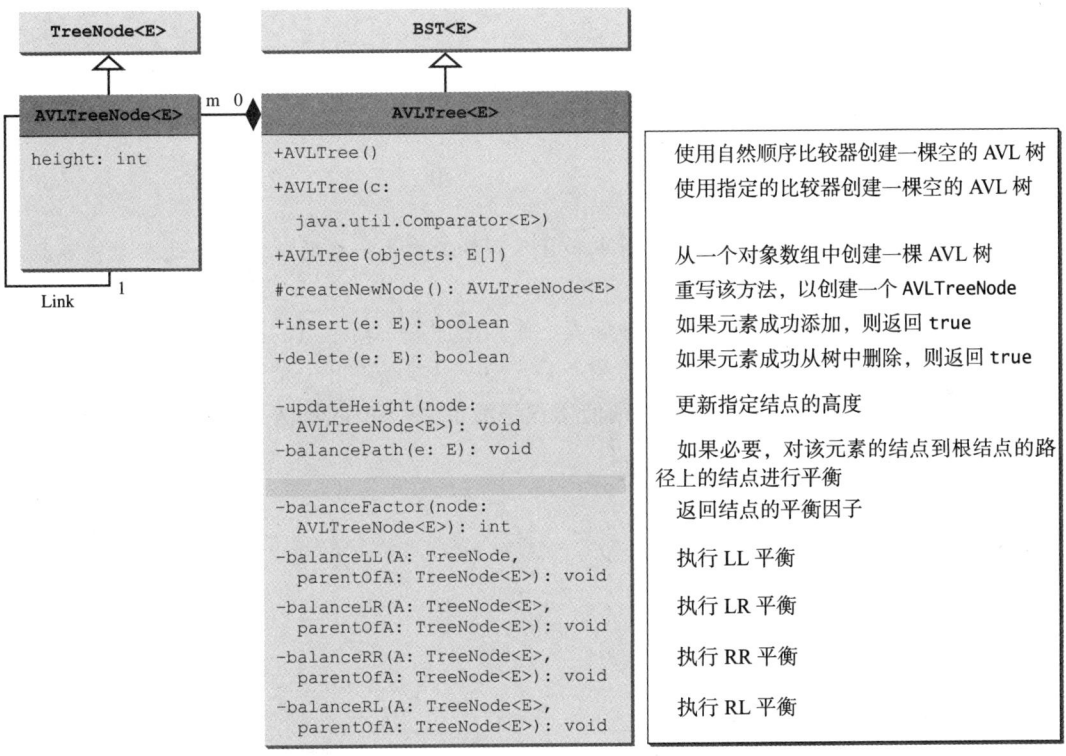

图 26-7 AVLTree 类继承自 BST，为 insert 和 delete 方法提供了新的实现

为了平衡一棵树，需要知道每个结点的高度。为方便起见，保存每个结点的高度在 AVLTreeNode 中，并定义 AVLTreeNode 为 BST.TreeNode 的子类。注意 TreeNode 定义为 BST 中的静态内部类。AVLTreeNode 将定义为 AVLTree 的静态内部类。TreeNode 包含了数据域 element、left、right，这些被 AVLTreeNode 所继承。因此，AVLTreeNode 包含了 4 个数据域，如图 26-8 所示。

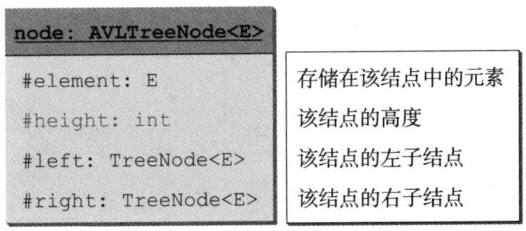

图 26-8 一个 AVLTreeNode 包含保护类型数据域 element、height、left、right

BST 类中的 createNewNode() 方法创建了一个 TreeNode 对象。该方法在 AVLTree 类中

被重写，用于创建一个 AVLTreeNode。注意，BST 类中 createNewNode() 方法返回类型为 TreeNode，而 AVLTree 类中 createNewNode() 方法返回类型为 AVLTreeNode。这是没有问题的，因为 AVLTreeNode 是 TreeNode 的子类。

在 AVLTree 中查找元素和在一个常规的二叉搜索树中查找是一样的，因此定义在 BST 类中的 search 方法同样可以应用于 AVLTree。

insert 和 delete 方法被重写，用于插入和删除一个元素，必要的时候执行重新平衡的操作，从而保证树是平衡的。

✔ 复习题

26.3.1 AVLTreeNode 类中的数据域是什么？
26.3.2 真或者假：AVLTreeNode 是 TreeNode 的子类。
26.3.3 真或者假：AVLTree 是 BST 的子类。

26.4 重写 insert 方法

🔑 **要点提示**：插入一个元素到 AVL 树中和插入到 BST 中是一样的，不同之处在于树可能需要重新平衡。

一个新的元素经常作为叶子结点被插入。作为增加一个新结点的结果，新的叶子结点的祖先结点的高度会增加。插入一个新的结点后，检查沿着新的叶子结点到根结点的路径上的结点，如果发现一个不平衡的结点，则使用程序清单 26-1 中的算法执行适当的旋转（见图 26-9）。

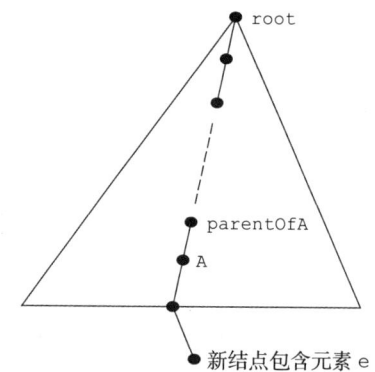

图 26-9 从新的叶子结点到根结点的路径上的结点可能变得不平衡

程序清单 26-1 平衡一条路径上的结点

```
1   balancePath(E e) {
2     Get the path from the node that contains element e to the root,
3       as illustrated in Figure 26.9;
4     for each node A in the path leading to the root {
5       Update the height of A;
6       Let parentOfA denote the parent of A,
7         which is the next node in the path, or null if A is the root;
8
9       switch (balanceFactor(A)) {
10        case -2:  if balanceFactor(A.left) == -1 or 0
11                    Perform LL rotation; // See Figure 26.3
12                  else
13                    Perform LR rotation; // See Figure 26.5
14                  break;
```

```
15          case +2:    if balanceFactor(A.right) == +1 or 0
16                        Perform RR rotation;  // See Figure 26.4
17                      else
18                        Perform RL rotation;  // See Figure 26.6
19        } // End of switch
20      } // End of for
21    } // End of method
```

算法考察从新的叶子结点到根结点的路径上的每个结点。更新路径上的结点的高度。如果结点是平衡的，则无须执行任何动作。如果结点是不平衡的，则执行适当的旋转操作。

✓ 复习题

26.4.1 针对图 26-1a 中的 AVL 树，显示添加元素 40 之后的新的 AVL 树。为了重新平衡该树，需要执行什么旋转操作？哪个结点是不平衡的？

26.4.2 针对图 26-1a 中的 AVL 树，显示添加元素 50 之后的新的 AVL 树。为了重新平衡该树，需要执行什么旋转操作？哪个结点是不平衡的？

26.4.3 针对图 26-1a 中的 AVL 树，显示添加元素 80 之后的新的 AVL 树。为了重新平衡该树，需要执行什么旋转操作？哪个结点是不平衡的？

26.4.4 针对图 26-1a 中的 AVL 树，显示添加元素 89 之后的新的 AVL 树。为了重新平衡该树，需要执行什么旋转操作？哪个结点是不平衡的？

26.5 实现旋转

要点提示：通过执行适当的旋转操作，将一棵不平衡的树变平衡。

26.2 节演示了如何在一个结点上执行旋转操作。程序清单 26-2 给出了 LL 旋转的算法，如图 26-3 所示。

程序清单 26-2 LL 旋转算法

```
1   balanceLL(TreeNode A, TreeNode parentOfA) {
2     Let B be the left child of A.
3
4     if (A is the root)
5       Let B be the new root
6     else {
7       if (A is a left child of parentOfA)
8         Let B be a left child of parentOfA;
9       else
10        Let B be a right child of parentOfA;
11    }
12
13    Make T2 the left subtree of A by assigning B.right to A.left;
14    Make A the right child of B by assigning A to B.right;
15    Update the height of node A and node B;
16  } // End of method
```

注意，结点 A 和 B 的高度可以被改变，而树中的其他结点的高度没有被改变。可以采用相似的方式实现 RR、LR 以及 RL 旋转。

✓ 复习题

26.5.1 以程序清单 26-2 作为模板，描述实现 RR、LR、RL 旋转的算法。

26.6 实现 delete 方法

要点提示：从 AVL 树中删除一个元素和从 BST 中删除是一样的，不同之处在于树可能需要重新平衡。

如 25.3 节中所讨论的，从一棵二叉树中删除一个元素，算法首先定位包含元素的结点。让 current 指向二叉树中包含该元素的结点，parent 指向 current 结点的父结点。current 结点可能是 parent 结点的左子结点或者右子结点。当删除一个元素的时候，可能出现两种情形：

情形 1：current 结点没有左子结点，如图 25-10a 所示。为了删除 current 结点，只需简单的连接 parent 结点和 current 结点的右子结点，如图 25-10b 所示。

从 parent 结点到 root 结点路径上的结点的高度可能减少。为了保证树是平衡的，调用

```
balancePath(parent.element); // Defined in Listing 26.1
```

情形 2：current 结点具有左子结点。让 rightMost 指向 current 结点的左子树中包含最大元素的结点，parentOfRightMost 指向 rightMost 结点的父结点，如图 25-12a 所示。rightMost 结点不会有右子结点，但是可能有左子结点。替换 current 结点中的元素值为 rightMost 结点中的元素值，并连接 parentOfRightMost 结点和 rightMost 结点的左子结点，删除 rightMost 结点，如图 25-12b 所示。

从 parentOfRightMost 结点到根结点路径上的结点的高度可能减少。为了保证树是平衡的，调用

```
balancePath(parentOfRightMost); // Defined in Listing 26.1
```

✓ 复习题

26.6.1 针对图 26-1a 中的 AVL 树，显示删除元素 107 之后的新的 AVL 树。为了重新平衡该树，需要执行什么旋转操作？哪个结点是不平衡的？

26.6.2 针对图 26-1a 中的 AVL 树，显示删除元素 60 之后的新的 AVL 树。为了重新平衡该树，需要执行什么旋转操作？哪个结点是不平衡的？

26.6.3 针对图 26-1a 中的 AVL 树，显示删除元素 55 之后的新的 AVL 树。为了重新平衡该树，需要执行什么旋转操作？哪个结点是不平衡的？

26.6.4 针对图 26-1b 中的 AVL 树，显示删除元素 67 和 87 之后的新的 AVL 树。为了重新平衡该树，需要执行什么旋转操作？哪个结点是不平衡的？

26.7 AVLTree 类

要点提示：AVLTree 类继承自 BST 类，并重写了 insert 和 delete 方法，以在必要的时候重新平衡该树。

程序清单 26-3 给出了 AVLTree 类的完整源代码。

程序清单 26-3 AVLTree.java

```
1  public class AVLTree<E> extends BST<E> {
2    /** Create an empty AVL tree using a natural comparator*/
3    public AVLTree() { // super() is implicitly called
4    }
5
6    /** Create a BST with a specified comparator */
7    public AVLTree(java.util.Comparator<E> c) {
8      super(c);
9    }
10
11   /** Create an AVL tree from an array of objects */
12   public AVLTree(E[] objects) {
13     super(objects);
14   }
```

```java
15
16    @Override /** Override createNewNode to create an AVLTreeNode */
17    protected AVLTreeNode<E> createNewNode(E e) {
18      return new AVLTreeNode<E>(e);
19    }
20
21    @Override /** Insert an element and rebalance if necessary */
22    public boolean insert(E e) {
23      boolean successful = super.insert(e);
24      if (!successful)
25        return false; // e is already in the tree
26      else {
27        balancePath(e); // Balance from e to the root if necessary
28      }
29
30      return true; // e is inserted
31    }
32
33    /** Update the height of a specified node */
34    private void updateHeight(AVLTreeNode<E> node) {
35      if (node.left == null && node.right == null) // node is a leaf
36        node.height = 0;
37      else if (node.left == null) // node has no left subtree
38        node.height = 1 + ((AVLTreeNode<E>)(node.right)).height;
39      else if (node.right == null) // node has no right subtree
40        node.height = 1 + ((AVLTreeNode<E>)(node.left)).height;
41      else
42        node.height = 1 +
43          Math.max(((AVLTreeNode<E>)(node.right)).height,
44          ((AVLTreeNode<E>)(node.left)).height);
45    }
46
47    /** Balance the nodes in the path from the specified
48     * node to the root if necessary
49     */
50    private void balancePath(E e) {
51      java.util.ArrayList<TreeNode<E>> path = path(e);
52      for (int i = path.size() - 1; i >= 0; i--) {
53        AVLTreeNode<E> A = (AVLTreeNode<E>)(path.get(i));
54        updateHeight(A);
55        AVLTreeNode<E> parentOfA = (A == root) ? null :
56          (AVLTreeNode<E>)(path.get(i - 1));
57
58        switch (balanceFactor(A)) {
59          case -2:
60            if (balanceFactor((AVLTreeNode<E>)A.left) <= 0) {
61              balanceLL(A, parentOfA); // Perform LL rotation
62            }
63            else {
64              balanceLR(A, parentOfA); // Perform LR rotation
65            }
66            break;
67          case +2:
68            if (balanceFactor((AVLTreeNode<E>)A.right) >= 0) {
69              balanceRR(A, parentOfA); // Perform RR rotation
70            }
71            else {
72              balanceRL(A, parentOfA); // Perform RL rotation
73            }
74        }
75      }
76    }
77
78    /** Return the balance factor of the node */
79    private int balanceFactor(AVLTreeNode<E> node) {
80      if (node.right == null) // node has no right subtree
```

```java
 81       return -node.height;
 82     else if (node.left == null) // node has no left subtree
 83       return +node.height;
 84     else
 85       return ((AVLTreeNode<E>)node.right).height -
 86         ((AVLTreeNode<E>)node.left).height;
 87   }
 88
 89   /** Balance LL (see Figure 26.3) */
 90   private void balanceLL(TreeNode<E> A, TreeNode<E> parentOfA) {
 91     TreeNode<E> B = A.left; // A is left-heavy and B is left-heavy
 92
 93     if (A == root) {
 94       root = B;
 95     }
 96     else {
 97       if (parentOfA.left == A) {
 98         parentOfA.left = B;
 99       }
100       else {
101         parentOfA.right = B;
102       }
103     }
104
105     A.left = B.right; // Make T2 the left subtree of A
106     B.right = A; // Make A the left child of B
107     updateHeight((AVLTreeNode<E>)A);
108     updateHeight((AVLTreeNode<E>)B);
109   }
110
111   /** Balance LR (see Figure 26.5) */
112   private void balanceLR(TreeNode<E> A, TreeNode<E> parentOfA) {
113     TreeNode<E> B = A.left; // A is left-heavy
114     TreeNode<E> C = B.right; // B is right-heavy
115
116     if (A == root) {
117       root = C;
118     }
119     else {
120       if (parentOfA.left == A) {
121         parentOfA.left = C;
122       }
123       else {
124         parentOfA.right = C;
125       }
126     }
127
128     A.left = C.right; // Make T3 the left subtree of A
129     B.right = C.left; // Make T2 the right subtree of B
130     C.left = B;
131     C.right = A;
132
133     // Adjust heights
134     updateHeight((AVLTreeNode<E>)A);
135     updateHeight((AVLTreeNode<E>)B);
136     updateHeight((AVLTreeNode<E>)C);
137   }
138
139   /** Balance RR (see Figure 26.4) */
140   private void balanceRR(TreeNode<E> A, TreeNode<E> parentOfA) {
141     TreeNode<E> B = A.right; // A is right-heavy and B is right-heavy
142
143     if (A == root) {
144       root = B;
145     }
146     else {
147       if (parentOfA.left == A) {
```

```java
148          parentOfA.left = B;
149        }
150        else {
151          parentOfA.right = B;
152        }
153      }
154
155      A.right = B.left; // Make T2 the right subtree of A
156      B.left = A;
157      updateHeight((AVLTreeNode<E>)A);
158      updateHeight((AVLTreeNode<E>)B);
159    }
160
161    /** Balance RL (see Figure 26.6) */
162    private void balanceRL(TreeNode<E> A, TreeNode<E> parentOfA) {
163      TreeNode<E> B = A.right; // A is right-heavy
164      TreeNode<E> C = B.left; // B is left-heavy
165
166      if (A == root) {
167        root = C;
168      }
169      else {
170        if (parentOfA.left == A) {
171          parentOfA.left = C;
172        }
173        else {
174          parentOfA.right = C;
175        }
176      }
177
178      A.right = C.left; // Make T2 the right subtree of A
179      B.left = C.right; // Make T3 the left subtree of B
180      C.left = A;
181      C.right = B;
182
183      // Adjust heights
184      updateHeight((AVLTreeNode<E>)A);
185      updateHeight((AVLTreeNode<E>)B);
186      updateHeight((AVLTreeNode<E>)C);
187    }
188
189    @Override /** Delete an element from the binary tree.
190     * Return true if the element is deleted successfully
191     * Return false if the element is not in the tree */
192    public boolean delete(E element) {
193      if (root == null)
194        return false; // Element is not in the tree
195
196      // Locate the node to be deleted and also locate its parent node
197      TreeNode<E> parent = null;
198      TreeNode<E> current = root;
199      while (current != null) {
200        if (c.compare(element, current.element) < 0) {
201          parent = current;
202          current = current.left;
203        }
204        else if (c.compare(element, current.element) > 0) {
205          parent = current;
206          current = current.right;
207        }
208        else
209          break; // Element is in the tree pointed by current
210      }
211
212      if (current == null)
213        return false; // Element is not in the tree
```

```
214
215      // Case 1: current has no left children (See Figure 23.6)
216      if (current.left == null) {
217        // Connect the parent with the right child of the current node
218        if (parent == null) {
219          root = current.right;
220        }
221        else {
222          if (c.compare(element, parent.element) < 0)
223            parent.left = current.right;
224          else
225            parent.right = current.right;
226
227          // Balance the tree if necessary
228          balancePath(parent.element);
229        }
230      }
231      else {
232        // Case 2: The current node has a left child
233        // Locate the rightmost node in the left subtree of
234        // the current node and also its parent
235        TreeNode<E> parentOfRightMost = current;
236        TreeNode<E> rightMost = current.left;
237
238        while (rightMost.right != null) {
239          parentOfRightMost = rightMost;
240          rightMost = rightMost.right; // Keep going to the right
241        }
242
243        // Replace the element in current by the element in rightMost
244        current.element = rightMost.element;
245
246        // Eliminate rightmost node
247        if (parentOfRightMost.right == rightMost)
248          parentOfRightMost.right = rightMost.left;
249        else
250          // Special case: parentOfRightMost is current
251          parentOfRightMost.left = rightMost.left;
252
253        // Balance the tree if necessary
254        balancePath(parentOfRightMost.element);
255      }
256
257      size--;
258      return true; // Element inserted
259    }
260
261    /** AVLTreeNode is TreeNode plus height */
262    protected static class AVLTreeNode<E> extends BST.TreeNode<E> {
263      protected int height = 0; // New data field
264
265      public AVLTreeNode(E o) {
266        super(o);
267      }
268    }
269  }
```

AVLTree 类继承自 BST。和 BST 类一样，AVLTree 类具有一个用于使用自然顺序比较器构建一棵空 AVLTree 的无参构造方法（第 3 和 4 行），一个用于使用指定的比较器构建一棵空 AVLTree 的构造方法（第 7～9 行），以及一个用于从一个元素数组构建一棵初始的 AVLTree 的构造方法（第 12～14 行）。

在 BST 类中定义的 createNewNode() 方法创建一个 TreeNode。该方法被重写以返回一个 AVLTreeNode（第 17～19 行）。

AVLTree 中的 insert 方法在第 22 ～ 31 行被重写。该方法首先调用 BST 中的 insert 方法，然后调用 balancePath(e)（第 27 行）来保证树是平衡的。

balancePath 方法首先得到从包含元素 e 的结点到根结点的路径上的所有结点（第 51 行）。对于路径上的每个结点，更新它的高度（第 54 行），检查它的平衡因子（第 58 行），如果必要的话执行适当的旋转（第 59 ～ 73 行）。

执行旋转的 4 个方法在第 90 ～ 187 行定义。每个方法带两个 TreeNode 类型的参数——A 和 parentOfA——来执行结点 A 处的适当的旋转。在图 26-3 ～ 图 26-6 中展示了每个旋转是如何执行的。旋转后，结点 A、B 以及 C 的高度被更新（第 107、134、157 和 184 行）。

AVLTree 中的 delete 方法在第 192 ～ 259 行被重写。该方法和 BST 类中的实现一样，不同之处在于需要在删除之后的两种情形中重新平衡结点（第 228 和 254 行）。

✓ 复习题

26.7.1 为什么 createNewNode 方法定义为受保护的？它在什么时候被调用？

26.7.2 updateHeight 方法什么时候被调用？balanceFactor 方法什么时候被调用？balancePath 方法什么时候被调用？如果把 AVLTree 类中第 66 行的 break 替换为 return，并在第 75 行增加一个 return，这个程序能正确运行吗？

26.7.3 AVLTree 类中的数据域是什么？

26.7.4 insert 和 delete 方法中，一旦执行了一个旋转来平衡树中的结点，那么可能还有不平衡的结点吗？

26.8 测试 AVLTree 类

要点提示：本节给出一个使用 AVLTree 类的例子。

程序清单 26-4 给出了一个测试程序。该程序创建了一个 AVLTree，使用整数数组 25、20 和 5 来进行初始化（第 4 ～ 5 行），在第 9 ～ 18 行插入元素，第 22 ～ 28 行删除元素。由于 AVLTree 是 BST 的子类，而 BST 中的元素是可以遍历的，该程序第 33 ～ 35 行使用了 foreach 循环来遍历所有的元素。

程序清单 26-4 TestAVLTree.java

```
1   public class TestAVLTree {
2     public static void main(String[] args) {
3       // Create an AVL tree
4       AVLTree<Integer> tree = new AVLTree<Integer>(new Integer[]{25,
5         20, 5});
6       System.out.print("After inserting 25, 20, 5:");
7       printTree(tree);
8
9       tree.insert(34);
10      tree.insert(50);
11      System.out.print("\nAfter inserting 34, 50:");
12      printTree(tree);
13
14      tree.insert(30);
15      System.out.print("\nAfter inserting 30");
16      printTree(tree);
17
18      tree.insert(10);
19      System.out.print("\nAfter inserting 10");
20      printTree(tree);
```

```
21
22       tree.delete(34);
23       tree.delete(30);
24       tree.delete(50);
25       System.out.print("\nAfter removing 34, 30, 50:");
26       printTree(tree);
27
28       tree.delete(5);
29       System.out.print("\nAfter removing 5:");
30       printTree(tree);
31
32       System.out.print("\nTraverse the elements in the tree: ");
33       for (int e: tree) {
34         System.out.print(e + " ");
35       }
36     }
37
38     public static void printTree(BST tree) {
39       // Traverse tree
40       System.out.print("\nInorder (sorted): ");
41       tree.inorder();
42       System.out.print("\nPostorder: ");
43       tree.postorder();
44       System.out.print("\nPreorder: ");
45       tree.preorder();
46       System.out.print("\nThe number of nodes is " + tree.getSize())
47       System.out.println();
48     }
49   }
```

```
After inserting 25, 20, 5:
Inorder (sorted): 5 20 25
Postorder: 5 25 20
Preorder: 20 5 25
The number of nodes is 3

After inserting 34, 50:
Inorder (sorted): 5 20 25 34 50
Postorder: 5 25 50 34 20
Preorder: 20 5 34 25 50
The number of nodes is 5

After inserting 30
Inorder (sorted): 5 20 25 30 34 50
Postorder: 5 20 30 50 34 25
Preorder: 25 20 5 34 30 50
The number of nodes is 6

After inserting 10
Inorder (sorted): 5 10 20 25 30 34 50
Postorder: 5 20 10 30 50 34 25
Preorder: 25 10 5 20 34 30 50
The number of nodes is 7

After removing 34, 30, 50:
Inorder (sorted): 5 10 20 25
Postorder: 5 20 25 10
Preorder: 10 5 25 20
The number of nodes is 4

After removing 5:
Inorder (sorted): 10 20 25
Postorder: 10 25 20
Preorder: 20 10 25
The number of nodes is 3
Traverse the elements in the tree: 10 20 25
```

图 26-10 显示了当元素添加到树上后，树是如何演化的。25 和 20 添加后，树如图 26-10a 所示。5 作为 20 的左子结点插入，如图 26-10b 所示。树是不平衡的，在结点 25 处左偏重。执行一个 LL 操作来获得一棵 AVL 树，如图 26-10c 所示。

插入 34 后，树如图 26-10d 所示。插入 50 后，树如图 26-10e 所示。树是不平衡的，在结点 25 处右偏重。执行一个 RR 操作来获得一棵 AVL 树，如图 26-10f 所示。

插入 30 后，树如图 26-10g 所示。树是不平衡的。执行一个 RL 操作来获得一棵 AVL 树，如图 26-10h 所示。

插入 10 后，树如图 26-10i 所示。树是不平衡的。执行一个 LR 操作来获得一棵 AVL 树，如图 26-10j 所示。

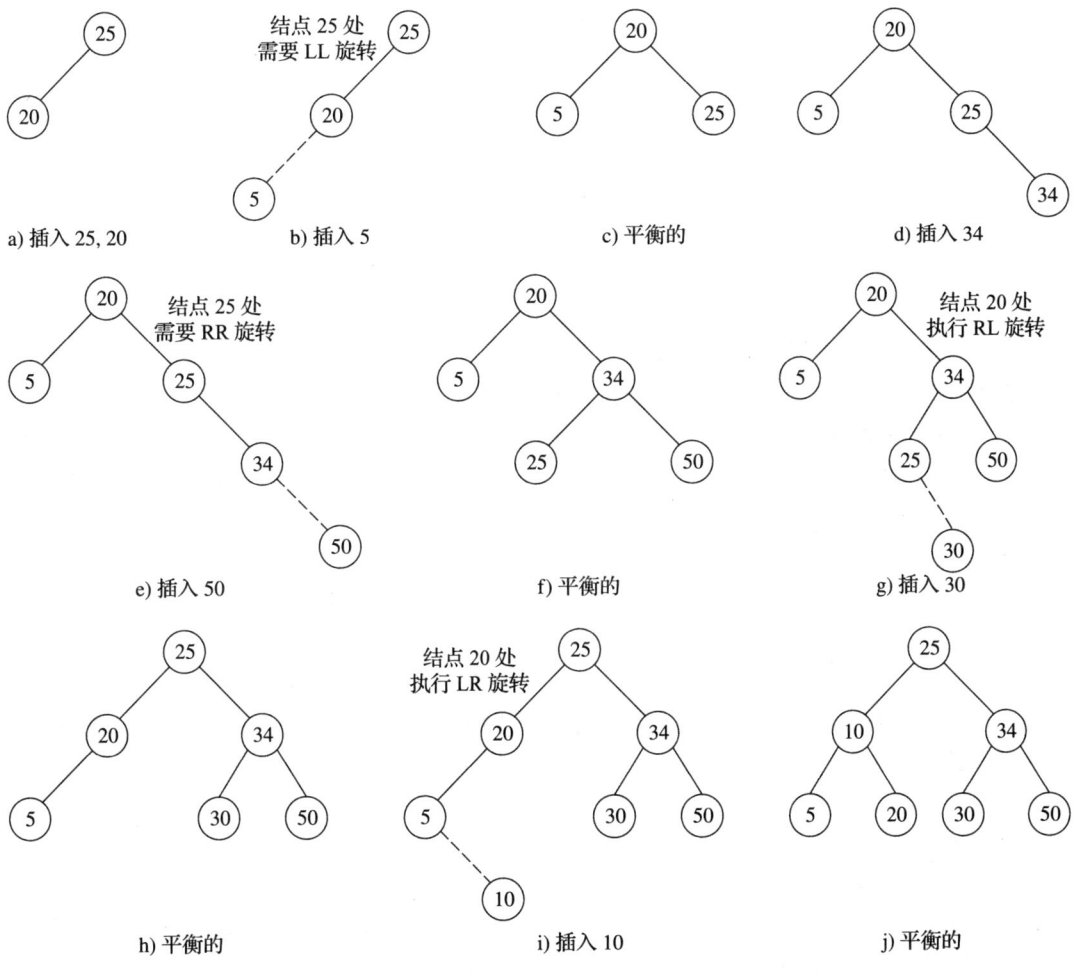

图 26-10 插入新的元素时树的演化

图 26-11 显示了当元素被删除后树是如何演化的。删除 34、30 以及 50 后，树如图 26-11b 所示。树是不平衡的。执行一个 LL 旋转来得到一棵 AVL 树，如图 26-11c 所示。

删除 5 后，树如图 26-11d 所示。树是不平衡的。执行一个 RL 旋转来得到一棵 AVL 树，如图 26-11e 所示。

✓ 复习题

26.8.1 顺序插入 1、2、3、4、10、9、7、5、8、6 到树中后，显示 AVL 树的变化。

26.8.2 针对前面问题所构建的树，顺序删除 1, 2, 3, 4, 10, 9, 7, 5, 8, 6 后，显示其变化。

26.8.3 可以使用 foreach 循环来遍历 AVL 树中的元素吗？

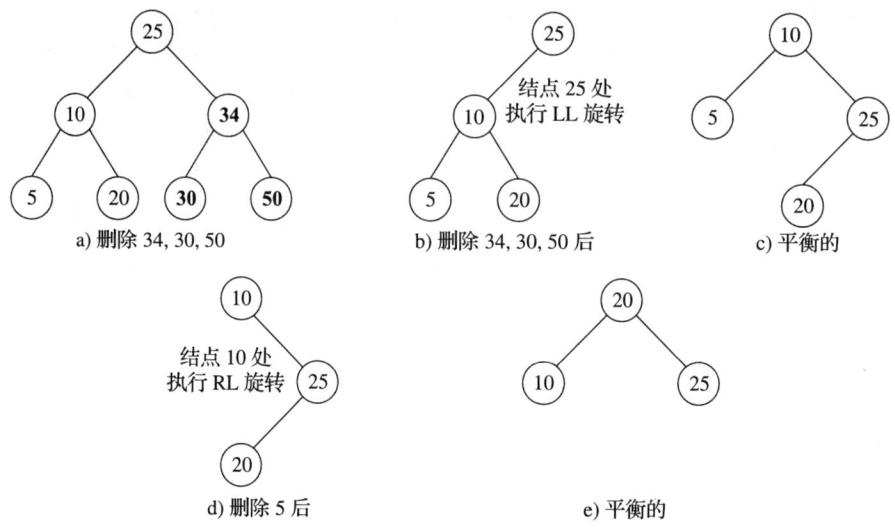

图 26-11 从树中删除元素后树的演化

26.9 AVL 树的时间复杂度分析

要点提示：由于一个 AVL 树的高度为 $O(\log n)$，所以 AVLTree 树中的 search、insert 以及 delete 方法的时间复杂度为 $O(\log n)$。

AVLTree 中的 search、insert 以及 delete 方法的时间复杂度依赖于树的高度。我们可以证明树的高度为 $O(\log n)$。

使用 $G(h)$ 表示一个高度为 h 的 AVL 树的最小结点数。二叉树的高度定义见 25.2 节。显然，$G(0)$ 为 1，$G(1)$ 为 2。高度为 $h \geq 2$ 的 AVL 树会有两棵最小子树结点数最小：一棵高度为 $h-1$，另一棵高度为 $h-2$。因此，

$$G(h) = G(h-1) + G(h-2) + 1$$

回顾一下，在下标为 i 处的斐波那契数可以使用递推关系 $F(i) = F(i-1) + F(i-2)$ 求得。因此，函数 $G(h)$ 实质上和 $F(i)$ 一样。可以证明

$$h < 1.4405 \log (n+2) - 1.3277$$

这里 n 是树中的结点数。因此，一棵 AVL 树的高度为 $O(\log n)$。

search、insert 以及 delete 方法只涉及沿着树中一条路径上的结点。updateHeight 和 balanceFactor 方法对于路径上的每个结点以常量时间执行。balancePath 方法对于路径上的结点也以常量时间执行。因此，search、insert 以及 delete 方法的时间复杂度为 $O(\log n)$。

复习题

26.9.1 对于具有 3 个结点、5 个结点以及 7 个结点的 AVL 树来说，最大 / 最小高度是多少？

26.9.2 如果一棵 AVL 树高度为 3，该树可以具有的最大结点数目为多少？该树可以具有的最小结点数目为多少？

26.9.3 如果一棵 AVL 树高度为 4，该树可以具有的最大结点数目为多少？该树可以具有的最小结点数目为多少？

关键术语

AVL tree（AVL 树）
balance factor（平衡因子）
LR rotation（LR 旋转）
perfectly balanced tree（完全平衡树）
right-heavy（右偏重）
RL rotation（RL 旋转）

left-heavy（左偏重）
LL rotation（LL 旋转）
rotation（旋转）
RR rotation（RR 旋转）
well-balanced tree（良好平衡树）

本章小结

1. AVL 树是良好平衡二叉树。在一棵 AVL 树中，对于每个结点而言，其两个子树的高度差为 0 或者 1。
2. 在一棵 AVL 树中插入或者删除元素的过程和在二叉搜索树中是一样的。不同之处在于可能需要在插入或者删除后重新平衡该树。
3. 插入和删除引起树的不平衡，可以通过不平衡结点处的子树的旋转重新获得平衡。
4. 一个结点重新平衡的过程称为*旋转*。有 4 种可能的旋转：LL *旋转*、LR *旋转*、RR *旋转*、RL *旋转*。
5. AVL 树的高度为 $O(\log n)$。因此，`search`、`insert` 以及 `delete` 方法的时间复杂度为 $O(\log n)$。

测试题

回答本书配套网站上的本章测试题。

编程练习题

*26.1 （图形化显示 AVL 树）编写一个程序，显示一棵 AVL 树，并显示每个结点的平衡因子。

26.2 （比较性能）编写一个测试程序，随机产生 500 000 个数字，并将它们插入 BST 中，重新打乱这 500 000 个数字并执行一次查找，然后将它们从树中删除前再次打乱这些数字。编写另外一个测试程序，为 AVLTree 执行同样的操作。比较两个程序的执行时间。

***26.3 （AVL 树的动画）编写一个程序，实现 AVL 树的 `insert`、`delete` 以及 `search` 方法的动画，如图 26-2 所示。

**26.4 （BST 的父引用）假设定义在 BST 中的 TreeNode 类包含了指向结点的父结点的引用，如编程练习题 25.15 所示。实现 AVLTree 类来支持这个改变。编写一个测试程序，添加数字 1, 2, …, 100 到该树中并显示所有叶子结点的路径。

**26.5 （第 *k* 小的元素）可以通过中序迭代器在 $O(n)$ 的时间内找到 BST 中第 *k* 小的元素。对于一棵 AVL 树而言，可以在 $O(\log n)$ 时间内找到。为了做到这点，在 AVLTreeNode 中添加一个新数据域 `size`，存储以该结点为根结点的子树中的结点数。注意，一个结点 *v* 比其两个子结点的大小的和多 1。图 26-12 显示了一棵 AVL 树，以及树中每个结点的 `size` 值。

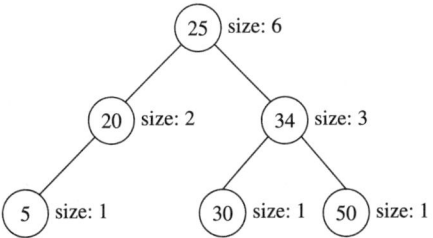

图 26-12 AVLTreeNode 中的 `size` 数据域存储以该结点为根结点的子树的结点数

在 AVLTree 类中添加以下方法，返回树中第 *k* 小的元素。

```
public E find(int k)
```

如果 k < 1 或者 k > 树的大小，方法返回 null。可以使用递归方法 find(k, root) 实现该方法，递归方法返回指定根结点的树中的第 k 小的元素。设 A 和 B 分别为该根结点的左子结点和右子结点。假设树不为空，并且 $k \leq root.size$，可以如下递归定义 find(k,root)：

$$find(k, root) = \begin{cases} root.element, & \text{if } A \text{ is null and } k \text{ is } 1; \\ B.element, & \text{if } A \text{ is null and } k \text{ is } 2; \\ find(k, A), & \text{if } k <= A.size; \\ root.elememt, & \text{if } k = A.size + 1; \\ find(k - A.size - 1, B), & \text{if } k > A.size + 1; \end{cases}$$

修改 AVLTree 树中的 insert 和 delete 方法，为每个结点中的 size 属性设置正确的值。insert 和 delete 方法将仍然为 $O(\log n)$ 时间。find(k) 方法可以在 $O(\log n)$ 时间内执行。因此，可以在 $O(\log n)$ 时间内找到 AVL 树中第 k 小的元素。

可以使用 liveexample.pearsoncmg.com/test/Exercise26_05Test.txt 中的代码来测试你的程序。

**26.6 （最近点对）22.8 节介绍了一种采用分治法在 $O(n\log n)$ 时间内得到最近点对的算法。该算法采用递归实现，开销很大。使用 AVL 树，可以在 $O(n\log n)$ 时间内解决同样的问题。采用 AVLTree 实现该算法。

**26.7 （测试 AVL 树）定义一个继承自 BST 类的名为 MyBST 的新类，它具有以下方法：

```
// Returns true if the tree is an AVL tree
public boolean isAVLTree()
```

使用 https://liveexample.pearsoncmg.com/test/Exercise26_07.txt 来测试你的代码。

第 27 章

Introduction to Java Programming and Data Structures, Comprehensive Version, Twelfth Edition

散　列

教学目标
- 理解什么是散列，以及散列用来做什么（27.2 节）。
- 获得一个对象的散列码，以及设计散列函数，将一个键映射到一个下标（27.3 节）。
- 使用开放地址法处理冲突（27.4 节）。
- 了解线性探测法、二次探测法和双重散列法的区别（27.4 节）。
- 使用分离链接法处理冲突（27.5 节）。
- 理解装填因子以及再散列的必要性（27.6 节）。
- 使用散列实现 MyHashMap（27.7 节）。
- 使用散列实现 MyHashSet（27.8 节）。

27.1 引言

☞ **要点提示**：散列非常高效。使用散列将耗费 $O(1)$ 时间来查找、插入以及删除一个元素。

前面章节介绍了二叉搜索树。在一个良好平衡的搜索树中，可以在 $O(\log n)$ 时间内找到一个元素。还有更加高效的方法在一个容器中查找元素吗？本章介绍一种称为散列（hashing）的技术。可以使用散列来实现一个映射（map）或者规则集（set），从而在 $O(1)$ 时间内查找、插入以及删除一个元素。

27.2 什么是散列

☞ **要点提示**：散列使用一个散列函数，将一个键映射到一个下标上。

介绍散列之前，我们回顾下映射（map）。映射是一种使用散列实现的数据结构。映射（21.5 节中介绍过）是一种存储条目的容器对象。每个条目包含两部分：键（key）和值（value）。键又称为搜索键，用于查找相应的值。例如，一个字典可以存储在一个映射中，其中单词作为键，而单词的定义作为值。

☞ **注意**：映射（map）又称为字典（dictionary）、散列表（hash table）或者关联数组（associate array）。

Java 集合框架定义了 `java.util.Map` 接口来对映射建模。三个具体的实现类为 `java.util.HashMap`、`java.util.LinkedHashMap` 以及 `java.util.TreeMap`。`java.util.HashMap` 使用散列实现，`java.util.LinkedHashMap` 使用 LinkedList 实现，`java.util.TreeMap` 使用红黑树实现（奖励章节第 41 章介绍红黑树）。本章中你将学习散列的概念，并使用它来实现一个散列映射。

如果知道一个数组中元素的下标，可以使用下标在 $O(1)$ 时间内获得元素。因此这是否意味着我们可以将值存储在一个数组中，然后用键作为下标来找到值呢？答案是可以——如果可以将键映射到一个下标上的话。存储了值的数组称为散列表（hash table）。将键映射到散列表中的下标的函数称为散列函数（hash function）。如图 27-1 所示，散列函数从一个键获得下标,并使用下标来获取该键的值。散列(hashing)是一种无须执行搜索即可通过从键得到的下标来获取值的技术。

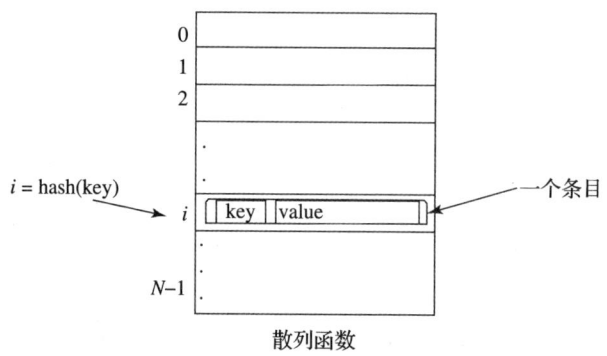

图27-1 散列函数将键映射到散列表中的下标上

如何设计一个散列函数,从而通过一个键得到一个下标呢?理想情况下,我们希望设计一个函数,将每个搜索键映射到散列表中的不同下标上。这样的函数称为完美散列函数(perfect hash function)。然而,很难找到一个完美散列函数。当两个或者更多的键映射到一个散列值上时,我们说产生了一个冲突(collision)。尽管有办法来处理冲突(这将在本章后面进行讨论),但是最好首先避免发生冲突。因此,应该设计一个快速以及易于计算的散列函数来最小化冲突。

✔ 复习题

27.2.1 什么是散列函数?什么是完美散列函数?什么是冲突?

27.3 散列函数和散列码

○━ 要点提示:典型的散列函数首先将搜索键转换成一个称为散列码的整数值,然后将散列码压缩为散列表中的下标。

散列码是从一个对象生成的数字,它使得一个对象可以快速地在散列表中存储/获取。Java的根类Object具有hashCode方法,该方法返回一个整数的散列码(hash code)。该方法默认返回该对象的内存地址。hashCode方法的一般约定如下:

1)当equals方法被重写时,应该重写hashCode方法,从而保证两个相等的对象返回同样的散列码。

2)程序执行过程中,如果对象的数据没有被修改,则多次调用hashCode将返回同样的整数。

3)两个不相等的对象可能具有同样的散列码,但是应该在实现hashCode方法时避免太多这样的情形出现。

27.3.1 基本数据类型的散列码

对于byte、short、int以及char类型的搜索键而言,简单地将它们转型为int。因此,这些类型中的任何两个不同的搜索键将有不同的散列码。

对于float类型的搜索键,使用Float.floatToIntBits(key)作为散列码。注意,floatToIntBits(float f)返回一个int值,该值的二进制位表示和浮点数f的二进制位表示相同。因此，两个不同的 float 类型的搜索键将具有不同的散列码。

对于 long 类型的搜索键，简单地将其类型转换为 int 不是很好的选择，因为只有前 32 位不同的键将具有相同的散列码。为了考虑前 32 位，将 64 位分为两部分，并执行异或操作将两部分结合。这个过程称为折叠（folding）。一个 long 类型键的散列码为：

```
int hashCode = (int)(key ^ (key >> 32));
```

注意，>> 为右移操作符，将二进制位向右移动 32 位。例如，1010110 >> 2 得到 0010101。^ 是按位异或操作符。它在双目操作数的相应位上执行操作。例如，1010110 ^ 0110111 得到 1100001。对于更多的位操作符，参见附录 G。

对于 double 类型的搜索键，首先使用 Double.doubleToLongBits 方法转化为 long 值。然后如下执行折叠操作：

```
long bits = Double.doubleToLongBits(key);
int hashCode = (int)(bits ^ (bits >> 32));
```

27.3.2 字符串的散列码

搜索键经常是字符串，因此为字符串设计好的散列函数非常重要。一个比较直观的方法是将所有字符的 Unicode 码求和作为字符串的散列码。这个方法在应用程序中的两个搜索键不包含同样字母的情况下可以工作，但如果搜索键包含同样字母，将产生许多冲突，例如 tod 和 dot。

一个更好的方法是生成散列码时考虑字符的位置。具体来说，令散列码为

$$s_0 \times b^{(n-1)} + s_1 \times b^{(n-2)} + \cdots + s_{n-1}$$

这里 s_i 为 s.charAt(i)。这个表达式为某个正数 b 的多项式，因此被称为多项式散列码 （polynomial hash code）。使用针对多项式求值的 Horner 规则（参见 6.7 节），可以如下高效地计算散列码：

$$(\cdots((s_0 \times b + s_1) \times b + s_2) \times b + \cdots + s_{n-2}) \times b + s_{n-1}$$

这个计算对于长的字符串来说会导致溢出，但是 Java 中会忽略算术溢出。应该选择一个合适的 b 值来最小化冲突。实验显示，b 的较好取值为 31、33、37、39 和 41。String 类中，hashCode 被重写为采用 b 值为 31 的多项式散列码。

27.3.3 压缩散列码

键的散列码可能是一个很大的整数，从而超过了散列表下标的范围，因此需要将它缩小到适合下标的范围内。假设散列表的下标处于 0 到 N-1 之间。将一个整数缩小到 0 到 N-1 之间的最通常做法是使用

```
index = hashCode % N;
```

理想情况下，应该为 N 选择一个素数来保证下标均匀展开。然而，选择一个大的素数将很耗时。Java API 针对 java.util.HashMap 的实现中，N 设置为一个 2 的整数次幂值。这样的选择具有合理性。当 N 为 2 的整数次幂值时，可以通过 & 操作符来把散列码压缩为散列表的下标值，如下所示：

```
index = hashCode & (N - 1);
```

index 将介于 0 与 N-1 之间。符号 & 是一个按位 AND 操作符（参见附录 G）。如果两个二进制位都为 1，则其 AND 操作结果为 1。例如，假设 N=4 以及 hashCode = 11，这样,11 & (4-1) =

1011 & 0011 = 0011。

为了保证散列码是均匀分布的，java.util.HashMap 的实现中使用主散列函数时还采用了补充的散列函数。该函数定义为：

```
private static int supplementalHash(int h) {
  h ^= (h >>> 20) ^ (h >>> 12);
  return h ^ (h >>> 7) ^ (h >>> 4);
}
```

^ 和 >>> 分别是比特异或和无符号右移操作（也在附录 G 中介绍）。位操作比乘、除以及求余操作要快许多，应该尽量使用位操作来代替这些操作。

完整的散列函数如下定义：

h(hashCode) = supplementalHash(hashCode) & (N - 1)

补充的散列函数帮助避免了两个低位相同的数之间的冲突。比如，11100101 & 00000111 和 11001101 & 00000111 都得到 00000111，但 supplementalHash(11100101) & 00000111 和 supplementalHash(11001101) & 00000111 得到的结果是不同的。使用补充的散列函数减少了这类冲突。

> **注意**：在 Java 中，int 是一个 32 位有符号整数。hashCode() 方法返回一个可能为负的 int。如果一个散列码为负，hashCode % N 也为负。但对于一个 int 值 N 来说，hashCode & (N-1) 将是非负的，因为 anyInt & aNonNegativeInt 总是非负的。

✓ **复习题**

27.3.1 什么是散列码？Byte、Short、Integer 以及 Character 的散列码是什么？
27.3.2 Float 对象的散列码是如何计算的？
27.3.3 Long 对象的散列码是如何计算的？
27.3.4 Double 对象的散列码是如何计算的？
27.3.5 String 对象的散列码是如何计算的？
27.3.6 一个散列码是如何压缩为一个散列表中下标的整数表示的？
27.3.7 如果 N 为 2 的整数次幂值，N/2 与 N>>1 一样吗？
27.3.8 如果 N 为 2 的整数次幂值，对于正整数 m，m% N 与 m & (N - 1) 等价吗？
27.3.9 Integer.value Of("-98").hashCode() 和 "ABCDEFGHIJK".hashCode() 的结果是什么？

27.4 使用开放地址法处理冲突

> **要点提示**：当两个键映射到散列表中的同一个下标上时，会产生冲突。通常，有两种方法处理冲突：开放地址法和分离链接法。

开放地址法（open addressing）是在冲突发生时，在散列表中找到一个开放位置的过程。开放地址法有几个变体：线性探测法、二次探测法和双重散列法。

27.4.1 线性探测法

当插入一个条目到散列表的过程中发生冲突时，线性探测法（linear probing）按顺序找到下一个可用的位置。例如，如果冲突发生在 hashTable[k % N]，则检查 hashTable[(k+1) % N] 是否可用。如果不可用，则检查 hashTable[(k+2) % N]，以此类推，直到找到一个可用单元，如图 27-2 所示。

> **注意**：当探测到表的终点时，则返回表的起点。因此，散列表被当成是循环的。

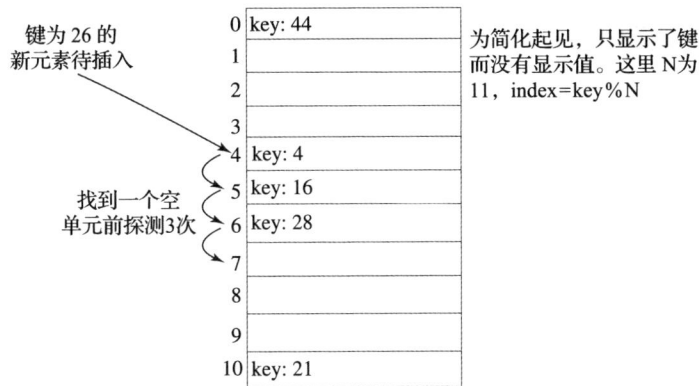

图 27-2 线性探测法按顺序找到下一个可用的位置

查找散列表中的条目时，从散列函数获得键相应的下标，比如说 k。检查 hashTable[k % N] 是否包含该条目。如果没有，检查 hashTable[(k+1) % N] 是否包含该条目，以此类推，直到找到，或者到达一个空的单元。

删除散列表中的条目时，查找匹配键的条目。如果条目找到，则放置一个特殊的标记表示该条目是可用的。散列表中的每个单元具有三个可能的状态：被占的、标记的或者空的。注意，一个被标记的单元对于插入同样是可用的。

线性探测法容易导致散列表中连续的单元组被占用。每个分组称为一个簇（cluster）。每个簇实际上是在获取、添加以及删除一个条目时必须查找的探测序列。当簇的大小增加时，它们可能合并为更大的簇，从而更加放慢查找的时间。这是线性探测法的一个较大的缺点。

○教学注意：参见网址 http://liveexample.pearsoncmg.com/dsanimation/LinearProbingeBook.html 获取线性探测法工作方式的交互式 GUI 演示，如图 27-3 所示。

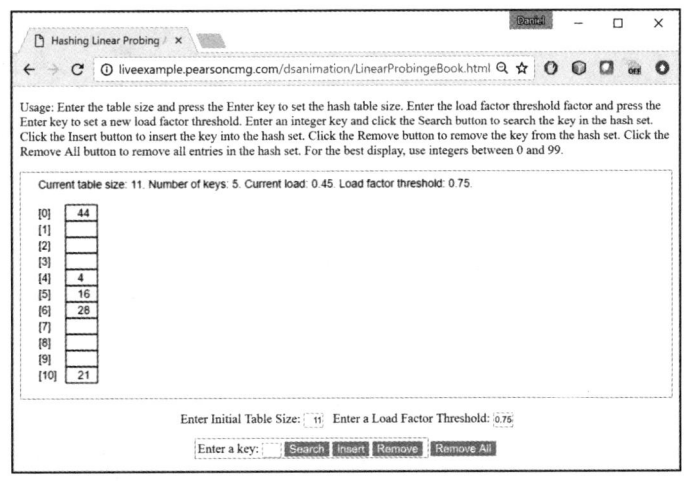

图 27-3 动画工具显示线性探测法如何工作

27.4.2 二次探测法

二次探测法（quadratic probing）可以避免线性探测法产生的成簇的问题。线性探测法从下标 k 位置审查连续单元。二次探测法则从下标为 $(k+j^2)$ % N 位置的单元开始审查，其中 $j \geqslant 0$。即 k % N，(k+1) % N，(k+4) % N，(k+9) % N，以此类推，如图 27-4 所示。

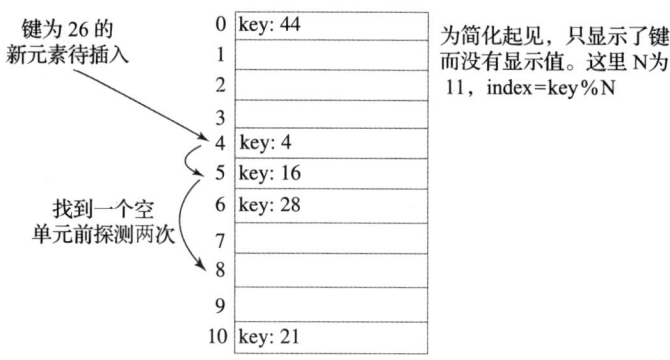

图 27-4　二次探测法以 j^2（$j=1,2,3,\cdots$）递增下一个下标

除修改了搜索序列外，二次探测法和线性探测法的工作机制相同。二次探测法避免了线性探测法的成簇问题，但是有自身的成簇问题，称为二次成簇（secondary clustering），即与一个被占据的条目产生冲突的条目将采用同样的探测序列。

线性探测法可以保证只要表不满，就总是可以找到一个可用的单元来插入新的元素。然而，二次探测法不能保证这一点。

教学注意：参见网址 http://liveexample.pearsoncmg.com/dsanimation/QuadraticProbingeBook.html 获取二次探测法工作方式的交互式 GUI 演示，如图 27-5 所示。

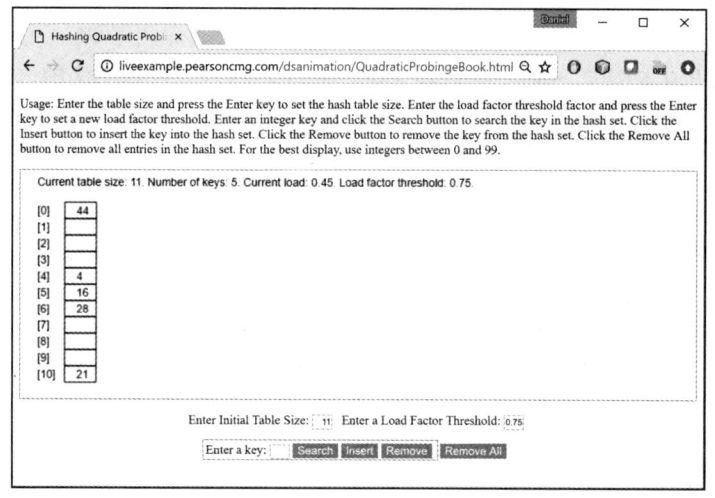

图 27-5　动画工具显示二次探测法如何工作

27.4.3　双重散列法

另外一个避免成簇问题的开放地址模式称为双重散列法（double hashing）。从初始下标 k 开始，线性探测法和二次探测法都对 k 加一个增量来定义搜索序列。对于线性探测法来说增量为 1，对于二次探测法来说增量为 j^2。这些增量都独立于键。双重散列法在键上应用第二个散列函数 $h'(key)$ 来确定增量，从而避免成簇问题。具体来说，双重散列法审查下标为 $(k+j*h'(key))\% N$ 处的单元，其中 $j \geq 0$，即 $k\%N$，$(k+h'(key))\%N$，$(k+2*h'(key))\%N$，$(k+3*h'(key))\%N$，以此类推。

例如，如下定义一个大小为 11 的散列表上的相关主散列函数 h 和二次散列函数 h'：

```
h(key) = key % 11;
h'(key) = 7 - key % 7;
```

对于搜索键 12，则有

```
h(12) = 12 % 11 = 1;
h'(12) = 7 - 12 % 7 = 2;
```

假设键为 45、58、4、28 以及 21 的元素已经位于散列表中，如图 27-6 所示。现在插入键为 12 的元素。对于键为 12 的探测序列从下标 1 处开始。因为下标为 1 的单元已经被占据，搜索下一个下标为 3(1+1*2) 处的单元。由于下标为 3 的单元已经被占据，搜索下一个下标为 5(1+2*2) 处的单元。由于下标为 5 的单元为空，键为 12 的元素插入该单元中。搜索过程在图 27-6 中展示。

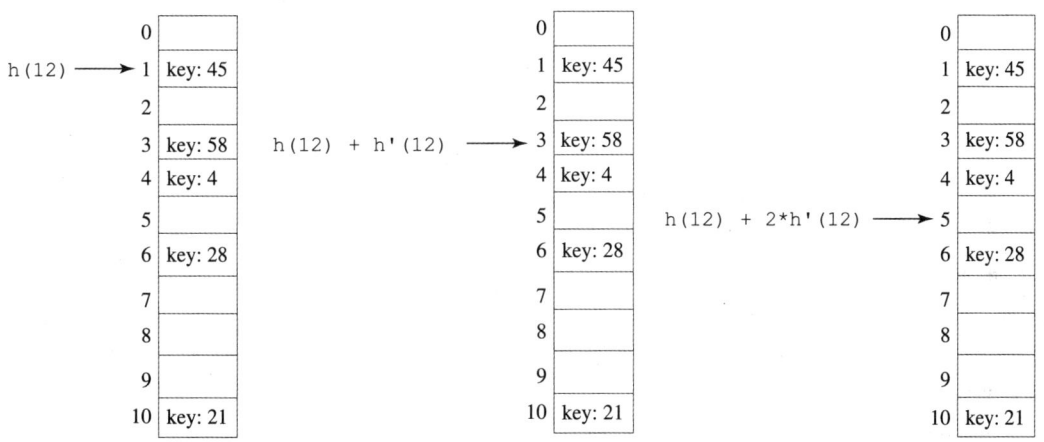

图 27-6 双重散列法中的第二个散列函数确定探测序列中下一个下标的增量

探测序列的下标如下：1、3、5、7、9、0、2、4、6、8、10。该序列覆盖了整个表。应该设计函数以产生一个到达整个表的探测序列。注意，二次函数不能具有一个为 0 的值，因为 0 不是增量。

教学注意：参见网址 http://liveexample.pearsoncmg.com/dsanimation/DoubleHashingeBook.html，获取二次散列法如何工作的交互式 GUI 演示。如图 27-7 所示。

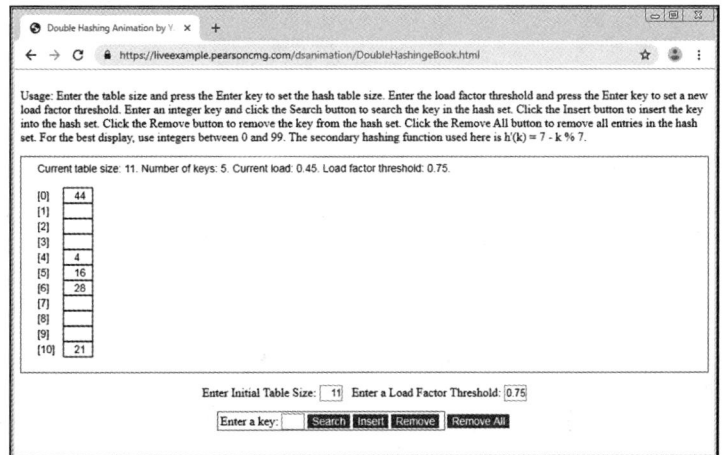

图 27-7 动画工具显示二次散列法如何工作

✓ 复习题

27.4.1 什么是开放地址法？什么是线性探测法？什么是二次探测法？什么是双重散列法？

27.4.2 描述线性探测产生的成簇问题。

27.4.3 什么是二次成簇？

27.4.4 显示在大小为 11 的散列表中使用线性探测法插入键为 34、29、53、44、120、39、45 以及 40 的条目后的情形。

27.4.5 显示在大小为 11 的散列表中使用二次探测法插入键为 34、29、53、44、120、39、45 以及 40 的条目后的情形。

27.4.6 显示在大小为 11 的散列表中使用双重散列法插入键为 34、29、53、44、120、39、45 以及 40 的条目后的情形，其中双重散列函数为：

```
h(k) = k % 11;
h'(k) = 7 - k % 7;
```

27.5 使用分离链接法处理冲突

要点提示：分离链接法将具有同样的散列下标的条目都放在同一个位置，而不是寻找一个新的位置。分离链接法的每个位置使用一个桶来放置多个条目。

前面小节介绍了使用开放地址法来处理冲突。开放地址模式在冲突发生的时候找到一个新的位置。本节介绍采用分离链接法处理冲突。分离链接模式将具有同样的散列下标的条目都放在同一个位置，而不是寻找一个新的位置。分离链接模式的每个位置称为一个桶。桶是一种放置多个条目的容器。

可以使用数组 ArrayList 或者 LinkedList 来实现一个桶。这里将使用 LinkedList 来演示。可以将散列表的每个单元视为指向一个链表头的引用，而链表中的元素从头部链接在一起，如图 27-8 所示。

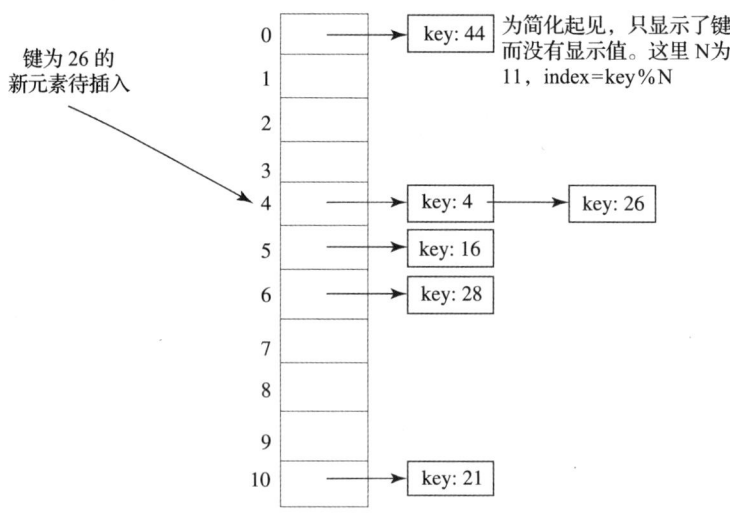

图 27-8 分离链接法将具有同样的散列下标的条目放在一个桶内

教学注意：参见网址 http://liveexample.pearsoncmg.com/dsanimation/SeparateChainingeBook.html，获取分离链接法如何工作的交互式 GUI 演示，如图 27-9 所示。

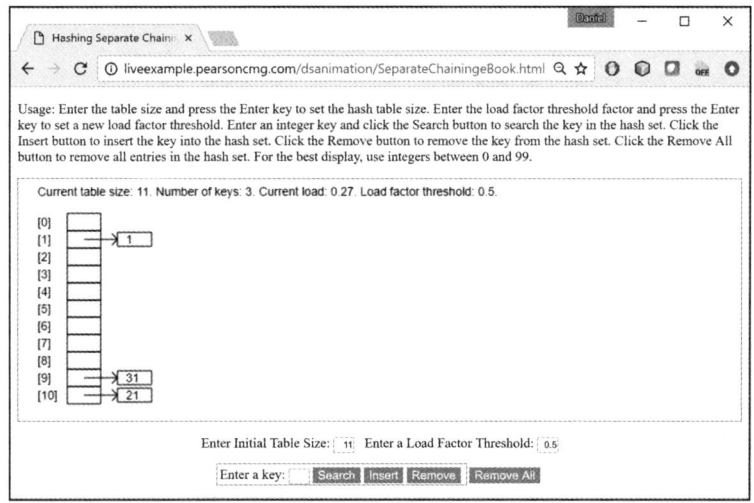

图 27-9 动画工具显示分离链接法如何工作

复习题

27.5.1 显示在大小为 11 的散列表中使用分离链接法插入键为 34、29、53、44、120、39、45 以及 40 的条目后的情形。

27.6 装填因子和再散列

要点提示：装填因子（load factor）衡量一个散列表有多满。如果装填因子溢出，则增加散列表的大小，并重新装载条目到一个更大的新散列表中。这称为再散列。

装填因子 λ（lamda）衡量一个散列表有多满。它是元素数目和散列表大小的比例，即 $\lambda = n/N$，这里 n 表示元素的数目，而 N 表示散列表中位置的数目。

注意，如果散列表为空则 λ 为 0。对于开放地址模式，λ 介于 0 和 1 之间。如果散列表满了，则 λ 为 1。对于分离链接法而言，λ 可能为任意值。当 λ 增加时，冲突的可能性增大。研究表明，对于开放地址法而言，需要维持装填因子在 0.5 以下，而对于分离链接法而言，维持在 0.9 以下。

将装填因子保持在一定的阈值下对于散列的性能是非常重要的。在 Java API 的 `java.util.HashMap` 类的实现中，采用了阈值 0.75。一旦装填因子超过阈值，就需要增加散列表的大小，并将映射中的所有条目再散列（rehash）到一个更大的新散列表中。注意，需要修改散列函数，因为散列表的大小改变了。由于再散列代价比较大，为了减少出现再散列的可能性，应该至少将散列表的大小翻倍。即使需要周期性的再散列，对于映射来说散列依然是一种高效的实现。

复习题

27.6.1 什么是装填因子？假设散列表初始大小为 4，它的装填因子为 0.5，显示在使用线性探测法插入键为 34、29、53、44、120、39、45 以及 40 的条目后的散列表。

27.6.2 假设散列表初始大小为 4，它的装填因子为 0.5，显示在使用二次探测法插入键为 34、29、53、44、120、39、45 以及 40 的条目后的散列表。

27.6.3 假设散列表初始大小为 4，它的装填因子为 0.5，显示在使用分离链接法插入键为 34、29、53、44、120、39、45 以及 40 的条目后的散列表。

27.7 使用散列实现映射

要点提示：可以使用散列来实现映射。

现在你理解了散列的概念，了解了如何设计一个好的散列函数来将一个键映射到散列表的下标上，了解了如何使用装填因子衡量性能，以及如何通过增加表的大小和再散列来保持性能。本节演示如何使用分离链接法来实现映射。

这里对照 java.util.Map 来设计我们自定义的 Map 接口，将接口命名为 MyMap，具体类命名为 MyHashMap，如图 27-10 所示。

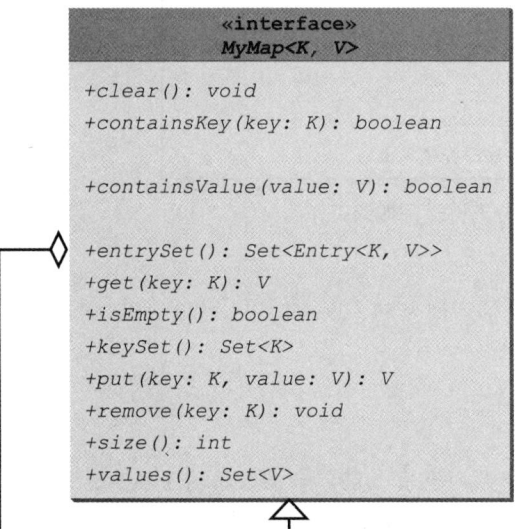

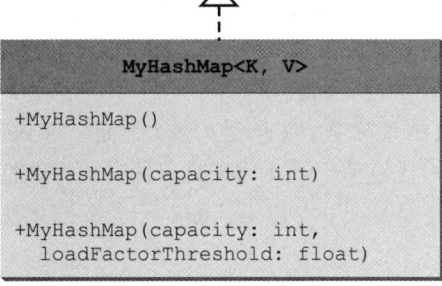

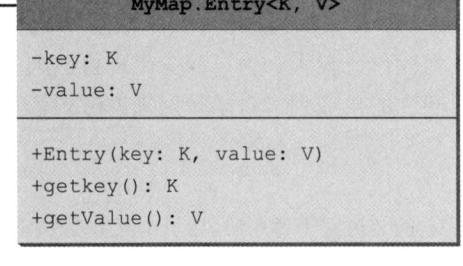

图 27-10 MyHashMap 实现 MyMap 接口

如何实现 MyHashMap 呢？我们将一个数组用于散列表，并且散列表中的每个元素都是一个桶。桶是一个 LinkedList。程序清单 27-1 给出了 MyMap 接口，程序清单 27-2 采用分离链接法实现了 MyHashMap。

程序清单 27-1 MyMap.java

```java
 1  public interface MyMap<K, V> {
 2    /** Remove all of the entries from this map */
 3    public void clear();
 4
 5    /** Return true if the specified key is in the map */
 6    public boolean containsKey(K key);
 7
 8    /** Return true if this map contains the specified value */
 9    public boolean containsValue(V value);
10
11    /** Return a set of entries in the map */
12    public java.util.Set<Entry<K, V>> entrySet();
13
14    /** Return the value that matches the specified key */
15    public V get(K key);
16
17    /** Return true if this map doesn't contain any entries */
18    public boolean isEmpty();
19
20    /** Return a set consisting of the keys in this map */
21    public java.util.Set<K> keySet();
22
23    /** Add an entry (key, value) into the map */
24    public V put(K key, V value);
25
26    /** Remove an entry for the specified key */
27    public void remove(K key);
28
29    /** Return the number of mappings in this map */
30    public int size();
31
32    /** Return a set consisting of the values in this map */
33    public java.util.Set<V> values();
34
35    /** Define an inner class for Entry */
36    public static class Entry<K, V> {
37      K key;
38      V value;
39
40      public Entry(K key, V value) {
41        this.key = key;
42        this.value = value;
43      }
44
45      public K getKey() {
46        return key;
47      }
48
49      public V getValue() {
50        return value;
51      }
52
53      @Override
54      public String toString() {
55        return "[" + key + ", " + value + "]";
56      }
57    }
58  }
```

程序清单 27-2 MyHashMap.java

```java
1  import java.util.LinkedList;
2
3  public class MyHashMap<K, V> implements MyMap<K, V> {
4    // Define the default hash-table size. Must be a power of 2
```

```java
  5    private static int DEFAULT_INITIAL_CAPACITY = 4;
  6
  7    // Define the maximum hash-table size. 1 << 30 is same as 2^30
  8    private static int MAXIMUM_CAPACITY = 1 << 30;
  9
 10    // Current hash-table capacity. Capacity is a power of 2
 11    private int capacity;
 12
 13    // Define default load factor
 14    private static float DEFAULT_MAX_LOAD_FACTOR = 0.75f;
 15
 16    // Specify a load factor used in the hash table
 17    private float loadFactorThreshold;
 18
 19    // The number of entries in the map
 20    private int size = 0;
 21
 22    // Hash table is an array with each cell being a linked list
 23    LinkedList<MyMap.Entry<K,V>>[] table;
 24
 25    /** Construct a map with the default capacity and load factor */
 26    public MyHashMap() {
 27      this(DEFAULT_INITIAL_CAPACITY, DEFAULT_MAX_LOAD_FACTOR);
 28    }
 29
 30    /** Construct a map with the specified initial capacity and
 31     * default load factor */
 32    public MyHashMap(int initialCapacity) {
 33      this(initialCapacity, DEFAULT_MAX_LOAD_FACTOR);
 34    }
 35
 36    /** Construct a map with the specified initial capacity
 37     * and load factor */
 38    public MyHashMap(int initialCapacity, float loadFactorThreshold) {
 39      if (initialCapacity > MAXIMUM_CAPACITY)
 40        this.capacity = MAXIMUM_CAPACITY;
 41      else
 42        this.capacity = trimToPowerOf2(initialCapacity);
 43
 44      this.loadFactorThreshold = loadFactorThreshold;
 45      table = new LinkedList[capacity];
 46    }
 47
 48    @Override /** Remove all of the entries from this map */
 49    public void clear() {
 50      size = 0;
 51      removeEntries();
 52    }
 53
 54    @Override /** Return true if the specified key is in the map */
 55    public boolean containsKey(K key) {
 56      if (get(key) != null)
 57        return true;
 58      else
 59        return false;
 60    }
 61
 62    @Override /** Return true if this map contains the value */
 63    public boolean containsValue(V value) {
 64      for (int i = 0; i < capacity; i++) {
 65        if (table[i] != null) {
 66          LinkedList<Entry<K, V>> bucket = table[i];
 67          for (Entry<K, V> entry: bucket)
 68            if (entry.getValue().equals(value))
 69              return true;
 70        }
```

```java
 71      }
 72
 73      return false;
 74    }
 75
 76    @Override /** Return a set of entries in the map */
 77    public java.util.Set<MyMap.Entry<K,V>> entrySet() {
 78      java.util.Set<MyMap.Entry<K, V>> set =
 79        new java.util.HashSet<>();
 80
 81      for (int i = 0; i < capacity; i++) {
 82        if (table[i] != null) {
 83          LinkedList<Entry<K, V>> bucket = table[i];
 84          for (Entry<K, V> entry: bucket)
 85            set.add(entry);
 86        }
 87      }
 88
 89      return set;
 90    }
 91
 92    @Override /** Return the value that matches the specified key */
 93    public V get(K key) {
 94      int bucketIndex = hash(key.hashCode());
 95      if (table[bucketIndex] != null) {
 96        LinkedList<Entry<K, V>> bucket = table[bucketIndex];
 97        for (Entry<K, V> entry: bucket)
 98          if (entry.getKey().equals(key))
 99            return entry.getValue();
100      }
101
102      return null;
103    }
104
105    @Override /** Return true if this map contains no entries */
106    public boolean isEmpty() {
107      return size == 0;
108    }
109
110    @Override /** Return a set consisting of the keys in this map */
111    public java.util.Set<K> keySet() {
112      java.util.Set<K> set = new java.util.HashSet<>();
113
114      for (int i = 0; i < capacity; i++) {
115        if (table[i] != null) {
116          LinkedList<Entry<K, V>> bucket = table[i];
117          for (Entry<K, V> entry: bucket)
118            set.add(entry.getKey());
119        }
120      }
121
122      return set;
123    }
124
125    @Override /** Add an entry (key, value) into the map */
126    public V put(K key, V value) {
127      if (get(key) != null) { // The key is already in the map
128        int bucketIndex = hash(key.hashCode());
129        LinkedList<Entry<K, V>> bucket = table[bucketIndex];
130        for (Entry<K, V> entry: bucket)
131          if (entry.getKey().equals(key)) {
132            V oldValue = entry.getValue();
133            // Replace old value with new value
134            entry.value = value;
135            // Return the old value for the key
136            return oldValue;
```

```java
137          }
138        }
139
140        // Check load factor
141        if (size >= capacity * loadFactorThreshold) {
142          if (capacity == MAXIMUM_CAPACITY)
143            throw new RuntimeException("Exceeding maximum capacity");
144
145          rehash();
146        }
147
148        int bucketIndex = hash(key.hashCode());
149
150        // Create a linked list for the bucket if not already created
151        if (table[bucketIndex] == null) {
152          table[bucketIndex] = new LinkedList<Entry<K, V>>();
153        }
154
155        // Add a new entry (key, value) to hashTable[index]
156        table[bucketIndex].add(new MyMap.Entry<K, V>(key, value));
157
158        size++; // Increase size
159
160        return value;
161      }
162
163      @Override /** Remove the entries for the specified key */
164      public void remove(K key) {
165        int bucketIndex = hash(key.hashCode());
166
167        // Remove the first entry that matches the key from a bucket
168        if (table[bucketIndex] != null) {
169          LinkedList<Entry<K, V>> bucket = table[bucketIndex];
170          for (Entry<K, V> entry: bucket)
171            if (entry.getKey().equals(key)) {
172              bucket.remove(entry);
173              size--; // Decrease size
174              break; // Remove just one entry that matches the key
175            }
176        }
177      }
178
179      @Override /** Return the number of entries in this map */
180      public int size() {
181        return size;
182      }
183
184      @Override /** Return a set consisting of the values in this map */
185      public java.util.Set<V> values() {
186        java.util.Set<V> set = new java.util.HashSet<>();
187
188        for (int i = 0; i < capacity; i++) {
189          if (table[i] != null) {
190            LinkedList<Entry<K, V>> bucket = table[i];
191            for (Entry<K, V> entry: bucket)
192              set.add(entry.getValue());
193          }
194        }
195
196        return set;
197      }
198
199      /** Hash function */
200      private int hash(int hashCode) {
201        return supplementalHash(hashCode) & (capacity - 1);
202      }
```

```java
203
204    /** Ensure the hashing is evenly distributed */
205    private static int supplementalHash(int h) {
206      h ^= (h >>> 20) ^ (h >>> 12);
207      return h ^ (h >>> 7) ^ (h >>> 4);
208    }
209
210    /** Return a power of 2 for initialCapacity */
211    private int trimToPowerOf2(int initialCapacity) {
212      int capacity = 1;
213      while (capacity < initialCapacity) {
214        capacity <<= 1; // Same as capacity *= 2. <= is more efficient
215      }
216
217      return capacity;
218    }
219
220    /** Remove all entries from each bucket */
221    private void removeEntries() {
222      for (int i = 0; i < capacity; i++) {
223        if (table[i] != null) {
224          table[i].clear();
225        }
226      }
227    }
228
229    /** Rehash the map */
230    private void rehash() {
231      java.util.Set<Entry<K, V>> set = entrySet(); // Get entries
232      capacity <<= 1; // Same as capacity *= 2. <= is more efficient
233      table = new LinkedList[capacity]; // Create a new hash table
234      size = 0; // Reset size to 0
235
236      for (Entry<K, V> entry: set) {
237        put(entry.getKey(), entry.getValue()); // Store to new table
238      }
239    }
240
241    @Override /** Return a string representation for this map */
242    public String toString() {
243      StringBuilder builder = new StringBuilder("[");
244
245      for (int i = 0; i < capacity; i++) {
246        if (table[i] != null && table[i].size() > 0)
247          for (Entry<K, V> entry: table[i])
248            builder.append(entry);
249      }
250
251      builder.append("]");
252      return builder.toString();
253    }
254  }
```

MyHashMap 类采用分离链接法实现 MyMap 接口。在类中定义了确定散列表大小和装填因子的参数。默认的初始容量为 4（第 5 行），最大容量为 2^{30}（第 8 行）。当前散列表容量设计为 2 的幂值（第 11 行）。默认的装填因子阈值为 0.75f（第 14 行）。可以在构建一个映射的时候指定一个自定义的装填因子阈值。自定义的装填因子阈值保存在 loadFactorThreshold 中（第 17 行）。数据域 size 表示映射中的条目数（第 20 行）。散列表是一个数组，数组中的每个单元是一个链表（第 23 行）。

提供了三个构造方法来构建一个映射。可以使用无参构造方法来构建具有默认容量和装填因子阈值的映射（第 26～28 行），可以构造具有指定容量和默认的装填因子阈值的映射（第

32～34行），以及构建具有指定的容量和装填因子阈值的映射（第38～46行）。

clear 方法从映射中删除所有的条目（第49～52行）。该方法调用 removeEntries()，这将删除桶中的所有条目（第221～227行）。removeEntries() 方法用 $O(capacity)$ 的时间来清除表中的所有条目。

containsKey(key) 方法通过调用 get 方法（第55～60行）检测指定的键是否在映射中。由于 get 方法耗费 $O(1)$ 时间，containsKey(key) 方法也耗费 $O(1)$ 时间。

containsValue(value) 方法检测某个值是否在映射中（第63～74行）。该方法耗费 $O(capacity + size)$ 时间。该时间实际上是 $O(capacity)$，因为 capacity > size。

entrySet() 方法返回一个包含映射中所有条目的规则集（第77～90行）。该方法需要 $O(capacity)$ 时间。

get(key) 方法返回具有指定键的第一个条目的值（第93～103行）。该方法需要 $O(1)$ 时间。

isEmpty() 方法在映射为空的情况下简单地返回 true（第106～108行）。该方法需要 $O(1)$ 时间。

keySet() 方法返回一个包含映射中所有键的规则集。该方法从每个桶中找到键并将它们加入一个规则集中（第111～123行）。该方法花费 $O(capacity)$ 时间。

put(key, value) 方法添加一个新的条目到映射中。该方法首先测试该键是否已经在映射中（第127行），如果是，它定位该条目，并将该键所在条目的旧值替换成新值（第134行），并返回旧值（第136行）。如果键不在映射中，则在映射中产生一个新的条目（第156行）。插入新条目之前，该方法检测大小是否超过了装填因子的阈值（第141行）。如果是，程序调用 rehash()（第145行）来增加容量，并将条目保存到更大的新散列表中。

rehash() 方法首先复制所有规则集中的条目（第231行），将容量翻倍（第232行），创建一个新的散列表（第233行），并将大小重置为 0（第234行）。该方法然后将所有条目复制到一个新散列表中（第236～238行）。rehash 方法花费 $O(capacity)$ 时间。如果不执行再散列，put 方法花费 $O(1)$ 时间来添加一个新的条目。

remove(key) 方法删除映射中指定键的条目（第164～177行）。该方法花费 $O(1)$ 时间。

size() 方法简单地返回映射的大小（第180～182行）。该方法花费 $O(1)$ 时间。

value() 方法返回映射中所有的值。该方法从所有的桶中检测每个条目，然后将其添加到一个规则集中（第185～197行）。该方法花费 $O(capacity)$ 时间。

hash() 方法调用 supplementalHash 方法来确保散列均匀分布，从而生成散列表的下标（第200～208行）。该方法花费 $O(1)$ 时间。

表 27-1 总结了 MyHashMap 中方法的时间复杂度。

表 27-1 MyHashMap 中方法的时间复杂度

方法	时间	方法	时间
clear()	$O(capacity)$	keySet()	$O(capacity)$
containsKey(key: Key)	$O(1)$	put(key: K, value: V)	$O(1)$
containsValue(value: V)	$O(capacity)$	remove(key: K)	$O(1)$
entrySet()	$O(capacity)$	size()	$O(1)$
get(key: K)	$O(1)$	values()	$O(capacity)$
isEmpty()	$O(1)$	rehash()	$O(capacity)$

由于再散列并不经常发生，put 方法的时间复杂度为 $O(1)$。注意，clear、entrySet、

keySet、values 以及 rehash 方法的复杂度依赖于 capacity，因此应该精心选择初始容量来避免这些方法的性能低下。

程序清单 27-3 给出了使用 MyHashMap 的测试程序。

程序清单 27-3 TestMyHashMap.java

```java
 1  public class TestMyHashMap {
 2    public static void main(String[] args) {
 3      // Create a map
 4      MyMap<String, Integer> map = new MyHashMap<>();
 5      map.put("Smith", 30);
 6      map.put("Anderson", 31);
 7      map.put("Lewis", 29);
 8      map.put("Cook", 29);
 9      map.put("Smith", 65);
10
11      System.out.println("Entries in map: " + map);
12
13      System.out.println("The age for Lewis is " +
14        map.get("Lewis"));
15
16      System.out.println("Is Smith in the map? " +
17        map.containsKey("Smith"));
18      System.out.println("Is age 33 in the map? " +
19        map.containsValue(33));
20
21      map.remove("Smith");
22      System.out.println("Entries in map: " + map);
23
24      map.clear();
25      System.out.println("Entries in map: " + map);
26    }
27  }
```

```
Entries in map: [[Anderson, 31][Smith, 65][Lewis, 29][Cook, 29]]
The age for Lewis is 29
Is Smith in the map? true
Is age 33 in the map? false
Entries in map: [[Anderson, 31][Lewis, 29][Cook, 29]]
Entries in map: []
```

该程序应用 MyHashMap 创建一个映射（第 4 行），并添加 5 个条目到映射中（第 5～9 行）。第 5 行添加键 Smith 和相应的值 30，第 9 行添加键 Smith 和相应的值 65。后者的值替换了前者的值。映射实际上只有 4 个条目。该程序显示了映射中的条目（第 11 行），得到一个键相应的值（第 14 行），检测映射中是否包含某个键（第 17 行）以及某个值（第 19 行），删除键 Smith 的条目（第 21 行），然后重新显示映射中的条目（第 22 行）。最后，程序清除映射（第 24 行）并显示一个空的映射（第 25 行）。

✓ **复习题**

27.7.1 程序清单 27-2 中，第 8 行的 1 << 30 是什么？1 << 1、1 << 2 以及 1 << 3 得到的整数是什么？

27.7.2 32 >> 1、32 >> 2、32 >> 3 以及 32 >> 4 得到的整数是什么？

27.7.3 程序清单 27-2 中，如果将 LinkedList 替换为 ArrayList，程序还能工作吗？程序清单 27-2 中，如何将第 56～59 行代码替换为一行代码？

27.7.4 描述 MyHashMap 类中 put(key, value) 方法是如何实现的。

27.7.5 程序清单 27-2 中，supplementalHash 方法声明为静态的，hash 方法可以声明为静态的吗？

27.7.6 给出下面代码的输出结果。

```java
MyMap<String, String> map = new MyHashMap<>();
map.put("Texas", "Dallas");
map.put("Oklahoma", "Norman");
map.put("Texas", "Austin");
map.put("Oklahoma", "Tulsa");

System.out.println(map.get("Texas"));
System.out.println(map.size());
```

27.7.7 如果 x 是一个负的 int 值，x & (N-1) 是负的吗？

27.8 使用散列实现规则集

要点提示：可以使用散列映射来实现散列集。

规则集（第 21 章中介绍过）是一种存储不同值的数据结构。Java 集合框架定义了 java.util.Set 接口来对规则集建模。三种具体的实现是 java.util.HashSet、java.util.LinkedHashSet 以及 java.util.TreeSet。java.util.HashSet 采用散列实现，java.util.LinkedHashSet 采用 LinkedList 实现，java.util.TreeSet 采用二叉搜索树实现。

可以采用与实现 MyHashMap 相同的方式来实现 MyHashSet。唯一不同之处在于键/值对存储在映射中，而元素存储在规则集中。

由于 HashSet 中的所有方法都继承自 Collection，我们通过实现 Collection 接口来设计自定义的 HashSet 类，如图 27-11 所示。

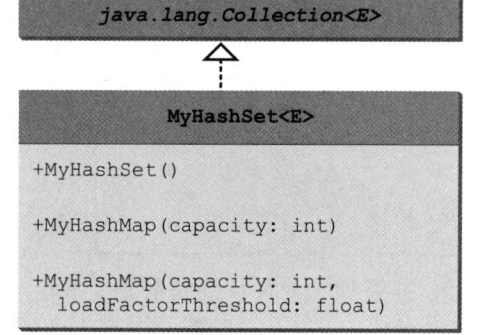

图 27-11 MyHashSet 实现 Collection 接口

程序清单 27-4 采用分离链接法实现了 MyHashSet。

程序清单 27-4 MyHashSet.java

```java
 1  import java.util.*;
 2
 3  public class MyHashSet<E> implements Collection<E> {
 4    // Define the default hash-table size. Must be a power of 2
 5    private static final int DEFAULT_INITIAL_CAPACITY = 4;
 6
 7    // Define the maximum hash-table size. 1 << 30 is same as 2^30
 8    private static final int MAXIMUM_CAPACITY = 1 << 30;
 9
10    // Current hash-table capacity. Capacity is a power of 2
11    private int capacity;
12
13    // Define default load factor
14    private static final float DEFAULT_MAX_LOAD_FACTOR = 0.75f;
15
```

```java
16    // Specify a load-factor threshold used in the hash table
17    private float loadFactorThreshold;
18
19    // The number of elements in the set
20    private int size = 0;
21
22    // Hash table is an array with each cell being a linked list
23    private LinkedList<E>[] table;
24
25    /** Construct a set with the default capacity and load factor */
26    public MyHashSet() {
27      this(DEFAULT_INITIAL_CAPACITY, DEFAULT_MAX_LOAD_FACTOR);
28    }
29
30    /** Construct a set with the specified initial capacity and
31     * default load factor */
32    public MyHashSet(int initialCapacity) {
33      this(initialCapacity, DEFAULT_MAX_LOAD_FACTOR);
34    }
35
36    /** Construct a set with the specified initial capacity
37     * and load factor */
38    public MyHashSet(int initialCapacity, float loadFactorThreshold) {
39      if (initialCapacity > MAXIMUM_CAPACITY)
40        this.capacity = MAXIMUM_CAPACITY;
41      else
42        this.capacity = trimToPowerOf2(initialCapacity);
43
44      this.loadFactorThreshold = loadFactorThreshold;
45      table = new LinkedList[capacity];
46    }
47
48    @Override /** Remove all elements from this set */
49    public void clear() {
50      size = 0;
51      removeElements();
52    }
53
54    @Override /** Return true if the element is in the set */
55    public boolean contains(E e) {
56      int bucketIndex = hash(e.hashCode());
57      if (table[bucketIndex] != null) {
58        LinkedList<E> bucket = table[bucketIndex];
59        return bucket.contains(e);
60      }
61
62      return false;
63    }
64
65    @Override /** Add an element to the set */
66    public boolean add(E e) {
67      if (contains(e))  // Duplicate element not stored
68        return false;
69
70      if (size + 1 > capacity * loadFactorThreshold) {
71        if (capacity == MAXIMUM_CAPACITY)
72          throw new RuntimeException("Exceeding maximum capacity");
73
74        rehash();
75      }
76
77      int bucketIndex = hash(e.hashCode());
78
79      // Create a linked list for the bucket if not already created
80      if (table[bucketIndex] == null) {
81        table[bucketIndex] = new LinkedList<E>();
```

```java
 82     }
 83
 84     // Add e to hashTable[index]
 85     table[bucketIndex].add(e);
 86
 87     size++; // Increase size
 88
 89     return true;
 90   }
 91
 92   @Override /** Remove the element from the set */
 93   public boolean remove(E e) {
 94     if (!contains(e))
 95       return false;
 96
 97     int bucketIndex = hash(e.hashCode());
 98
 99     // Create a linked list for the bucket if not already created
100     if (table[bucketIndex] != null) {
101       LinkedList<E> bucket = table[bucketIndex];
102       bucket.removed(e);
103     }
104
105     size--; // Decrease size
106
107     return true;
108   }
109
110   @Override /** Return true if the set contain no elements */
111   public boolean isEmpty() {
112     return size == 0;
113   }
114
115   @Override /** Return the number of elements in the set */
116   public int size() {
117     return size;
118   }
119
120   @Override /** Return an iterator for the elements in this set */
121   public java.util.Iterator<E> iterator() {
122     return new MyHashSetIterator(this);
123   }
124
125   /** Inner class for iterator */
126   private class MyHashSetIterator implements java.util.Iterator<E> {
127     // Store the elements in a list
128     private java.util.ArrayList<E> list;
129     private int current = 0; // Point to the current element in list
130     private MyHashSet<E> set;
131
132     /** Create a list from the set */
133     public MyHashSetIterator(MyHashSet<E> set) {
134       this.set = set;
135       list = setToList();
136     }
137
138     @Override /** Next element for traversing? */
139     public boolean hasNext() {
140       return current < list.size();
141     }
142
143     @Override /** Get current element and move cursor to the next */
144     public E next() {
145       return list.get(current++);
146     }
147
```

```java
148       /** Remove the current element returned by the last next() */
149       public void remove() {
150         // Left as an exercise
151         // You need to remove the element from the set
152         // You also need to remove it from the list
153       }
154     }
155
156     /** Hash function */
157     private int hash(int hashCode) {
158       return supplementalHash(hashCode) & (capacity - 1);
159     }
160
161     /** Ensure the hashing is evenly distributed */
162     private static int supplementalHash(int h) {
163       h ^= (h >>> 20) ^ (h >>> 12);
164       return h ^ (h >>> 7) ^ (h >>> 4);
165     }
166
167     /** Return a power of 2 for initialCapacity */
168     private int trimToPowerOf2(int initialCapacity) {
169       int capacity = 1;
170       while (capacity < initialCapacity) {
171         capacity <<= 1; // Same as capacity *= 2. <= is more efficient
172       }
173
174       return capacity;
175     }
176
177     /** Remove all e from each bucket */
178     private void removeElements() {
179       for (int i = 0; i < capacity; i++) {
180         if (table[i] != null) {
181           table[i].clear();
182         }
183       }
184     }
185
186     /** Rehash the set */
187     private void rehash() {
188       java.util.ArrayList<E> list = setToList(); // Copy to a list
189       capacity <<= 1; // Same as capacity *= 2. <= is more efficient
190       table = new LinkedList[capacity]; // Create a new hash table
191       size = 0;
192
193       for (E element: list) {
194         add(element); // Add from the old table to the new table
195       }
196     }
197
198     /** Copy elements in the hash set to an array list */
199     private java.util.ArrayList<E> setToList() {
200       java.util.ArrayList<E> list = new java.util.ArrayList<>();
201
202       for (int i = 0; i < capacity; i++) {
203         if (table[i] != null) {
204           for (E e: table[i]) {
205             list.add(e);
206           }
207         }
208       }
209
210       return list;
211     }
212
213     @Override /** Return a string representation for this set */
```

```java
214   public String toString() {
215     java.util.ArrayList<E> list = setToList();
216     StringBuilder builder = new StringBuilder("[");
217
218     // Add the elements except the last one to the string builder
219     for (int i = 0; i < list.size() - 1; i++) {
220       builder.append(list.get(i) + ", ");
221     }
222
223     // Add the last element in the list to the string builder
224     if (list.size() == 0)
225       builder.append("]");
226     else
227       builder.append(list.get(list.size() - 1) + "]");
228
229     return builder.toString();
230   }
231
232   @Override
233   public boolean addAll(Collection<? extends E> arg0) {
234     // Left as an exercise
235     return false;
236   }
237
238   @Override
239   public boolean containsAll(Collection<?> arg0) {
240     // Left as an exercise
241     return false;
242   }
243
244   @Override
245   public boolean removeAll(Collection<?> arg0) {
246     // Left as an exercise
247     return false;
248   }
249
250   @Override
251   public boolean retainAll(Collection<?> arg0) {
252     // Left as an exercise
253     return false;
254   }
255
256   @Override
257   public Object[] toArray() {
258     // Left as an exercise
259     return null;
260   }
261
262   @Override
263   public <T> T[] toArray(T[] arg0) {
264     // Left as an exercise
265     return null;
266   }
267 }
```

MyHashSet 类使用分离链接法实现了 MySet 接口。实现 MyHashSet 类类似于实现 MyHashMap，不过有以下不同：

1）对于 MyHashSet 来说，元素存储在散列表中，而对于 MyHashMap 来说，条目（键/值对）存储在散列表中。

2）MyHashSet 实现了 Collection。由于 Collection 接口继承自 Iterable 接口，所以 MyHashSet 中的元素是可遍历的。

提供了三个构造方法来构建一个规则集。可以使用无参构造方法来构建具有默认容量和装填因子阈值的默认规则集（第 26～28 行），可以构造具有指定容量和默认装填因子阈值的规则集（第 32～34 行），以及构建具有指定容量和装填因子阈值的规则集（第 38～46 行）。

clear 方法从规则集中删除所有的条目（第 49～52 行）。该方法调用 removeElements()，这将删除表中所有的单元（第 181 行）。表中的每个单元是一个存储了具有相同散列表下标的元素的链表。removeElements() 方法花费 $O(capacity)$ 时间。

contains(element) 方法通过审查指定的桶里是否包含元素来检测指定的键是否在规则集中（第 59 行）。该方法花费 $O(1)$ 时间，因为桶的大小被认为非常小。

add(element) 方法添加一个新元素到规则集中。该方法首先检测该元素是否已经在规则集中（第 67 行）。如果是，该方法返回 false。接着该方法检测是否大小超出了装填因子的阈值（第 70 行）。如果是，该程序调用 rehash()（第 74 行）来增加容量并将元素存储到更大的新散列表中。

rehash() 方法首先复制所有元素到一个线性表中（第 188 行），将容量翻倍（第 189 行），创建一个新的散列表（第 190 行），并将大小重置为 0（第 191 行）。该方法然后将所有元素复制到一个更大的新散列表中（第 193～195 行）。rehash 方法花费 $O(capacity)$ 时间。如果不执行再散列，add 方法花费 $O(1)$ 时间来添加一个新的元素。

remove(element) 方法删除规则集中的指定元素（第 93～108 行）。该方法花费 $O(1)$ 时间。

size() 方法简单地返回规则集中元素的数目（第 116～118 行）。该方法花费 $O(1)$ 时间。

iterator() 方法返回一个 java.util.Iterator 的实例。MyHashSetIterator 类实现了 java.util.Iterator 以创建一个前向迭代器。当构建一个 MyHashSetIterator 时，复制规则集中所有的元素到一个线性表中（第 135 行）。变量 current 指向线性表中的元素。初始时，current 为 0（第 129 行），这表示指向线性表中的第一个元素。MyHashSetIterator 实现了 java.util.Iterator 中的 hasNext()、next() 以及 remove() 方法。如果 current < list.size()，则调用 hasNext() 会返回 true。调用 next() 返回当前元素并移动 current 以指向下一个元素（第 145 行）。调用 remove() 删除最后一个 next() 调用的元素。

hash() 方法调用 supplementalHash 方法来确保散列均匀分布，从而生成散列表的下标（第 157～159 行）。该方法花费 $O(1)$ 时间。

定义在 Collection 接口中的 containsAll、addAll、removeAll、retainAll、toArray() 和 toArray(T[]) 方法在 MyHashSet 中被重写。它们的实现留作编程练习题 27.11。

表 27-2 总结了 MyHashSet 中方法的时间复杂度。

表 27-2 MyHashSet 中方法的时间复杂度

方法	时间
clear()	$O(capacity)$
contains(e: E)	$O(1)$
add(e: E)	$O(1)$
remove(e: E)	$O(1)$
isEmpty()	$O(1)$
size()	$O(1)$
iterator()	$O(capacity)$
rehash()	$O(capacity)$

程序清单 27-5 给出了使用 MyHashSet 的测试程序。

程序清单 27-5 TestMyHashSet.java

```java
 1  public class TestMyHashSet {
 2    public static void main(String[] args) {
 3      // Create a MyHashSet
 4      java.util.Collection<String> set = new MyHashSet<>();
 5      set.add("Smith");
 6      set.add("Anderson");
 7      set.add("Lewis");
 8      set.add("Cook");
 9      set.add("Smith");
10
11      System.out.println("Elements in set: " + set);
12      System.out.println("Number of elements in set: " + set.size());
13      System.out.println("Is Smith in set? " + set.contains("Smith"));
14
15      set.remove("Smith");
16      System.out.print("Names in set in uppercase are ");
17      for (String s: set)
18        System.out.print(s.toUpperCase() + " ");
19
20      set.clear();
21      System.out.println("\nElements in set: " + set);
22    }
23  }
```

```
Elements in set: [Cook, Anderson, Smith, Lewis]
Number of elements in set: 4
Is Smith in set? true
Names in set in uppercase are COOK ANDERSON LEWIS
Elements in set: []
```

该程序应用 MyHashSet 创建一个规则集（第 4 行），并添加 5 个元素到规则集中（第 5～9 行）。第 5 行添加 Smith，第 9 行再次添加 Smith。由于只有不重复的元素可以存储在规则集中，Smith 只在规则集中出现一次。规则集中实际上有 4 个元素。程序显示了这些元素（第 11 行），得到它的大小（第 12 行），检测规则集是否包含某个指定的元素（第 13 行），删除一个元素（第 15 行）。由于规则集中的元素是可遍历的，该程序使用了 foreach 循环来遍历规则集中的所有元素（第 17～18 行）。最后，程序清除规则集（第 20 行）并显示一个空的规则集（第 21 行）。

✔ **复习题**

27.8.1 为什么可以使用 foreach 循环来遍历规则集中的元素？

27.8.2 描述 MyHashSet 类中的 add(e) 方法是如何实现的。

27.8.3 程序清单 27-4 中第 100～103 行可以删去吗？

27.8.4 实现程序清单 27-4 中第 150～152 行中的 remove() 方法。

关键术语

associative array（关联数组）
cluster（簇）
dictionary（字典）
double hashing（双重散列）
hash code（散列码）
hash function（散列函数）
hash map（散列映射）
hash set（散列规则集）
hash table（散列表）
linear probing（线性探测）

load factor（装填因子）
open addressing（开放地址法）
perfect hash function（完美散列函数）
separate chaining（分离链接法）

quadratic probing（二次探测法）
rehashing（再散列）
polynomial hash code（多项式散列码）

本章小结

1. 映射是一种存储条目的数据结构。每个条目包含两部分：键和值。键也称为搜索键，用于查找相应的值。可以使用散列技术来实现映射，实现使用 $O(1)$ 的时间复杂度来查找、获取、插入以及删除条目。
2. 规则集是一种存储元素的数据结构。可以使用散列技术来实现规则集，实现使用 $O(1)$ 的时间复杂度来查找、获取、插入以及删除元素。
3. 散列是一种无须执行搜索即可通过从键得到的下标来获取值的技术。典型的散列函数首先将搜索键转化为一个称为散列码的整数值，然后将散列码压缩为散列表中的一个下标。
4. 当两个键映射到散列表中的同样下标上时，产生冲突。通常有两种方法处理冲突：开放地址法和分离链接法。
5. 开放地址法是在发生冲突时，在散列表中找到一个开放位置的过程。开放地址法有几种变体：线性探测、二次探测以及双重散列。
6. 分离链接法将具有同样散列下标的条目放到相同的位置，而不是寻找新的位置。分离链接法中每个位置称为一个桶。桶是容纳多个条目的容器。

测试题

回答位于本书配套网站上的本章测试题。

编程练习题

**27.1 （应用开放地址法的线性探测法来实现 MyMap）应用开放地址法的线性探测法创建一个实现 MyMap 的新的具体类。简单起见，使用 f(key) = key % size 作为散列函数，这里 size 是散列表的大小。初始时，散列表的大小为 4。当装填因子超过阈值 (0.5) 时，表的大小翻倍。使用位于 https://liveexample.pearsoncmg.com/test/Exercise27_01.txt 的代码测试你的新 MyHashMap 类。

**27.2 （应用开放地址法的二次探测法来实现 MyMap）应用开放地址法的二次探测法创建一个实现 MyMap 的新的具体类。简单起见，使用 f(key) = key % size 作为散列函数，这里 size 是散列表的大小。初始时，散列表的大小为 4。当装填因子超过阈值 (0.5) 时，表的大小翻倍。

**27.3 （应用开放地址法的双重散列法来实现 MyMap）应用开放地址法的双重散列法创建一个实现 MyMap 的新的具体类。简单起见，使用 f(key) = key % size 作为散列函数，这里 size 是散列表的大小。初始时，散列表的大小为 4。当装填因子超过阈值 (0.5) 时，表的大小翻倍。

**27.4 （修改 MyHashMap 使得可以有重复的键）修改 MyHashMap 从而允许条目可以有重复的键。需要修改 put(key,value) 的实现。同时，添加一个名为 getAll(key) 的新方法，返回一个匹配映射中键的值的规则集。

**27.5 （使用 MyHashMap 实现 MyHashSet）使用 MyHashMap 实现 MyHashSet。注意，可以使用 (key, key) 创建条目，而不是使用 (key, value)。

**27.6 （实现线性探测法的动画）编写程序，实现线性探测法的动画，如图 27-3 所示。可以在程序中修改散列表的初始大小。假设装填因子阈值为 0.75。

**27.7 （实现分离链接法的动画）编写程序，实现 MyHashMap 的动画，如图 27-9 所示。可以在程序中修改散列表的初始大小。假设装填因子阈值为 0.75。

**27.8 （实现二次探测法的动画）编写程序，实现二次探测法的动画，如图 27-5 所示。可以在程序中修改散列表的初始大小。假设装填因子阈值为 0.75。

**27.9 （实现字符串的散列码）编写一个方法，使用 27.3.2 节中描述的方法返回字符串的散列码，其中 b 取值 31。方法头如下：

```
public static int hashCodeForString(String s)
```

**27.10 （比较 MyHashSet 和 MyArrayList）程序清单 24-2 定义了 MyArrayList。编写一个程序，产生 0 到 999999 之间的 1000000 个随机双精度值，并将它们存储在 MyArrayList 和 MyHashSet 中。产生一个包含 0 到 1999999 之间的 1000000 个随机双精度值的线性表。对于线性表中的每个数字，检测是否在数组线性表中以及是否在散列规则集中。运行程序，显示针对数组线性表和散列规则集的总体测试时间。

**27.11 （实现 MyHashSet 中的规则集操作）在 MyHashSet 类中省略了 addAll、removeAll、retainAll、toArray() 和 toArray(T[]) 方法的实现，实现这些方法。同时，在 MyHashSet 类中添加一个新的构造方法 MyHashSet(E[] list)。用 liveexample.pearsoncmg.com/test/Exercise27_11.txt 上的代码来测试你的新 MyHashSet 类。

**27.12 （SetToList）编写以下方法，从一个规则集中返回 ArrayList。

```
public static <E> ArrayList<E> setToList(Set<E> s)
```

*27.13 （Date 类）设计一个 Date 类，满足下列要求：
- 用三个数据域 year、month、day 来表示一个日期。
- 一个构造方法，它以指定的年、月、日为参数构造一个日期。
- 重写 equals 方法。
- 重写 hashCode 方法（可以参考 Java API 中 Date 类的实现）。

*27.14 （Point 类）设计一个 Point 类，满足下列要求：
- 两个带有 getter 方法的数据域 x、y，用来表示一个点。
- 一个无参的构造方法，生成一个代表 (0,0) 的点。
- 一个构造方法，以指定 x、y 值构造点。
- 重写 equals 方法。对于点 p1 和点 p2，当 p1.x == p2.x 且 p1.y == p2.y 时，认为 p1 == p2。
- 重写 hashCode 方法（可以参考 Java API 中 Point2D 类的实现）。

*27.15 （改写程序清单 27-4）文中用 LinkedList 作为桶。用 AVLTree 来代替 LinkedList。假设 E 是 Comparable 类型的。如下重定义 MyHashSet：

```
public class MyHashSet<E extends Comparable<E>> implements
    Collection<E> {
  ...
}
```

用程序清单 27-5 中的 main 方法来测试你的程序。

第 28 章

图及其应用

教学目标
- 使用图对真实世界中的问题进行建模并解释哥尼斯堡七桥问题（28.1 节）。
- 描述图中的术语：顶点、边、简单图、加权／非加权图以及有向／无向图（28.2 节）。
- 使用线性表、边数组、边对象、邻接矩阵和邻接线性表来表示顶点和边（28.3 节）。
- 使用 Graph 接口和 UnweightedGraph 类对图建模（28.4 节）。
- 可视化显示图（28.5 节）。
- 使用 UnweightedGraph.SearchTree 类表示图的遍历（28.6 节）。
- 设计并且实现深度优先搜索（28.7 节）。
- 使用深度优先搜索解决连通圆问题（28.8 节）。
- 设计并且实现广度优先搜索（28.9 节）。
- 使用广度优先搜索解决 9 枚硬币反面的问题（28.10 节）。

28.1 引言

要点提示：真实世界中的许多问题可以使用图算法解决。

图对现实世界中的问题的建模和求解非常有用。例如，可以使用图对求两座城市之间最少飞行次数的问题进行建模，其中顶点代表城市，边代表两座相邻城市之间的航班，如图 28-1 所示。求两座城市之间最少飞行次数的问题就简化为寻找图中两个顶点之间最短路径的问题。在 UPS（United Parcel Service，联合包裹服务）公司，平均每位司机每天会经过 120 站。有很多方式对这些站排序。UPS 在 10 年间花了上亿美元开发了一个叫作 Orion(On-Road Integrated Optimization and Navigation，行车集成优化和导航) 的系统，它使用图算法来为每位司机规划出性价比最高的路线。本章学习非加权图的算法，下一章将学习加权图的算法。

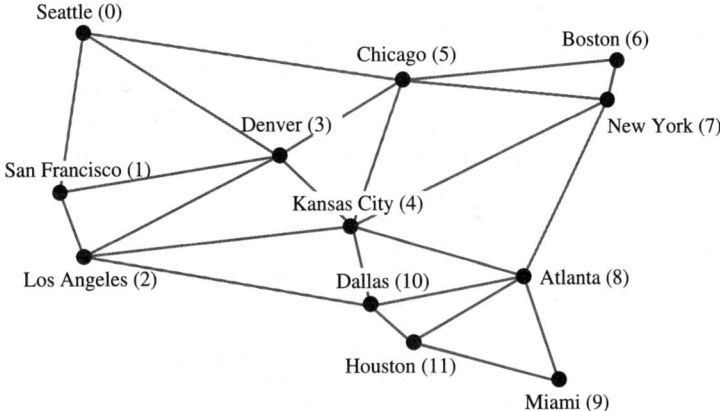

图 28-1　图可以用来对城市之间的飞行次数进行建模

对图的研究也称为图论（graph theory）。1736年莱昂哈德·欧拉创立了图论，当时他用"图"术语来解决著名的哥尼斯堡七桥问题。位于普鲁士的哥尼斯堡（现俄罗斯的加里宁格勒）被普累格河分开，该河流经两座岛，这座城市和岛由七座桥相连，如图28-2a所示。问题在于，一个人可以经过每座桥一次且只经过一次，然后返回起点吗？欧拉证明了这是不可能的。

为了证明这个结论，欧拉首先通过删除所有的街道来抽象出哥尼斯堡的城市地图，得到了如图28-2a所示的草图。然后，他将每一块陆地用一个点来替换，这个点称为顶点（vertex）或者结点（node），并且将每一座桥用一条线来替换，这条线称为边（edge），如图28-2b所示。这种有顶点和边的结构称为图（graph）。

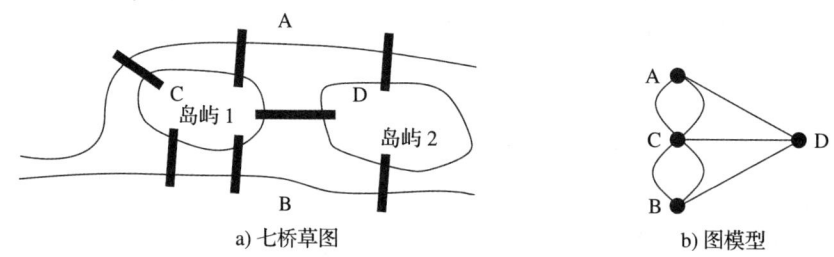

图28-2 七座桥连接了岛屿和陆地

对于这样的图，我们询问是否存在一条从任意顶点出发的路径，这条路径遍历所有的边一次且只有一次，然后返回起始顶点。欧拉证明了这种路径存在的条件是，每个顶点必须拥有偶数条边。因此，哥尼斯堡七桥问题无解。

图算法广泛应用于不同的领域，例如计算机科学、数学、生物学、工程学、经济学、遗传学和社会科学。本章讲述深度优先搜索和广度优先搜索以及它们的应用。下一章将讲述在加权图中找到最小生成树和最短路径的算法，以及它们的应用。

28.2 基本的图术语

○┳ 要点提示：图由顶点以及连接顶点的边所组成。

本章不假定读者对图论或者离散数学有任何预备知识。下面使用简单明了的术语来定义图。

什么是图？图（graph）是一种数学结构，它表示真实世界中实体之间的关系。例如，图28-1中的图代表了城市间的航班，图28-2b中的图代表了陆地之间的桥梁。

一个图包含了非空的顶点（也称为结点或者点），以及一个连接顶点的边的集合。为方便起见，我们定义一个图为$G=(V, E)$，其中V代表顶点的集合，E代表边的集合。例如，图28-1中图的V和E分别如下所示：

```
V = {"Seattle", "San Francisco", "Los Angeles",
     "Denver", "Kansas City", "Chicago", "Boston", "New York",
     "Atlanta", "Miami", "Dallas", "Houston"};

E = {{"Seattle", "San Francisco"},{"Seattle", "Chicago"},
     {"Seattle", "Denver"}, {"San Francisco", "Denver"},
     ...
    };
```

图可以是有向的，也可以是无向的。在有向图（directed graph）中，每条边都有一个方向，表明可以沿着这条边从一个顶点移到另一个顶点。可以使用有向图来对父/子之间的关系进行建模，其中从顶点 A 到 B 的边表示 A 是 B 的父结点。图 28-3a 显示了一个有向图。

在无向图（undirected graph）中，可以在顶点之间双向移动。图 28-1 中的图是无向的。

边可以是加权的，也可以是非加权的。例如，你可以给图 28-1 中图的每条边分配一个权重，表示两个城市之间的飞行时间。

如果图中的两个顶点被同一条边连接，那么它们被称为邻接的（adjacent）。类似地，如果两条边连接到同一个顶点，它们也被称为邻接的。在图中，连接两个顶点的边被认为关联（incident）到这两个顶点。顶点的度（degree）就是与该顶点关联的边的数目。

如果两个顶点是邻接的，那么它们互为邻居（neighbor）。类似地，两条邻接的边也互为邻居。

环（loop）是一条将顶点连接到它自身的边。如果两个顶点可通过两条或者多条边相连，这些边就称为平行边（parallel edge）。简单图（simple graph）是指没有环和平行边的图。完全图（complete graph）是指每一对顶点都相连的图，如图 28-3b 所示。

如果图中任意两个顶点之间存在一条路径，称该图为连通的（connected）。一个图 G 的子图（subgraph）是如下的图：其顶点集合是 G 的子集，其边的集合是 G 的子集。例如，图 28-3c 中的图是 28-3b 中图的子图。

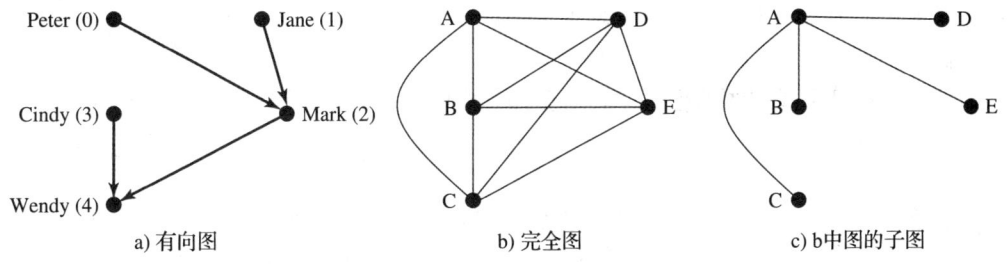

a) 有向图　　　　b) 完全图　　　　c) b中图的子图

图 28-3　图可以呈现为各种形式

假设图是连通且无向的。回路（cycle）是指始于一个顶点然后终于同一顶点的封闭路径。没有回路的连通图是一棵树（tree）。图 G 的生成树（spanning tree）是一个 G 的连通子图，该子图是包含 G 中所有顶点的树。

教学注意：在开始介绍图算法及其应用之前，通过网址 liveexample.pearsoncmg.com/dsanimation/GraphLearningTooleBook.html 提供的交互式工具来了解图是很有帮助的，如图 28-4 所示。该工具可以让你通过鼠标操作添加/删除/移动顶点以及绘制边。也可以找到深度优先搜索（DFS）树和广度优先搜索（BFS）树，以及两个顶点之间的最短路径。

✓ **复习题**

28.2.1　什么是著名的哥尼斯堡七桥问题？

28.2.2　什么是图？解释下列术语：无向图、有向图、加权图、顶点的度、平行边、简单图、完全图、连通图、回路、子图、树以及生成树。

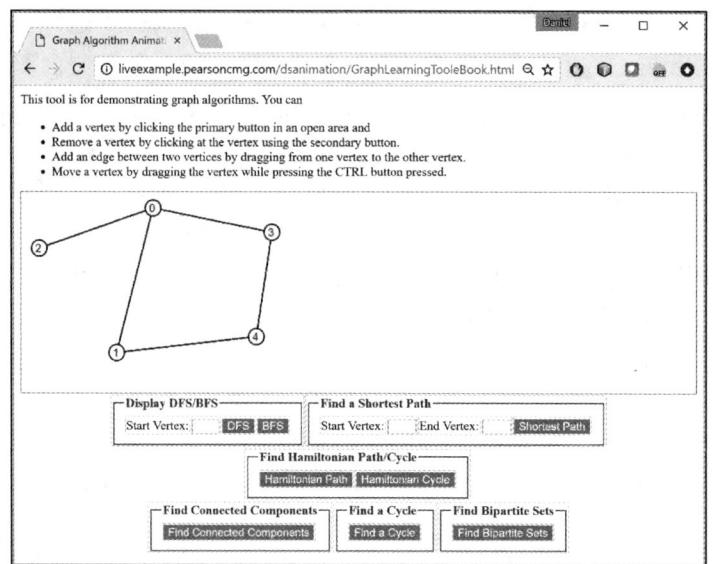

图 28-4 可以使用工具通过鼠标操作来创建图，以及显示 DFS/BFS 树和最短路径

28.2.3 具有 5 个顶点的完全图中有几条边？具有 5 个结点的树中有几条边？

28.2.4 具有 n 个顶点的完全图中有几条边？具有 n 个结点的树中有几条边？

28.3 表示图

☞ 要点提示：表示一个图是指在程序中存储它的顶点和边。存储图的数据结构是数组或者线性表。

为了编写处理和操作图的程序，必须在计算机中存储和表示图。

28.3.1 表示顶点

顶点可以存储在数组或线性表中。例如，可以用下面的数组来存储图 28-1 中的所有城市名：

```
String[] vertices = {"Seattle", "San Francisco", "Los Angeles",
  "Denver", "Kansas City", "Chicago", "Boston", "New York",
  "Atlanta", "Miami", "Dallas", "Houston"};
```

☞ 注意：顶点可以是任意类型的对象。例如，可以将城市考虑为包含其名字、人口和市长等信息的对象。于是，可以将顶点定义为：

```
City city0 = new City("Seattle", 608660, "Mike McGinn");
...
City city11 = new City("Houston", 2099451, "Annise Parker");
City[] vertices = {city0, city1, . . . , city11};

public class City {
  private String cityName;
  private int population;
  private String mayor;

  public City(String cityName, int population, String mayor) {
    this.cityName = cityName;
    this.population = population;
    this.mayor = mayor;
  }
```

```java
  public String getCityName() {
    return cityName;
  }

  public int getPopulation() {
    return population;
  }

  public String getMayor() {
    return mayor;
  }

  public void setMayor(String mayor) {
    this.mayor = mayor;
  }

  public void setPopulation(int population) {
    this.population = population;
  }
}
```

对于一个拥有 n 个顶点的图，这 n 个顶点可以使用自然数 0, 1, 2, …, n−1 来标注。于是，vertices[0] 表示 "Seattle"，vertices[1] 表示 "San Francisco"，等等，如图 28-5 所示。

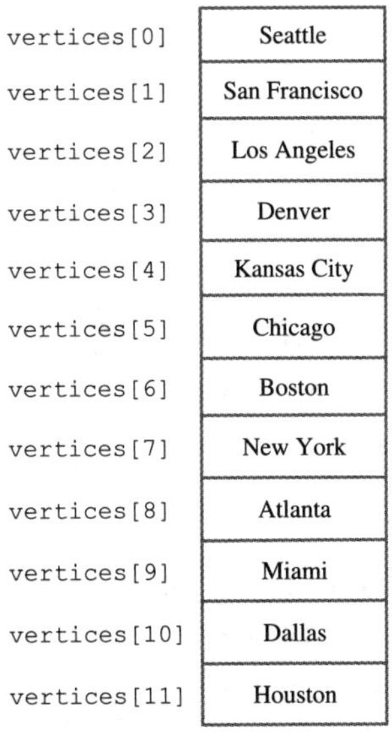

vertices[0]	Seattle
vertices[1]	San Francisco
vertices[2]	Los Angeles
vertices[3]	Denver
vertices[4]	Kansas City
vertices[5]	Chicago
vertices[6]	Boston
vertices[7]	New York
vertices[8]	Atlanta
vertices[9]	Miami
vertices[10]	Dallas
vertices[11]	Houston

图 28-5 存储顶点名字的数组

☛ **注意**：可以通过顶点的名字或者下标来引用顶点，就看哪一种方式使用起来更方便。显然，在程序中通过下标访问顶点比较容易。

28.3.2 表示边：边数组

边可以使用二维数组来表示。例如，可以使用下面的数组来存储图 28-1 中图的所有边：

```java
int[][] edges = {
  {0, 1}, {0, 3}, {0, 5},
  {1, 0}, {1, 2}, {1, 3},
  {2, 1}, {2, 3}, {2, 4}, {2, 10},
  {3, 0}, {3, 1}, {3, 2}, {3, 4}, {3, 5},
  {4, 2}, {4, 3}, {4, 5}, {4, 7}, {4, 8}, {4, 10},
  {5, 0}, {5, 3}, {5, 4}, {5, 6}, {5, 7},
```

```
    {6, 5}, {6, 7},
    {7, 4}, {7, 5}, {7, 6}, {7, 8},
    {8, 4}, {8, 7}, {8, 9}, {8, 10}, {8, 11},
    {9, 8}, {9, 11},
    {10, 2}, {10, 4}, {10, 8}, {10, 11},
    {11, 8}, {11, 9}, {11, 10}
};
```

这种表示称为边数组（edge array）。图 28-3a 中的顶点和边可以如下表示：

```
String[] vertices = {"Peter", "Jane", "Mark", "Cindy", "Wendy"};

int[][] edges = {{0, 2}, {1, 2}, {2, 4}, {3, 4}};
```

28.3.3 表示边：Edge 对象

另外一种表示边的方法就是将边定义为对象，并存储在 java.util.ArrayList 中。Edge 类可以如程序清单 28-1 所示定义：

程序清单 28-1 Edge.java

```java
public class Edge {
  int u;
  int v;

  public Edge(int u, int v) {
    this.u = u;
    this.v = v;
  }

  public boolean equals(Object o) {
    return u == ((Edge)o).u && v == ((Edge)o).v;
  }
}
```

例如，可以使用下面的线性表来存储图 28-1 中图的所有边：

```java
java.util.ArrayList<Edge> list = new java.util.ArrayList<>();
list.add(new Edge(0, 1));
list.add(new Edge(0, 3));
list.add(new Edge(0, 5));
...
```

如果事先不知道所有的边，那么将 Edge 对象存储在一个 ArrayList 中比较好。

使用边数组或 Edge 对象来表示边对输入来说是很直观的，但是内部处理的效率不高。接下来的两节将介绍使用邻接矩阵（adjacency matrix）和邻接线性表（adjacency list）来表示图，使用这两种数据结构处理图很高效。

28.3.4 表示边：邻接矩阵

假设图有 n 个顶点，那么可以使用名为 adjacencyMatrix 的二维 n × n 矩阵来表示边。矩阵中的每一个元素或者为 0 或者为 1。如果从顶点 i 到顶点 j 存在一条边，那么 adjacencyMatrix[i][j] 为 1；否则，adjacencyMatrix[i][j] 为 0。如果图是无向的，那么该矩阵是对称的，因为 adjacencyMatrix[i][j] 与 adjacencyMatrix[j][i] 是相同的。例如，图 28-1 中图的边可以使用邻接矩阵表示为：

```
int[][] adjacencyMatrix = {
  {0, 1, 0, 1, 0, 1, 0, 0, 0, 0, 0, 0}, // Seattle
  {1, 0, 1, 1, 0, 0, 0, 0, 0, 0, 0, 0}, // San Francisco
  {0, 1, 0, 1, 1, 1, 0, 0, 0, 0, 0, 0}, // Los Angeles
```

```
    {1, 1, 1, 0, 1, 1, 0, 0, 0, 0, 0, 0}, // Denver
    {0, 0, 1, 1, 0, 1, 0, 1, 1, 0, 1, 0}, // Kansas City
    {1, 0, 0, 1, 1, 0, 1, 1, 0, 0, 0, 0}, // Chicago
    {0, 0, 0, 0, 0, 1, 0, 1, 0, 0, 0, 0}, // Boston
    {0, 0, 0, 0, 1, 1, 1, 0, 1, 0, 0, 0}, // New York
    {0, 0, 0, 1, 1, 0, 0, 1, 0, 1, 1, 1}, // Atlanta
    {0, 0, 0, 0, 0, 0, 0, 0, 1, 0, 0, 1}, // Miami
    {0, 0, 1, 0, 1, 0, 0, 0, 1, 0, 0, 1}, // Dallas
    {0, 0, 0, 0, 0, 0, 0, 0, 1, 1, 1, 0}  // Houston
};
```

注意：由于对于无向图来说，矩阵是对称的，因此可以用不规则矩阵来存储它。

图 28-3a 中的有向图的邻接矩阵可以如下表示：

```
int[][] a = {{0, 0, 1, 0, 0}, // Peter
             {0, 0, 1, 0, 0}, // Jane
             {0, 0, 0, 0, 1}, // Mark
             {0, 0, 0, 0, 1}, // Cindy
             {0, 0, 0, 0, 0}  // Wendy
            };
```

28.3.5 表示边：邻接线性表

可以使用邻接顶点线性表（adjacency vertex list）或邻接边线性表（adjacency edge list）来表示边。顶点 i 的邻接顶点线性表包含了所有与 i 邻接的顶点；顶点 i 的邻接边线性表包含了所有与 i 邻接的边。可以定义一个线性表数组。该数组具有 n 个条目，每个条目是一个线性表。顶点 i 的邻接顶点线性表包含了所有的顶点 j，其中顶点 i 和 j 之间存在一条边。例如，为了表示图 28-1 中的图，可以如下创建一个线性表数组：

```
java.util.List<Integer>[] neighbors = new java.util.List[12];
```

neighbors[0] 包含顶点 0（即 Seattle）的所有邻接顶点，neighbors[1] 包含顶点 1（即 San Francisco）的所有邻接顶点，以此类推，如图 28-6 所示。

Seattle	neighbors[0]	1	3	5		
San Francisco	neighbors[1]	0	2	3		
Los Angeles	neighbors[2]	1	3	4	10	
Denver	neighbors[3]	0	1	2	4	5
Kansas City	neighbors[4]	2	3	5	7	8	10
Chicago	neighbors[5]	0	3	4	6	7	
Boston	neighbors[6]	5	7				
New York	neighbors[7]	4	5	6	8		
Atlanta	neighbors[8]	4	7	9	10	11	
Miami	neighbors[9]	8	11				
Dallas	neighbors[10]	2	4	8	11		
Houston	neighbors[11]	8	9	10			

图 28-6 使用邻接顶点线性表来表示图 28-1 中图的边

为了表示图 28-1 中图的邻接边线性表，可以如下创建一个线性表数组：

```
java.util.List<Edge>[] neighbors = new java.util.List[12];
```

neighbors[0] 包含顶点 0（即 Seattle）的所有邻接边，neighbors[1] 包含顶点 1（即 San Francisco）的所有邻接边，以此类推，如图 28-7 所示。

Seattle	neighbors[0]	Edge(0, 1)	Edge(0, 3)	Edge(0, 5)			
San Francisco	neighbors[1]	Edge(1, 0)	Edge(1, 2)	Edge(1, 3)			
Los Angeles	neighbors[2]	Edge(2, 1)	Edge(2, 3)	Edge(2, 4)	Edge(2, 10)		
Denver	neighbors[3]	Edge(3, 0)	Edge(3, 1)	Edge(3, 2)	Edge(3, 4)	Edge(3, 5)	
Kansas City	neighbors[4]	Edge(4, 2)	Edge(4, 3)	Edge(4, 5)	Edge(4, 7)	Edge(4, 8)	Edge(4, 10)
Chicago	neighbors[5]	Edge(5, 0)	Edge(5, 3)	Edge(5, 4)	Edge(5, 6)	Edge(5, 7)	
Boston	neighbors[6]	Edge(6, 5)	Edge(6, 7)				
New York	neighbors[7]	Edge(7, 4)	Edge(7, 5)	Edge(7, 6)	Edge(7, 8)		
Atlanta	neighbors[8]	Edge(8, 4)	Edge(8, 7)	Edge(8, 9)	Edge(8, 10)	Edge(8, 11)	
Miami	neighbors[9]	Edge(9, 8)	Edge(9, 11)				
Dallas	neighbors[10]	Edge(10, 2)	Edge(10, 4)	Edge(10, 8)	Edge(10, 11)		
Houston	neighbors[11]	Edge(11, 8)	Edge(11, 9)	Edge(11, 10)			

图 28-7 使用邻接边线性表来表示图 28-1 中图的边

> **注意**：可以使用邻接矩阵或者邻接线性表来表示一个图。哪种方法更好呢？如果图很稠密（也就是说，存在大量的边），那么建议使用邻接矩阵。如果图很稀疏（也就是说，存在很少的边），由于使用邻接矩阵会浪费大量的存储空间，因此最好使用邻接线性表。

邻接矩阵和邻接线性表都可以用在程序中，以提高算法的效率。例如，使用邻接矩阵来检查两个顶点是否相连只需要 $O(1)$ 常量时间，使用邻接线性表来打印图中所有的边需要线性时间 $O(m)$，这里的 m 表示边的条数。

> **注意**：用邻接顶点线性表表示非加权图更加简单。然而，对于许多应用来说，邻接边线性表更加灵活。使用邻接边线性表更易于在边上添加额外的约束。为此，本书将用邻接边线性表来表示图。

可以使用数组、数组线性表或者链表来存储邻接线性表。我们将使用线性表而不使用数组，因为线性表更易于扩充以添加新的顶点。而且我们使用数组线性表而不是链表，因为我们的算法仅要求搜索线性表中的邻接顶点。对于我们的算法而言，使用数组线性表更加高效。使用数组线性表，图 28-6 中的邻接边线性表可以如下构建：

```
List<ArrayList<Edge>> neighbors = new ArrayList<>();
neighbors.add(new ArrayList<Edge>());
neighbors.get(0).add(new Edge(0, 1));
neighbors.get(0).add(new Edge(0, 3));
neighbors.get(0).add(new Edge(0, 5));
neighbors.add(new ArrayList<Edge>());
neighbors.get(1).add(new Edge(1, 0));
neighbors.get(1).add(new Edge(1, 2));
neighbors.get(1).add(new Edge(1, 3));
...
...
neighbors.get(11).add(new Edge(11, 8));
```

```
neighbors.get(11).add(new Edge(11, 9));
neighbors.get(11).add(new Edge(11, 10));
```

✓ **复习题**

28.3.1 如何表示图中的顶点？如何使用边数组表示边？如何使用边对象表示边？如何使用邻接矩阵表示边？如何使用邻接线性表表示边？

28.3.2 分别使用边数组、边对象线性表、邻接矩阵、邻接顶点线性表、邻接边线性表来表示下面的图。

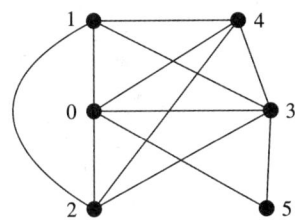

28.4 图的建模

要点提示：Graph接口定义了图的常用操作。

Java集合框架是设计复杂数据结构的良好示例。数据结构的常用特征在接口中定义（例如，Collection、Set、List、Queue），如图20-1所示。这种设计模式对图的建模非常有用。我们将定义一个名为Graph的接口来包含图的所有常用操作，以及一个名为AbstractGraph的抽象类来部分地实现Graph接口。可以添加许多具体的图到这个设计中。例如，我们将定义名为UnweightedGraph和WeightedGraph的图。这些接口和类的关系如图28-8所示。

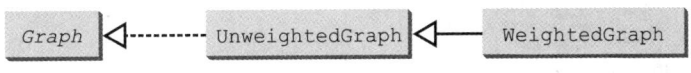

图28-8 图的常用操作定义在接口中，具体类定义具体的图

什么是针对图的常用操作？一般来说，是指要得到图中顶点的个数，得到图中所有的顶点，得到指定下标的顶点对象，得到指定名字的顶点下标，得到顶点的邻居，得到顶点的度，清除图，添加新的顶点，添加新的边，执行深度优先搜索及广度优先搜索。深度优先搜索及广度优先搜索将在下一节中介绍。图28-9在UML图中列举了这些方法。

UnweightedGraph没有引入任何新方法。在UnweightedGraph类中定义了一个顶点的线性表和一个边的邻接线性表。有了这些数据域，就足以实现所有定义在Graph接口中的方法。为方便起见，假设图是简单图，即顶点没有连接到自身的边，并且没有从顶点u到v的平行边。

注意：可以使用任意类型的顶点来创建图。每个顶点与一个下标相关联，该下标同顶点线性表中的顶点下标是一样的。如果创建图时没有指定顶点，则顶点与其下标一样。

假设Graph接口和UnweightedGraph类都是可用的。程序清单28-2给出了一个测试程序，它创建了图28-1中的图，并且为图28-3a创建了一个图。

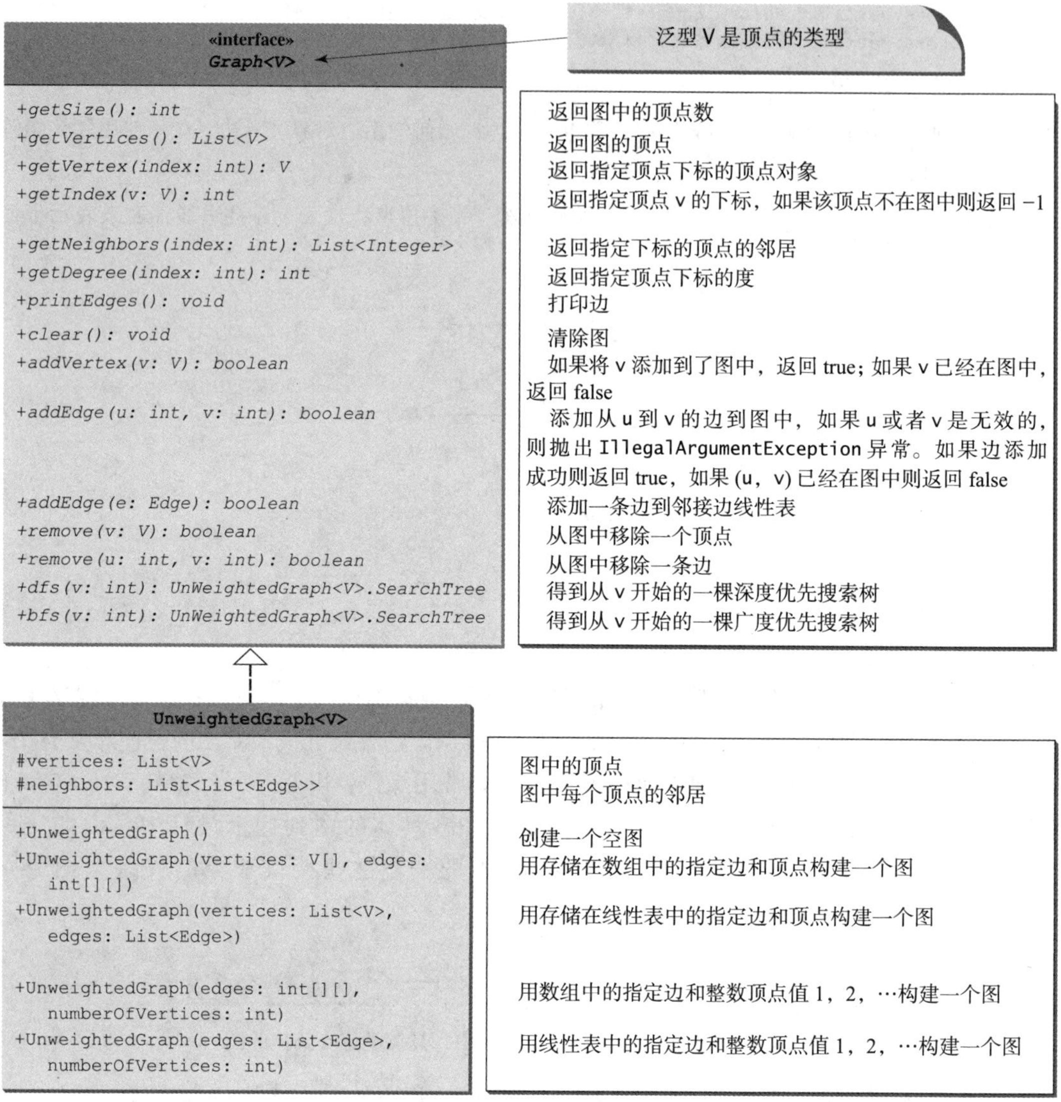

第 28 章

```
         «interface»
          Graph<V>                           ← 泛型 V 是顶点的类型

+getSize(): int                              返回图中的顶点数
+getVertices(): List<V>                      返回图的顶点
+getVertex(index: int): V                    返回指定顶点下标的顶点对象
+getIndex(v: V): int                         返回指定顶点 v 的下标，如果该顶点不在图中则返回 –1

+getNeighbors(index: int): List<Integer>     返回指定下标的顶点的邻居
+getDegree(index: int): int                  返回指定顶点下标的度
+printEdges(): void                          打印边
+clear(): void                               清除图
+addVertex(v: V): boolean                    如果将 v 添加到了图中，返回 true；如果 v 已经在图中，
                                             返回 false
+addEdge(u: int, v: int): boolean            添加从 u 到 v 的边到图中，如果 u 或者 v 是无效的，
                                             则抛出 IllegalArgumentException 异常。如果边添加
                                             成功则返回 true，如果 (u, v) 已经在图中则返回 false

+addEdge(e: Edge): boolean                   添加一条边到邻接边线性表
+remove(v: V): boolean                       从图中移除一个顶点
+remove(u: int, v: int): boolean             从图中移除一条边
+dfs(v: int): UnWeightedGraph<V>.SearchTree  得到从 v 开始的一棵深度优先搜索树
+bfs(v: int): UnWeightedGraph<V>.SearchTree  得到从 v 开始的一棵广度优先搜索树


       UnweightedGraph<V>

#vertices: List<V>                           图中的顶点
#neighbors: List<List<Edge>>                 图中每个顶点的邻居

+UnweightedGraph()                           创建一个空图
+UnweightedGraph(vertices: V[], edges:       用存储在数组中的指定边和顶点构建一个图
  int[][])
+UnweightedGraph(vertices: List<V>,          用存储在线性表中的指定边和顶点构建一个图
  edges: List<Edge>)

+UnweightedGraph(edges: int[][],             用数组中的指定边和整数顶点值 1，2，…构建一个图
  numberOfVertices: int)
+UnweightedGraph(edges: List<Edge>,          用线性表中的指定边和整数顶点值 1，2，…构建一个图
  numberOfVertices: int)
```

图 28-9　Graph 接口定义了所有类型的图的常用操作

程序清单 28-2　TestGraph.java

```java
 1  public class TestGraph {
 2    public static void main(String[] args) {
 3      String[] vertices = {"Seattle", "San Francisco", "Los Angeles",
 4        "Denver", "Kansas City", "Chicago", "Boston", "New York",
 5        "Atlanta", "Miami", "Dallas", "Houston"};
 6
 7      // Edge array for graph in Figure 28.1
 8      int[][] edges = {
 9        {0, 1}, {0, 3}, {0, 5},
10        {1, 0}, {1, 2}, {1, 3},
11        {2, 1}, {2, 3}, {2, 4}, {2, 10},
12        {3, 0}, {3, 1}, {3, 2}, {3, 4}, {3, 5},
13        {4, 2}, {4, 3}, {4, 5}, {4, 7}, {4, 8}, {4, 10},
14        {5, 0}, {5, 3}, {5, 4}, {5, 6}, {5, 7},
15        {6, 5}, {6, 7},
16        {7, 4}, {7, 5}, {7, 6}, {7, 8},
```

```java
17        {8, 4}, {8, 7}, {8, 9}, {8, 10}, {8, 11},
18        {9, 8}, {9, 11},
19        {10, 2}, {10, 4}, {10, 8}, {10, 11},
20        {11, 8}, {11, 9}, {11, 10}
21      };
22
23      Graph<String> graph1 = new UnweightedGraph<>(vertices, edges);
24      System.out.println("The number of vertices in graph1: "
25        + graph1.getSize());
26      System.out.println("The vertex with index 1 is "
27        + graph1.getVertex(1));
28      System.out.println("The index for Miami is " +
29        graph1.getIndex("Miami"));
30      System.out.println("The edges for graph1:");
31      graph1.printEdges();
32
33      // List of Edge objects for graph in Figure 28.3a
34      String[] names = {"Peter", "Jane", "Mark", "Cindy", "Wendy"};
35      java.util.ArrayList<Edge> edgeList
36        = new java.util.ArrayList<>();
37      edgeList.add(new Edge(0, 2));
38      edgeList.add(new Edge(1, 2));
39      edgeList.add(new Edge(2, 4));
40      edgeList.add(new Edge(3, 4));
41      // Create a graph with 5 vertices
42      Graph<String> graph2 = new UnweightedGraph<>
43        (java.util.Arrays.asList(names), edgeList);
44      System.out.println("\nThe number of vertices in graph2: "
45        + graph2.getSize());
46      System.out.println("The edges for graph2:");
47      graph2.printEdges();
48    }
49  }
```

```
The number of vertices in graph1: 12
The vertex with index 1 is San Francisco
The index for Miami is 9
The edges for graph1:
Seattle (0): (0, 1) (0, 3) (0, 5)
San Francisco (1): (1, 0) (1, 2) (1, 3)
Los Angeles (2): (2, 1) (2, 3) (2, 4) (2, 10)
Denver (3): (3, 0) (3, 1) (3, 2) (3, 4) (3, 5)
Kansas City (4): (4, 2) (4, 3) (4, 5) (4, 7) (4, 8) (4, 10)
Chicago (5): (5, 0) (5, 3) (5, 4) (5, 6) (5, 7)
Boston (6): (6, 5) (6, 7)
New York (7): (7, 4) (7, 5) (7, 6) (7, 8)
Atlanta (8): (8, 4) (8, 7) (8, 9) (8, 10) (8, 11)
Miami (9): (9, 8) (9, 11)
Dallas (10): (10, 2) (10, 4) (10, 8) (10, 11)
Houston (11): (11, 8) (11, 9) (11, 10)

The number of vertices in graph2: 5
The edges for graph2:
Peter (0): (0, 2)
Jane (1): (1, 2)
Mark (2): (2, 4)
Cindy (3): (3, 4)
Wendy (4):
```

该程序在第 3～23 行为图 28-1 中的图创建 graph1。graph1 中的顶点在第 3～5 行定义。graph1 的边在第 8～21 行定义。这里使用二维数组来表示边。对于数组中的每一行 i，edges[i][0] 和 edges[i][1] 表示存在从顶点 edges[i][0] 到顶点 edges[i][1] 的一条边。例如，第一行 {0,1} 表示从顶点 0(edges[0][0]) 到顶点 1(edges[0][1]) 的边，{0,5} 表示从顶点 0(edges[2][0]) 到顶点 5(edges[2][1]) 的边。第 23 行创建图。第 31 行调用 graph1 上的方法 printEdges() 来显示 graph1 中的所有边。

该程序在第 34～43 行为图 28-3a 中的图创建 graph2。第 37～40 行定义 graph2 中的边。第 43 行使用 Edge 对象线性表创建 graph2。第 47 行调用 graph2 上的方法 printEdges() 来显示 graph2 中的所有边。

注意，graph1 和 graph2 都包含字符串顶点。这些顶点与下标 0,1,…,n-1 相关联。下标是顶点在 vertices 中的位置。例如，顶点 Miami 的下标为 9。

现在将注意力放在接口和类的实现上。程序清单 28-3 和程序清单 28-4 分别给出了 Graph 接口以及 UnweightedGraph 类。

程序清单 28-3 Graph.java

```java
 1  public interface Graph<V> {
 2    /** Return the number of vertices in the graph */
 3    public int getSize();
 4
 5    /** Return the vertices in the graph */
 6    public java.util.List<V> getVertices();
 7
 8    /** Return the object for the specified vertex index */
 9    public V getVertex(int index);
10
11    /** Return the index for the specified vertex object */
12    public int getIndex(V v);
13
14    /** Return the neighbors of vertex with the specified index */
15    public java.util.List<Integer> getNeighbors(int index);
16
17    /** Return the degree for a specified vertex */
18    public int getDegree(int v);
19
20    /** Print the edges */
21    public void printEdges();
22
23    /** Clear the graph */
24    public void clear();
25
26    /** Add a vertex to the graph */
27    public boolean addVertex(V vertex);
28
29    /** Add an edge to the graph */
30    public boolean addEdge(int u, int v);
31
32    /** Add an edge to the graph */
33    public boolean addEdge(Edge e);
34
35    /** Remove a vertex v from the graph, return true if successful */
36    public boolean remove(V v);
37
38    /** Remove an edge (u, v) from the graph, return true if successful */
39    public boolean remove(int u, int v);
40
41    /** Obtain a depth-first search tree */
42    public UnweightedGraph<V>.SearchTree dfs(int v);
43
44    /** Obtain a breadth-first search tree */
45    public UnweightedGraph<V>.SearchTree bfs(int v);
46  }
```

程序清单28-4 UnweightedGraph.java

```java
 1  import java.util.*;
 2
 3  public class UnweightedGraph<V> implements Graph<V> {
 4    protected List<V> vertices = new ArrayList<>(); // Store vertices
 5    protected List<List<Edge>> neighbors
 6      = new ArrayList<>(); // Adjacency Edge lists
 7
 8    /** Construct an empty graph */
 9    protected UnweightedGraph() {
10    }
11
12    /** Construct a graph from vertices and edges stored in arrays */
13    protected UnweightedGraph(V[] vertices, int[][] edges) {
14      for (int i = 0; i < vertices.length; i++)
15        addVertex(vertices[i]);
16
17      createAdjacencyLists(edges, vertices.length);
18    }
19
20    /** Construct a graph from vertices and edges stored in List */
21    protected UnweightedGraph(List<V> vertices, List<Edge> edges) {
22      for (int i = 0; i < vertices.size(); i++)
23        addVertex(vertices.get(i));
24
25      createAdjacencyLists(edges, vertices.size());
26    }
27
28    /** Construct a graph for integer vertices 0, 1, 2 and edge list */
29    protected UnweightedGraph(List<Edge> edges, int numberOfVertices) {
30      for (int i = 0; i < numberOfVertices; i++)
31        addVertex((V)(Integer.valueOf(i))); // vertices is {0, 1, ... }
32
33      createAdjacencyLists(edges, numberOfVertices);
34    }
35
36    /** Construct a graph from integer vertices 0, 1, and edge array */
37    protected UnweightedGraph(int[][] edges, int numberOfVertices) {
38      for (int i = 0; i < numberOfVertices; i++)
39        addVertex((V)(Integer.valueOf(i))); // vertices is {0, 1, ... }
40
41      createAdjacencyLists(edges, numberOfVertices);
42    }
43
44    /** Create adjacency lists for each vertex */
45    private void createAdjacencyLists(
46        int[][] edges, int numberOfVertices) {
47      for (int i = 0; i < edges.length; i++) {
48        addEdge(edges[i][0], edges[i][1]);
49      }
50    }
51
52    /** Create adjacency lists for each vertex */
53    private void createAdjacencyLists(
54        List<Edge> edges, int numberOfVertices) {
55      for (Edge edge: edges) {
56        addEdge(edge.u, edge.v);
57      }
58    }
59
60    @Override /** Return the number of vertices in the graph */
61    public int getSize() {
62      return vertices.size();
63    }
64
```

```java
 65    @Override /** Return the vertices in the graph */
 66    public List<V> getVertices() {
 67      return vertices;
 68    }
 69
 70    @Override /** Return the object for the specified vertex */
 71    public V getVertex(int index) {
 72      return vertices.get(index);
 73    }
 74
 75    @Override /** Return the index for the specified vertex object */
 76    public int getIndex(V v) {
 77      return vertices.indexOf(v);
 78    }
 79
 80    @Override /** Return the neighbors of the specified vertex */
 81    public List<Integer> getNeighbors(int index) {
 82      List<Integer> result = new ArrayList<>();
 83      for (Edge e: neighbors.get(index))
 84        result.add(e.v);
 85
 86      return result;
 87    }
 88
 89    @Override /** Return the degree for a specified vertex */
 90    public int getDegree(int v) {
 91      return neighbors.get(v).size();
 92    }
 93
 94    @Override /** Print the edges */
 95    public void printEdges() {
 96      for (int u = 0; u < neighbors.size(); u++) {
 97        System.out.print(getVertex(u) + " (" + u + "): ");
 98        for (Edge e: neighbors.get(u)) {
 99          System.out.print("(" + getVertex(e.u) + ", " +
100            getVertex(e.v) + ") ");
101        }
102        System.out.println();
103      }
104    }
105
106    @Override /** Clear the graph */
107    public void clear() {
108      vertices.clear();
109      neighbors.clear();
110    }
111
112    @Override /** Add a vertex to the graph */
113    public boolean addVertex(V vertex) {
114      if (!vertices.contains(vertex)) {
115        vertices.add(vertex);
116        neighbors.add(new ArrayList<Edge>());
117        return true;
118      }
119      else {
120        return false;
121      }
122    }
123
124    @Override /** Add an edge to the graph */
125    public boolean addEdge(Edge e) {
126      if (e.u < 0 || e.u > getSize() - 1)
127        throw new IllegalArgumentException("No such index: " + e.u);
128
129      if (e.v < 0 || e.v > getSize() - 1)
130        throw new IllegalArgumentException("No such index: " + e.v);
```

```java
131      if (!neighbors.get(e.u).contains(e)) {
132        neighbors.get(e.u).add(e);
133        return true;
134      }
135      else {
136        return false;
137      }
138    }
139
140    @Override /** Add an edge to the graph */
141    public boolean addEdge(int u, int v) {
142      return addEdge(new Edge(u, v));
143    }
144
145    @Override /** Obtain a DFS tree starting from vertex v */
146    /** To be discussed in Section 28.7 */
147    public SearchTree dfs(int v) {
148      List<Integer> searchOrder = new ArrayList<>();
149      int[] parent = new int[vertices.size()];
150      for (int i = 0; i < parent.length; i++)
151        parent[i] = -1; // Initialize parent[i] to -1
152
153      // Mark visited vertices
154      boolean[] isVisited = new boolean[vertices.size()];
155
156      // Recursively search
157      dfs(v, parent, searchOrder, isVisited);
158
159      // Return a search tree
160      return new SearchTree(v, parent, searchOrder);
161    }
162
163    /** Recursive method for DFS search */
164    private void dfs(int v, int[] parent, List<Integer> searchOrder,
165        boolean[] isVisited) {
166      // Store the visited vertex
167      searchOrder.add(v);
168      isVisited[v] = true; // Vertex v visited
169
170      for (Edge e : neighbors.get(v)) {// e.u is v
171        int w = e.v; // e.v is w in Listing 28.8
172        if (!isVisited[w]) {
173          parent[w] = v; // The parent of vertex w is v
174          dfs(w, parent, searchOrder, isVisited); // Recursive search
175        }
176      }
177    }
178
179    @Override /** Starting bfs search from vertex v */
180    /** To be discussed in Section 28.9 */
181    public SearchTree bfs(int v) {
182      List<Integer> searchOrder = new ArrayList<>();
183      int[] parent = new int[vertices.size()];
184      for (int i = 0; i < parent.length; i++)
185        parent[i] = -1; // Initialize parent[i] to -1
186
187      java.util.LinkedList<Integer> queue =
188        new java.util.LinkedList<>(); // list used as a queue
189      boolean[] isVisited = new boolean[vertices.size()];
190      queue.offer(v); // Enqueue v
191      isVisited[v] = true; // Mark it visited
192
193      while (!queue.isEmpty()) {
194        int u = queue.poll(); // Dequeue to u
195        searchOrder.add(u); // u searched
```

```java
197        for (Edge e: neighbors.get(u)) {// Note that e.u is u
198          int w = e.v; // e.v is w in Listing 28.8
199          if (!isVisited[w]) {// e.v is w in Listing 28.11
200            queue.offer(w); // Enqueue w
201            parent[w] = u; // The parent of w is u
202            isVisited[w] = true; // Mark it visited
203          }
204        }
205      }
206
207      return new SearchTree(v, parent, searchOrder);
208    }
209
210    /** Tree inner class inside the UnweightedGraph class */
211    /** To be discussed in Section 28.6 */
212    public class SearchTree {
213      private int root; // The root of the tree
214      private int[] parent; // Store the parent of each vertex
215      private List<Integer> searchOrder; // Store the search order
216
217      /** Construct a tree with root, parent, and searchOrder */
218      public SearchTree(int root, int[] parent,
219          List<Integer> searchOrder) {
220        this.root = root;
221        this.parent = parent;
222        this.searchOrder = searchOrder;
223      }
224
225      /** Return the root of the tree */
226      public int getRoot() {
227        return root;
228      }
229
230      /** Return the parent of vertex v */
231      public int getParent(int v) {
232        return parent[v];
233      }
234
235      /** Return an array representing search order */
236      public List<Integer> getSearchOrder() {
237        return searchOrder;
238      }
239
240      /** Return number of vertices found */
241      public int getNumberOfVerticesFound() {
242        return searchOrder.size();
243      }
244
245      /** Return the path of vertices from a vertex to the root */
246      public List<V> getPath(int index) {
247        ArrayList<V> path = new ArrayList<>();
248
249        do {
250          path.add(vertices.get(index));
251          index = parent[index];
252        }
253        while (index != -1);
254
255        return path;
256      }
257
258      /** Print a path from the root to vertex v */
259      public void printPath(int index) {
260        List<V> path = getPath(index);
261        System.out.print("A path from " + vertices.get(root) + " to " +
262          vertices.get(index) + ": ");
```

```
263        for (int i = path.size() - 1; i >= 0; i--)
264          System.out.print(path.get(i) + " ");
265      }
266
267      /** Print the whole tree */
268      public void printTree() {
269        System.out.println("Root is: " + vertices.get(root));
270        System.out.print("Edges: ");
271        for (int i = 0; i < parent.length; i++) {
272          if (parent[i] != -1) {
273            // Display an edge
274            System.out.print("(" + vertices.get(parent[i]) + ", " +
275              vertices.get(i) + ") ");
276          }
277        }
278        System.out.println();
279      }
280    }
281
282    @Override /** Remove vertex v and return true if successful */
283    public boolean remove(V v) {
284      return true; // Implementation left as an exercise
285    }
286
287    @Override /** Remove edge (u, v) and return true if successful */
288    public boolean remove(int u, int v) {
289      return true; // Implementation left as an exercise
290    }
291  }
```

程序清单 28-3 中 Graph 接口的代码很直接，下面我们消化一下程序清单 28-4 中 UnweightedGraph 类的代码。

UnweightedGraph 类定义了数据域 vertices（第 4 行）来存储顶点，定义了 neighbors（第 5 行）来在邻接边线性表中存储边。neighbors.get(i) 存储顶点 i 的所有邻接边。在第 9 ~ 42 行定义了 4 个重载的构造方法，可以创建默认图，或者从边和顶点的数组或线性表来创建图。createAdjacencyLists(int[][]edges,int numberOfVertices) 方法从一个数组中的边来创建邻接线性表（第 45 ~ 50 行）。createAdjacencyLists (List<Edge>edges,int numberOfVertices) 方法从一个线性表中的边来创建邻接线性表（第 53 ~ 58 行）。

getNeighbors(u) 方法（第 81 ~ 87 行）返回顶点 u 的邻接顶点线性表。clear() 方法（第 106 ~ 110 行）从图中移除所有顶点和边。addVertex(u) 方法（第 112 ~ 122 行）添加一个新的顶点到 vertices 中并返回 true。如果顶点已经在图中了则返回 false（第 120 行）。

addEdge(e) 方法（第 124 ~ 139 行）添加一条新的边到邻接边线性表中并返回 true。如果边已经在图中了则返回 false。如果边是无效的，则该方法可能抛出 IllegalArgumentException（第 126 ~ 130 行）。

addEdge(u, v) 方法（第 141 ~ 144 行）添加一条边 (u, v) 到图中。如果图是无向的，应该调用 addEdge(u, v) 和 addEdge(v, u) 在顶点 u 和 v 之间添加一条边。

printEdges() 方法（第 95 ~ 104 行）显示了所有顶点以及每一个顶点的邻接边。

第 148 ~ 208 行代码给出了查找深度优先搜索树和广度优先搜索树的方法，这将分别在 28.7 节和 28.9 节中介绍。

✓ 复习题

28.4.1 描述 Graph 和 UnweightedGraph 中的方法。

28.4.2 对于程序清单 28-2 中的代码来说，graph1.getIndex("Seattle") 的结果是什么？graph1.

getDegree(5) 的结果是什么？graph1.getVertex(4) 的结果是什么？

28.4.3 给出下列代码的输出：

```java
public class Test {
  public static void main(String[] args) {
    Graph<Character> graph = new UnweightedGraph<>();
    graph.addVertex('U');
    graph.addVertex('V');
    int indexForU = graph.getIndex('U');
    int indexForV = graph.getIndex('V');
    System.out.println("indexForU is " + indexForU);
    System.out.println("indexForV is " + indexForV);
    graph.addEdge(indexForU, indexForV);
    System.out.println("Degree of U is " +
      graph.getDegree(indexForU));
    System.out.println("Degree of V is " +
      graph.getDegree(indexForV));
  }
}
```

28.4.4 如果 v 不在图中，getIndex(v) 会返回什么？如果下标不在图中，getVertex(index) 会返回什么？如果 v 已经在图中了，addVertex(v) 会返回什么？如果 u 或者 v 不在图中，addEdge(u, v) 会返回什么？

28.5 图的可视化

要点提示：为了可视化地显示图，每个顶点必须被赋予一个位置。

前一节介绍了 Graph 接口和 UnweightedGraph 类。本节介绍如何图形化地显示图。为了显示一个图，需要知道每个顶点显示的位置以及每个顶点的名字。为了确保图可以显示，我们在程序清单 28-5 中定义了一个名为 Displayable 的接口，该接口具有获取 x 和 y 坐标以及顶点名字的方法，并且让顶点作为 Displayable 的实例。

程序清单 28-5 Displayable.java

```java
1 public interface Displayable {
2   public double getX(); // Get x-coordinate of the vertex
3   public double getY(); // Get y-coordinate of the vertex
4   public String getName(); // Get display name of the vertex
5 }
```

现在，一个具有 Displayable 顶点的图可以显示在一个名为 GraphView 的面板上了，如程序清单 28-6 所示。

程序清单 28-6 GraphView.java

```java
1  import javafx.scene.Group;
2  import javafx.scene.layout.BorderPane;
3  import javafx.scene.shape.Circle;
4  import javafx.scene.shape.Line;
5  import javafx.scene.text.Text;
6
7  public class GraphView extends BorderPane {
8    private Graph<? extends Displayable> graph;
9    private Group group = new Group();
10
11   public GraphView(Graph<? extends Displayable> graph) {
12     this.graph = graph;
13     this.setCenter(group); // Center the group
14     repaintGraph();
15   }
```

```
16
17    private void repaintGraph() {
18      group.getChildren().clear(); // Clear group for a new display
19
20      // Draw vertices and text for vertices
21      java.util.List<? extends Displayable> vertices
22        = graph.getVertices();
23      for (int i = 0; i < graph.getSize(); i++) {
24        double x = vertices.get(i).getX();
25        double y = vertices.get(i).getY();
26        String name = vertices.get(i).getName();
27
28        group.getChildren().add(new Circle(x, y, 16));
29        group.getChildren().add(new Text(x - 8, y - 18, name));
30      }
31
32      // Draw edges for pairs of vertices
33      for (int i = 0; i < graph.getSize(); i++) {
34        java.util.List<Integer> neighbors = graph.getNeighbors(i);
35        double x1 = graph.getVertex(i).getX();
36        double y1 = graph.getVertex(i).getY();
37        for (int v: neighbors) {
38          double x2 = graph.getVertex(v).getX();
39          double y2 = graph.getVertex(v).getY();
40
41          // Draw an edge for (i, v)
42          group.getChildren().add(new Line(x1, y1, x2, y2));
43        }
44      }
45    }
46  }
```

要在一个面板上显示图，只需通过将图作为参数传入构造方法来创建一个 GraphView 的实例（第 11 行）。要显示顶点，图顶点的类必须实现 Displayable 接口（第 21～44 行）。对于每个顶点的下标 i，调用 graph.getNeighbors(i) 方法返回它的邻接线性表（第 34 行）。通过这个线性表，可以找到所有与顶点 i 相邻的顶点，并且绘出一条将顶点 i 与其相邻顶点相连的线（第 35～42 行）。

程序清单 28-7 给出了一个例子，它显示图 28-1 中的图，如图 28-10 所示。

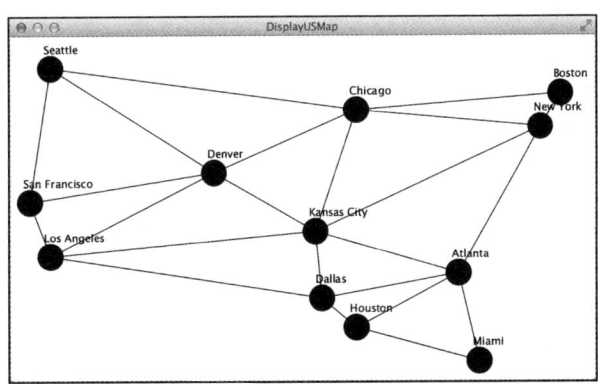

图 28-10 在一个面板中显示图

程序清单 28-7 DisplayUSMap.java

```
1  import javafx.application.Application;
2  import javafx.scene.Scene;
3  import javafx.stage.Stage;
```

```java
 4
 5  public class DisplayUSMap extends Application {
 6    @Override // Override the start method in the Application class
 7    public void start(Stage primaryStage) {
 8      City[] vertices = {new City("Seattle", 75, 50),
 9        new City("San Francisco", 50, 210),
10        new City("Los Angeles", 75, 275), new City("Denver", 275, 175),
11        new City("Kansas City", 400, 245),
12        new City("Chicago", 450, 100), new City("Boston", 700, 80),
13        new City("New York", 675, 120), new City("Atlanta", 575, 295),
14        new City("Miami", 600, 400), new City("Dallas", 408, 325),
15        new City("Houston", 450, 360) };
16
17      // Edge array for graph in Figure 28.1
18      int[][] edges = {
19        {0, 1}, {0, 3}, {0, 5}, {1, 0}, {1, 2}, {1, 3},
20        {2, 1}, {2, 3}, {2, 4}, {2, 10},
21        {3, 0}, {3, 1}, {3, 2}, {3, 4}, {3, 5},
22        {4, 2}, {4, 3}, {4, 5}, {4, 7}, {4, 8}, {4, 10},
23        {5, 0}, {5, 3}, {5, 4}, {5, 6}, {5, 7},
24        {6, 5}, {6, 7}, {7, 4}, {7, 5}, {7, 6}, {7, 8},
25        {8, 4}, {8, 7}, {8, 9}, {8, 10}, {8, 11},
26        {9, 8}, {9, 11}, {10, 2}, {10, 4}, {10, 8}, {10, 11},
27        {11, 8}, {11, 9}, {11, 10}
28      };
29
30      Graph<City> graph = new UnweightedGraph<>(vertices, edges);
31
32      // Create a scene and place it in the stage
33      Scene scene = new Scene(new GraphView(graph), 750, 450);
34      primaryStage.setTitle("DisplayUSMap"); // Set the stage title
35      primaryStage.setScene(scene); // Place the scene in the stage
36      primaryStage.show(); // Display the stage
37    }
38
39    static class City implements Displayable {
40      private double x, y;
41      private String name;
42
43      City(String name, double x, double y) {
44        this.name = name;
45        this.x = x;
46        this.y = y;
47      }
48
49      @Override
50      public double getX() {
51        return x;
52      }
53
54      @Override
55      public double getY() {
56        return y;
57      }
58
59      @Override
60      public String getName() {
61        return name;
62      }
63    }
64  }
```

定义 City 类对具有坐标和名字的顶点建模（第 39 ~ 63 行）。该程序创建一个顶点为 City 类型的图（第 30 行）。由于 City 实现了 Displayable，所以为图创建的 GraphView 对象在面板上显示图（第 33 行）。

作为熟悉图的类和接口的练习，以恰当的边添加一个城市（例如 Savannah）到图中。

复习题

28.5.1 如果程序清单 28-6 中第 38 ~ 42 行的代码替换为以下代码，程序清单 28-7 还能运行吗？

```
if (i < v) {
  double x2 = graph.getVertex(v).getX();
  double y2 = graph.getVertex(v).getY();

  // Draw an edge for (i, v)
  getChildren().add(new Line(x1, y1, x2, y2));
}
```

28.5.2 对于程序清单 28-1 中创建的 graph1 对象，可以如下创建一个 GraphView 对象吗？

```
GraphView view = new GraphView(graph1);
```

28.6 图的遍历

要点提示：深度优先和广度优先是遍历图的两个常用方法。

图的遍历（graph traversal）是指仅访问一次图中的每一个顶点的过程。存在两种流行的遍历图的方法：深度优先遍历（或深度优先搜索）和广度优先遍历（或广度优先搜索）。这两种遍历方法都会产生一个生成树，它可以用类来建模，如图 28-11 所示。注意，SearchTree 是定义在 UnweightedGraph 类中的一个内部类。UnweightedGraph<V>.SearchTree 和定义在 25.2.5 节中的 Tree 接口不同。UnweightedGraph<V>.SearchTree 是一个特定的类，设计为描述结点的父子关系，而 Tree 接口定义诸如树的搜索、插入和删除等常用的操作。因为没有必要对生成树执行这些操作，所以 UnweightedGraph<V>.SearchTree 没有定义为 Tree 的子类型。

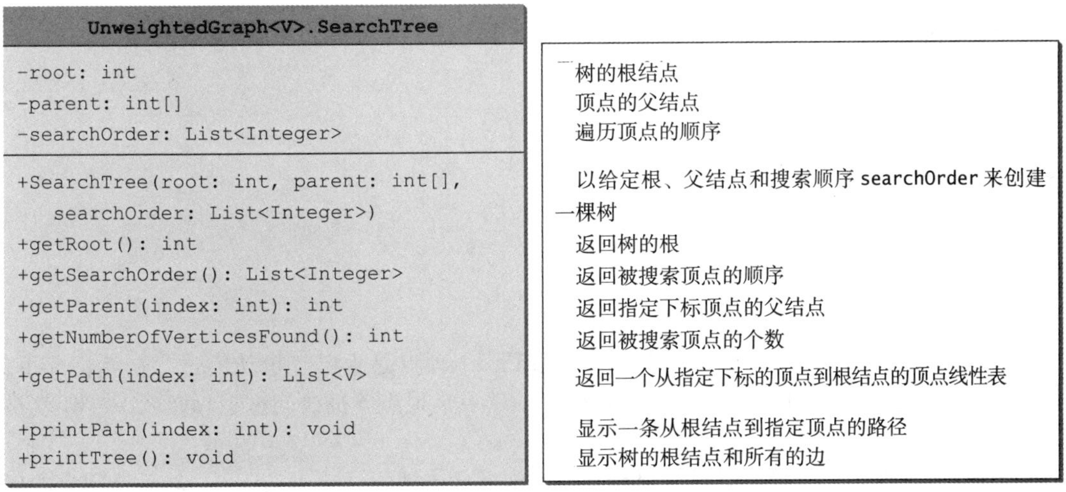

UnweightedGraph<V>.SearchTree	
-root: int	树的根结点
-parent: int[]	顶点的父结点
-searchOrder: List<Integer>	遍历顶点的顺序
+SearchTree(root: int, parent: int[], searchOrder: List<Integer>)	以给定根、父结点和搜索顺序 searchOrder 来创建一棵树
+getRoot(): int	返回树的根
+getSearchOrder(): List<Integer>	返回被搜索顶点的顺序
+getParent(index: int): int	返回指定下标顶点的父结点
+getNumberOfVerticesFound(): int	返回被搜索顶点的个数
+getPath(index: int): List<V>	返回一个从指定下标的顶点到根结点的顶点线性表
+printPath(index: int): void	显示一条从根结点到指定顶点的路径
+printTree(): void	显示树的根结点和所有的边

图 28-11 SearchTree 类描述具有父子关系的结点

在程序清单 28-4 中的第 210 ~ 278 行，将 SearchTree 定义为 UnweightedGraph 类中的一个内部类。构造方法通过根、边和搜索顺序来创建一棵树。

SearchTree 类定义了 7 个方法。getRoot() 方法返回树的根。可以通过调用 getSearchOrder() 方法来获取被搜索的顶点的顺序。可以调用 getParent(v) 来找出顶点 v 在这个搜索中的父结点。调用方法 getNumberOfVerticesFound() 返回搜索到的顶点个数。getPath(index) 方法返回一个从指定下标的顶点到根结点的顶点线性表。调用 printPath(v) 显示一条从根结点到顶点 v 的路径。可以使用 printTree() 方法来显示树中所有的边。

28.7 节和 28.9 节将分别介绍深度优先搜索和广度优先搜索。两种搜索都将产生一个 SearchTree 类的实例。

✔ 复习题

28.6.1 UnweightedGraph<V>.SearchTree 实现了程序清单 25-3 中定义的 Tree 接口吗？

28.6.2 使用什么方法找到树中一个顶点的父结点？

28.7 深度优先搜索

> **要点提示**：图的深度优先搜索（DFS）从图中的一个结点出发，在回溯前尽可能地访问图中的所有结点。

图的深度优先搜索（DFS）和 25.2.4 节中讨论的树的深度优先搜索很相似。对于树，搜索从根结点开始；对于图，搜索可以从任意一个顶点开始。

树的深度优先搜索首先访问根结点，然后递归地访问根结点的子树。类似地，图的深度优先搜索首先访问一个顶点，然后递归地访问和这个顶点相连的所有顶点。不同之处在于图可能包含环，这可能会导致无限的递归。为了避免这个问题，需要跟踪已经访问过的顶点。

这种搜索之所以称为深度优先（depth-first），是因为它尽可能地搜索图中的"更深处"。搜索从某个顶点 v 开始，然后访问顶点 v 的第一个未被访问的邻居。如果顶点 v 没有未被访问的邻居，返回到到达顶点 v 的那个顶点。我们假定图是连通的并且从任意结点开始的搜索可以到达所有的结点。如果不是这种情况，参见编程练习题 28.4 以找到图中连通的部分。

28.7.1 DFS 算法

程序清单 28-8 描述了深度优先搜索算法。

程序清单 28-8 深度优先搜索算法

```
Input: G = (V, E) and a starting vertex v
Output: a DFS tree rooted at v

1  SearchTree dfs(vertex v) {
2    visit v;
3    for each neighbor w of v
4      if (w has not been visited) {
5        set v as the parent for w in the tree;
6        dfs(w);
7      }
8  }
```

可以使用一个名为 isVisited 的数组，标识一个顶点是否已经被访问过。初始情况下，每个顶点 i 对应的 isVisited[i] 都为 false。一旦一个顶点 v 被访问过，isVisited[v] 就被设置为 true。

考虑图 28-12a 中的图。假设从顶点 0 开始深度优先搜索。首先访问 0，然后访问它的任意一个邻居，比如顶点 1。现在，1 已经被访问，如图 28-12b 所示。顶点 1 有三个邻居——顶点 0、2 和 4。由于顶点 0 已经被访问，因而将访问顶点 2 或者顶点 4。我们选择顶点 2，现在顶点 2 已经被访问过，如图 28-12c 所示。顶点 2 有三个邻居，分别为顶点 0、1 和 3。由于顶点 0 和 1 已经被访问，因此选取顶点 3。现在顶点 3 已经被访问过，如图 28-12d 所示。此时，顶点已经被以如下的顺序访问过：

0, 1, 2, 3

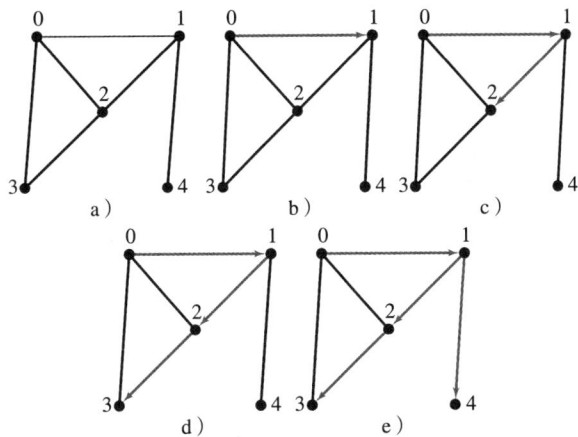

图 28-12 深度优先搜索递归地访问一个结点和它的邻居

由于顶点 3 的所有邻居都已被访问，因此回溯到顶点 2。由于顶点 2 的所有邻居也都已经被访问，因此回溯到顶点 1。顶点 4 与顶点 1 相连，但是顶点 4 还没有被访问过，因此访问顶点 4，如图 28-12e 所示。由于顶点 4 的所有邻居都已被访问，因此回溯到顶点 1。由于顶点 1 的所有邻居都已经被访问，因此返回到顶点 0。由于顶点 0 的所有邻居都已被访问，搜索终止。

由于每条边和每个顶点仅被访问一次，所以 dfs 方法的时间复杂度为 O(|E|+|V|)，其中 |E| 表示边的条数，|V| 表示顶点的个数。

28.7.2 DFS 的实现

在程序清单 28-8 中描述的 DFS 算法使用的是递归，很自然地，在实现它的时候也应使用递归。还可以使用栈来实现（参见编程练习题 28.3）。

`dfs(int v)` 方法在程序清单 28-4 中的第 148～178 行实现，它返回一个将顶点 v 作为根结点的 `SearchTree` 类的实例。该方法将搜索过的顶点存储在一个线性表 `searchOrder` 中（第 149 行），每个顶点的父结点存储在数组 `parent` 中（第 150 行），使用数组 `isVisited` 来表示顶点是否已经被访问过（第 155 行）。调用辅助方法 `dfs(v, parent, searchOrder, isVisited)` 来执行深度优先搜索（第 159 行）。

在递归的辅助方法中，搜索从顶点 v 开始。在第 168 行顶点 v 被添加到 `searchOrder` 中，并且被标记为已访问过（第 169 行）。对于顶点 v 的每一个未被访问的邻居，递归地调用该方法来执行深度优先搜索。当顶点 w（w 为程序清单 28-4 中第 172 行的 e.v）被访问，顶点 w 的父结点被存储在 `parent[e.v]` 中（第 174 行）。对于一个连通的图或者一个连通组件，所有顶点都被访问过时，该方法返回。

程序清单 28-9 给出了一个测试程序，用来显示图 28-1 中的图由 Chicago 开始的深度优先搜索。由 Chicago 开始的深度优先搜索的图示如图 28-13 所示。

程序清单 28-9 TestDFS.java

```
1  public class TestDFS {
2    public static void main(String[] args) {
3      String[] vertices = {"Seattle", "San Francisco", "Los Angeles",
4        "Denver", "Kansas City", "Chicago", "Boston", "New York",
5        "Atlanta", "Miami", "Dallas", "Houston"};
6
7      int[][] edges = {
8        {0, 1}, {0, 3}, {0, 5},
```

```
 9        {1, 0}, {1, 2}, {1, 3},
10        {2, 1}, {2, 3}, {2, 4}, {2, 10},
11        {3, 0}, {3, 1}, {3, 2}, {3, 4}, {3, 5},
12        {4, 2}, {4, 3}, {4, 5}, {4, 7}, {4, 8}, {4, 10},
13        {5, 0}, {5, 3}, {5, 4}, {5, 6}, {5, 7},
14        {6, 5}, {6, 7},
15        {7, 4}, {7, 5}, {7, 6}, {7, 8},
16        {8, 4}, {8, 7}, {8, 9}, {8, 10}, {8, 11},
17        {9, 8}, {9, 11},
18        {10, 2}, {10, 4}, {10, 8}, {10, 11},
19        {11, 8}, {11, 9}, {11, 10}
20      };
21
22      Graph<String> graph = new UnweightedGraph<>(vertices, edges);
23      UnweightedGraph<String>.SearchTree dfs =
24        graph.dfs(graph.getIndex("Chicago"));
25
26      java.util.List<Integer> searchOrders = dfs.getSearchOrder();
27      System.out.println(dfs.getNumberOfVerticesFound() +
28        " vertices are searched in this DFS order:");
29      for (int i = 0; i < searchOrders.size(); i++)
30        System.out.print(graph.getVertex(searchOrders.get(i)) + " ");
31      System.out.println();
32
33      for (int i = 0; i < searchOrders.size(); i++)
34        if (dfs.getParent(i) != -1)
35          System.out.println("parent of " + graph.getVertex(i) +
36            " is " + graph.getVertex(dfs.getParent(i)));
37    }
38  }
```

```
12 vertices are searched in this DFS order:
  Chicago Seattle San Francisco Los Angeles Denver
  Kansas City New York Boston Atlanta Miami Houston Dallas
parent of Seattle is Chicago
parent of San Francisco is Seattle
parent of Los Angeles is San Francisco
parent of Denver is Los Angeles
parent of Kansas City is Denver
parent of Boston is New York
parent of New York is Kansas City
parent of Atlanta is New York
parent of Miami is Atlanta
parent of Dallas is Houston
parent of Houston is Miami
```

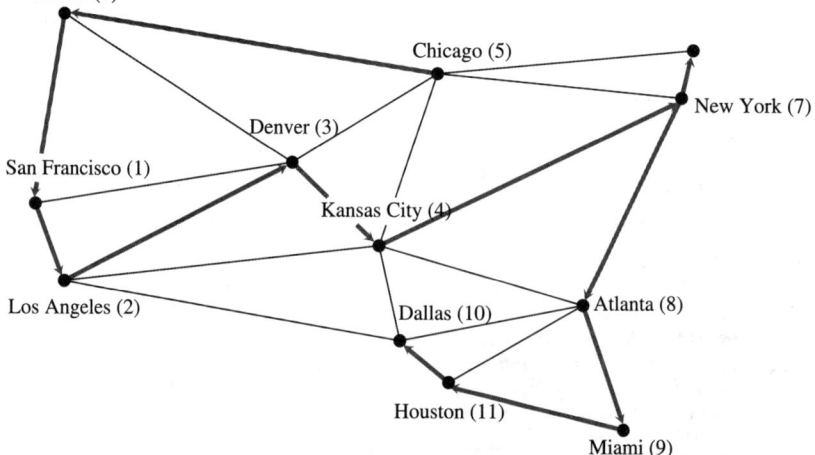

图 28-13 由 Chicago 开始的深度优先搜索（来源：©Mozilla Firefox）

28.7.3 DFS 的应用

深度优先搜索可以用来解决如下所示的许多问题：
- 检测图是否连通。由任何一个顶点开始搜索图，如果搜索的顶点的个数与图中顶点的个数一致，那么图是连通的；否则，图就不是连通的。（参见编程练习题 28.1）。
- 检测两个顶点之间是否存在路径（参见编程练习题 28.5）。
- 找出两个顶点之间的路径（参见编程练习题 28.5）。
- 找出所有连通的部分。一个连通的部分是指一个最大的连通子图，其中每一个顶点对都有路径连接（参见编程练习题 28.4）。
- 检测图中是否存在回路（参见编程练习题 28.6）。
- 找出图中的回路（参见编程练习题 28.7）。
- 找出哈密尔顿路径/回路。图的哈密尔顿路径（Hamiltonian path）是指可以访问图中每个顶点正好一次的路径。哈密尔顿回路（Hamiltonian cycle）是指访问图中每个顶点正好一次并且返回到出发顶点的路径（参见编程练习题 28.17）。

前 6 个问题可以通过程序清单 28-4 中的 dfs 方法轻松解决。要找到哈密尔顿路径/回路，需要探索所有可能的 DFS 来找到导致最长路径的那个。哈密尔顿路径/回路具有许多应用，包括解决著名的骑士巡游问题，这个问题在配套网站的补充材料 VI.E 中给出了。

✔ 复习题

28.7.1 什么是深度优先搜索？

28.7.2 为图 28-3b 中的图从结点 A 开始绘制一个 DFS 树。

28.7.3 为图 28-1 中的图从结点 Atlanta 开始绘制一个 DFS 树。

28.7.4 调用 dfs(v) 的返回类型是什么？

28.7.5 程序清单 28-8 中描述的深度优先搜索算法使用了递归。另外，也可以使用栈来实现，如下所示。指出下面算法的错误之处并给出正确的算法。

```
// Wrong version
SearchTree dfs(vertex v) {
  push v into the stack;
  mark v visited;

  while (the stack is not empty) {
    pop a vertex, say u, from the stack
    visit u;
    for each neighbor w of u
      if (w has not been visited)
        push w into the stack;
  }
}
```

28.8 示例学习：连通圆问题

> **要点提示**：连通圆问题是确定在一个二维平面上的所有圆是否连通的。这个问题可以使用深度优先遍历方法解决。

DFS 算法有许多的应用。本节应用 DFS 算法来解决连通圆问题。

在连通圆问题中，要确定在一个二维平面上的所有圆是否连通。如果所有圆是连通的，则以填充方式绘制，如图 28-14a 所示。否则，就不填充，如图 28-14b 所示。

我们将编写一个程序，让用户通过在没有被圆占据的空白区域点击鼠标来创建一个圆。当添加圆后，这些圆如果是连通的则以填充方式绘制，否则不被填充。

我们将创建一个图来对此问题建模。每个圆是图中的一个顶点。如果两个圆交叉则是连通的。我们在图上应用 DFS，如果深度优先搜索找到了所有的顶点，则图是连通的。

在程序清单 28-10 中给出了该程序。

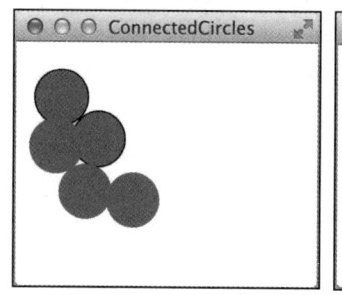
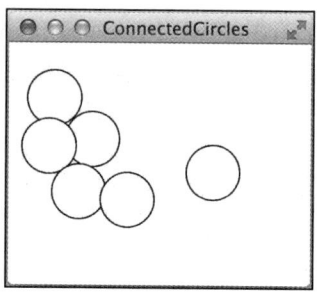

a) 连通的圆　　　　　　　b) 不连通的圆

图 28-14　可以使用 DFS 方法来确定是否圆是连通的（来源：Oracle 或其附属公司版权所有 © 1995 ~ 2016，经授权使用）

程序清单 28-10　ConnectedCircles.java

```java
 1  import javafx.application.Application;
 2  import javafx.geometry.Point2D;
 3  import javafx.scene.Node;
 4  import javafx.scene.Scene;
 5  import javafx.scene.layout.Pane;
 6  import javafx.scene.paint.Color;
 7  import javafx.scene.shape.Circle;
 8  import javafx.stage.Stage;
 9
10  public class ConnectedCircles extends Application {
11    @Override // Override the start method in the Application class
12    public void start(Stage primaryStage) {
13      // Create a scene and place it in the stage
14      Scene scene = new Scene(new CirclePane(), 450, 350);
15      primaryStage.setTitle("ConnectedCircles"); // Set the stage title
16      primaryStage.setScene(scene); // Place the scene in the stage
17      primaryStage.show(); // Display the stage
18    }
19
20    /** Pane for displaying circles */
21    class CirclePane extends Pane {
22      public CirclePane() {
23        this.setOnMouseClicked(e -> {
24          if (!isInsideACircle(new Point2D(e.getX(), e.getY()))) {
25            // Add a new circle
26            getChildren().add(new Circle(e.getX(), e.getY(), 20));
27            colorIfConnected();
28          }
29        });
30      }
31
32      /** Returns true if the point is inside an existing circle */
33      private boolean isInsideACircle(Point2D p) {
34        for (Node circle: this.getChildren())
35          if (circle.contains(p))
36            return true;
37
38        return false;
39      }
40
41      /** Color all circles if they are connected */
42      private void colorIfConnected() {
```

```
43        if (getChildren().size() == 0)
44          return; // No circles in the pane
45
46        // Build the edges
47        java.util.List<Edge> edges
48          = new java.util.ArrayList<>();
49        for (int i = 0; i < getChildren().size(); i++)
50          for (int j = i + 1; j < getChildren().size(); j++)
51            if (overlaps((Circle)(getChildren().get(i)),
52                (Circle)(getChildren().get(j)))) {
53              edges.add(new Edge(i, j));
54              edges.add(new Edge(j, i));
55            }
56
57        // Create a graph with circles as vertices
58        Graph<Node> graph = new UnweightedGraph<>
59            ((java.util.List<Node>)getChildren(), edges);
60        UnweightedGraph<Node>.SearchTree tree = graph.dfs(0);
61        boolean isAllCirclesConnected = getChildren().size() == tree
62          .getNumberOfVerticesFound();
63
64        for (Node circle: getChildren()) {
65          if (isAllCirclesConnected) {  // All circles are connected
66            ((Circle)circle).setFill(Color.RED);
67          }
68          else {
69            ((Circle)circle).setStroke(Color.BLACK);
70            ((Circle)circle).setFill(Color.WHITE);
71          }
72        }
73      }
74    }
75
76    public static boolean overlaps(Circle circle1, Circle circle2) {
77      return new Point2D(circle1.getCenterX(), circle1.getCenterY()).
78        distance(circle2.getCenterX(), circle2.getCenterY())
79        <= circle1.getRadius() + circle2.getRadius();
80    }
81  }
```

JavaFX 的 Circle 类包含了数据域 x、y 和 radius，这些数据域给出了圆的圆心位置以及半径。同时还定义了 contains 方法来检测一个点是否在圆内。overlaps 方法（第 76～80 行）检测两个圆是否交叉。

当用户在任何已经存在的圆外点击鼠标，一个新的圆在以鼠标点击处为中心的位置创建并添加到 circles 列表中（第 26 行）。

为了检测圆是否连通，该程序构建了一个图（第 46～59 行）。圆作为图的顶点。边在第 47～55 行构建。如果两个圆交叉则代表它们的顶点是连通的（第 51 行）。对图执行 DFS 的结果为一棵树（第 60 行）。该树的 getNumberOfVerticesFound() 返回搜索到的顶点数。如果它等于圆的个数，则所有的圆是连通的（第 61～62 行）。

✓ 复习题

28.8.1 解决连通圆问题的图是如何创建的？
28.8.2 当你在圆内点击鼠标时，程序会创建一个新的圆吗？
28.8.3 程序是如何知道所有圆是连通的？

28.9 广度优先搜索

✪ 要点提示：图的广度优先搜索（BFS）逐层访问顶点。第一层由起始顶点组成，每个下一层由与前一层邻接的顶点组成。

图的广度优先遍历与25.2.4节中讨论的树的广度优先遍历类似。对于树的广度优先遍历而言，将逐层访问结点。首先访问根结点，然后是根结点的所有子结点，接着是根结点的孙子结点，依此类推。同样，图的广度优先搜索首先访问一个顶点，然后是所有与其邻接的顶点，最后是所有与这些顶点邻接的顶点，依此类推。为了确保每个顶点只被访问一次，如果一个顶点已经被访问过，那么就跳过这个顶点。

28.9.1 BFS 算法

从图中顶点 v 开始的广度优先搜索算法，可以描述为如程序清单 28-11 所示。

程序清单 28-11 广度优先搜索算法

```
Input: G = (V, E) and a starting vertex v
Output: a BFS tree rooted at v

1  SearchTree bfs(vertex v) {
2    create an empty queue for storing vertices to be visited;
3    add v into the queue;
4    mark v visited;
5
6    while (the queue is not empty) {
7      dequeue a vertex, say u, from the queue;
8      add u into a list of traversed vertices;
9      for each neighbor w of u
10       if w has not been visited {
11         add w into the queue;
12         set u as the parent for w in the tree;
13         mark w visited;
14       }
15   }
16 }
```

考虑图 28-15a 中的图。假设从顶点 0 开始广度优先搜索，首先访问顶点 0，然后访问它的所有邻居顶点 1、2、3，如图 28-15b 所示。顶点 1 有三个邻居，顶点 0、2、4。由于顶点 0 和 2 已经被访问，现在只能访问顶点 4，如图 28-15c 所示。顶点 2 有三个邻居，顶点 0、1 和 3，它们都被访问过。顶点 3 有三个邻居，顶点 0、2 和 4，它们也都被访问过。顶点 4 有两个邻居，顶点 1 和 3，它们都被访问过。因此，搜索终止。

由于每条边和每个顶点只访问一次，所以 bfs 方法的时间复杂度为 O(|E|+|V|)，其中 |E| 表示边的条数，|V| 表示顶点的个数。

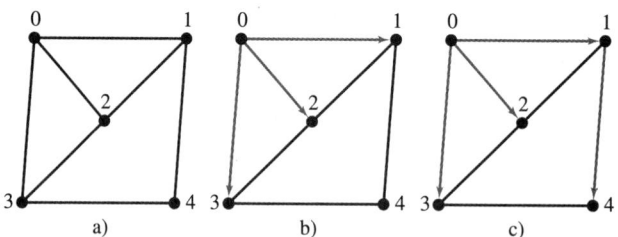

图 28-15 广度优先搜索访问一个结点，然后是其邻居，接着是其邻居的邻居，依此类推

28.9.2 BFS 的实现

方法 bfs(int v) 在 Graph 接口定义，并且在程序清单 28-4 中的 UnweightedGraph 类中实现（第 182～208 行）。它返回一个将顶点 v 作为根结点的 SearchTree 类的实例。该方法将搜索到的顶点存储在线性表 searchOrder 中（第 183 行），将每个顶点的父结点存储在一

个名为 parent 的数组中（第 184 行），为队列使用一个链表（第 188～189 行），并使用数组 isVisited 来表明顶点是否已经访问过（第 190 行）。该搜索从顶点 v 开始。顶点 v 在第 191 行被添加到队列中，并且被标记为已访问（第 192 行）。该方法现在检测队列中的每一个顶点 u（第 194 行）并且将它添加到 searchOrder 中（第 196 行）。该方法将顶点 u 的每一个未被访问的邻居顶点 w（w 为程序清单 28-4 中第 198 行的 e.v）添加到队列中（第 200 行），然后设置它的父结点为 u（第 201 行），并将其标记为已访问（第 202 行）。

程序清单 28-12 给出了一个测试程序，用来显示图 28-1 中的图从 chicago 开始的 BFS。由 Chicago 开始的 BFS 的图示如图 28-16 所示。

程序清单 28-12 TestBFS.java

```
1  public class TestBFS {
2    public static void main(String[] args) {
3      String[] vertices = {"Seattle", "San Francisco", "Los Angeles",
4        "Denver", "Kansas City", "Chicago", "Boston", "New York",
5        "Atlanta", "Miami", "Dallas", "Houston"};
6
7      int[][] edges = {
8        {0, 1}, {0, 3}, {0, 5},
9        {1, 0}, {1, 2}, {1, 3},
10       {2, 1}, {2, 3}, {2, 4}, {2, 10},
11       {3, 0}, {3, 1}, {3, 2}, {3, 4}, {3, 5},
12       {4, 2}, {4, 3}, {4, 5}, {4, 7}, {4, 8}, {4, 10},
13       {5, 0}, {5, 3}, {5, 4}, {5, 6}, {5, 7},
14       {6, 5}, {6, 7},
15       {7, 4}, {7, 5}, {7, 6}, {7, 8},
16       {8, 4}, {8, 7}, {8, 9}, {8, 10}, {8, 11},
17       {9, 8}, {9, 11},
18       {10, 2}, {10, 4}, {10, 8}, {10, 11},
19       {11, 8}, {11, 9}, {11, 10}
20     };
21
22     Graph<String> graph = new UnweightedGraph<>(vertices, edges);
23     UnweightedGraph<String>.SearchTree bfs =
24       graph.bfs(graph.getIndex("Chicago"));
25
26     java.util.List<Integer> searchOrders = bfs.getSearchOrder();
27     System.out.println(bfs.getNumberOfVerticesFound() +
28       " vertices are searched in this order:");
29     for (int i = 0; i < searchOrders.size(); i++)
30       System.out.println(graph.getVertex(searchOrders.get(i)));
31
32     for (int i = 0; i < searchOrders.size(); i++)
33       if (bfs.getParent(i) != -1)
34         System.out.println("parent of " + graph.getVertex(i) +
35           " is " + graph.getVertex(bfs.getParent(i)));
36   }
37 }
```

```
12 vertices are searched in this order:
  Chicago Seattle Denver Kansas City Boston New York
  San Francisco Los Angeles Atlanta Dallas Miami Houston
parent of Seattle is Chicago
parent of San Francisco is Seattle
parent of Los Angeles is Denver
parent of Denver is Chicago
parent of Kansas City is Chicago
parent of Boston is Chicago
parent of New York is Chicago
parent of Atlanta is Kansas City
parent of Miami is Atlanta
parent of Dallas is Kansas City
parent of Houston is Atlanta
```

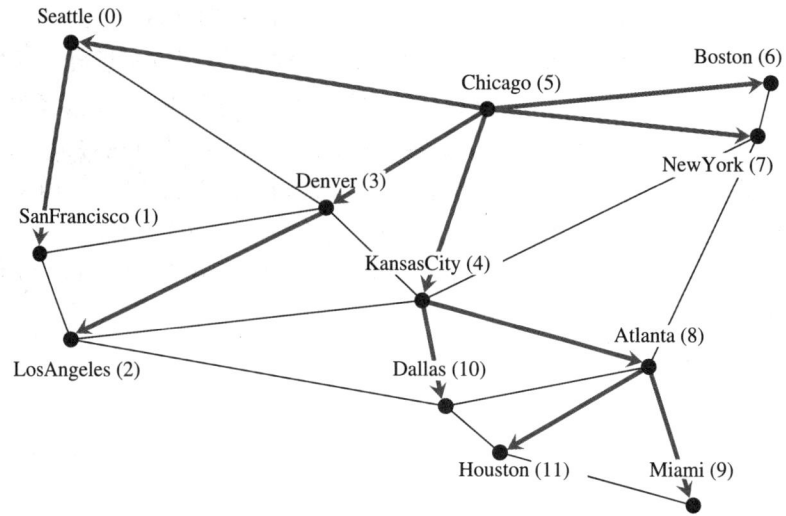

图 28-16　由 Chicago 开始的广度优先搜索（来源：©Mozilla Firefox）

28.9.3　BFS 的应用

许多由 DFS 解决的问题也可以由 BFS 解决。具体而言，BFS 可以用来解决以下问题：

- 检测图是否是连通的。如果在图中任意两个顶点之间都存在一条路径，那么该图是连通的。
- 检测在两个顶点之间是否存在一条路径。
- 找出两个顶点之间的最短路径。可以证明根结点和广度优先搜索树中的任意一个结点之间的路径是根结点和该结点之间的最短路径（参见复习题 28.9.5）。
- 找出所有的连通组件。一个连通组件是指一个最大的连通子图，其中的每个顶点对都有路径相连接。
- 检测图中是否存在回路（参见编程练习题 28.6）。
- 找出图中的回路（参见编程练习题 28.7）。
- 检测一个图是否是二分图：如果图的顶点可以分为两个不相交的集合，而且同一个集合中的顶点之间不存在边，那么这个图就是二分图（参见编程练习题 28.8）。

✔ 复习题

28.9.1　调用 `bfs(v)` 的返回类型是什么？

28.9.2　什么是广度优先搜索？

28.9.3　绘制图 28-3b 中由结点 A 开始的图的广度优先搜索树。

28.9.4　绘制图 28-1 中由顶点 `Atlanta` 开始的图的广度优先搜索树。

28.9.5　证明广度优先搜索树中的根结点和任意结点之间的路径是它们之间的最短路径。

28.10　示例学习：9 枚硬币反面问题

🔑 **要点提示**：9 枚硬币反面的问题可以简化为最短路径问题。

9 枚硬币反面问题如下：将 9 枚硬币放在一个 3×3 的矩阵中，其中一些正面朝上，另一些正面朝下。一个合法的移动是指翻转任何一个正面朝上的硬币以及与它相邻的硬币（不包括对角线相邻的）。任务就是找到最少次数的移动，使得所有硬币正面朝下。例如，从如图 28-17a 所示的 9 枚硬币开始。当翻动最后一行的第二个硬币之后，9 枚硬币将如图 28-17b 所示。当翻动第一行的第二个硬币之后，9 枚硬币都将正面朝下，如图 28-17c 所示。在网址 liveexample.pearsoncmg.com/dsanimation/NineCoin.html 上可以观看交互式演示。

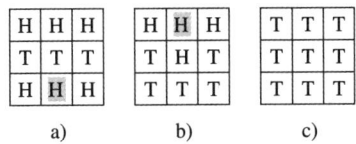

图 28-17 当所有硬币都正面朝下，问题得到解决

下面编写一个程序，提示用户输入 9 枚硬币的初始状态，然后显示解决方案，如下面的运行示例所示。

```
Enter the initial nine coins Hs and Ts: HHHTTTHHH  ↵Enter
The steps to flip the coins are
HHH
TTT
HHH

HHH
THT
TTT

TTT
TTT
TTT
```

9 枚硬币的每一个状态代表图中的一个结点。例如，图 28-17 中的三个状态对应图中的三个结点。为了方便起见，使用一个 3×3 的矩阵来表示所有的结点，其中 0 表示正面，1 表示背面。由于存在 9 个格子，并且每个格子不是 0 就是 1，因此一共有 2^9（512）个结点，分别标记为 0,1,…,511，如图 28-18 所示。

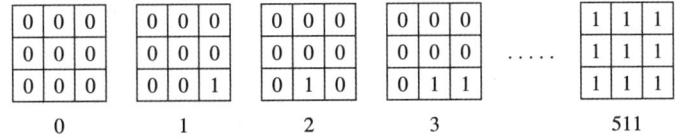

图 28-18 一共有 512 个结点，以 0,1,2,…,511 的顺序标记

如果存在一个由结点 u 到结点 v 的合法移动，我们就分配一个由结点 v 到结点 u 的边。图 28-20 显示了图的一部分。注意这里有一条边从 511 到 47，因为可以翻转结点 47 的一个单元格从而成为结点 511。

图 28-18 中的最后一个结点代表 9 枚硬币正面朝下的状态。为方便起见，我们称最后一个结点为目标结点（target node），这样，目标结点被标记为 511。假设 9 枚硬币反面的问题的初始状态对应到结点 s，那么问题就简化为搜索结点 s 和目标结点之间的最短路径，这就等价于在一个以目标结点为根结点的广度优先搜索树中搜索从结点 s 到目标结点的路径。

现在的任务是创建一个标记为 0,1,2,…,511 的包含 512 个结点的有向图，并且顶点之间有边相连。一旦图被创建，就得到以结点 511 为根结点的一个广度优先搜索树。从该广度优先搜索树可以得到从根结点到任意一个结点的最短路径。我们将创建一个名为 NineTailModel 的类，其包含了获取从目标结点到任意其他结点之间最短路径的方法。类的 UML 图如图 28-19 所示。

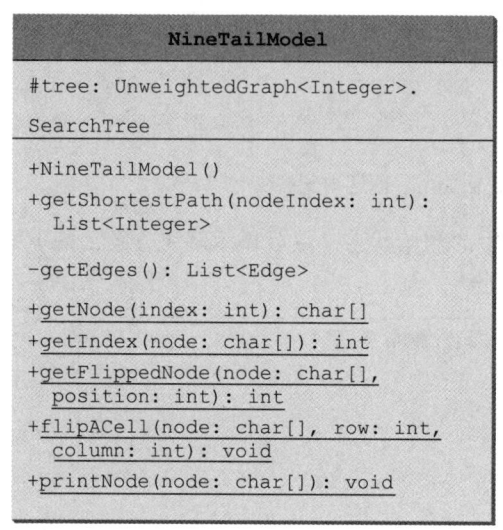

图 28-19　NineTailModel 类使用一个图建模 9 枚硬币反面的问题

结点被可视化地表示为一个包含字母 H 和 T 的 3×3 的矩阵。在程序中，我们使用一个包含 9 个字符的一维数组来表示一个结点。例如，图 28-18 中代表顶点 1 的结点在数组中表示为 {'H','H','H','H','H','H','H','H','T'}。

getEdges() 方法返回一个包含 Edge 对象的线性表。

getNode(index) 方法返回指定下标的结点。例如，getNode(0) 返回包含 9 个 H 的结点，getNode(511) 返回包含 9 个 T 的结点。getIndex(node) 方法返回结点的下标。

注意，数据域 tree 被定义为保护的，因此它可以被下一章中的 WeightedNineTail 子类访问。

getFlippedNode(char[] node,int position) 方法翻转指定位置和其邻接位置的结点。该方法返回新结点的下标。位置是从 0 到 8 的一个值，指向了结点中的一个硬币，如下图所示。

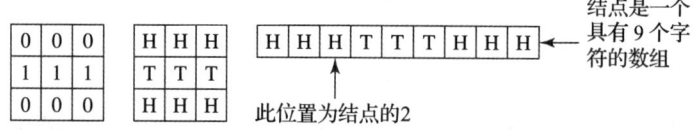

例如，对于图 28-20 中的结点 56，在位置 0 处翻转，那么将会得到结点 51。如果在位置 1 处翻转结点 56，将会得到结点 47。

方法 flipACell(char[] node,int row,int column) 翻转指定行和列的结点。例如，如果在第 0 行第 0 列翻转结点 56，那么新结点为 408。如果在第 2 行第 0 列翻转结点 56，那么新结点为 30。

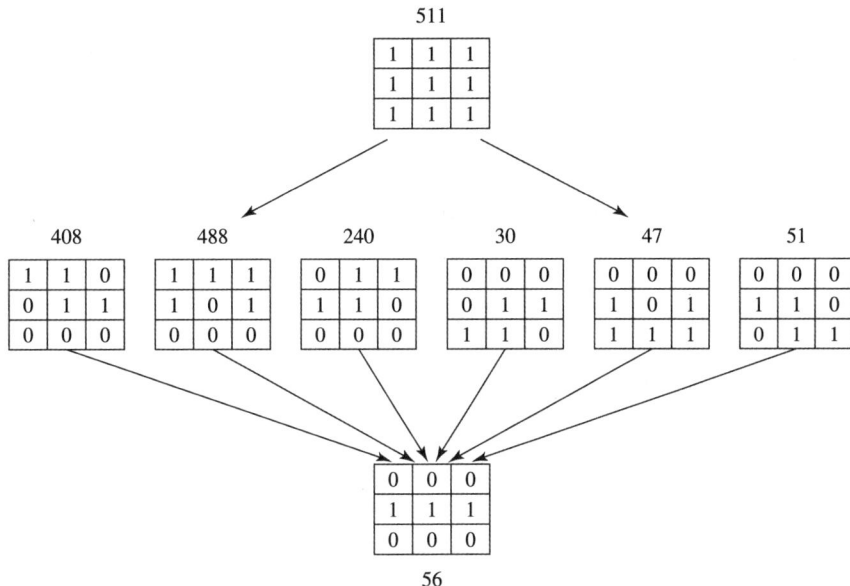

图 28-20 如果翻转单元格后结点 u 变为结点 v，就分配一条由结点 v 到结点 u 的边

程序清单 28-13 给出了 NineTailModel.java 的源代码。

程序清单 28-13 NineTailModel.java

```java
1  import java.util.*;
2
3  public class NineTailModel {
4    public final static int NUMBER_OF_NODES = 512;
5    protected UnweightedGraph<Integer>.SearchTree tree;
6
7    /** Construct a model */
8    public NineTailModel() {
9      // Create edges
10     List<Edge> edges = getEdges();
11
12     // Create a graph
13     UnweightedGraph<Integer> graph = new UnweightedGraph<>(
14       edges, NUMBER_OF_NODES);
15
16     // Obtain a BSF tree rooted at the target node
17     tree = graph.bfs(511);
18   }
19
20   /** Create all edges for the graph */
21   private List<Edge> getEdges() {
22     List<Edge> edges =
23       new ArrayList<>(); // Store edges
24
25     for (int u = 0; u < NUMBER_OF_NODES; u++) {
26       for (int k = 0; k < 9; k++) {
27         char[] node = getNode(u); // Get the node for vertex u
28         if (node[k] == 'H') {
29           int v = getFlippedNode(node, k);
30           // Add edge (v, u) for a legal move from node u to node v
31           edges.add(new Edge(v, u));
32         }
33       }
34     }
35
36     return edges;
```

```java
37      }
38
39      public static int getFlippedNode(char[] node, int position) {
40        int row = position / 3;
41        int column = position % 3;
42
43        flipACell(node, row, column);
44        flipACell(node, row - 1, column);
45        flipACell(node, row + 1, column);
46        flipACell(node, row, column - 1);
47        flipACell(node, row, column + 1);
48
49        return getIndex(node);
50      }
51
52      public static void flipACell(char[] node, int row, int column) {
53        if (row >= 0 && row <= 2 && column >= 0 && column <= 2) {
54          // Within the boundary
55          if (node[row * 3 + column] == 'H')
56            node[row * 3 + column] = 'T'; // Flip from H to T
57          else
58            node[row * 3 + column] = 'H'; // Flip from T to H
59        }
60      }
61
62      public static int getIndex(char[] node) {
63        int result = 0;
64
65        for (int i = 0; i < 9; i++)
66          if (node[i] == 'T')
67            result = result * 2 + 1;
68          else
69            result = result * 2 + 0;
70
71        return result;
72      }
73
74      public static char[] getNode(int index) {
75        char[] result = new char[9];
76
77        for (int i = 0; i < 9; i++) {
78          int digit = index % 2;
79          if (digit == 0)
80            result[8 - i] = 'H';
81          else
82            result[8 - i] = 'T';
83          index = index / 2;
84        }
85
86        return result;
87      }
88
89      public List<Integer> getShortestPath(int nodeIndex) {
90        return tree.getPath(nodeIndex);
91      }
92
93      public static void printNode(char[] node) {
94        for (int i = 0; i < 9; i++)
95          if (i % 3 != 2)
96            System.out.print(node[i]);
97          else
98            System.out.println(node[i]);
99
100       System.out.println();
101     }
102   }
```

构造方法（第 8 ～ 18 行）创建一个有 512 个结点的图，其中每一条边对应着从一个结点到另一个结点的移动（第 10 行）。从这个图，可以得到一棵以目标结点 511 为根结点的广度优先搜索树（第 17 行）。

为创建边，getEdges 方法（第 21 ～ 37 行）检测每一个结点 u，查看它是否可以翻转成为另一个结点 v。如果可以，将 (v, u) 添加到 Edge 线性表（第 31 行）。getFlippedNode(node,position) 方法通过在一个结点中翻转一个 H 格子和其邻居来找到翻转的结点（第 43 ～ 47 行）。flipACell (node,row,column) 方法真正在一个结点中翻转一个 H 格子和其邻居（第 52 ～ 60 行）。

getIndex(node) 方法实现的方式与将二进制数转换为十进制数的方式一样（第 62 ～ 72 行）。getNode(index) 方法返回一个包含字母 H 和 T 的结点（第 74 ～ 87 行）。

getShortestpath(nodeIndex) 方法调用 getPath(nodeIndex) 方法来获取从指定的结点到目标结点之间的最短路径上的顶点（第 89 ～ 91 行）。

printNode(node) 方法在控制台上显示一个结点（第 93 ～ 101 行）。

程序清单 28-14 给出了一个程序，提示用户输入一个初始结点，并且显示到达目标结点的步骤。

程序清单 28-14 NineTail.java

```java
1  import java.util.Scanner;
2
3  public class NineTail {
4    public static void main(String[] args) {
5      // Prompt the user to enter nine coins' Hs and Ts
6      System.out.print("Enter the initial nine coins Hs and Ts: ");
7      Scanner input = new Scanner(System.in);
8      String s = input.nextLine();
9      char[] initialNode = s.toCharArray();
10
11     NineTailModel model = new NineTailModel();
12     java.util.List<Integer> path =
13       model.getShortestPath(NineTailModel.getIndex(initialNode));
14
15     System.out.println("The steps to flip the coins are ");
16     for (int i = 0; i < path.size(); i++)
17       NineTailModel.printNode(
18         NineTailModel.getNode(path.get(i).intValue()));
19   }
20 }
```

该程序第 8 行提示用户输入一个包含 9 个 H 和 T 字母的字符串作为初始结点，从字符串中得到一个字符数组（第 9 行），创建一个图的模型以得到广度优先搜索树（第 11 行），得到一个从初始结点到目标结点的最短路径（第 12 ～ 13 行），然后显示该路径上的结点（第 16 ～ 18 行）。

✓ 复习题

28.10.1 NineTailModel 中图的结点是如何创建的？

28.10.2 NineTailModel 中图的边是如何创建的？

28.10.3 在程序清单 28-13 中调用 getIndex("HTHTTTHHH".toCharArray()) 会返回什么？在程序清单 28-13 中调用 getNode(46) 会返回什么？

28.10.4 如果将程序清单 28-13 中的第 26 行和第 27 行交换，程序还能工作吗？为什么？

关键术语

adjacency list（邻接线性表）
adjacency matrix（邻接矩阵）
adjacent vertices（邻接顶点）
breadth-first search（广度优先搜索）
complete graph（完全图）
cycle（回路）
degree（度）
depth-first search（深度优先搜索）
directed graph（有向图）
graph（图）
incident edges（关联边）

parallel edge（平行边）
Seven Bridges of Königsberg（哥尼斯堡七桥问题）
simple graph（简单图）
spanning tree（生成树）
strongly connected graph（强连通图）
tree（树）
undirected graph（无向图）
unweighted graph（非加权图）
weakly connected graph（弱连通图）
weighted graph（加权图）

本章小结

1. 图是一种有用的数学结构，可以表示现实世界中实体之间的联系。学习了如何使用类和接口来对图建模，如何使用数组和链表来表示顶点和边，以及如何实现图的操作。
2. 图的遍历是指访问图中的每个顶点正好一次的过程。学习了两种遍历图的常用方法：深度优先搜索（DFS）和广度优先搜索（BFS）。
3. DFS 和 BFS 可以解决许多问题，如检测图是否连通，检测图中是否存在环，以及找出两个顶点之间的最短路径等。

测试题

回答位于本书配套网站上的本章测试题。

编程练习题

28.6～28.10 节

*28.1 （测试一个图是否是连通的）编写一个程序，从文件读入图并且检测该图是否是连通的。文件中的第一行包含了一个表示顶点个数的数字（n）。顶点被标记为 0,1,…,n-1。接下来的每一行，以 u v1 v2… 的形式描述边（u,v1）、（u,v2），以此类推。图 28-21 给出了对应图的两个文件例子。

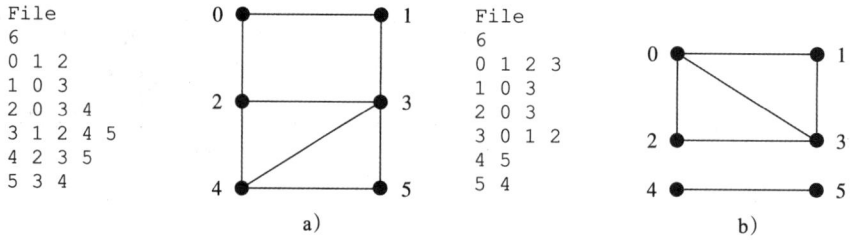

图 28-21　图的顶点和边可以存储在一个文件中

你的程序应该提示用户输入文件的 URL，然后从该文件中读取数据，创建 UnweightedGraph 的一个实例 g，调用 g.printEdges() 来显示所有的边，并调用 dfs() 来获取 UnweightedGraph<V>.SearchTree 的一个实例 tree。如果 tree.getNumberOfVerticeFound() 与图中的顶点数目相同，那么该图就是连通的。下面是这个程序的运行示例：

```
Enter a URL:
https://liveexample.pearsoncmg.com/test/GraphSample1.txt  ↵Enter
The number of vertices is 6
Vertex 0: (0, 1) (0, 2)
Vertex 1: (1, 0) (1, 3)
Vertex 2: (2, 0) (2, 3) (2, 4)
Vertex 3: (3, 1) (3, 2) (3, 4) (3, 5)
Vertex 4: (4, 2) (4, 3) (4, 5)
Vertex 5: (5, 3) (5, 4)
The graph is connected
```

> **提示**：使用 new UnweightedGraph(list,numberOfVertices) 来创建一个图，其中 list 包含一个 Edge 对象的线性表。使用 new Edge(u,v) 来创建一条边。读取第一行来获取顶点的数目。将接下来的每一行读入一个字符串 s 中，并且使用 s.split("[\\s+]") 来从字符串中提取顶点并从顶点创建边。

*28.2 （为图创建文件）修改程序清单 28-2 来创建一个文件表示 graph1。该文件格式在编程练习题 28.1 中描述。从程序清单 28-2 中第 8～21 行定义的数组创建该文件。图的顶点数为 12，它存储在文件的第一行。文件的内容应该如下所示：

```
12
0 1 3 5
1 0 2 3
2 1 3 4 10
3 0 1 2 4 5
4 2 3 5 7 8 10
5 0 3 4 6 7
6 5 7
7 4 5 6 8
8 4 7 9 10 11
9 8 11
10 2 4 8 11
11 8 9 10
```

*28.3 （使用栈实现深度优先搜索）程序清单 28-8 中描述的深度优先搜索使用的是递归。设计一个不使用递归的新算法。使用伪代码描述该算法。通过定义一个名为 UnweightedGraphWithNonrecursiveDFS 的新类来实现它，该类继承自 UnweightedGraph 并且重写 dfs 方法。按照程序清单 28-9 写一个相同的测试程序，除了用 UnweightedGraphWithNonrecursiveDFS 代替 UnweightedGraph。

*28.4 （寻找连通组件）创建一个名为 MyGraph 的新类作为 UnweightedGraph 的子类，其中包含找到图中所有连通组件的方法，使用的方法头如下：

public List<List<Integer>> getConnectedComponents();

该方法返回一个 List<List<Integer>>。线性表中的每个元素是另一个线性表，它包含了连通组件的所有顶点。例如，对于图 28-21b 中的图而言，getConnectedComponents() 返回 [[0,1,2,3],[4,5]]。

*28.5 （寻找路径）定义一个继承自 UnweightedGraph 类的名为 UnweightedGraphWithGetPath 的新类，它有一个找出两个顶点之间路径的新方法，其方法头如下：

public List<Integer> getPath(int u, int v);

该方法返回一个 List<Integer>，它包含从顶点 u 到顶点 v 的路径上的所有顶点。使用 BFS 方法，可以获取从顶点 u 到顶点 v 的最短路径。如果顶点 u 和顶点 v 之间不存在路径，该方法返回 null。写一个测试程序，为图 28-1 创建一个图。该程序提示用户输入两个城市，然后输出它们之间的路径。使用 https://liveexample.pearsoncmg.com/test/Exercise28_05.txt 测试你的代码。下面是一个运行示例：

```
Enter a starting city: Seattle  ↵Enter
Enter an ending city: Miami  ↵Enter
The path is Seattle Denver Kansas City Atlanta Miami
```

*28.6 （探测回路）定义一个继承自 UnweightedGraph 类的名为 UnweightedGraphDetectCycle 的新类，它有一个判定图中是否存在环的新方法，其方法头如下：

`public boolean isCyclic();`

用伪代码描述这个算法并实现它。注意图可能是有向图。

*28.7 （找出回路）定义一个继承自 UnweightedGraph 类的名为 UnweightedGraphFindCycle 的新类，它有一个判定图中是否存在环的新方法，其方法头如下：

`public List<Integer> getACycle(int u);`

该方法返回一个 List，它包含从顶点 u 开始的回路上的所有顶点。如果图中没有回路，方法返回 null。用伪代码描述这个算法并实现它。

**28.8 （测试二分图）回顾一下，如果图的顶点可以分为两个不相交的集合，而且同一个集合中的顶点之间不存在边，那么这个图是二分图。定义一个名为 UnweightedGraphTestBipartite 的类，该类有以下方法来检测图是否是二分图：

`public boolean isBipartite();`

**28.9 （得到二分集合）在 UnweightedGraph 中添加一个新方法，如果图是二分图，返回这两个二分集合，其方法头如下：

`public List<List<Integer>> getBipartite();`

该方法返回一个包含两个子线性表的 List，每一个都包含了一个顶点集合。如果图不是二分的，方法返回 null。

28.10 （找出最短路径）编写一个程序，从文件中读取一个连通图。图存储在一个文件中，使用和编程练习题 28.1 中给定的相同格式。你的程序应该提示用户输入文件名和两个顶点，然后应该显示两个顶点之间的最短路径。例如，对于图 28-21a 中的图，顶点 0 和顶点 5 之间的最短路径可以显示为 0 1 3 5。

下面是该程序的一个运行示例：

```
Enter a file name: c:\exercise\GraphSample1.txt  ↵Enter
Enter two vertices (integer indexes): 0 5  ↵Enter
The number of vertices is 6
Vertex 0: (0, 1) (0, 2)
Vertex 1: (1, 0) (1, 3)
Vertex 2: (2, 0) (2, 3) (2, 4)
Vertex 3: (3, 1) (3, 2) (3, 4) (3, 5)
Vertex 4: (4, 2) (4, 3) (4, 5)
Vertex 5: (5, 3) (5, 4)
The path is 0 1 3 5
```

**28.11 （修改程序清单 28-14）程序清单 28-14 中的程序允许用户在控制台上为 9 枚硬币反面问题输入数据并且在控制台上显示结果。编写一个程序，让用户设置 9 枚硬币的初始状态（如图 28-22a 所示），然后单击 Solve 按钮来显示解决方案，如图 28-22b 所示。初始情况下，用户可以通过单击鼠标来翻转硬币。将翻转的单元设置为红色。

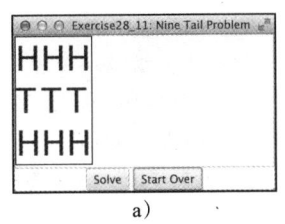
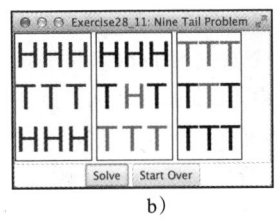

a) b)

图 28-22 解决 9 枚硬币反面问题的程序（来源：Oracle 或其附属公司版权所有 © 1995 ～ 2016，经授权使用）

**28.12 （9 枚硬币反面问题的变体）在 9 枚硬币反面问题中，当翻转一个正面的硬币时，水平和垂直方向上的邻居也都被翻转。重新编写程序，假设对角线上的邻居也都被翻转。

**28.13 （4×4 的 16 枚硬币反面问题）程序清单 28-14 提供了 9 枚硬币反面问题的解答。修改该程序，成为一个 4×4 的 16 枚硬币反面问题。注意可能对于一个开始的模式并不存在解答。如果是这样，报告没有解答存在。

**28.14 （4×4 的 16 枚硬币反面问题的分析）本书中的 9 枚硬币反面问题使用的是 3×3 的矩阵。假设在一个 4×4 的矩阵中放置了 16 枚硬币。编写一个程序，找出不存在解答的开始模式的数目。

*28.15 （4×4 的 16 枚硬币反面问题的 GUI）修改编程练习题 28.14，使得用户可以设置 4×4 的 16 枚硬币反面问题的初始化模式（参见图 28-23a）。用户可以单击 Solve 按钮来显示解答，如图 28-23b 所示。开始时，用户可以点击鼠标按钮来翻转硬币。如果解答不存在，显示一个消息对话框来报告该消息。

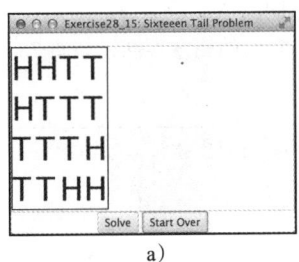
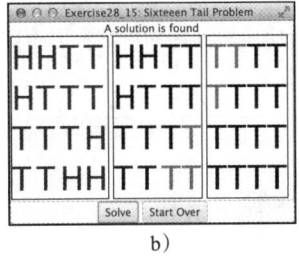

a) b)

图 28-23 解决 16 枚硬币反面问题的程序（来源：Oracle 或其附属公司版权所有 © 1995 ～ 2016，经授权使用）

**28.16 （诱导子图）给定一个无向图 $G=(V, E)$ 和一个整数 k，找出 G 的一个最大的诱导子图 H，H 中的所有结点的度 $\geq k$，或者得出这样的子图不存在的结论。使用下面的方法头实现该方法：

`public static <V> Graph<V> maxInducedSubgraph(Graph<V> g, int k)`

如果这样的子图不存在，方法返回一个空图。

提示：一个直观的方法是删除那些度小于 k 的顶点。随着顶点及其邻接边被删除，其他顶点的度可能会减小。继续这个过程直到没有顶点被删除，或者所有的顶点都被删除。

***28.17 （哈密尔顿环）补充材料 VI.E 给出了哈密尔顿路径算法的实现。在 Graph 接口中添加以下 getHamiltonianCycle 方法，并且在 UnweightedGraph 类中实现它：

```
/** Return a Hamiltonian cycle
 * Return null if the graph doesn't contain a Hamiltonian cycle */
public List<Integer> getHamiltonianCycle()
```

***28.18 （骑士巡游回路）改写补充材料 VI.E 中示例学习的 KnightTourApp.java 程序，找出骑士访问棋盘的每个方块并且返回到起始方块的路径。将骑士巡游回路问题简化为寻找哈密尔顿环的问题。

28.19 (显示一个图中的深度优先搜索/广度优先搜索树) 修改程序清单 28-6 中的 `GraphView`，使用 setter 方法添加一个数据域 `tree`。树的边显示为红色。编写一个程序，显示图 28-1 中的图，以及从一个指定城市出发的深度优先搜索/广度优先搜索树，如图 28-13 和图 28-16 所示。如果输入了一个地图中没有的城市，程序用一个标签中给出错误信息。

*28.20 **(显示图)** 编写一个程序，从一个文件中读取一个图并显示它。文件的第一行包含了表示顶点个数的数字（n）。顶点被标记为 0,1,…,n-1。接下来的每一行以 u x y v1 v2… 的格式描述了 u 的位置 (x, y) 以及边 (u, v1) 和 (u, v2)，依此类推。图 28-24a 给出了对应图的文件的例子。你的程序提示用户输入文件名，从文件中读取数据并且使用 `GraphView` 在面板上显示图，如图 28-24b 所示。

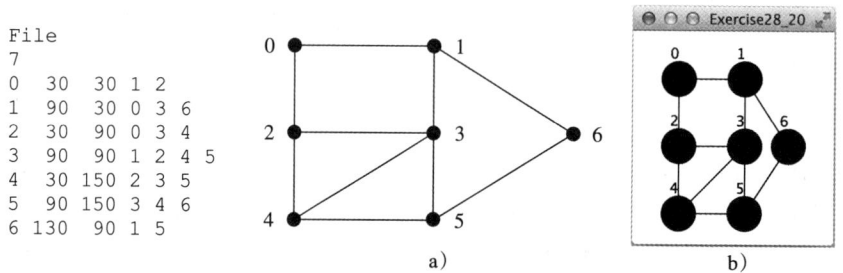

图 28-24 该程序读取关于图的信息并且可视化地显示它（来源：Oracle 或其附属公司版权所有 © 1995 ～ 2016，经授权使用）

28.21 (显示连通圆集合) 修改程序清单 28-10，以不同颜色显示连通圆的集合。也就是说，如果两个圆是连通的，则使用相同的颜色显示。否则，它们的颜色不同，如图 28-25 所示。（提示：参见编程练习题 28.4。）

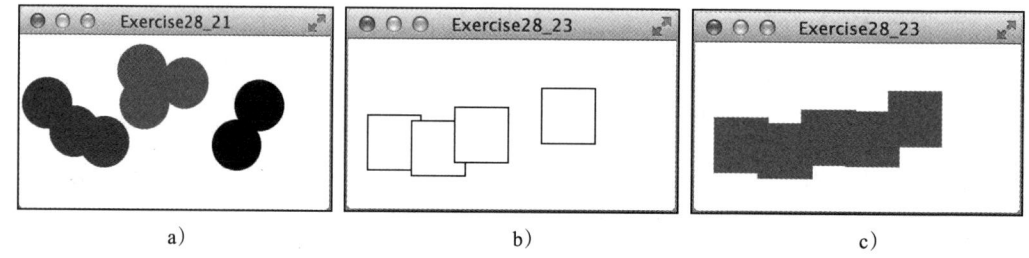

图 28-25 a) 连通圆以相同的颜色显示；b) 如果矩形不是连通的，则不使用颜色填充；c) 如果矩形是连通的，则使用颜色填充（来源：Oracle 或其附属公司版权所有 © 1995 ～ 2016，经授权使用）

28.22 (移动圆) 修改程序清单 28-10，使得用户可以拖放和移动圆。

28.23 (连通矩形) 程序清单 28-10 允许用户创建圆并确定它们是否是连通的。为矩形重写该程序。该程序使用户可以在没有被矩形占据的空白区域点击鼠标来创建矩形。当矩形被添加时，如果一些矩形是连通的则以填充方式绘制，否则不填充。如图 28-25b ～图 28-25c 所示。

*28.24 **(移除圆)** 修改程序清单 28-10，使得用户可以通过在圆内点击移除圆。

*28.25 **(实现 remove(V, v))** 修改程序清单 28-4，重写定义在 `Graph` 接口中的 `remove(V, v)` 方法。

*28.26 **(实现 remove(int u, int v))** 修改程序清单 28-4，重写定义在 `Graph` 接口中的 `remove(int u, int v)` 方法。

第 29 章

Introduction to Java Programming and Data Structures, Comprehensive Version, Twelfth Edition

加权图及其应用

教学目标

- 使用邻接矩阵和邻接线性表来表示加权边（29.2 节）。
- 使用继承自 UnweightedGraph 类的 WeightedGraph 类来对加权图建模（29.3 节）。
- 设计并实现得到最小生成树的算法（29.4 节）。
- 定义继承自 SearchTree 类的 MST 类（29.4 节）。
- 设计并实现得到单源最短路径的算法（29.5 节）。
- 定义继承自 SearchTree 类的 ShortestPathTree 类（29.5 节）
- 用最短路径算法来解决加权 9 枚硬币反面的问题（29.6 节）。

29.1 引言

🔑 **要点提示**：如果图的每条边都赋予了一个权重，则该图是一个加权图。加权图有很多实际的应用。

图 28-1 假设图表示了城市之间的飞行次数。可以应用广度优先搜索（BFS）来找到两个城市之间的最小飞行次数。假设边代表了城市之间的驾驶距离，如图 29-1 所示。如何找到连接所有城市的最小总距离呢？又如何找到两个城市之间的最短路径？本章讨论这些问题。前者称为最小生成树（MST）问题，后者是最短路径问题。

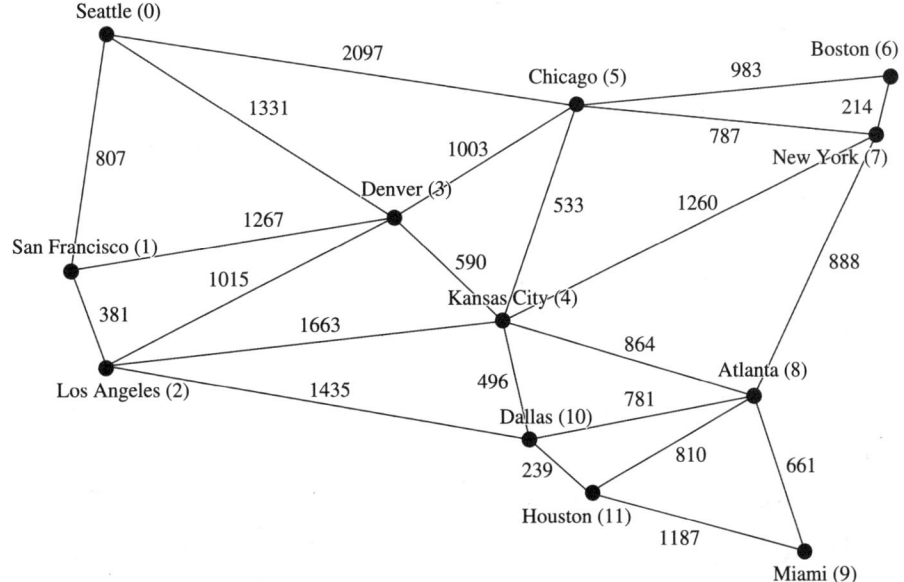

图 29-1 该图对城市之间的距离进行了建模

前面章节中介绍了图的概念。你学会了如何使用边数组、边线性表、邻接矩阵和邻接线性表来表示边，以及如何使用 Graph 接口和 UnweightedGraph 类来对图建模。前面章节中还介绍了两种重要的遍历图的方法：深度优先搜索和广度优先搜索，并将其应用于解决实际的问题。本章将介绍加权图。29.4 节介绍找出最小生成树的算法，29.5 节介绍得到最短路径的算法。

> **教学注意**：在开始介绍加权图的算法和应用之前，通过网址 liveexample.pearsoncmg.com/dsanimation/WeightedGraphLearningTooleBook.html 提供的交互式工具来了解加权图是很有帮助的，如图 29-2 所示。该工具可以让你输入顶点，创建加权边，查看图，从单一源中找到一个 MST 以及所有的最短路径。

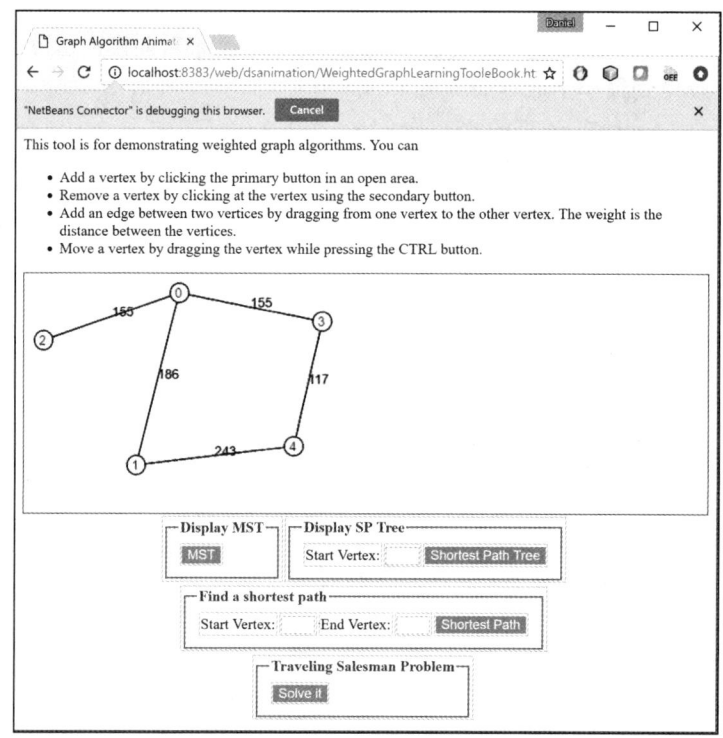

图 29-2　可以使用该工具，通过鼠标操作来创建一个加权图，显示 MST 和最短路径，并执行其他操作（来源：©Mozilla Firefox）

29.2　加权图的表示

> **要点提示**：加权边可以存储在邻接线性表中。

加权图有两种类型：顶点加权和边加权。在顶点加权图中，每个顶点都分配了一个权重。在边加权图中，每条边都分配了一个权重。这两种类型中，边加权图应用更广泛，本章主要介绍边加权图。

除开需要表示边的权重，加权图与非加权图的表示方法一样。与非加权图一样，加权图的顶点可以存储在一个数组中。本节介绍表示加权图的边的三种方法。

29.2.1　加权边的表示：边数组

可以使用一个二维数组来表示加权边。例如，可以使用 29-3b 所示的数组存储图 29-3a 中图的所有边。

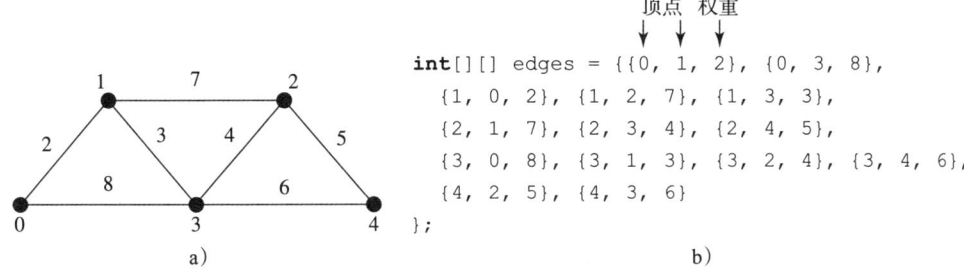

图 29-3 加权图中每条边都分配了一个权重

> **注意**：权重可以为任何类型：Integer, Double, BigDecimal, 等等。可以采用一个如下所示的 Object 类型的二维数组来表示加权边：

```
Object[][] edges = {
  {Integer.valueOf(0), Integer.valueOf(1), new SomeTypeForWeight(2)},
  {Integer.valueOf(0), Integer.valueOf(3), new SomeTypeForWeight(8)},
  ...
};
```

29.2.2 加权邻接矩阵

假设图有 n 个顶点。可以使用一个 $n \times n$ 的二维矩阵 weights 来表示边上的权重。weights[i][j] 表示边 (i,j) 上的权重。如果顶点 i 和 j 不相连，weights[i][j] 为 null。例如，在图 29-3a 中图的权重可以使用邻接矩阵表示，如下所示：

```
Integer[][] adjacencyMatrix = {
  {null, 2, null, 8, null},
  {2, null, 7, 3, null},
  {null, 7, null, 4, 5},
  {8, 3, 4, null, 6},
  {null, null, 5, 6, null}
};
```

	0	1	2	3	4
0	null	2	null	8	null
1	2	null	7	3	null
2	null	7	null	4	5
3	8	3	4	null	6
4	null	null	5	6	null

29.2.3 邻接线性表

另一种表示边的方法是将边定义为对象。程序清单 28-3 中 Edge 类用来表示非加权的边。对于加权图，我们定义如程序清单 29-1 所示的 WeightedEdge 类。

程序清单 29-1 WeightedEdge.java

```java
 1  public class WeightedEdge extends Edge
 2      implements Comparable<WeightedEdge> {
 3    public double weight;  // The weight on edge (u, v)
 4
 5    /** Create a weighted edge on (u, v) */
 6    public WeightedEdge(int u, int v, double weight) {
 7      super(u, v);
 8      this.weight = weight;
 9    }
10
11    @Override /** Compare two edges on weights */
12    public int compareTo(WeightedEdge edge) {
13      if (weight > edge.weight)
14        return 1;
15      else if (weight == edge.weight)
16        return 0;
```

```
17        else
18          return -1;
19      }
20    }
```

一个 Edge 对象表示一条由顶点 u 到顶点 v 的边。WeightedEdge 继承自 Edge，添加了一个新的属性 weight。

为了创建一个 WeightedEdge 对象，使用 new WeightedEdge(i,j,w)，其中 w 是边 (i,j) 上的权重。通常你会需要比较边的权重。因此，WeightedEdge 类实现了 Comparable 接口。

对于非加权图，我们使用邻接线性表来表示边。对于加权图，我们依然使用邻接线性表，图 29-3a 中图的顶点的邻接线性表可以表示为：

```
java.util.List<WeightedEdge>[] list = new java.util.List[5];
```

list[0]	WeightedEdge(0, 1, 2)	WeightedEdge(0, 3, 8)		
list[1]	WeightedEdge(1, 0, 2)	WeightedEdge(1, 3, 3)	WeightedEdge(1, 2, 7)	
list[2]	WeightedEdge(2, 3, 4)	WeightedEdge(2, 4, 5)	WeightedEdge(2, 1, 7)	
list[3]	WeightedEdge(3, 1, 3)	WeightedEdge(3, 2, 4)	WeightedEdge(3, 4, 6)	WeightedEdge(3, 0, 8)
list[4]	WeightedEdge(4, 2, 5)	WeightedEdge(4, 3, 6)		

list[i] 存储了所有与顶点 i 相邻的边。

为了灵活性，我们使用一个数组线性表而不是固定大小的数组来表示 list，如下所示：

```
List<List<WeightedEdge>> list = new java.util.ArrayList<>();
```

✔ 复习题

29.2.1 对于代码 WeightedEdge edge = new WeightedEdge(1, 2, 3.5) 而言，edge.u、edge.v 以及 edge.weight 是什么？

29.2.2 下面代码的输出是什么？

```
List<WeightedEdge> list = new ArrayList<>();
list.add(new WeightedEdge(1, 2, 3.5));
list.add(new WeightedEdge(2, 3, 4.5));
WeightedEdge e = java.util.Collections.max(list);
System.out.println(e.u);
System.out.println(e.v);
System.out.println(e.weight);
```

29.3 WeightedGraph 类

⛰ 要点提示：WeightedGraph 类继承自 UnweightedGraph。

前面章节设计了 Graph 接口和 UnweightedGraph 类来对图建模。我们现在设计 WeightedGraph 作为 UnweightedGraph 的子类，如图 29-4 所示。

WeightedGraph 简单地继承自 UnweightedGraph，采用 5 个构造方法来创建具体的 WeightedGraph 实例。WeightedGraph 继承了 UnweightedGraph 的所有方法，重写了 clear 和 addVertex 方法，实现了一个新方法 addEdge 用于添加一条加权边，并且还引入了新的方法用于获取最小生成树并找出所有的单源最短路径。最小生成树和最短路径将分别在 29.4 节和 29.5 节中介绍。

程序清单 29-2 实现了 WeightedGraph。内部使用边的邻接线性表（第 38～63 行）来存储每个顶点的邻接边。当创建一个 WeightedGraph 时，就会创建其边的邻接线性表（第 47 和 57 行）。方法 getMinimumSpanningTree()（第 99 ～ 138 行）和 getShortestPaths()（第 156 ～ 197 行）将在后面小节中介绍。

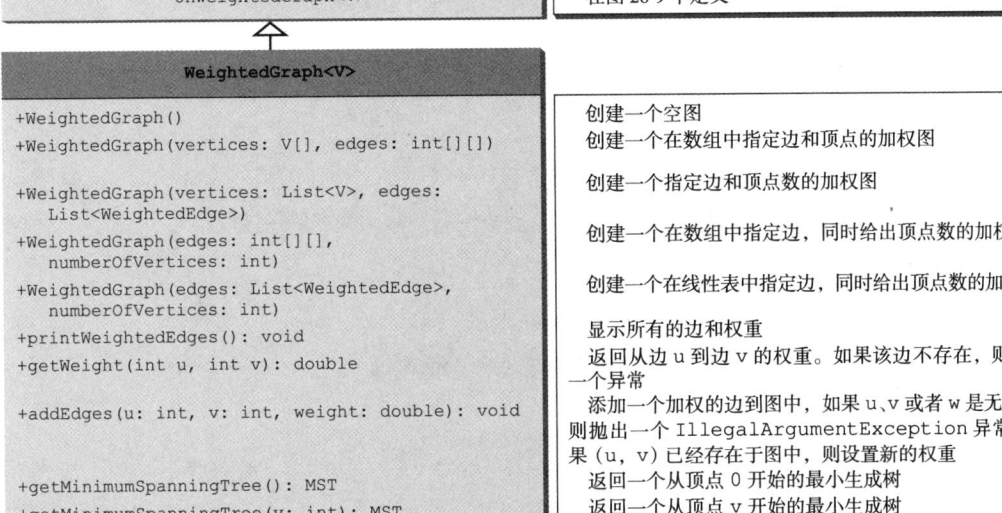

图 29-4　WeightedGraph 继承自 AbstractGraph

程序清单 29-2 WeightedGraph.java

```java
 1  import java.util.*;
 2
 3  public class WeightedGraph<V> extends UnweightedGraph<V> {
 4    /** Construct an empty */
 5    public WeightedGraph() {
 6    }
 7
 8    /** Construct a WeightedGraph from vertices and edged in arrays */
 9    public WeightedGraph(V[] vertices, int[][] edges) {
10      createWeightedGraph(java.util.Arrays.asList(vertices), edges);
11    }
12
13    /** Construct a WeightedGraph from vertices and edges in list */
14    public WeightedGraph(int[][] edges, int numberOfVertices) {
15      List<V> vertices = new ArrayList<>();
16      for (int i = 0; i < numberOfVertices; i++)
17        vertices.add((V)(Integer.valueOf(i)));
18
19      createWeightedGraph(vertices, edges);
20    }
21
22    /** Construct a WeightedGraph for vertices 0, 1, 2 and edge list */
23    public WeightedGraph(List<V> vertices, List<WeightedEdge> edges) {
24      createWeightedGraph(vertices, edges);
25    }
26
27    /** Construct a WeightedGraph from vertices 0, 1, and edge array */
28    public WeightedGraph(List<WeightedEdge> edges,
29        int numberOfVertices) {
30      List<V> vertices = new ArrayList<>();
31      for (int i = 0; i < numberOfVertices; i++)
```

```java
32          vertices.add((V)(Integer.valueOf(i)));
33
34      createWeightedGraph(vertices, edges);
35    }
36
37    /** Create adjacency lists from edge arrays */
38    private void createWeightedGraph(List<V> vertices, int[][] edges) {
39      this.vertices = vertices;
40
41      for (int i = 0; i < vertices.size(); i++) {
42        neighbors.add(new ArrayList<Edge>()); // Create a list for vertices
43      }
44
45      for (int i = 0; i < edges.length; i++) {
46        neighbors.get(edges[i][0]).add(
47          new WeightedEdge(edges[i][0], edges[i][1], edges[i][2]));
48      }
49    }
50
51    /** Create adjacency lists from edge lists */
52    private void createWeightedGraph(
53        List<V> vertices, List<WeightedEdge> edges) {
54      this.vertices = vertices;
55
56      for (int i = 0; i < vertices.size(); i++) {
57        neighbors.add(new ArrayList<Edge>()); // Create a list for vertices
58      }
59
60      for (WeightedEdge edge: edges) {
61        neighbors.get(edge.u).add(edge); // Add an edge into the list
62      }
63    }
64
65    /** Return the weight on the edge (u, v) */
66    public double getWeight(int u, int v) throws Exception {
67      for (Edge edge : neighbors.get(u)) {
68        if (edge.v == v) {
69          return ((WeightedEdge)edge).weight;
70        }
71      }
72
73      throw new Exception("Edge does not exit");
74    }
75
76    /** Display edges with weights */
77    public void printWeightedEdges() {
78      for (int i = 0; i < getSize(); i++) {
79        System.out.print(getVertex(i) + " (" + i + "): ");
80        for (Edge edge : neighbors.get(i)) {
81          System.out.print("(" + edge.u +
82            ", " + edge.v + ", " + ((WeightedEdge)edge).weight + ") ");
83        }
84        System.out.println();
85      }
86    }
87
88    /** Add edges to the weighted graph */
89    public boolean addEdge(int u, int v, double weight) {
90      return addEdge(new WeightedEdge(u, v, weight));
91    }
92
93    /** Get a minimum spanning tree rooted at vertex 0 */
94    public MST getMinimumSpanningTree() {
95      return getMinimumSpanningTree(0);
96    }
```

```java
 97
 98    /** Get a minimum spanning tree rooted at a specified vertex */
 99    public MST getMinimumSpanningTree(int startingVertex) {
100      // cost[v] stores the cost by adding v to the tree
101      double[] cost = new double[getSize()];
102      for (int i = 0; i < cost.length; i++) {
103        cost[i] = Double.POSITIVE_INFINITY; // Initial cost
104      }
105      cost[startingVertex] = 0; // Cost of source is 0
106
107      int[] parent = new int[getSize()]; // Parent of a vertex
108      parent[startingVertex] = -1; // startingVertex is the root
109      double totalWeight = 0; // Total weight of the tree thus far
110
111      List<Integer> T = new ArrayList<>();
112
113      // Expand T
114      while (T.size() < getSize()) {
115        // Find smallest cost u in V - T
116        int u = -1; // Vertex to be determined
117        double currentMinCost = Double.POSITIVE_INFINITY;
118        for (int i = 0; i < getSize(); i++) {
119          if (!T.contains(i) && cost[i] < currentMinCost) {
120            currentMinCost = cost[i];
121            u = i;
122          }
123        }
124
125        if (u == -1) break; else T.add(u); // Add a new vertex to T
126        totalWeight += cost[u]; // Add cost[u] to the tree
127
128        // Adjust cost[v] for v that is adjacent to u and v in V - T
129        for (Edge e: neighbors.get(u)) {
130          if (!T.contains(e.v) && cost[e.v] > ((WeightedEdge)e).weight) {
131            cost[e.v] = ((WeightedEdge)e).weight;
132            parent[e.v] = u;
133          }
134        }
135      } // End of while
136
137      return new MST(startingVertex, parent, T, totalWeight);
138    }
139
140    /** MST is an inner class in WeightedGraph */
141    public class MST extends SearchTree {
142      private double totalWeight; // Total weight of all edges in the tree
143
144      public MST(int root, int[] parent, List<Integer> searchOrder,
145          double totalWeight) {
146        super(root, parent, searchOrder);
147        this.totalWeight = totalWeight;
148      }
149
150      public double getTotalWeight() {
151        return totalWeight;
152      }
153    }
154
155    /** Find single-source shortest paths */
156    public ShortestPathTree getShortestPath(int sourceVertex) {
157      // cost[v] stores the cost of the path from v to the source
158      double[] cost = new double[getSize()];
159      for (int i = 0; i < cost.length; i++) {
160        cost[i] = Double.POSITIVE_INFINITY; // Initial cost set to infinity
161      }
```

```java
162      cost[sourceVertex] = 0; // Cost of source is 0
163
164      // parent[v] stores the previous vertex of v in the path
165      int[] parent = new int[getSize()];
166      parent[sourceVertex] = -1; // The parent of source is set to -1
167
168      // T stores the vertices whose path found so far
169      List<Integer> T = new ArrayList<>();
170
171      // Expand T
172      while (T.size() < getSize()) {
173        // Find smallest cost u in V - T
174        int u = -1; // Vertex to be determined
175        double currentMinCost = Double.POSITIVE_INFINITY;
176        for (int i = 0; i < getSize(); i++) {
177          if (!T.contains(i) && cost[i] < currentMinCost) {
178            currentMinCost = cost[i];
179            u = i;
180          }
181        }
182
183        if (u == -1) break; else T.add(u); // Add a new vertex to T
184
185        // Adjust cost[v] for v that is adjacent to u and v in V - T
186        for (Edge e: neighbors.get(u)) {
187          if (!T.contains(e.v)
188              && cost[e.v] > cost[u] + ((WeightedEdge)e).weight) {
189            cost[e.v] = cost[u] + ((WeightedEdge)e).weight;
190            parent[e.v] = u;
191          }
192        }
193      } // End of while
194
195      // Create a ShortestPathTree
196      return new ShortestPathTree(sourceVertex, parent, T, cost);
197    }
198
199    /** ShortestPathTree is an inner class in WeightedGraph */
200    public class ShortestPathTree extends SearchTree {
201      private double[] cost; // cost[v] is the cost from v to source
202
203      /** Construct a path */
204      public ShortestPathTree(int source, int[] parent,
205          List<Integer> searchOrder, double[] cost) {
206        super(source, parent, searchOrder);
207        this.cost = cost;
208      }
209
210      /** Return the cost for a path from the root to vertex v */
211      public double getCost(int v) {
212        return cost[v];
213      }
214
215      /** Print paths from all vertices to the source */
216      public void printAllPaths() {
217        System.out.println("All shortest paths from " +
218          vertices.get(getRoot()) + " are:");
219        for (int i = 0; i < cost.length; i++) {
220          printPath(i); // Print a path from i to the source
221          System.out.println("(cost: " + cost[i] + ")"); // Path cost
222        }
223      }
224    }
225  }
```

WeightedGraph 类继承自 UnweightedGraph（第 3 行）。UnweightedGraph 中的属性 vertices

和 neighbors 被 WeightedGraph 所继承。neighbors 是一个线性表。线性表中的每个元素是包含了边的另一个线性表。对于非加权图来说，每条边是 Edge 的一个实例。对于加权图来说，每条边是 WeightedEdge 的一个实例。WeightedEdge 是 Edge 的子类型。因此，对于加权图来说，可以添加一个加权边到 neighbors.get(i) 中（第 47 行）。

addEdge(u, v, weight) 方法（第 88～91 行）添加一条边 (u, v, weight) 到图中。如果图是无向的，应该调用 addEdge(u, v, weight) 和 addEdge(v, u, weight) 添加一条位于顶点 u 和 v 之间的边。

程序清单 29-3 给出了一个测试程序，按图 29-1 所示创建一个图以及按图 29-3a 所示创建另外一个图。

程序清单 29-3 TestWeightedGraph.java

```java
 1  public class TestWeightedGraph {
 2    public static void main(String[] args) {
 3      String[] vertices = {"Seattle", "San Francisco", "Los Angeles",
 4        "Denver", "Kansas City", "Chicago", "Boston", "New York",
 5        "Atlanta", "Miami", "Dallas", "Houston"};
 6
 7      int[][] edges = {
 8        {0, 1, 807}, {0, 3, 1331}, {0, 5, 2097},
 9        {1, 0, 807}, {1, 2, 381}, {1, 3, 1267},
10        {2, 1, 381}, {2, 3, 1015}, {2, 4, 1663}, {2, 10, 1435},
11        {3, 0, 1331}, {3, 1, 1267}, {3, 2, 1015}, {3, 4, 599},
12          {3, 5, 1003},
13        {4, 2, 1663}, {4, 3, 599}, {4, 5, 533}, {4, 7, 1260},
14          {4, 8, 864}, {4, 10, 496},
15        {5, 0, 2097}, {5, 3, 1003}, {5, 4, 533},
16          {5, 6, 983}, {5, 7, 787},
17        {6, 5, 983}, {6, 7, 214},
18        {7, 4, 1260}, {7, 5, 787}, {7, 6, 214}, {7, 8, 888},
19        {8, 4, 864}, {8, 7, 888}, {8, 9, 661},
20          {8, 10, 781}, {8, 11, 810},
21        {9, 8, 661}, {9, 11, 1187},
22        {10, 2, 1435}, {10, 4, 496}, {10, 8, 781}, {10, 11, 239},
23        {11, 8, 810}, {11, 9, 1187}, {11, 10, 239}
24      };
25
26      WeightedGraph<String> graph1 =
27        new WeightedGraph<>(vertices, edges);
28      System.out.println("The number of vertices in graph1: "
29        + graph1.getSize());
30      System.out.println("The vertex with index 1 is "
31        + graph1.getVertex(1));
32      System.out.println("The index for Miami is " +
33        graph1.getIndex("Miami"));
34      System.out.println("The edges for graph1:");
35      graph1.printWeightedEdges();
36
37      edges = new int[][] {
38        {0, 1, 2}, {0, 3, 8},
39        {1, 0, 2}, {1, 2, 7}, {1, 3, 3},
40        {2, 1, 7}, {2, 3, 4}, {2, 4, 5},
41        {3, 0, 8}, {3, 1, 3}, {3, 2, 4}, {3, 4, 6},
42        {4, 2, 5}, {4, 3, 6}
43      };
44      WeightedGraph<Integer> graph2 = new WeightedGraph<>(edges, 5);
45      System.out.println("\nThe edges for graph2:");
46      graph2.printWeightedEdges();
47    }
48  }
```

```
The number of vertices in graph1: 12
The vertex with index 1 is San Francisco
The index for Miami is 9
The edges for graph1:
Vertex 0: (0, 1, 807) (0, 3, 1331) (0, 5, 2097)
Vertex 1: (1, 2, 381) (1, 0, 807) (1, 3, 1267)
Vertex 2: (2, 1, 381) (2, 3, 1015) (2, 4, 1663) (2, 10, 1435)
Vertex 3: (3, 4, 599) (3, 5, 1003) (3, 1, 1267)
   (3, 0, 1331) (3, 2, 1015)
Vertex 4: (4, 10, 496) (4, 8, 864) (4, 5, 533) (4, 2, 1663)
   (4, 7, 1260) (4, 3, 599)
Vertex 5: (5, 4, 533) (5, 7, 787) (5, 3, 1003)
   (5, 0, 2097) (5, 6, 983)
Vertex 6: (6, 7, 214) (6, 5, 983)
Vertex 7: (7, 6, 214) (7, 8, 888) (7, 5, 787) (7, 4, 1260)
Vertex 8: (8, 9, 661) (8, 10, 781) (8, 4, 864)
   (8, 7, 888) (8, 11, 810)
Vertex 9: (9, 8, 661) (9, 11, 1187)
Vertex 10: (10, 11, 239) (10, 4, 496) (10, 8, 781) (10, 2, 1435)
Vertex 11: (11, 10, 239) (11, 9, 1187) (11, 8, 810)

The edges for graph2:
Vertex 0: (0, 1, 2) (0, 3, 8)
Vertex 1: (1, 0, 2) (1, 2, 7) (1, 3, 3)
Vertex 2: (2, 3, 4) (2, 1, 7) (2, 4, 5)
Vertex 3: (3, 1, 3) (3, 4, 6) (3, 2, 4) (3, 0, 8)
Vertex 4: (4, 2, 5) (4, 3, 6)
```

该程序在第 3 ~ 27 行为图 29-1 中的图创建 graph1。第 3 ~ 5 行定义 graph1 的顶点。第 7 ~ 24 行定义 graph1 的边。边采用二维数组表示。对于数组中的每一行 i，edges[i][0] 和 edges[i][1] 表明存在一条由顶点 edges[i][0] 到顶点 edges[i][1] 的边，并且这条边的权重为 edges[i][2]。例如，{0,1,807}（第 8 行）表示由顶点 0(edges[0][0]) 到顶点 1(edges[0][1]) 的边，并且权重为 807(edges[0][2])。{0,5,2097}（第 8 行）表示由顶点 0(edges[2][0]) 到顶点 5(edges[2][1]) 的边，并且权重为 2097(edges[2][2])。第 35 行调用 graph1 中的方法 printWeightedEdges() 来显示 graph1 中的所有边。

该程序在第 37 ~ 44 行为图 29-3a 中的图 graph2 创建边。第 46 行调用 graph2 中的方法 printWeightedEdges() 来显示 graph2 中的所有边。

复习题

29.3.1 如果使用优先队列来存储加权边，下面代码的输出是什么？

```
PriorityQueue<WeightedEdge> q = new PriorityQueue<>();
q.offer(new WeightedEdge(1, 2, 3.5));
q.offer(new WeightedEdge(1, 6, 6.5));
q.offer(new WeightedEdge(1, 7, 1.5));
System.out.println(q.poll().weight);
System.out.println(q.poll().weight);
System.out.println(q.poll().weight);
```

29.3.2 如果使用优先队列来存储加权边，下面代码中有什么错误？修改错误并显示输出。

```
List<PriorityQueue<WeightedEdge>> queues = new ArrayList<>();
queues.get(0).offer(new WeightedEdge(0, 2, 3.5));
queues.get(0).offer(new WeightedEdge(0, 6, 6.5));
queues.get(0).offer(new WeightedEdge(0, 7, 1.5));
queues.get(1).offer(new WeightedEdge(1, 0, 3.5));
queues.get(1).offer(new WeightedEdge(1, 5, 8.5));
queues.get(1).offer(new WeightedEdge(1, 8, 19.5));
System.out.println(queues.get(0).peek()
   .compareTo(queues.get(1).peek()));
```

29.3.3 下面代码的输出是什么?

```java
public class Test {
  public static void main(String[] args) {
    WeightedGraph<Character> graph = new WeightedGraph<>();
    graph.addVertex('U');
    graph.addVertex('V');
    int indexForU = graph.getIndex('U');
    int indexForV = graph.getIndex('V');
    System.out.println("indexForU is " + indexForU);
    System.out.println("indexForV is " + indexForV);
    graph.addEdge(indexForU, indexForV, 2.5);
    System.out.println("Degree of U is " +
      graph.getDegree(indexForU));
    System.out.println("Degree of V is " +
      graph.getDegree(indexForV));
    System.out.println("Weight of UV is " +
      graph.getWeight(indexForU, indexOfV));
  }
}
```

29.4 最小生成树

要点提示：图的最小生成树是一棵具有最小总权重的生成树。

一个图可能有很多生成树。假设边具有权重，最小生成树拥有最小的权重和。例如，图29-5b、c、d中的树都是图29-5a中图的生成树。图29-3c和图29-3d中的树是最小生成树。

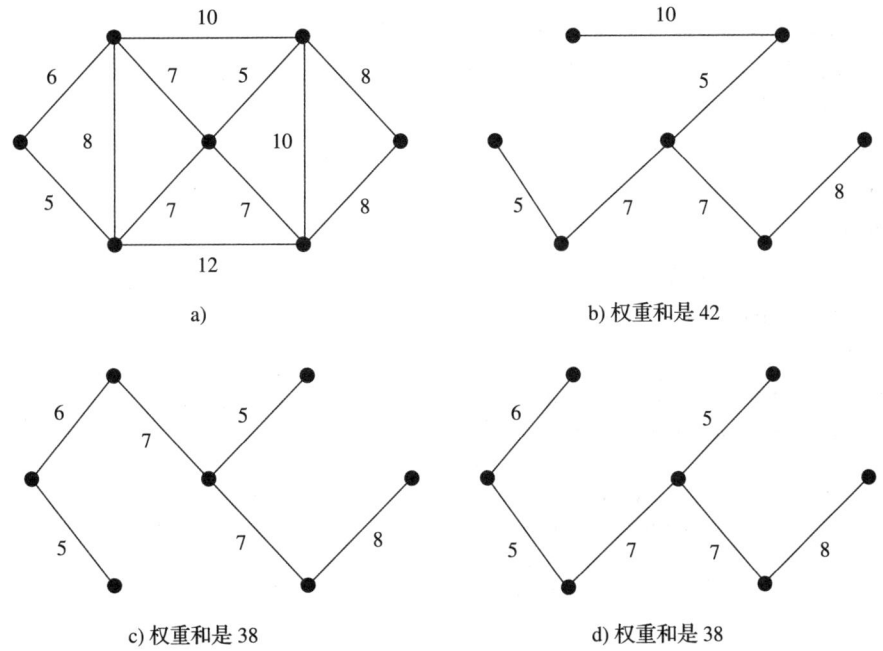

a)

b) 权重和是 42

c) 权重和是 38

d) 权重和是 38

图 29-5　c 和 d 中的树都是 a 中图的最小生成树

找到最小生成树的问题有很多应用。考虑一个在很多城市拥有分公司的公司，公司想要架设电话线来连接所有的分公司。电话公司对不同城市之间的连接要价不同。存在很多种方法可以将所有分公司连接在一起，最便宜的方法就是找出一棵费用总和最小的生成树。

29.4.1 最小生成树算法

如何找出最小生成树？对于这个问题，有几个著名的算法。本节将介绍 Prim 算法。Prim 算法从包含任意顶点的生成树 T 开始。算法通过反复添加与已经在树中的顶点相连的具有最短边的顶点来对树进行扩展。Prim 算法是一种贪婪算法，其描述在程序清单 29-4 中。

程序清单 29-4 Prim 的最小生成树算法

```
1   MST getMinimumSpanningTree(s) {
2     Let T be a set for the vertices in the spanning tree;
3     Initially, add the starting vertex, s, to T;
4
5     while (size of T < n) {
6       Find x in T and y in V - T with the smallest weight
7         on the edge (x, y), as shown in Figure 29.6;
8       Add y to T and set parent[y] = x;
9     }
10  }
```

该算法从将起始顶点添加到 T 中开始，然后持续地将顶点（比方说 y）从 V-T 添加到 T 中。y 是与 T 中顶点相邻的顶点中边权重最小的那个顶点。例如，存在 5 条边连接着 T 和 V-T 之间的顶点，如图 29-6 所示，其中 (x,y) 就是权重最小的那个。考虑图 29-7 中的图。算法以如下的顺序将顶点添加到 T 中：

1）将顶点 0 添加到 T 中。

2）将顶点 5 添加到 T 中，因为 WeightedEdge(5,0,5) 在所有与 T 中的顶点相连的边中具有最小权重，如图 29-7a 所示。从 0 指向 5 的箭头线表示 0 是 5 的父结点。

3）将顶点 1 添加到 T 中，因为 WeightedEdge(1,0,6) 在所有与 T 中的顶点相连的边中具有最小权重，如图 29-7b 所示。

4）将顶点 6 添加到 T 中，因为 WeightedEdge(6,1,7) 在所有与 T 中的顶点相连的边中具有最小权重，如图 29-7c 所示。

5）将顶点 2 添加到 T 中，因为 WeightedEdge(2,6,5) 在所有与 T 中的顶点相连的边中具有最小权重，如图 29-7d 所示。

6）将顶点 4 添加到 T 中，因为 WeightedEdge(4,6,7) 在所有与 T 中的顶点相连的边中具有最小权重，如图 29-7e 所示。

7）将顶点 3 添加到 T 中，因为 WeightedEdge(3,2,8) 在所有与 T 中的顶点相连的边中具有最小权重，如图 29-7f 所示。

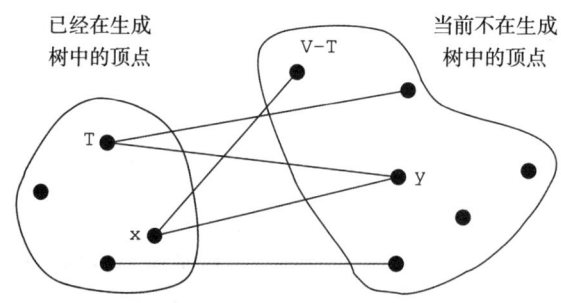

图 29-6 找到 T 中的顶点 x，该顶点以最小权重连接 V-T 中的顶点 y

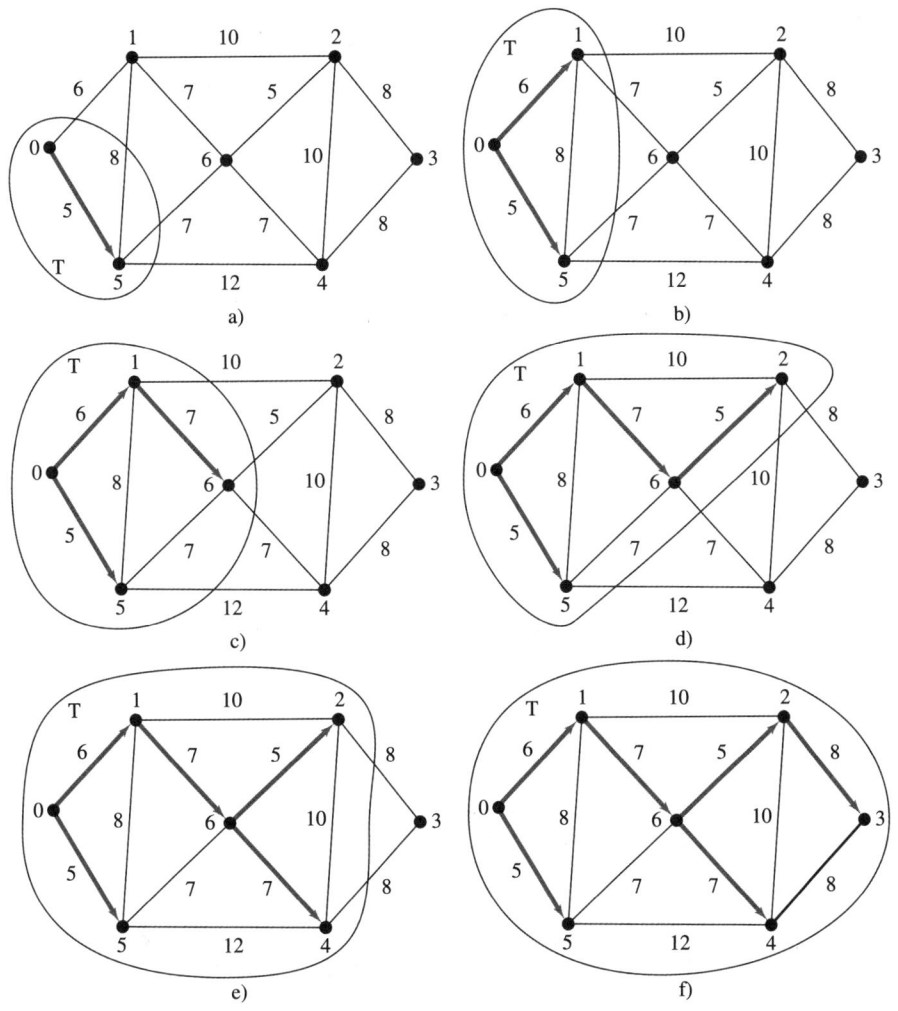

图 29-7 具有最小权重的邻接顶点被不断地添加到 T 中

> **注意**：最小生成树不是唯一的。例如，图 29-5 中的 c 和 d 都是图 29-5a 中图的最小生成树。然而，如果权重是不同的，则图只有唯一的最小生成树。

> **注意**：这里假设图是连通且无向的。如果一个图不是连通的或者是有向的，这里的算法是无效的。可以修改算法，为任何无向图找出生成森林。生成森林是一种图，该图中每个连通的组件是一棵树。

29.4.2 完善 Prim 的 MST 算法

为了易于确定加入树中的下一个顶点，使用 cost[v] 存储加入顶点 v 到生成树 T 中的开销。初始时，对于起始顶点来说 cost[s] 为 0，而对于其他顶点设置 cost[v] 值为无穷大。算法重复地在 V-T 中找到具有最小 cost[u] 的顶点 u，并将其移到 T 中。改进的算法在程序清单 29-5 中给出。

程序清单 29-5 改进的 Prim 算法

```
Input: A connected undirected weighted G = (V, E) with nonnegative weights
Output: a minimum spanning tree with the starting vertex s as the root
```

```
1  MST getMinimumSpanningTree(s) {
2    Let T be a set that contains the vertices in the spanning tree;
3    Initially T is empty;
4    Set cost[s] = 0 and cost[v] = infinity for all other vertices in V;
5
6    while (size of T < n) {
7      Find u not in T with the smallest cost[u];
8      Add u to T;
9      for (each v not in T and (u, v) in E)
10       if (cost[v] > w(u, v)) { // Adjust cost[v]
11         cost[v] = w(u, v); parent[v] = u;
12       }
13   }
14 }
```

参见位于网址 liveexample.pearsoncmg.com/dsanimation/RefinedPrim.html 的改进 Prim 算法的交互式演示。

29.4.3 MST 算法的实现

getMinimumSpanningTree(int v) 方法定义在 WeightedGraph 类中,参见图 29-4。它返回一个 MST 类的实例。MST 类定义为 WeightedGraph 类中的一个内部类,WeightedGraph 类继承自 SearchTree 类,如图 29-8 所示。SearchTree 类在图 28-11 中给出。MST 类在程序清单 29-2 中的第 141～153 行实现。

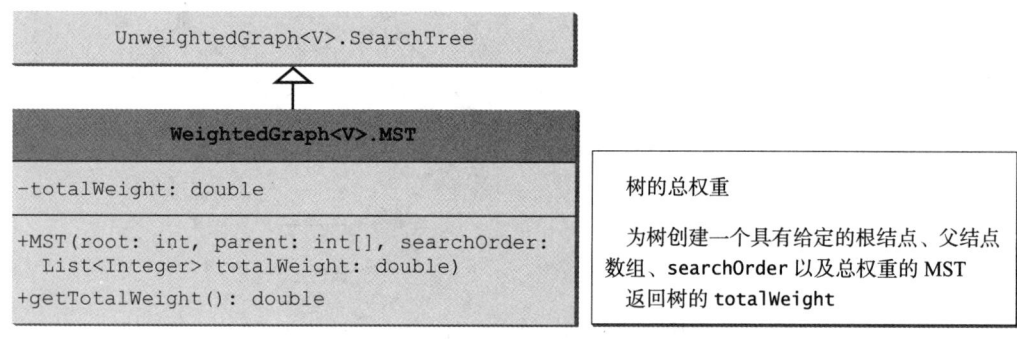

图 29-8 MST 类继承自 SearchTree 类

改进的 Prim 算法大大简化了实现。getMinimumSpanningTree 方法使用了改进的 Prim 算法实现,在程序清单 29-2 中的第 99～138 行中。getMinimumSpanningTree(int startingVertex) 方法设置 cost[startingVertex] 为 0(第 105 行),为其他顶点设置 cost[v] 为无穷大(第 102～104 行)。startingVertex 的父结点设为 -1(第 108 行)。T 是存储添加到生成树的顶点的线性表(第 111 行)。使用线性表而不是集合来表示 T 中是因为要记录加入 T 中的顶点的次序。

初始时 T 为空。为了扩充 T,该方法执行以下操作:

1) 找到具有最小 cost[u] 的顶点 u(第 118～123 行)。

2) 如果找到了 u,则将其加入 T 中(第 125 行)。注意,如果没有找到 u(u == -1),那么这个图是非连通的。在这种情况下,使用 break 语句退出 while 循环。

3) 添加 u 到 T 中后,如果 cost[v]>w(u,v),则对 V-T 中的顶点 u 的每个邻接顶点 v 更新 cost[v] 和 parent[v](第 129～134 行)。

在一个新顶点被添加到 T 中之后,更新 totalWeight(第 126 行)。一旦所有的顶点都被添加到 T 中，就创建了一个 MST 的实例（第 137 行）。注意，如果图不是连通的，那么这个方法就不起作用。然而，可以修改它来获得一个局部的 MST。

MST 类继承了 SearchTree 类（第 141 行）。为了创建一个 MST 实例，传递 root、parent、T 和 totalWeight（第 144～145 行）。数据域 root、parent 和 searchOrder 在 SearchTree 类中定义，SearchTree 类是定义在 UnweightedGraph 中的一个内部类。

注意，因为 T 是一个线性表，通过调用 T.contains(i) 检测顶点 i 是否在 T 中需要 $O(n)$ 的时间。因此，该实现的总体时间复杂度为 $O(n^3)$。有兴趣的读者可以参考编程练习题 29.20 来改善实现，降低复杂度到 $O(n^2)$。

程序清单 29-6 给出了一个测试程序，分别显示图 29-1 中图的最小生成树和图 29-3a 中的图。

程序清单 29-6 TestMinimumSpanningTree.java

```java
public class TestMinimumSpanningTree {
  public static void main(String[] args) {
    String[] vertices = {"Seattle", "San Francisco", "Los Angeles",
      "Denver", "Kansas City", "Chicago", "Boston", "New York",
      "Atlanta", "Miami", "Dallas", "Houston"};

    int[][] edges = {
      {0, 1, 807}, {0, 3, 1331}, {0, 5, 2097},
      {1, 0, 807}, {1, 2, 381}, {1, 3, 1267},
      {2, 1, 381}, {2, 3, 1015}, {2, 4, 1663}, {2, 10, 1435},
      {3, 0, 1331}, {3, 1, 1267}, {3, 2, 1015}, {3, 4, 599},
        {3, 5, 1003},
      {4, 2, 1663}, {4, 3, 599}, {4, 5, 533}, {4, 7, 1260},
        {4, 8, 864}, {4, 10, 496},
      {5, 0, 2097}, {5, 3, 1003}, {5, 4, 533},
        {5, 6, 983}, {5, 7, 787},
      {6, 5, 983}, {6, 7, 214},
      {7, 4, 1260}, {7, 5, 787}, {7, 6, 214}, {7, 8, 888},
      {8, 4, 864}, {8, 7, 888}, {8, 9, 661},
        {8, 10, 781}, {8, 11, 810},
      {9, 8, 661}, {9, 11, 1187},
      {10, 2, 1435}, {10, 4, 496}, {10, 8, 781}, {10, 11, 239},
      {11, 8, 810}, {11, 9, 1187}, {11, 10, 239}
    };

    WeightedGraph<String> graph1 =
      new WeightedGraph<>(vertices, edges);
    WeightedGraph<String>.MST tree1 = graph1.getMinimumSpanningTree();
    System.out.println("tree1: Total weight is " +
      tree1.getTotalWeight());
    tree1.printTree();

    edges = new int[][] {
      {0, 1, 2}, {0, 3, 8},
      {1, 0, 2}, {1, 2, 7}, {1, 3, 3},
      {2, 1, 7}, {2, 3, 4}, {2, 4, 5},
      {3, 0, 8}, {3, 1, 3}, {3, 2, 4}, {3, 4, 6},
      {4, 2, 5}, {4, 3, 6}
    };

    WeightedGraph<Integer> graph2 = new WeightedGraph<>(edges, 5);
    WeightedGraph<Integer>.MST tree2 =
      graph2.getMinimumSpanningTree(1);
    System.out.println("\ntree2: Total weight is " +
      tree2.getTotalWeight());
    tree2.printTree();
```

```
47
48          System.out.println("\nShow the search order for tree1:");
49          for (int i: tree1.getSearchOrder())
50            System.out.print(graph1.getVertex(i) + " ");
51       }
52  }
```

```
Total weight is 6513.0
Root is: Seattle
Edges: (Seattle, San Francisco) (San Francisco, Los Angeles)
  (Los Angeles, Denver) (Denver, Kansas City) (Kansas City, Chicago)
  (New York, Boston) (Chicago, New York) (Dallas, Atlanta)
  (Atlanta, Miami) (Kansas City, Dallas) (Dallas, Houston)

Total weight is 14.0
Root is: 1
Edges: (1, 0) (3, 2) (1, 3) (2, 4)

Show the search order for tree1:
Seattle San Francisco Los Angeles Denver Kansas City Dallas

Houston Chicago Atlanta Miami New York Boston
```

该程序在第 27 行为图 29-1 创建了一个加权图。接着它调用 getMinimumSpanningTree()（第 28 行）来返回一个表示图的最小生成树的 MST。调用 MST 对象的 printTree()（第 31 行）显示树中的边。注意，MST 是 SearchTree 类的子类。方法 printTree() 定义在 SearchTree 类中。

最小生成树的图示如图 29-9 所示。顶点以如下的顺序添加到树中：Seattle（西雅图）、San Francisco（圣弗朗西斯科）、Los Angeles（洛杉矶）、Denver（丹佛）、Kansas City（堪萨斯城）、Dallas（达拉斯）、Houston（休斯敦）、Chicago（芝加哥）、Atlanta（亚特兰大）、Miami（迈阿密）、New York（纽约）和 Boston（波士顿）。

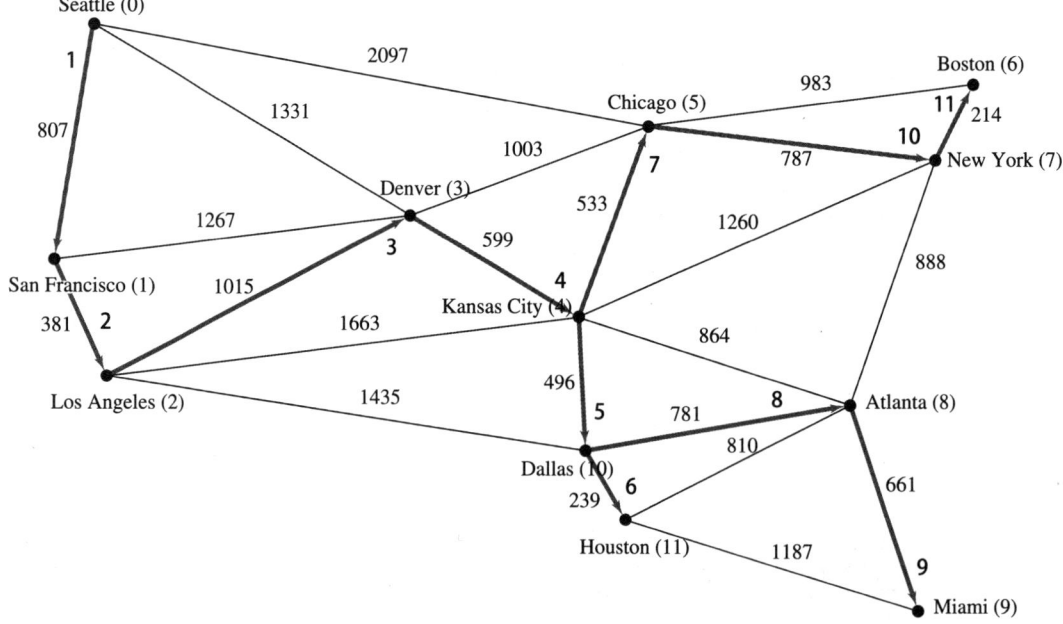

图 29-9　代表城市的最小生成树中的边在图中加粗显示

复习题

29.4.1 找出下图的一棵最小生成树。

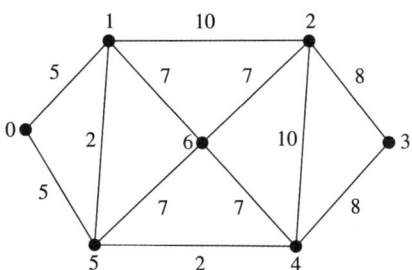

29.4.2 如果所有边的权重不同，那么最小生成树是唯一的吗？

29.4.3 如果使用邻接矩阵来表示加权边，Prim 算法的时间复杂度为多少？

29.4.4 如果图不是连通的，那么 WeightedGraph 中的方法 getMinimumSpanningTree() 将会怎样？通过编写一个测试程序，创建一个非连通图并且调用方法 getMinimumSpanningTree() 来验证你的答案。如何通过获取局部 MST 来解决这个问题？

29.4.5 给出以下代码的输出：

```java
public class Test {
  public static void main(String[] args) {
    WeightedGraph<Character> graph = new WeightedGraph<>();
    graph.addVertex('U');
    graph.addVertex('V');
    graph.addVertex('X');
    int indexForU = graph.getIndex('U');
    int indexForV = graph.getIndex('V');
    int indexForX = graph.getIndex('X');
    System.out.println("indexForU is " + indexForU);
    System.out.println("indexForV is " + indexForV);
    System.out.println("indexForX is " + indexForV);
    graph.addEdge(indexForU, indexForV, 3.5);
    graph.addEdge(indexForV, indexForU, 3.5);
    graph.addEdge(indexForU, indexForX, 2.1);
    graph.addEdge(indexForX, indexForU, 2.1);
    graph.addEdge(indexForV, indexForX, 3.1);
    graph.addEdge(indexForX, indexForV, 3.1);
    WeightedGraph<Character>.MST mst
      = graph.getMinimumSpanningTree();
    graph.printWeightedEdges();
    System.out.println(mst.getTotalWeight());
    mst.printTree();
  }
}
```

29.5 寻找最短路径

☞ 要点提示：两个顶点之间的最短路径，是指具有最小总权重的路径。

给定一个边的权重非负的图，一个著名的找出两个顶点间最短路径的算法是由荷兰计算机科学家 Edsger Dijkstra 发现的。为了找到从顶点 s 到顶点 v 的最短路径，Dijkstra 算法寻找从 s 到所有顶点的最短路径。因此 Dijkstra 的算法被称为单源最短路径算法。算法使用 cost[v] 来存储从顶点 v 到源顶点 s 的最短路径的开销。cost[s] 为 0。初始时，为所有其他顶点设置 cost[v] 为无穷大。该算法重复找出 V-T 中的一个具有最小 cost[u] 的顶点 u，并将 u 移到 T 中。

该算法在程序清单 29-7 中描述。

程序清单 29-7 Dijkstra 的单源最短路径算法

Input: a graph G = (V, E) with nonnegative weights
Output: a shortest-path tree with the source vertex s as the root

```
1   ShortestPathTree getShortestPath(s) {
2     Let T be a set that contains the vertices whose
3       paths to s are known; Initially T is empty;
4     Set cost[s] = 0; and cost[v] = infinity for all other vertices in V;
5
6     while (size of T < n) {
7       Find u not in T with the smallest cost[u];
8       Add u to T;
9       for (each v not in T and (u, v) in E)
10        if (cost[v] > cost[u] + w(u, v)) {
11          cost[v] = cost[u] + w(u, v); parent[v] = u;
12        }
13    }
14  }
```

这个算法与 Prim 的寻找最小生成树算法非常相似，它们都将顶点分为两个集合 T 和 V-T。在 Prim 的算法中，集合 T 包含已经添加到树中的顶点。在 Dijkstra 的算法中，集合 T 包含那些已经找到的与源顶点之间距离最短的顶点。这两种算法都重复地从 V-T 中寻找一个顶点，然后将其添加到 T 中。在 Prim 算法中，该顶点以最小权重的边邻接到集合中某个顶点。在 Dijkstra 算法中，该顶点邻接到集合中某个顶点并具有到源顶点的最小总开销。

该算法开始时将 cost[s] 设置为 0（第 4 行），并为所有其他顶点设置 cost[v] 为无穷大。然后不断地将顶点（称为 u）从 V-T 添加到 T 中，该顶点具有最小的 cost[u]（第 7～8 行），如图 29-10a 所示。在顶点 u 被添加到 T 中后，对于每个不在 T 中的 v 顶点，如果 (u,v) 在 T 中并且 cost[v] > cost[u] + w(u,v)，该算法更新 cost[v] 和 parent[v]（第 10～12 行）。

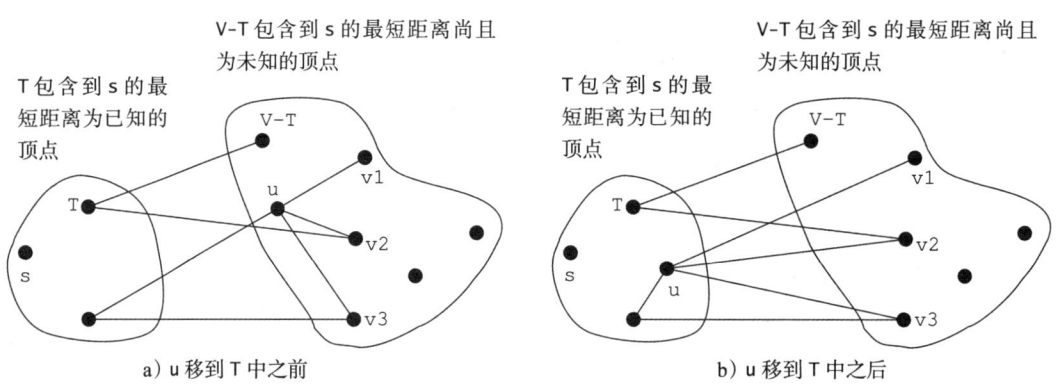

图 29-10 a) 在 V-T 中找到一个具有最小 cost[u] 的顶点 u；b) 为每个在 V-T 中并且和 u 邻接的顶点 v 更新 cost[v]

我们使用图 29-11a 中的图来解释 Dijkstra 算法。假设源顶点为顶点 1。因此，cost[1]=0，其他顶点的初始开销为 ∞，如图 29-11b 所示。我们使用 parent[i] 来表示路径中顶点 i 的父顶点。为方便起见，设置源结点的父结点为 -1。

初始时，设置集合 T 为空。该算法选择具有最小开销的顶点。这种情况下，顶点为 1。算法将 1 添加到 T 中，如图 29-12a 所示。之后，算法为每个和 1 相邻的顶点调整开销值。现在顶点 2、0、6 和 3 的开销，以及它们的父顶点被更新，如图 29-12b 所示。

顶点 2、0、6 和 3 与源顶点相邻，而顶点 2 为 V-T 中具有最小开销的顶点，于是将顶点 2 添加到 T 中，如图 29-13 所示。更新 V-T 中与 2 相邻的顶点的开销及其父顶点。cost[0] 现在更新为 6，并且其父顶点设为 2。从 1 到 2 的箭头线表明 2 添加到 T 中之后，1 是 2 的父顶点。

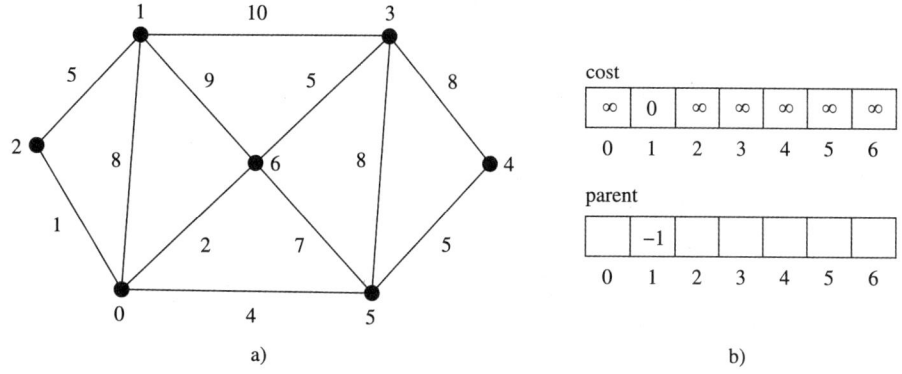

图 29-11 算法将找到从源顶点 1 开始的所有最短路径

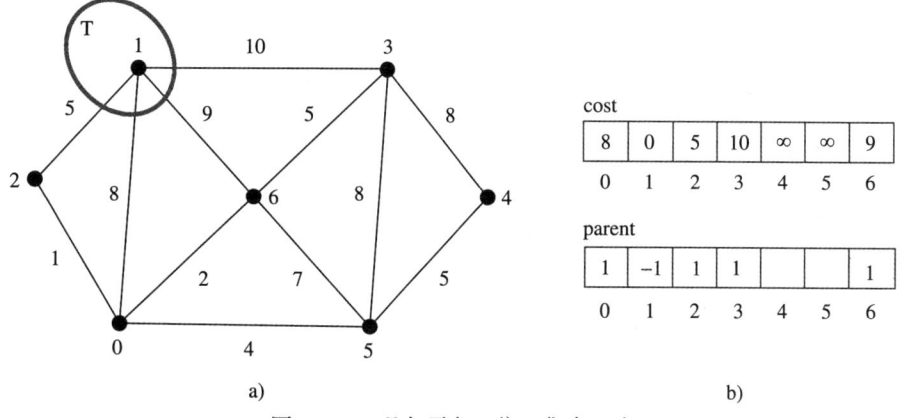

图 29-12 现在顶点 1 位于集合 T 中

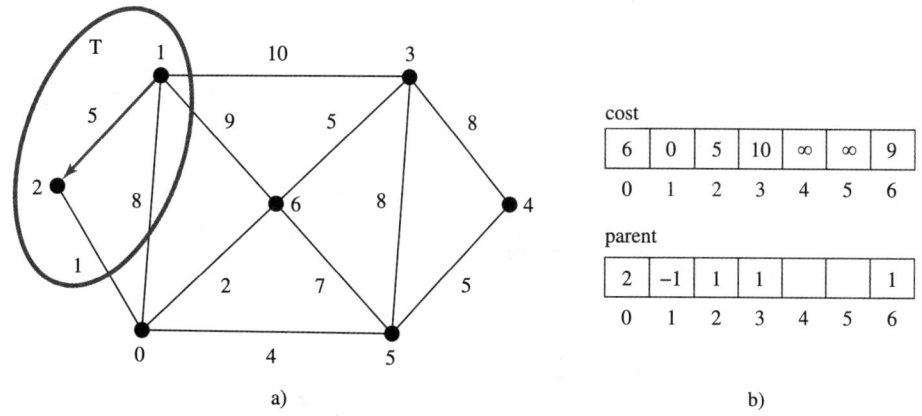

图 29-13 现在顶点 1 和 2 位于集合 T 中

现在，T 包含 {1,2}。在 V-T 中，顶点 0 具有到源顶点 1 的开销最小的路径，于是将顶点 0 添加到 T 中，如图 29-14 所示。如果可行的话，更新 V-T 中与 0 相邻的顶点的开销和父顶点。cost[5] 现在更新为 10，并且它的父顶点设为 0；cost[6] 现在更新为 8，并且它的父顶点设为 0。

现在，T 包含 {1,2,0}。在 V-T 中，顶点 6 具有到源顶点 1 的开销最小的路径，于是将顶点 6 添加到 T 中，如图 29-15 所示。更新 V-T 中与 6 相邻的顶点的开销和父顶点。

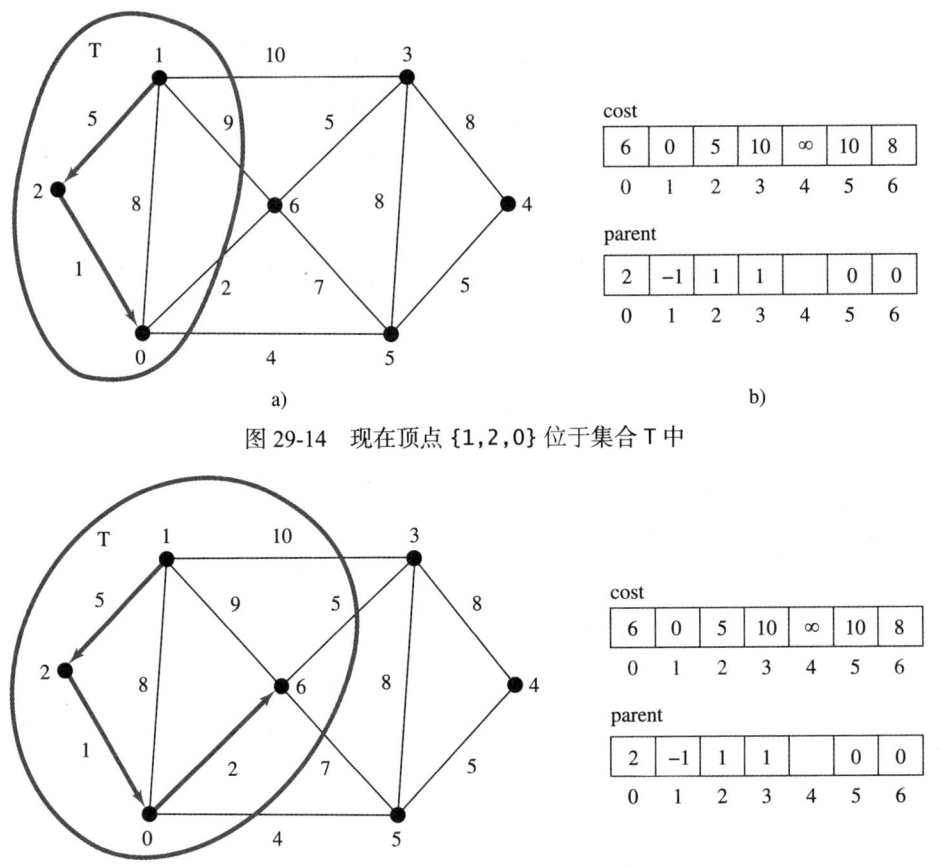

图 29-14 现在顶点 {1,2,0} 位于集合 T 中

图 29-15 现在顶点 {1,2,0,6} 位于集合 T 中

现在，T 包含 {1,2,0,6}。在 V-T 中，顶点 3 或者 5 具有最小开销。既可以选择顶点 3，也可以选择顶点 5 放到 T 中。我们将顶点 3 添加到 T 中，如图 29-16 所示。如果可行的话，更新 V-T 中与 3 相邻的顶点的开销和其父顶点。cost[4] 现在更新为 18，并且它的父顶点设为 3。

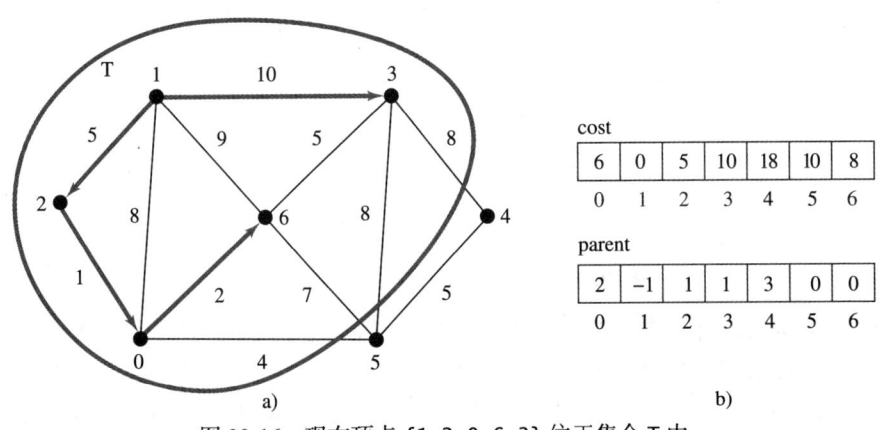

图 29-16 现在顶点 {1,2,0,6,3} 位于集合 T 中

现在，T 包含 {1,2,0,6,3}。在 V-T 中，顶点 5 具有最小开销，因此将顶点 5 添加到 T 中，如图 29-17 所示。更新 V-T 中与 5 相邻的顶点的开销和其父顶点。如果可行的话，cost[4] 现在更新为 15，并且它的父顶点设为 5。

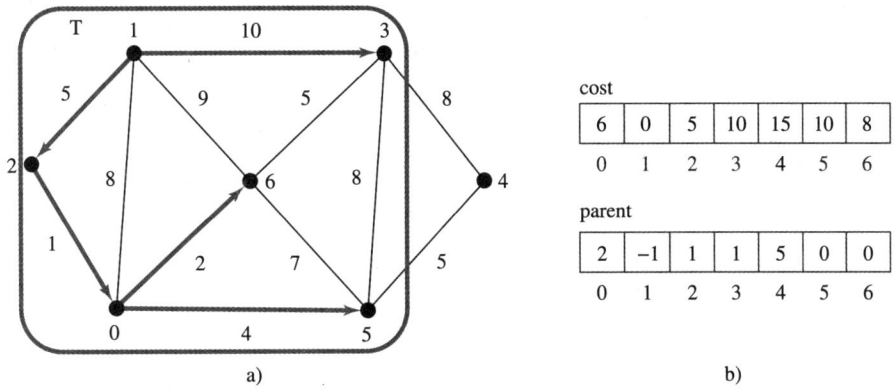

图 29-17 现在顶点 {1,2,0,6,3,5} 位于集合 T 中（来源：Oracle 或其附属公司版权所有 © 1995 ～ 2016，经授权使用）

现在，T 包含 {1,2,0,6,3,5}。在 V-T 中，顶点 4 具有最小开销，因此将顶点 4 添加到 T 中，如图 29-18 所示。

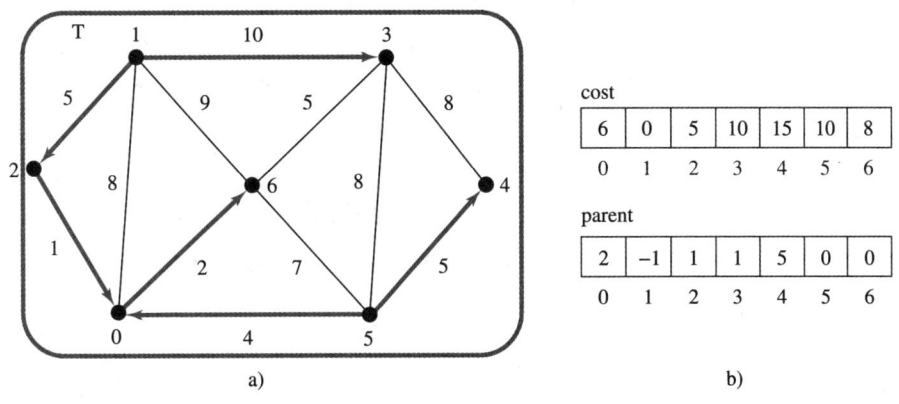

图 29-18 现在顶点 {1,2,0,6,3,5,4} 位于集合 T 中（来源：Oracle 或其附属公司版权所有 © 1995 ～ 2016，经授权使用）

正如你所看到的，该算法本质上就是找出从源顶点出发的所有最短路径，它将产生一个以源顶点为根结点的树。我们称这棵树为单源所有最短路径树（或简称最短路径树）。为了对这棵树建模，定义一个继承自 SearchTree 类的名为 ShortestPathTree 的类，如图 29-19 所示。ShortestPathTree 在程序清单 29-2 的第 200 ～ 224 行中定义为 WeightedGraph 的一个内部类。

getShortestPath(int sourceVertex) 方法在程序清单 29-2 的第 156 ～ 197 行中实现。该方法设置 cost[sourceVertex] 为 0（第 162 行），其他顶点设置 cost[v] 为无穷大（第 159 ～ 161 行）。sourceVertex 的父顶点设置为 -1（第 166 行）。T 为存储添加到最短路径树的顶点的线性表（第 169 行）。为 T 使用线性表而不是集合是为了记录顶点添加到 T 中的次序。

图 29-19 WeightedGraph<V>.ShortestPathTree 继承自 UnweightGraph<V>.SearchTree
（来源：Oracle 或其附属公司版权所有 © 1995 ～ 2016，经授权使用）

初始时，T 为空。为了扩充 T，该方法执行以下操作：

1）找到具有最小 cost[u] 值的顶点 u（第 175 ～ 181 行）。

2）如果找到了 u，则将其添加到 T 中（第 183 行）。注意，如果找不到 u（u == -1），那么该图是非连通的。在这种情况下，使用 break 语句退出 while 循环。

3）将 u 添加到 T 后，对于 V-T 中每个和 u 相邻的 v 顶点，如果 cost[v] > cost[u] + w(u,v)，则更新 cost[v] 和 parent[v]（第 186 ～ 192 行）。

一旦 s 的所有顶点都添加到 T 中，就会创建一个 ShortestPathTree 的实例（第 196 行）。ShortestPathTree 类继承自 SearchTree 类（第 200 行）。为了创建一个 ShortestPathTree 的实例，传递 sourceVertex、parent、T 和 cost（第 204 ～ 205 行）。sourceVertex 成为树的根结点。数据域 root、parent 和 searchOrder 在 SearchTree 类中定义，SearchTree 类是定义在 UnweightedGraph 中的一个内部类。

注意，因为 T 是一个线性表，所以通过调用 T.contains(i) 检测顶点 i 是否在 T 中需要 $O(n)$ 的时间。因此，该实现的总体时间复杂度为 $O(n^3)$。有兴趣的读者可以参考编程练习题 29.20 来改进实现，降低复杂度到 $O(n^2)$。

Dijkstra 算法是贪婪算法和动态编程的结合。它总是添加到源顶点具有最短距离的新顶点，从这个意义上来说，它是一种贪婪算法。它存储每个已知顶点到源顶点的最短距离，并使用它避免之后的重复计算，因此 Dijkstra 算法也使用了动态编程。

程序清单 29-8 给出了一个测试程序，分别显示图 29-1 中从 Chicago（芝加哥）出发到所有其他城市的最短路径，以及图 29-3a 中从顶点 3 到所有顶点的最短路径。

程序清单 29-8 TestShortestPath.java

```
1  public class TestShortestPath {
2    public static void main(String[] args) {
3      String[] vertices = {"Seattle", "San Francisco", "Los Angeles",
4        "Denver", "Kansas City", "Chicago", "Boston", "New York",
5        "Atlanta", "Miami", "Dallas", "Houston"};
6
7      int[][] edges = {
8        {0, 1, 807}, {0, 3, 1331}, {0, 5, 2097},
9        {1, 0, 807}, {1, 2, 381}, {1, 3, 1267},
10       {2, 1, 381}, {2, 3, 1015}, {2, 4, 1663}, {2, 10, 1435},
11       {3, 0, 1331}, {3, 1, 1267}, {3, 2, 1015}, {3, 4, 599},
12         {3, 5, 1003},
13       {4, 2, 1663}, {4, 3, 599}, {4, 5, 533}, {4, 7, 1260},
```

```
14            {4, 8, 864}, {4, 10, 496},
15          {5, 0, 2097}, {5, 3, 1003}, {5, 4, 533},
16            {5, 6, 983}, {5, 7, 787},
17          {6, 5, 983}, {6, 7, 214},
18          {7, 4, 1260}, {7, 5, 787}, {7, 6, 214}, {7, 8, 888},
19          {8, 4, 864}, {8, 7, 888}, {8, 9, 661},
20            {8, 10, 781}, {8, 11, 810},
21          {9, 8, 661}, {9, 11, 1187},
22          {10, 2, 1435}, {10, 4, 496}, {10, 8, 781}, {10, 11, 239},
23          {11, 8, 810}, {11, 9, 1187}, {11, 10, 239}
24        };
25
26        WeightedGraph<String> graph1 =
27          new WeightedGraph<>(vertices, edges);
28        WeightedGraph<String>.ShortestPathTree tree1 =
29          graph1.getShortestPath(graph1.getIndex("Chicago"));
30        tree1.printAllPaths();
31
32        // Display shortest paths from Houston to Chicago
33        System.out.print("Shortest path from Houston to Chicago: ");
34        java.util.List<String> path
35          = tree1.getPath(graph1.getIndex("Houston"));
36        for (String s: path) {
37          System.out.print(s + " ");
38        }
39
40        edges = new int[][] {
41          {0, 1, 2}, {0, 3, 8},
42          {1, 0, 2}, {1, 2, 7}, {1, 3, 3},
43          {2, 1, 7}, {2, 3, 4}, {2, 4, 5},
44          {3, 0, 8}, {3, 1, 3}, {3, 2, 4}, {3, 4, 6},
45          {4, 2, 5}, {4, 3, 6}
46        };
47        WeightedGraph<Integer> graph2 = new WeightedGraph<>(edges, 5);
48        WeightedGraph<Integer>.ShortestPathTree tree2 =
49          graph2.getShortestPath(3);
50        System.out.println("\n");
51        tree2.printAllPaths();
52      }
53    }
```

```
All shortest paths from Chicago are:
A path from Chicago to Seattle: Chicago Seattle (cost: 2097.0)
A path from Chicago to San Francisco: Chicago Denver San Francisco
  (cost: 2270.0)
A path from Chicago to Los Angeles: Chicago Denver Los Angeles
  (cost: 2018.0)
A path from Chicago to Denver: Chicago Denver (cost: 1003.0)
A path from Chicago to Kansas City: Chicago Kansas City (cost: 533.0)
A path from Chicago to Chicago: Chicago (cost: 0.0)
A path from Chicago to Boston: Chicago Boston (cost: 983.0)
A path from Chicago to New York: Chicago New York (cost: 787.0)
A path from Chicago to Atlanta: Chicago Kansas City Atlanta
  (cost: 1397.0)
A path from Chicago to Miami:
  Chicago Kansas City Atlanta Miami (cost: 2058.0)
A path from Chicago to Dallas:
  Chicago Kansas City Dallas (cost: 1029.0)
A path from Chicago to Houston:
  Chicago Kansas City Dallas Houston (cost: 1268.0)
Shortest path from Houston to Chicago:
  Houston Dallas Kansas City Chicago

All shortest paths from 3 are:
A path from 3 to 0: 3 1 0 (cost: 5.0)
```

```
A path from 3 to 1: 3 1 (cost: 3.0)
A path from 3 to 2: 3 2 (cost: 4.0)
A path from 3 to 3: 3 (cost: 0.0)
A path from 3 to 4: 3 4 (cost: 6.0)
```

该程序在第 27 行为图 29-1 创建了一个加权图。然后，调用方法 getShortestPath(graph1.getIndex("Chicago")) 来返回一个 Path 对象，该对象包含从芝加哥出发的所有最短路径。调用 ShortestPathTree 对象上的 printAllPaths() 显示所有的路径（第 30 行）。

图 29-20 展示了从芝加哥出发的所有最短路径。从芝加哥出发到其他城市的最短路径按以下顺序被找到：Kansas City（堪萨斯城），New York（纽约），Boston（波士顿），Denver（丹佛），Dallas（达拉斯），Houston（休斯敦），Atlanta（亚特兰大），Los Angeles（洛杉矶），Miami（迈阿密），Seattle（西雅图）和 San Francisco（圣弗朗西斯科）。

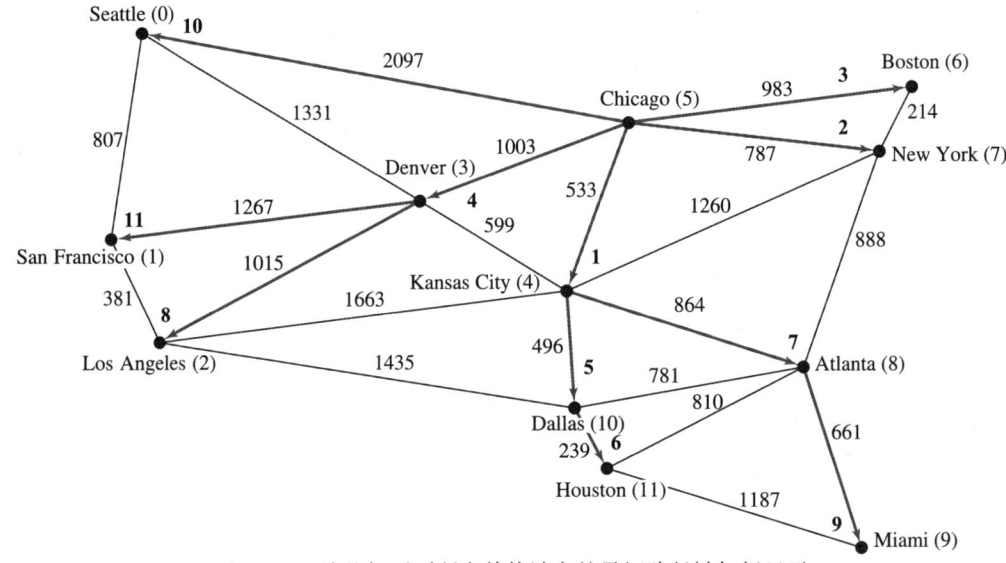

图 29-20 从芝加哥到所有其他城市的最短路径被加粗显示

✓ 复习题

29.5.1 追踪求解图 29-1 中从波士顿到所有其他城市的最短路径的 Dijkstra 算法。

29.5.2 如果所有的边都有不同的权重，那么两个顶点之间的最短路径是唯一的吗？

29.5.3 如果使用邻接矩阵来表示加权边，Dijkstra 算法的时间复杂度是多少？

29.5.4 如果源顶点不能到达图中的所有顶点，那么运行 WeightedGraph 中的 getShortestPath() 方法将会怎样？编写一个测试程序，创建一个非连通图并且调用 getShortestPath() 方法来验证你的答案。如何通过获得一个局部的最短路径树来修改这个问题？

29.5.5 如果没有从顶点 v 到源顶点的路径，cost[v] 将是什么？

29.5.6 假设图是连通的；如果删除掉 WeightedGraph 中的第 159～161 行，getShortestPath 方法可以正确地找到最短路径吗？

29.5.7 给出下列代码的输出：

```java
public class Test {
  public static void main(String[] args) {
    WeightedGraph<Character> graph = new WeightedGraph<>();
    graph.addVertex('U');
    graph.addVertex('V');
```

```java
      graph.addVertex('X');
      int indexForU = graph.getIndex('U');
      int indexForV = graph.getIndex('V');
      int indexForX = graph.getIndex('X');
      System.out.println("indexForU is " + indexForU);
      System.out.println("indexForV is " + indexForV);
      System.out.println("indexForX is " + indexForV);
      graph.addEdge(indexForU, indexForV, 3.5);
      graph.addEdge(indexForV, indexForU, 3.5);
      graph.addEdge(indexForU, indexForX, 2.1);
      graph.addEdge(indexForX, indexForU, 2.1);
      graph.addEdge(indexForV, indexForX, 3.1);
      graph.addEdge(indexForX, indexForV, 3.1);
      WeightedGraph<Character>.ShortestPathTree tree =
        graph.getShortestPath(1);
      graph.printWeightedEdges();
      tree.printTree();
   }
}
```

29.6 示例学习：加权的 9 枚硬币反面问题

要点提示：加权的 9 枚硬币反面问题可以简化为加权最短路径问题。

28.10 节中给出了 9 枚硬币反面问题，并且使用广度优先搜索算法解决了这个问题。本节中给出这个问题的变体，并使用最短路径算法解决它。

9 枚硬币反面的问题就是找出使所有的硬币正面朝下的最小数目的移动。每次移动时，翻转一个正面朝上的硬币及其邻居。加权的 9 枚硬币反面问题在每次移动上将翻转的次数指定为权重。例如，可以通过翻转第一行的第一枚硬币和它的两个邻居，将图 29-21a 中的硬币转移成图 29-21b 中的状态，因此这次移动的权重为 3。可以通过翻转位于中央的硬币和它的 4 个邻居，将图 29-21c 中的硬币转移成图 29-21d 中的状态，因此这次移动的权重为 5。

H	H	H		T	T	H		T	T	H		T	H	H
T	T	T		H	T	T		H	H	T		T	T	H
H	H	H		H	H	H		H	H	H		H	T	H
a)				b)				c)				d)		

图 29-21 每次移动的权重为该移动的翻转硬币数目

加权的 9 枚硬币反面问题可以简化为在一个边加权图中找出从一个起始结点到目标结点的最短路径。这个图包含 512 个结点。如果存在一个从结点 u 到结点 v 的转移，那么创建一条从结点 v 到结点 u 的边，将翻转的次数指定为边的权重。

回顾一下，在 28.10 节我们定义了一个 `NineTailModel` 类来对 9 枚硬币反面的问题进行建模。现在，我们定义一个名为 `WeightedNineTailModel` 的新类继承自 `NineTailModel`，如图 29-22 所示。

`NineTailModel` 类创建一个 `Graph` 并且获取一个以目标结点 511 为根结点的 `Tree`。除创建了一个 `WeightedGraph` 和获取一个以目标结点 511 为根结点的 `ShortestPathTree` 外，`WeightedNineTailModel` 与 `NineTailModel` 是一样的。`getEdges()` 方法找出图中所有的边。`getNumberOfFlips(int u, int v)` 方法返回从结点 u 到结点 v 的翻转次数。`getNumberOfFlips(int u)` 方法返回从结点 u 到目标结点的翻转次数。

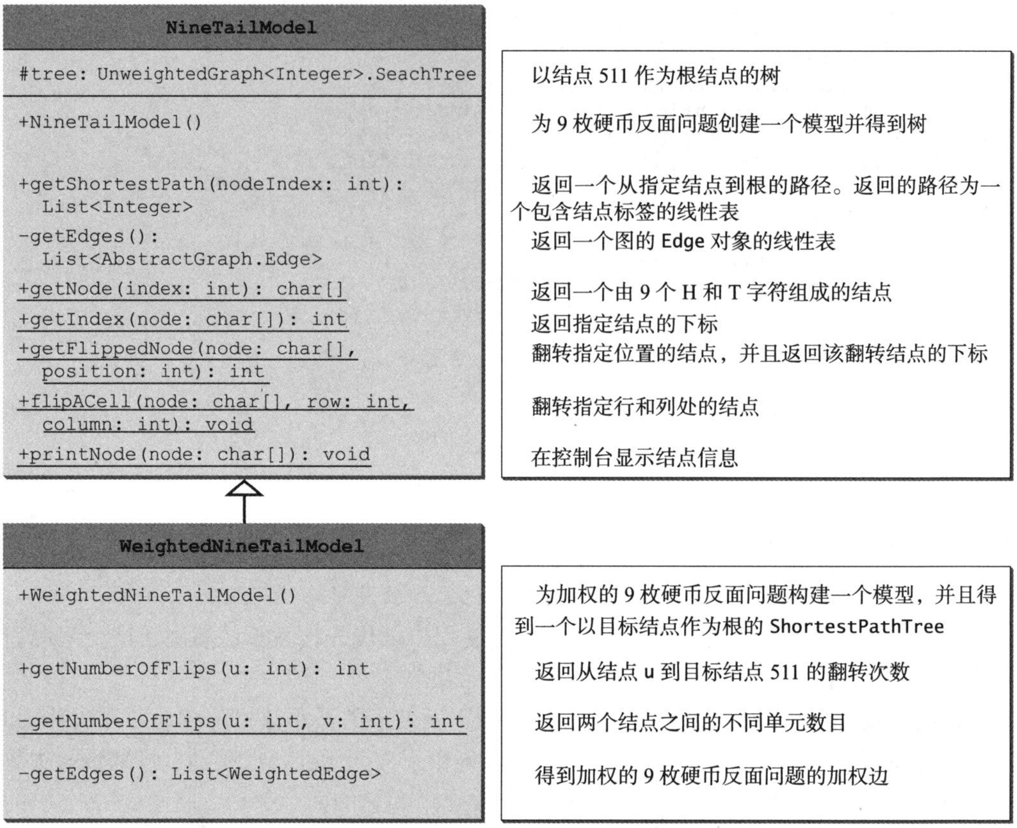

图 29-22 WeightedNineTailModel 类继承自 NineTailModel

NineTailModel
#tree: UnweightedGraph<Integer>.SeachTree
+NineTailModel()
+getShortestPath(nodeIndex: int): List<Integer>
-getEdges(): List<AbstractGraph.Edge>
+getNode(index: int): char[]
+getIndex(node: char[]): int
+getFlippedNode(node: char[], position: int): int
+flipACell(node: char[], row: int, column: int): void
+printNode(node: char[]): void

说明:
- 以结点 511 作为根结点的树
- 为 9 枚硬币反面问题创建一个模型并得到树
- 返回一个从指定结点到根的路径。返回的路径为一个包含结点标签的线性表
- 返回一个图的 Edge 对象的线性表
- 返回一个由 9 个 H 和 T 字符组成的结点
- 返回指定结点的下标
- 翻转指定位置的结点，并且返回该翻转结点的下标
- 翻转指定行和列处的结点
- 在控制台显示结点信息

WeightedNineTailModel
+WeightedNineTailModel()
+getNumberOfFlips(u: int): int
-getNumberOfFlips(u: int, v: int): int
-getEdges(): List<WeightedEdge>

说明:
- 为加权的 9 枚硬币反面问题构建一个模型，并且得到一个以目标结点作为根的 ShortestPathTree
- 返回从结点 u 到目标结点 511 的翻转次数
- 返回两个结点之间的不同单元数目
- 得到加权的 9 枚硬币反面问题的加权边

图 29-22 WeightedNineTailModel 类继承自 NineTailModel

程序清单 29-9 实现了 WeightedNineTailModel。

程序清单 29-9 WeightedNineTailModel.java

```java
import java.util.*;

public class WeightedNineTailModel extends NineTailModel {
  /** Construct a model */
  public WeightedNineTailModel() {
    // Create edges
    List<WeightedEdge> edges = getEdges();

    // Create a graph
    WeightedGraph<Integer> graph = new WeightedGraph<Integer>(
      edges, NUMBER_OF_NODES);

    // Obtain a shortest-path tree rooted at the target node
    tree = graph.getShortestPath(511);
  }

  /** Create all edges for the graph */
  private List<WeightedEdge> getEdges() {
    // Store edges
    List<WeightedEdge> edges = new ArrayList<>();

    for (int u = 0; u < NUMBER_OF_NODES; u++) {
      for (int k = 0; k < 9; k++) {
        char[] node = getNode(u); // Get the node for vertex u
        if (node[k] == 'H') {
          int v = getFlippedNode(node, k);
```

```
27          int numberOfFlips = getNumberOfFlips(u, v);
28
29          // Add edge (v, u) for a legal move from node u to node v
30          edges.add(new WeightedEdge(v, u, numberOfFlips));
31        }
32      }
33    }
34
35    return edges;
36  }
37
38  private static int getNumberOfFlips(int u, int v) {
39    char[] node1 = getNode(u);
40    char[] node2 = getNode(v);
41
42    int count = 0; // Count the number of different cells
43    for (int i = 0; i < node1.length; i++)
44      if (node1[i] != node2[i]) count++;
45
46    return count;
47  }
48
49  public int getNumberOfFlips(int u) {
50    return (int)((WeightedGraph<Integer>.ShortestPathTree)tree)
51      .getCost(u);
52  }
53 }
```

WeightedNineTailModel 继承自 NineTailModel，创建了一个 WeightedGraph 对加权的 9 枚硬币反面问题进行建模（第 10～11 行）。对于每个结点 u，getEdges() 方法找出一个被翻转的结点 v，然后将翻转的次数指定为边 (v,u) 的权重（第 30 行）。getNumberOfFlips(int u,int v) 方法返回从结点 u 到结点 v 的翻转次数（第 38～47 行）。翻转次数是指两个结点之间的不同格子的个数（第 44 行）。

WeightedNineTailModel 获取一个以目标结点 511 为根结点的 ShortestPathTree（第 14 行）。注意，tree 是一个定义在 NineTailModel 中的被保护的数据域，并且 ShortestPathTree 是 Tree 的子类。NineTailModel 中定义的方法使用属性 tree。

getNumberOfFlips(int u) 方法（第 49～52 行）返回从结点 u 到目标结点的翻转次数，即从结点 u 到目标结点的路径开销。可以调用定义在 ShortestPathTree 类中的方法 getCost(u) 得到这个开销（第 51 行）。

程序清单 29-10 给出了一个程序，提示用户输入一个初始结点并且显示到达目标结点的最小翻转次数。

程序清单 29-10 WeightedNineTail.java

```
1  import java.util.Scanner;
2
3  public class WeightedNineTail {
4    public static void main(String[] args) {
5      // Prompt the user to enter the nine coins' Hs and Ts
6      System.out.print("Enter an initial nine coins' Hs and Ts: ");
7      Scanner input = new Scanner(System.in);
8      String s = input.nextLine();
9      char[] initialNode = s.toCharArray();
10
11     WeightedNineTailModel model = new WeightedNineTailModel();
12     java.util.List<Integer> path =
13       model.getShortestPath(NineTailModel.getIndex(initialNode));
14
15     System.out.println("The steps to flip the coins are ");
```

```
16      for (int i = 0; i < path.size(); i++)
17        NineTailModel.printNode(NineTailModel.getNode(path.get(i)));
18
19      System.out.println("The number of flips is " +
20        model.getNumberOfFlips(NineTailModel.getIndex(initialNode)));
21    }
22  }
```

```
Enter an initial nine coins Hs and Ts: HHHTTTHHH  ↵Enter
The steps to flip the coins are
HHH
TTT
HHH

HHH
THT
TTT

TTT
TTT
TTT

The number of flips is 8
```

该程序在第8行提示用户将一个由9个H和T字母组成的字符串作为初始结点输入，从该字符串获取一个字符数组（第9行），创建一个模型（第11行），获取从初始结点到目标结点的最短路径（第12～13行），显示路径中的结点（第16～17行），并调用 `getNumberOfFlips` 来获取到达目标结点所需的翻转次数（第20行）。

✓ 复习题

29.6.1 为什么程序清单28-13中 `NineTailModel` 的 `tree` 数据域定义为受保护的？

29.6.2 `WeightedNineTailModel` 中是如何创建图的结点的？

29.6.3 `WeightedNineTailModel` 中是如何创建图的边的？

关键术语

Dijkstra'algorithm（Dijkstra算法）　　　　　shortest path（最短路径）
edge-weighted graph（边加权图）　　　　　　single-source shortest path（单源最短路径）
minimum spanning tree（最小生成树）　　　　vertex-weighted graph（顶点加权图）
Prim's algorithm（Prim算法）

本章小结

1. 可以使用邻接矩阵或者线性表来存储图中的加权边。
2. 图的生成树是一个树结构的子图，并连接着图中所有的顶点。
3. Prim算法找出最小生成树的工作机制如下：算法首先从包含任意一个结点的生成树T开始。算法通过添加与已在树中的顶点具有最小权重边的结点来扩展这棵树。
4. Dijkstra算法从源顶点开始搜索，然后不断寻找与源顶点具有最短路径的结点，直到所有结点被找到。

测试题

回答位于本书配套网站上的本章测试题。

编程练习题

*29.1 （Kruskal 算法）本书中介绍了找出最小生成树的 Prim 算法。Kruskal 算法是另一个找出最小生成树的著名算法。该算法重复地找出最小权重边，如果不会形成环，就将它添加到树中。当所有顶点都在树中时，终止这个过程。使用 Kruskal 算法设计和实现一个找出 MST 的算法。

*29.2 （使用邻接矩阵实现 Prim 算法）教材在邻接边上使用线性表实现 Prim 算法。对于加权图，使用邻接矩阵实现该算法。

*29.3 （使用邻接矩阵实现 Dijkstra 算法）教材在邻接边上使用线性表来实现 Dijkstra 算法。对于加权图，使用邻接矩阵实现该算法。

*29.4 （修改 9 枚硬币反面问题中的权重）教材中我们将翻转的次数作为每次移动的权重。假设权重是翻转次数的 3 倍，修改这个程序。

*29.5 （证明或反证）猜想 NineTailModel 和 WeightedNineTailModel 可能会得到相同的最短路径。编写程序去证明或者反证这个观点。（提示：令 tree1 和 tree2 分别表示从 NineTailModel 和 WeightedNineTailModel 获取的根结点为 511 的树。如果一个结点 u 在 tree1 中的深度和在 tree2 中的深度一样，那么，从结点 u 到目标结点的路径长度是相同的。）

**29.6 （加权 4×4 的 16 枚硬币反面问题的模型）教材中加权的 9 枚硬币反面问题使用的是 3×3 的矩阵。假设有 16 枚放在 4×4 的矩阵中的硬币。创建一个名为 WeightedTailModel16 的新的模型类，然后创建模型的一个实例并且将这个对象存入一个名为 WeightedTailModel16.dat 的文件中。

**29.7 （加权 4×4 的 16 枚硬币反面问题）为加权 4×4 的 16 枚硬币反面的问题修改程序清单 29-9。你的程序应该读取前一个编程练习题创建的模型对象。

**29.8 （旅行商人问题）旅行商人问题（traveling salesman problem, TSP）就是找出往返的最短路径，使得可以访问每个城市一次且只能访问一次，最后返回到起始城市。这个问题等价于编程练习题 28.17 中的寻找一条最短的哈密尔顿环。在 WeightedGraph 类中添加下面的方法：

```
// Return a shortest cycle
// Return null if no such cycle exists
public List<Integer> getShortestHamiltonianCycle()
```

*29.9 （得到最小生成树）编写一个程序，从文件中读取一个连通图并且显示它的最小生成树。文件中的第一行是表明顶点个数（n）的数字。顶点被标记为 0,1,…,n-1。接下来的每一行以 u1,v1,w1|u2,v2,w2|… 的形式来描述边。采用这种形式的每个三元组描述一条边和它的权重。图 29-23 显示了与图对应的文件的例子。注意，我们假设图是无向的。如果图有一条边 (u,v)，那么它也有一条边 (v,u)，但文件中只表示了一条边。当构建一个图时，两条边都需要添加。

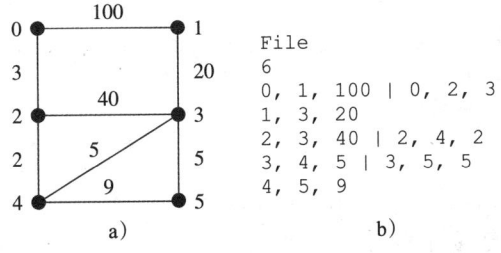

图 29-23 加权图的顶点和边可以存储在一个文件中

你的程序应该提示用户输入文件的 URL，然后从文件中读取数据，创建 WeightedGraph 的一个实例 g，调用 g.printWeightedEdges() 来显示所有的边，调用 getMinimumSpanningTree() 来获取一个 WeightedGraph.MST 的实例 tree，调用 tree.getTotalWeight() 来显示最小生成树的权重，以及调用 tree.printTree() 来显示这棵树。下面是这个程序的运行示例：

```
Enter a URL:
   https://liveexample.pearsoncmg.com/test/WeightedGraphSample.txt  ↵Enter
The number of vertices is 6
Vertex 0: (0, 2, 3) (0, 1, 100)
Vertex 1: (1, 3, 20) (1, 0, 100)
Vertex 2: (2, 4, 2) (2, 3, 40) (2, 0, 3)
Vertex 3: (3, 4, 5) (3, 5, 5) (3, 1, 20) (3, 2, 40)
Vertex 4: (4, 2, 2) (4, 3, 5) (4, 5, 9)
Vertex 5: (5, 3, 5) (5, 4, 9)
Total weight in MST is 35
Root is: 0
Edges: (3, 1) (0, 2) (4, 3) (2, 4) (3, 5)
```

提示：使用 new WeightedGraph(list,numberOfVertices) 来创建一个图，其中 list 包含一个 WeightedEdge 对象的线性表。使用 new WeightedEdges(u,v,w) 来创建一条边。读取第一行以获取顶点个数。将接下来的每一行读入一个字符串 s 中，并且使用 s.split("[\\|]") 来提取三元组。对于每个三元组，使用 triplet.split("[,]") 提取顶点和权重。

*29.10 （为图创建文件）修改程序清单 29-3，创建一个表示 graph1 的文件。文件格式在编程练习题 29.9 中描述。从程序清单 29-3 中第 7 ～ 24 行定义的数组来创建文件。图的顶点个数为 12，它将存储在文件的第一行。如果 u<v，那么存储边 (u,v)。文件的内容应该如下所示：

```
12
0, 1, 807 | 0, 3, 1331 | 0, 5, 2097
1, 2, 381 | 1, 3, 1267
2, 3, 1015 | 2, 4, 1663 | 2, 10, 1435
3, 4, 599 | 3, 5, 1003
4, 5, 533 | 4, 7, 1260 | 4, 8, 864 | 4, 10, 496
5, 6, 983 | 5, 7, 787
6, 7, 214
7, 8, 888
8, 9, 661 | 8, 10, 781 | 8, 11, 810
9, 11, 1187
10, 11, 239
```

*29.11 （得到最短路径）编写一个程序，从文件中读取一个连通图。该图存储在一个文件中，使用与编程练习题 29.9 一样的指定格式。你的程序应该提示用户输入文件的 URL、两个顶点，然后显示这两个顶点之间的最短路径。例如，对于图 29-23 中的图，顶点 0 和顶点 1 之间的最短路径可以显示为 0 2 4 3 1。

下面是该程序的一个运行示例：

```
Enter a URL:
   https://liveexample.pearsoncmg.com/test/WeightedGraphSample2.txt  ↵Enter
Enter two vertices (integer indexes): 0 1  ↵Enter
The number of vertices is 6
Vertex 0: (0, 2, 3) (0, 1, 100)
Vertex 1: (1, 3, 20) (1, 0, 100)
Vertex 2: (2, 4, 2) (2, 3, 40) (2, 0, 3)
Vertex 3: (3, 4, 5) (3, 5, 5) (3, 1, 20) (3, 2, 40)
Vertex 4: (4, 2, 2) (4, 3, 5) (4, 5, 9)
Vertex 5: (5, 3, 5) (5, 4, 9)
A path from 0 to 1: 0 2 4 3 1
```

*29.12 （显示加权图）修改程序清单 28-6 中的 GraphView 以显示加权图。编写一个程序，显示图 29-1

中的图,如图 29-24 所示。(教师可以要求学生扩展该程序,添加具有合适的边的新城市到该图中。)

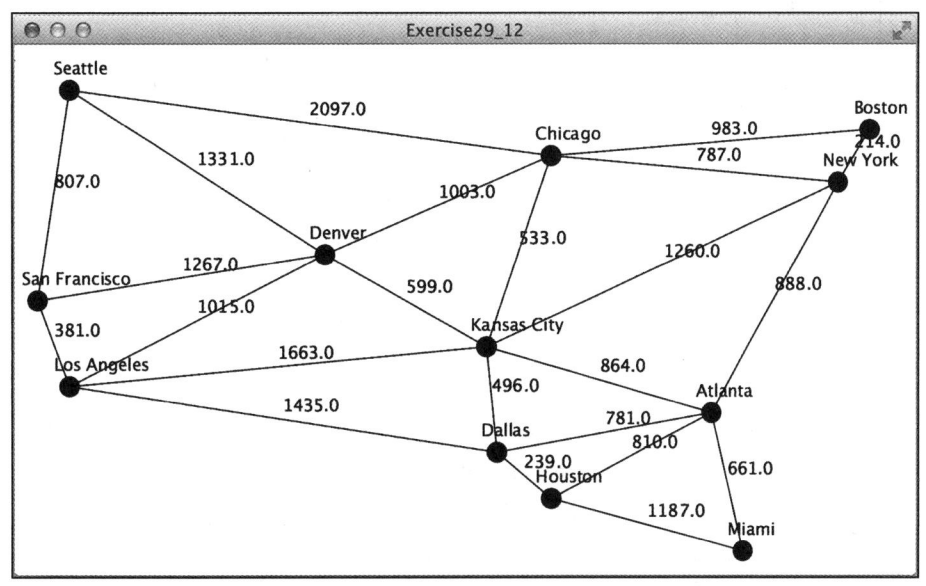

图 29-24 编程练习题 29.12 显示一个加权图

*29.13 (显示最短路径)修改程序清单 28-6 中的 GraphView,以显示一个加权图和两个指定城市之间的最短路径,如图 29-25 所示。需要在 GraphView 中添加一个数据域 path。如果 path 不为空,路径中的边显示为红色。如果输入了一个图中没有的城市,程序显示一个对话框来提醒用户。

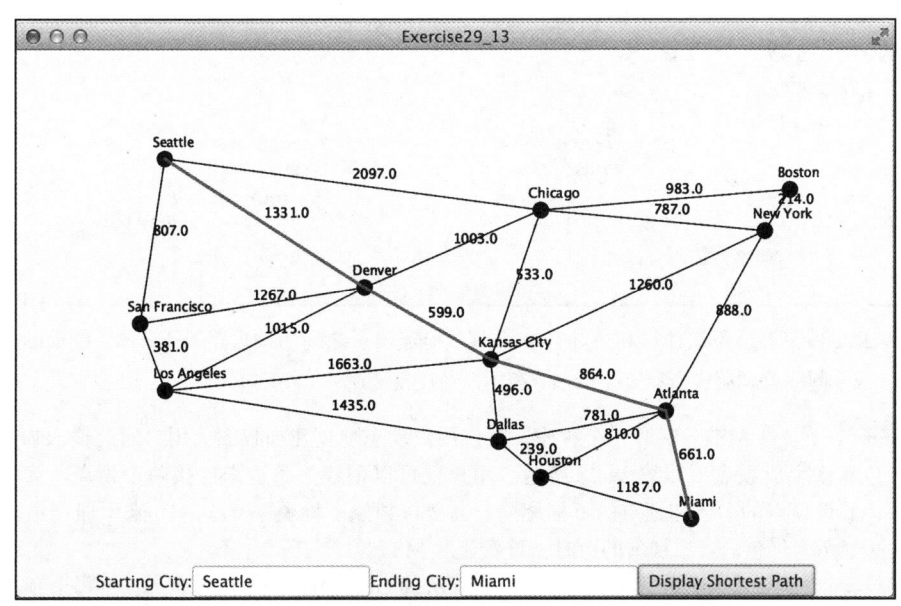

图 29-25 编程练习题 29.13 显示一条最短路径

*29.14 (显示最小生成树)修改程序清单 28-6 中的 GraphView,为图 29-1 中的图显示其加权图和最小生成树,如图 29-26 所示。MST 的路径显示为红色。

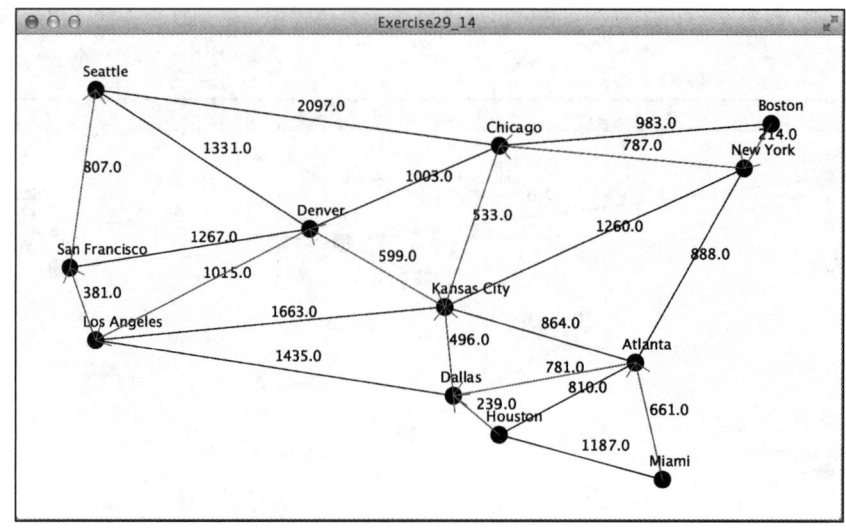

图 29-26 编程练习题 29.14 显示一个 MST

***29.15 （动态图）编写一个程序，允许用户动态创建一个加权图。用户通过输入顶点的名字和位置来创建顶点，如图 29-27 所示。用户也可以创建一条边来连接两个顶点。为了简化程序，假设顶点的名字和顶点的下标相同。需要以顶点下标顺序 0,1,…,n 来添加顶点。用户可以指定两个顶点并且让程序以红色来显示它们之间的最短路径。

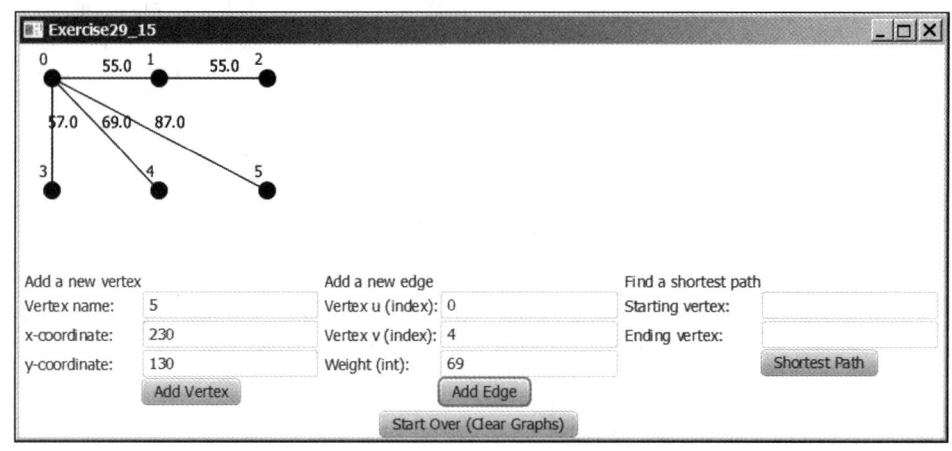

图 29-27 程序可以添加顶点和边并且显示两个指定顶点之间的最短路径（来源：Oracle 或其附属公司版权所有 © 1995 ～ 2016，经授权使用）

***29.16 （显示一个动态 MST）编写一个程序，允许用户动态地创建加权图。用户通过输入顶点的名字和位置来生成顶点，如图 29-28 所示。用户也可以创建一条边来连接两个顶点。为了简化程序，假设顶点的名字和顶点的下标相同。需要以顶点下标顺序 0,1,…,n 来添加顶点。MST 中的边显示为红色。当添加新的边时，重新显示 MST。

***29.17 （加权图可视化工具）开发一个如图 29-2 所示的 GUI 程序，要求如下：（1）每个顶点的半径为 20 像素。（2）用户点击鼠标左键时，如果鼠标点没有在一个已经存在的顶点内部或者过于接近，则放置一个位于鼠标点的顶点。（3）在一个已经存在的顶点内部右击鼠标来删除该顶点。（4）在一个顶点内部按下鼠标键并且拖放到另外一个顶点处释放，则产生一条边，并且显示两个顶点之间的距离。（5）用户在按下 CTRL 键的同时拖放一个顶点，则移动该顶点。

（6）顶点是从 0 开始的数字。当移动一个顶点时，顶点被重新标号。（7）可以单击 Show MST 或者 Show ALL SP From the Source 按钮来显示一个起始顶点的 MST 或者 SP 树。（8）可以单击 Show Shortest Path 按钮来显示两个指定顶点之间的最短路径。

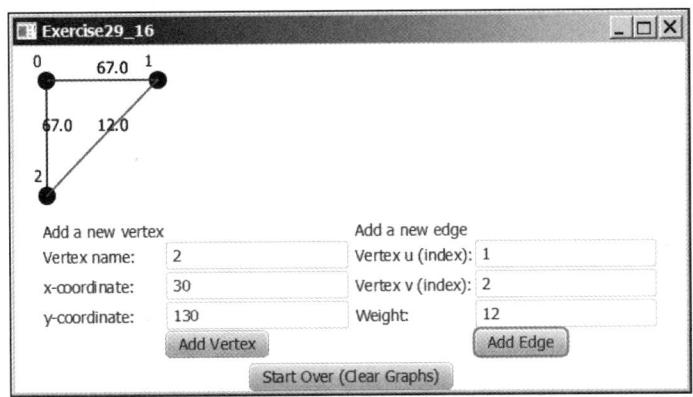

图 29-28　程序可以动态添加顶点和边并且显示 MST（来源：Oracle 或其附属公司版权所有 © 1995 ～ 2016，经授权使用）

***29.18　(Dijkstra 算法的替代版本) 一个 Dijkstra 算法的替代版本可以如下描述：

```
Input: a weighted graph G = (V, E) with nonnegative weights
Output: A shortest-path tree from a source vertex s

1  ShortestPathTree getShortestPath(s) {
2    Let T be a set that contains the vertices whose
3      paths to s are known;
4    Initially T contains source vertex s with cost[s] = 0;
5    for (each u in V - T)
6      cost[u] = infinity;
7
8    while (size of T < n) {
9      Find v in V - T with the smallest cost[u] + w(u, v) value
10       among all u in T;
11     Add v to T and set cost[v] = cost[u] + w(u, v);
12     parent[v] = u;
13   }
14 }
```

该算法使用 cost[v] 来存储从顶点 v 到源顶点 s 的最短路径的开销。cost[s] 为 0。初始时，设置 cost[v] 为无穷大，表示没有找到从 v 到 s 的路径。让 V 表示图中的所有顶点，T 表示已经知道开销的顶点集合。初始时，源顶点 s 位于 T 中。该算法重复地找到 T 中的顶点 u 和 V-T 中的顶点 v，使得 cost[u] + w(u + v) 最小，并将 v 移至 T 中。教材中给出的最短路径算法不断为 V-T 中的顶点更新开销和父顶点。该算法初始时为每个顶点的开销设置为无穷大，然后仅在顶点被加入 T 中的时候修改一次该顶点的开销。实现这个算法，并使用程序清单 29-7 来测试你的新算法。

***29.19　(高效地找到具有最小 cost[u] 的顶点 u) getShortestPath 方法使用线性搜索，找到具有最小 cost[u] 的顶点 u。这将使用 O(|V|) 的时间。使用一棵 AVL 树可以将搜索时间缩减为 O(log|V|)。修改该方法，使用一个 AVL 树来存储 V-T 中的顶点。使用程序清单 29-7 来测试你的新实现。

***29.20　(高效地测试一个顶点 u 是否在 T 中) 由于在程序清单 29-2 中方法 getMinimumSpanningTree 和 getShortestPath 采用线性表实现了 T，这样通过调用 T.contains(u) 来测试一个顶点 u 是否在 T 中需要 O(n) 的时间。通过引入一个名为 isInT 的数组来修改这两个方法。当一个顶

点 u 被加入 T 中时设置 isInT[u] 为 true。测试一个顶点 u 是否在 T 中现在可以在 O(1) 的时间内完成。使用下面的代码编写一个测试程序,其中 graph1 是从图 29-1 中创建的。

```
WeightedGraph<String> graph1 = new WeightedGraph<>(edges, vertices);
WeightedGraph<String>.MST tree1 = graph1.getMinimumSpanningTree();
System.out.println("Total weight is " + tree1.getTotalWeight());
tree1.printTree();
WeightedGraph<String>.ShortestPathTree tree2 =
  graph1.getShortestPath(graph1.getIndex("Chicago"));
tree2.printAllPaths();
```

第 30 章

Introduction to Java Programming and Data Structures, Comprehensive Version, Twelfth Edition

集合流的聚合操作

教学目标
- 对集合流使用聚合操作来简化代码和提高性能（30.1 节）。
- 在流上创建一个流管道，使用惰性中间方法（skip, limit, filter, distinct, sorted, map, mapToInt）和终止方法（count, sum, average, max, min, forEach, findFirst, firstAny, anyMatch, allMatch, noneMatch, toArray）（30.2 节）。
- 使用 IntStream、LongStream、DoubleStream 来处理基本数据值（30.3 节）。
- 创建并行流以快速执行（30.4 节）。
- 使用 reduce 方法将流中的元素缩减为单一结果（30.5 节）。
- 使用 collect 方法把流中元素放入可变集合（30.6 节）。
- 将流中元素分组并对组中的元素使用聚合方法（30.7 节）。
- 使用各种例子演示怎样通过流简化代码（30.8 节）。

30.1 引言

要点提示：对集合流使用聚合操作可以简化代码并提高性能。

通常，需要处理在数组或集合中的数据。比如，假定需要计算一个数组中大于 60 的数字个数，可以使用 foreach 循环编写代码，如下所示：

```
Double[] numbers = {2.4, 55.6, 90.12, 26.6};
Set<Double> set = new HashSet<>(Arrays.asList(numbers));
int count = 0;
for (double e: set)
  if (e > 60)
    count++;
System.out.println("Count is " + count);
```

这段代码没有问题，但是 Java 为这种任务提供了一种更好、更简单的方式。可以使用聚合操作重写代码，如下所示：

```
System.out.println("Count is "
  + set.stream().filter(e -> e > 60).count());
```

在规则集上调用 stream() 方法会为规则集中的元素返回一个 Stream。filter 方法指定了"选择其值大于 60 的元素"这一条件。count() 方法返回了流中满足这一条件的元素个数。

集合流（collection stream），或简称流（stream），是一个元素序列。对流的操作称为聚合操作（aggregate operation），也称为流操作（stream operation），因为这些操作适用于流中的所有数据。filter 和 count 方法是聚合操作的示例。使用 foreach 循环编写的代码描述了获得计数的过程，即如果元素大于 60，则增加计数。使用聚合操作编写的代码告诉程序返回大于 60 的元素的计数，但不指定如何获得计数。显然，使用集合操作将细节的实现留给了计算机，这使代码变得简洁和简单。此外，可以利用多个处理器并行地执行流上的聚合操作。因此，使用聚合操作编写的代码通常比使用 foreach 循环的代码运行得更快。

Java 提供了许多聚合操作和许多不同的使用聚合操作的方法。本章全面介绍了聚合操作和流。

✓ 复习题

30.1.1 用集合流的聚合操作处理数据有什么好处？

30.2 流管道

要点提示：流管道由一个从数据源创建的流、零个或多个中间方法和一个最终的终止方法组成。

数组或集合是用于存储数据的对象。流是处理数据的临时对象。当数据处理完后，流就被销毁。Java 8 在 Collection 接口中引入了一个新的缺省 stream() 方法，用以返回一个 Stream 对象。Stream 接口继承了 BaseStream 接口并包括如图 30-1 所示的聚合方法和效用方法。

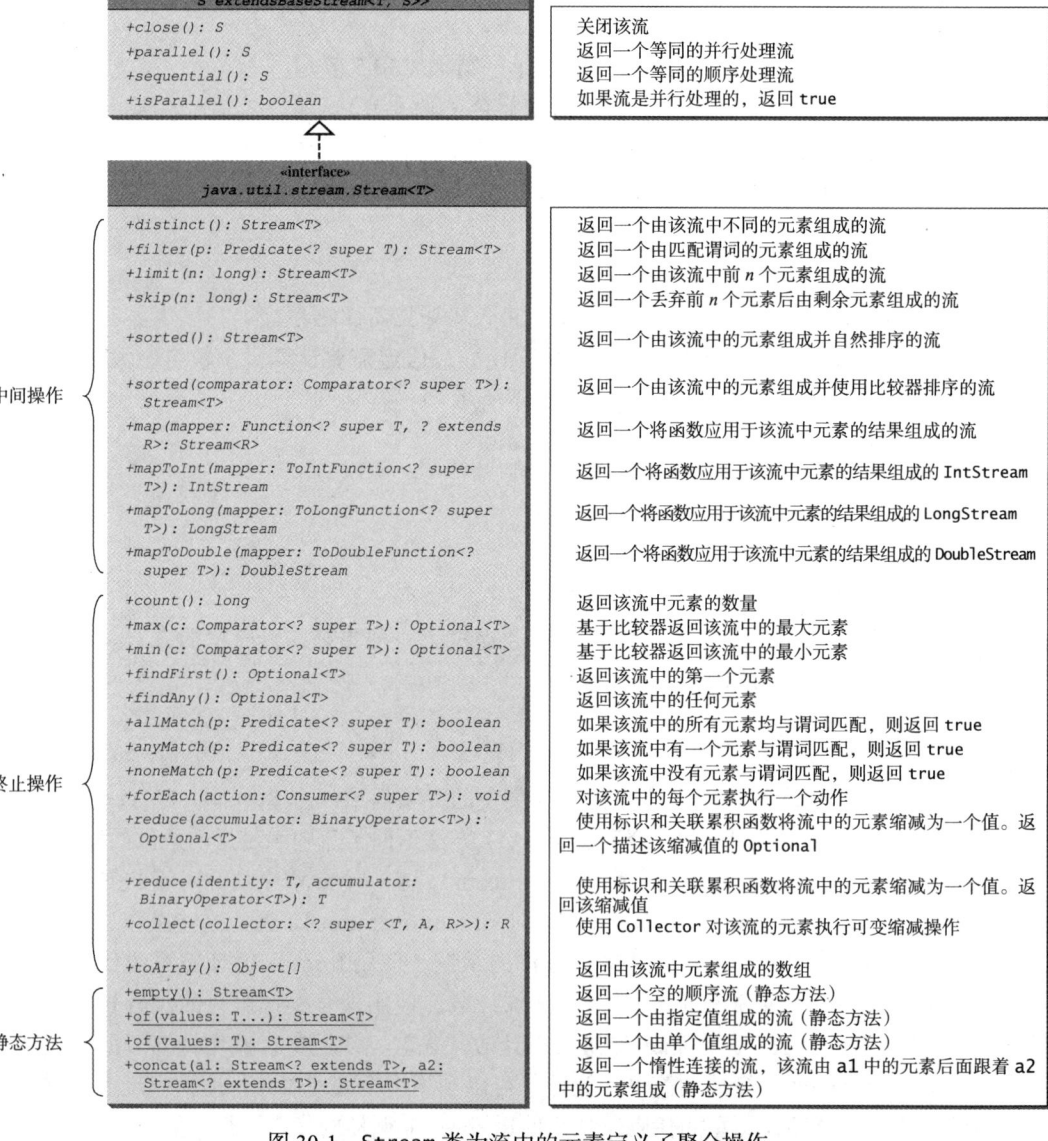

图 30-1　Stream 类为流中的元素定义了聚合操作

Stream 接口中的方法分为三类：中间方法，终止方法，静态方法。中间方法将流转换为另一个流。终止方法返回结果或执行操作。在执行终止方法后，流会自动关闭。静态方法创建一个流。

这些方法通过流管道调用。流管道（stream pipeline）由一个源（例如，一个线性表，一个规则集，或一个数组）、一个创建流的方法、零个或多个中间方法、一个最终的终止方法组成。以下是流管道的一个示例：

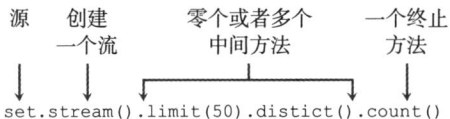

这里，set 是数据源，调用 stream() 方法为源中的数据创建了一个流，调用 limit(50) 返回由流中前 50 个元素组成的流，调用 distinct() 获得由这个流中不同的元素组成的流，而调用 count() 返回了最终流中元素的个数。

流是惰性的，这意味着仅仅当终止操作开始时才会进行计算。这允许 JVM 优化计算。

大多数流方法的参数是函数接口的实例。所以，可以用 lambda 表达式或方法引用来创建参数。程序清单 30-1 给出了一个示例，它展示了创建一个流并在流上使用方法。

程序清单 30-1 StreamDemo.java

```java
 1  import java.util.stream.Stream;
 2
 3  public class StreamDemo {
 4    public static void main(String[] args) {
 5      String[] names = {"John", "Peter", "Susan", "Kim", "Jen",
 6        "George", "Alan", "Stacy", "Michelle", "john"};
 7
 8      // Display the first four names sorted
 9      Stream.of(names).limit(4).sorted()
10        .forEach(e -> System.out.print(e + " "));
11
12      // Skip four names and display the rest sorted ignore case
13      System.out.println();
14      Stream.of(names).skip(4)
15        .sorted((e1, e2) -> e1.compareToIgnoreCase(e2))
16        .forEach(e -> System.out.print(e + " "));
17
18      System.out.println();
19      Stream.of(names).skip(4)
20        .sorted(String::compareToIgnoreCase)
21        .forEach(e -> System.out.print(e + " "));
22
23      System.out.println("\nLargest string with length > 4: "
24        + Stream.of(names)
25          .filter(e -> e.length() > 4)
26          .max(String::compareTo).get());
27
28      System.out.println("Smallest string alphabetically: "
29        + Stream.of(names).min(String::compareTo).get());
30
31      System.out.println("Stacy is in names? "
32        + Stream.of(names).anyMatch(e -> e.equals("Stacy")));
33
34      System.out.println("All names start with a capital letter? "
35        + Stream.of(names)
36          .allMatch(e -> Character.isUpperCase(e.charAt(0))));
37
38      System.out.println("No name begins with Ko? "
```

```
39          + Stream.of(names).noneMatch(e -> e.startsWith("Ko")));
40
41      System.out.println("Number of distinct case-insensitive strings: "
42          + Stream.of(names).map(e -> e.toUpperCase())
43            .distinct().count());
44
45      System.out.println("First element in this stream in lowercase: "
46          + Stream.of(names).map(String::toLowerCase).findFirst().get());
47
48      System.out.println("Skip 4 and get any element in this stream:"
49          + Stream.of(names).skip(4).sorted().findAny().get());
50
51      Object[] namesInLowerCase =
52          Stream.of(names).map(String::toLowerCase).toArray();
53      System.out.println(java.util.Arrays.toString(namesInLowerCase));
54    }
55  }
```

```
John Kim Peter Susan
Alan George Jen john Michelle Stacy
Alan George Jen john Michelle Stacy
Largest string with length > 4: Susan
Smallest string alphabetically: Alan
Stacy is in names? true
All names start with a capital letter? false
No name begins with Ko? true
Number of distinct case-insensitive strings: 9
First element in this stream in lowercase: john
Skip 4 and get any element in this stream: Alan
[john, peter, susan, kim, jen, george, alan, stacy, michelle, john]
```

30.2.1 Stream.of、limit 和 forEach 方法

该程序创建了一个字符串数组（第 5 ~ 6 行）。在第 9 ~ 10 行，调用静态方法 Stream.of(names) 返回一个由 names 数组中字符串组成的 Stream，调用 limit(4) 返回一个由流中前 4 个元素组成的新 Stream，调用 sorted() 给流排序，并调用 forEach 方法显示流中的每个元素。传入 forEach 方法的参数是一个 lambda 表达式。正如 15.6 节中介绍的，lambda 表达式是一种简洁的语法，用于替换实现函数接口的匿名内部类。传入 forEach 方法的参数是带有抽象方法 accept(T, t) 的函数接口 Consumer<? super T> 实例。下面图 a 中使用 lambda 表达式的语句（即程序清单中第 10 行）和图 b 中使用匿名内部类是等价的。lambda 表达式不仅简化了代码，而且简化了方法的概念。现在可以简单地说：对于流中的每个元素，执行表达式给定的动作。

```
forEach(e -> System.out.print(e + " "))
```

```
forEach(
  new java.util.function.Consumer<String>() {
    public void accept(String e) {
      System.out.print(e + " ");
    }
  }
)
```

a) 使用 lambda 表达式   b) 使用匿名内部类

30.2.2 sorted 方法

第 15 行中的 sorted 方法使用了 Comparator 对流中的数据排序。Comparator 是一个函

数接口。使用 lambda 表达式实现该接口，并指定两个字符串忽略大小写进行比较。下面图 a 中使用 lambda 表达式和图 b 中使用匿名内部类是等价的。在这个示例中，lambda 表达式只是简单地调用了一个方法。所以，它可以使用第 20 行（也可以参见图 c）中的方法引用进一步简化。方法引用在 20.6 节中进行了介绍。

```
sorted((e1, e2) ->
  e1.compareToIgnoreCase(e2))
```
a) 使用 lambda 表达式

```
sorted(String::compareToIgnoreCase)
```
c) 使用方法引用

```
sorted(
  new java.util.Comparator<String>() {
    public int compare(String e1, String e2) {
      return e1.compareToIgnoreCase(e2);
    }
  }
)
```
b) 使用匿名内部类

30.2.3 filter 方法

filter 方法传入一个 Predicate<? super T> 类型的参数，Predicate<? super T> 是一个函数接口，它有一个返回布尔值的抽象方法 test(T t)。该方法从流中选出满足谓词的元素。第 25 行使用 lambda 表达式来实现 Predicate 接口，如图 a 中所示，这和图 b 中使用匿名内部类是等价的。

```
filter(e -> e.length() > 4)
```
a) 使用 lambda 表达式

```
filter(
  new java.util.function.Predicate<String>() {
    public boolean test(String e) {
      return e.length() > 4;
    }
  }
)
```
b) 使用匿名内部类

30.2.4 max 和 min 方法

max 和 min 方法传入一个 Comparator<? Super T> 类型的参数。该参数指定了为获得最大值或最小值将如何比较元素。该程序使用方法引用 String::compareTo 来简化创建 Comparator 的代码（第 26 和 29 行）。max 和 min 方法返回一个描述元素的 Optional<T>。需要调用 Optional 类中的 get() 方法来返回该元素。

30.2.5 anyMatch、allMatch 和 noneMatch 方法

anyMatch、allMatch 和 noneMatch 方法传入一个 Predicate<? super T> 类型的参数，分别测试该流中有一个元素、所有元素或没有元素满足谓词的情况。该程序测试了名字 Stacy 是否在流中（第 32 行），是否所有名字以大写字母开头（第 36 行），以及是否有以字符串 Ko 开头的名字（第 39 行）。

30.2.6 map、distinct 和 count 方法

map 方法通过将流中元素映射为新元素而返回一个新流。因此，第 42 行中的 map 方法返回了全是大写字母的新流。distinct() 方法生成了一个包含不同元素的新流。count() 方法计算流中元素的个数。所以，第 43 行流管道计算了数组 names 中不同字符串的个数。

map 方法传入一个 Function<? super T,? super R> 类型的参数，返回一个 Stream<R>

的实例。Function 是带有抽象方法 apply(T t) 的函数接口,将 t 映射为 R 类型的值。第 42 行使用 lambda 表达式来实现 Function 接口,如图 a 中所示,这和图 b 中使用匿名内部类是等价的。可以使用图 c 中所示的方法引用进一步简化。

```
map(e -> e.toUpperCase())
```
a) 使用 lambda 表达式

```
map(
  new java.util.function.Function<String, String>() {
    public String apply(String e) {
      return e.toUpperCase();
    }
  }
)
```
b) 使用匿名内部类

```
map(String::toUpperCase)
```
c) 使用方法引用

30.2.7 findFirst、findAny 和 toArray 方法

findFirst() 方法(第 46 行)返回包装在一个 Optional<T> 实例中的流的第一个元素。实际的元素值则通过调用 Optional<T> 类中的 get() 方法返回。findAny() 方法返回流中的任何一个元素(第 49 行),具体选择哪个元素取决于流的内部状态。findAny() 方法比 findFirst() 方法更高效。

toArray() 方法(第 52 行)返回流中对象的数组。

> **注意**:BaseStream 接口定义了 close() 方法,调用它可以关闭流。但没有必要使用它,因为流在执行完终止方法后会自动关闭。

✔ 复习题

30.2.1 给出下列代码的输出:

```
Character[] chars = {'D', 'B', 'A', 'C'};
System.out.println(Stream.of(chars).sorted().findFirst().get());
System.out.println(Stream.of(chars).sorted(
  java.util.Comparator.reverseOrder()).findFirst().get());
System.out.println(Stream.of(chars)
  .limit(2).sorted().findFirst().get());
System.out.println(Stream.of(chars).distinct()
  .skip(2).filter(e -> e > 'A').findFirst().get());
System.out.println(Stream.of(chars)
  .max(Character::compareTo).get());
System.out.println(Stream.of(chars)
  .max(java.util.Comparator.reverseOrder()).get());
System.out.println(Stream.of(chars)
  .filter(e -> e > 'A').findFirst().get());
System.out.println(Stream.of(chars)
  .allMatch(e -> e >= 'A'));
System.out.println(Stream.of(chars)
  .anyMatch(e -> e > 'F'));
System.out.println(Stream.of(chars)
  .noneMatch(e -> e > 'F'));
Stream.of(chars).map(e -> e + "").map(e -> e.toLowerCase())
  .forEach(System.out::println);
Object[] temp = Stream.of(chars).map(e -> e + "Y")
  .map(e -> e.toLowerCase()).sorted().toArray();
System.out.println(java.util.Arrays.toString(temp));
```

30.2.2 下列代码有什么错误?

```
Character[] chars = {'D', 'B', 'A', 'C'};
Stream<Character> stream = Stream.of(chars).sorted();
System.out.println(stream.findFirst());
System.out.println(stream.skip(2).findFirst());
```

30.2.3 使用方法引用和匿名内部类重写 (a)，并使用 lambda 表达式和匿名内部类重写 (b)：

(a) `sorted((s1, s2) -> s1.compareToIgnoreCase(s2))`
(b) `forEach(System.out::println)`

30.2.4 给定一个 `Map<String, Double>` 类型的映射 `map`，写一个表达式，它返回 `map` 中所有值的和。比如，如果 `map` 包含 `{"john", 1.5}` 和 `{"Peter", 1.1}`，和就是 2.6。

30.3 IntStream、LongStream 和 DoubleStream

🔑 要点提示：IntStream、LongStream 和 DoubleStream 是特殊类型的处理基本数据类型 int、long 和 double 值序列的流。

Stream 表示一个对象序列。除了 Stream 之外，Java 提供了 IntStream、LongStream 和 DoubleStream 来表示 int、long 和 double 值序列。这些流也是 BaseStream 的子接口，可以像使用 Stream 一样使用这些流。此外，可以使用 sum()、average() 和 summaryStatistics() 方法来返回流中元素的和、平均值以及各种统计。可以使用 mapToInt 方法将 Stream 转化为 IntStream，或使用 map 方法将包括 IntStream 在内的任何流变成 Stream。

程序清单 30-2 给出了一个使用 IntStream 的示例。

程序清单 30-2 IntstreamDemo.java

```java
import java.util.IntSummaryStatistics;
import java.util.stream.IntStream;
import java.util.stream.Stream;

public class IntStreamDemo {
  public static void main(String[] args) {
    int[] values = {3, 4, 1, 5, 20, 1, 3, 3, 4, 6};

    System.out.println("The average of distinct even numbers > 3: " +
      IntStream.of(values).distinct()
        .filter(e -> e > 3 && e %  2 == 0).average().getAsDouble());

    System.out.println("The sum of the first 4 numbers is " +
      IntStream.of(values).limit(4).sum());

    IntSummaryStatistics stats =
    IntStream.of(values).summaryStatistics();

    System.out.printf("The summary of the stream is\n%-10s%10d\n" +
      "%-10s%10d\n%-10s%10d\n%-10s%10d\n%-10s%10.2f\n",
      " Count:", stats.getCount(), " Max:", stats.getMax(),
      " Min:", stats.getMin(), " Sum:", stats.getSum(),
      " Average:", stats.getAverage());

    String[] names = {"John", "Peter", "Susan", "Kim", "Jen",
      "George", "Alan", "Stacy", "Michelle", "john"};

    System.out.println("Total character count for all names is "
      + Stream.of(names).mapToInt(e -> e.length()).sum());

    System.out.println("The number of digits in array values is " +
      Stream.of(values).map(e -> e + "")
        .mapToInt(e -> e.length()).sum());
  }
}
```

```
The average of distinct even numbers > 3: 10.0
The sum of the first 4 numbers is 13
The summary of the stream is

  Count:           10
  Max:             20
  Min:              1
  Sum:             50
  Average:       5.00

Total character count for all names is 47
The number of digits in array values is 11
```

该程序创建了一个 int 值的数组（第7行）。第10行的流管道使用了中间方法 distinct 和 filter 以及终止方法 average。average() 方法以 OptionalDouble 对象返回流的平均值（第11行），通过调用 getAsDouble() 方法得到了实际的平均值。

第14行中的流管道使用了中间方法 limit 和终止方法 sum。sum() 方法返回流中所有值的和。

如果需要获得流中的多个汇总值，使用 summaryStatistics() 更高效。该方法（第17行）返回 IntSummaryStatistics 的一个实例，它包含了计数值、最小值、最大值、和以及平均值等汇总值（第19～23行）。注意，sum()、average() 和 summaryStatistics() 方法只能应用于 IntStream、LongStream 和 DoubleStream。

mapToInt 方法通过将流中每个值映射为一个 int 值来返回一个 IntStream。第29行流管道中的 mapToInt 方法将每个字符串映射为一个字符串长度的 int 值，而 sum 方法得到 IntStream 中所有 int 值的和。第29行的流管道得到了流中所有字符的总数。

mapToInt 方法有一个 ToIntFunction<? super T> 类型的参数，并返回一个 IntStream 的实例。ToIntFunction 是带有抽象方法 applyAsInt(T t) 的函数接口，该方法将 t 映射为 int 类型的值。第33行使用 lambda 表达式来实现 ToIntFunction 接口，如以下图 a 中所示，这和图 b 中使用匿名内部类是等价的。可以使用图 c 中所示的方法引用进一步简化。

```
mapToInt(e -> e.length())
```
a) 使用 lambda 表达式

```
mapToInt(
  new java.util.function.ToIntFunction<String>() {
    public int applyAsInt(String e) {
      return e.length();
    }
  }
)
```
b) 使用匿名内部类

```
mapToInt(String::length)
```
c) 使用方法引用

第32行的 map 方法返回一个新的字符串流。每个字符串由 values 数组中的整数转化而来。第33行的 mapToInt 方法返回一个新的整数流，每个整数代表字符串的长度。sum() 方法返回最终流中所有 int 值的和。所以第32～33行的流管道得到了 values 数组中数位的个数。

复习题

30.3.1 给出下列代码的输出：

```
int[] numbers = {1, 4, 2, 3, 1};
System.out.println(IntStream.of(numbers)
  .sorted().findFirst().getAsInt());
```

```
System.out.println(IntStream.of(numbers)
    .limit(2).sorted().findFirst().getAsInt());
System.out.println(IntStream.of(numbers).distinct()
    .skip(1).filter(e -> e > 2).sum());
System.out.println(IntStream.of(numbers).distinct()
    .skip(1).filter(e -> e > 2).average().getAsDouble());
System.out.println(IntStream.of(numbers).max().getAsInt());
System.out.println(IntStream.of(numbers).max().getAsInt());
System.out.println(IntStream.of(numbers)
    .filter(e -> e > 1).findFirst().getAsInt());
System.out.println(IntStream.of(numbers)
    .allMatch(e -> e >= 1));
System.out.println(IntStream.of(numbers)
    .anyMatch(e -> e > 4));
System.out.println(IntStream.of(numbers).noneMatch(e -> e > 4));
IntStream.of(numbers).mapToObj(e -> (char)(e + 50))
    .forEach(System.out::println);

Object[] temp = IntStream.of(numbers)
    .mapToObj(e -> (char)(e + 'A')).toArray();
System.out.println(java.util.Arrays.toString(temp));
```

30.3.2 下列代码有什么错误？

```
int[] numbers = {1, 4, 2, 3, 1};
DoubleSummaryStatistics stats =
    DoubleStream.of(numbers).summaryStatistics();
System.out.printf("The summary of the stream is\n%-10s%10d\n" +
    "%-10s%10.2f\n%-10s%10.2f\n%-10s%10.2f\n%-10s%10.2f\n",
    " Count:", stats.getCount(), " Max:", stats.getMax(),
    " Min:", stats.getMin(), " Sum:", stats.getSum(),
    " Average:", stats.getAverage());
```

30.3.3 用匿名内部类重写下列代码，将 int 映射为 Character：

```
mapToObj(e -> (char)(e + 50))
```

30.3.4 给出下列代码的输出：

```
int[][] m = {{1, 2}, {3, 4}, {5, 6}};
System.out.println(Stream.of(m)
    .mapToInt(e -> IntStream.of(e).sum()).sum());
```

30.3.5 给定程序清单 30-1 中的 names 数组，编写代码显示 names 中的字符总数。

30.4 并行流

要点提示：流可以并行执行以改进性能。

多核系统的广泛应用引发了软件革命。为了从多核处理器中受益，软件需要并行运行。所有的流操作可以利用多核处理器并行执行。Collection 接口中的 stream() 方法返回顺序流。为了并行执行操作，使用 Collection 接口中的 parallelStream() 方法来获得并行流。所有的流都可以通过调用在 BaseStream 接口中定义的 parallel() 方法来转变为并行流。类似地，可以调用 sequential() 方法把并行流转变为顺序流。

中间方法可以进一步分为无状态方法（stateless method）和有状态方法（stateful method）。如 filter 和 map 等无状态方法可以独立于流中的其他元素来执行。如 distinct 和 sorted 等有状态方法必须考虑整个流来执行。比如，distinct 方法必须考虑流中所有元素才能得到结果。无状态方法在内部机制上是可以并行化的，并可以一次性并行执行。有状态方法必须并行执行多次。

程序清单 30-3 给出的示例展示了使用并行流的优势。

程序清单 30-3 ParallelStreamDemo.java

```java
1  import java.util.Arrays;
2  import java.util.Random;
3  import java.util.stream.IntStream;
4
5  public class ParallelStreamDemo {
6    public static void main(String[] args) {
7      Random random = new Random();
8      int[] list = random.ints(200_000_000).toArray();
9
10     System.out.println("Number of processors: " +
11       Runtime.getRuntime().availableProcessors());
12
13     long startTime = System.currentTimeMillis();
14     int[] list1 = IntStream.of(list).filter(e -> e > 0).sorted()
15       .limit(5).toArray();
16     System.out.println(Arrays.toString(list1));
17     long endTime = System.currentTimeMillis();
18     System.out.println("Sequential execution time is " +
19       (endTime - startTime) + " milliseconds");
20
21     startTime = System.currentTimeMillis();
22     int[] list2 = IntStream.of(list).parallel().filter(e -> e > 0)
23       .sorted().limit(5).toArray();
24     System.out.println(Arrays.toString(list2));
25     endTime = System.currentTimeMillis();
26     System.out.println("Parallel execution time is " +
27       (endTime - startTime) + " milliseconds");
28   }
29 }
```

```
Number of processors: 8
[4, 9, 38, 42, 52]
Sequential execution time is 12362 milliseconds
[4, 9, 38, 42, 52]
Parallel execution time is 3448 milliseconds
```

9.6.2 节中介绍的 Random 类可以用来生成随机数字。可以使用 ints(n) 方法来生成由 n 个随机 int 值组成的 IntStream（第 8 行）。也可以使用 ints(n, r1, r2) 来生成 r1（包含）到 r2（不包含）范围中 n 个元素组成的 IntStream，用 doubles(n) 和 doubles(n, r1, r2) 来生成随机浮点数构成的 DoubleStream。在 IntStream 上调用 toArray() 方法（第 8 行）从流中返回一个 int 值数组。回顾一下，可以在整数 200_000_000 中用下划线来提升可读性（参见 2.10.1 节）。

调用 Runtime.getRuntime() 返回 Runtime 对象（第 11 行）。调用 Runtime 的 availableProcessors() 返回对于 JVM 可用的处理器数量。在该示例中，系统有 8 个处理器。

使用 IntStream.of(list) 来创建 IntStream（第 14 行）。中间方法 filter(e -> e > 0) 选择流中正数。中间方法 sorted() 将过滤后的流排序。中间方法 limit(5) 选择排序后的流中的前 5 个整数（第 15 行）。最后，终止方法 toArray() 返回流中 5 个整数组成的数组。这是个顺序流，为了将它转变为并行流，只需要简单地调用 parallel() 方法（第 22 行），就可以将流设置为并行处理。正如你从示例运行中看到的，并行执行比顺序执行快很多。

有几个有趣的问题。

1)中间方法是惰性的,它在启动终止方法时执行。这可以从下列代码中确认:

```
1   long startTime = System.currentTimeMillis();
2   IntStream stream = IntStream.of(list).filter(e -> e > 0).sorted()
3     .limit(5);
4   System.out.println("The time for the preceding method is " +
5     (System.currentTimeMillis() - startTime) + " milliseconds");
6   int[] list1 = stream.toArray();
7   System.out.println("The execution time is " +
8     (System.currentTimeMillis() - startTime) + " milliseconds");
```

当你运行这段代码时,可以看到第 2~3 行几乎不花任何时间,因为中间方法还没被执行。当终止方法 toArray() 被调用时,执行针对流管道的所有方法。所以,该流管道的实际执行时间在第 6 行。

2)流管道中的中间方法的顺序有影响吗?是的,会有影响。比如,如果方法 limit(5) 和 sorted() 交换顺序,结果将不一样。即使结果相同,运行性能也会不同。比如,如果 sorted() 方法放在 filter(e -> e > 0) 前,结果是相同的,但将花更多的时间来执行流,因为对大量元素排序将花费更多的时间。将 filter 放在 sorted 前可以去掉约一半的排序元素。

3)并行流是否总是更快?不一定。并行执行需要同步,这将带来一定的开销。如果将第 14 行和第 22 行中的 IntStream.of(list) 替换为 random.ints(200_000_000),第 22~23 行中的并行流会比第 14~15 行中的顺序流花更多的时间。原因是生成伪随机数序列的算法是高度顺序的,并行流的开销比并行处理节约下来的时间更多。所以,在选择并行流部署前,应该对顺序流和并行流都进行测试。

4)并行执行流方法时,流中的元素可能以任意顺序处理。所以,下列代码会以随机顺序显示流中的数字:

```
IntStream.of(1, 2, 3, 4, 5).parallel()
  .forEach(e -> System.out.print(e + " "));
```

然而,如果顺序执行的话,数字将被显示为 1 2 3 4 5。

✓ 复习题

30.4.1 什么是无状态方法?什么是有状态方法?

30.4.2 如何创建并行流?

30.4.3 假设 names 是字符串规则集,下列两个流哪个更好?

```
Object[] s = set.parallelStream().filter(e -> e.length() > 3)
  .sorted().toArray();

Object[] s = set.parallelStream().sorted()
  .filter(e -> e.length() > 3).toArray();
```

30.4.4 下列代码的输出是什么?

```
int[] values = {3, 4, 1, 5, 20, 1, 3, 3, 4, 6};
System.out.print("The values are ");
  IntStream.of(values)
    .forEach(e -> System.out.print(e + " "));
```

30.4.5 下列代码的输出是什么?

```
int[] values = {3, 4, 1, 5, 20, 1, 3, 3, 4, 6};
System.out.print("The values are ");
  IntStream.of(values).parallel()
    .forEach(e -> System.out.print(e + " "));
```

30.4.6 编写一条语句,生成由 1000 个 0.0 到 1.0 之间的随机双精度浮点数组成的数组,不包括 1.0。

30.5 使用 reduce 方法进行流的归约

要点提示：可以使用 reduce 方法将流中的元素归约为单一值。

通常，需要处理一个集合中的所有元素来产生一个汇总值，如总和、最大值或最小值。例如，下列代码获得了规则集 s 中所有元素的总和：

```
int total = 0;
for (int e: s) {
  total += e;
}
```

这个代码很简单，但它指定了获得总和的确切步骤，并且是高度顺序的。流中的 reduce 方法可以用来编写高层次的代码以便并行执行。

归约（reduction）通过重复应用二元运算，例如加法、乘法，或在两个元素之间找到最大值，读取流中的元素并生成单一的值。通过使用归约，可以写出如下求规则集中元素总和的代码：

```
int sum = s.parallelStream().reduce(0, (e1, e2) -> e1 + e2);
```

这里，reduce 方法读入两个参数。第一个参数是一个标识，即初始值。第二个参数是函数接口 IntBinaryOperator 的一个对象。该接口包含一个应用二元运算后返回 int 值的抽象方法 applyAsInt(int e1, int e2)。下面图 a 中的上述 lambda 表达式与图 b 中使用匿名内部类的代码是等价的。

```
reduce(0, e -> (e1, e2) -> e1 + e2)
```

```
reduce(0,
  new java.util.function.IntBinaryOperator() {
    public int applyAsInt(int e1, int e2) {
      return e1 + e2;
    }
  }
)
```

a) 使用 lambda 表达式　　　　　　　　　　b) 使用匿名内部类

上述 reduce 方法在语义上等同于一个命令式的代码，如下所示：

```
int total = identity (i.e., 0, in this case);
for (int e: s) {
  total = applyAsInt(total, e);
}
```

reduce 方法使得代码简洁。此外，代码可以并行化，因为多个处理器可以同时对两个整数不断调用 applyAsInt 方法。

通过使用 reduce 方法，可以写出如下求规则集中元素最大值的代码：

```
int result = s.parallelStream()
  .reduce(Integer.MIN_VALUE, (e1, e2) -> Math.max(e1, e2));
```

事实上，sum、max 和 min 方法是通过 reduce 方法实现的。

程序清单 30-4 给出了使用 reduce 方法的一个示例。

程序清单 30-4 StreamReductionDemo.java

```
1  import java.util.stream.IntStream;
2  import java.util.stream.Stream;
3
4  public class StreamReductionDemo {
```

```
5    public static void main(String[] args) {
6      int[] values = {3, 4, 1, 5, 20, 1, 3, 3, 4, 6};
7
8      System.out.print("The values are ");
9        IntStream.of(values).forEach(e -> System.out.print(e + " "));
10
11     System.out.println("\nThe result of multiplying all values is " +
12       IntStream.of(values).parallel().reduce(1, (e1, e2) -> e1 * e2));
13
14     System.out.print("The values are " +
15       IntStream.of(values).mapToObj(e -> e + "")
16         .reduce((e1, e2) -> e1 + ", " + e2).get());
17
18     String[] names = {"John", "Peter", "Susan", "Kim", "Jen",
19       "George", "Alan", "Stacy", "Michelle", "john"};
20     System.out.print("\nThe names are: ");
21     System.out.println(Stream.of(names)
22       .reduce((x, y) -> x + ", " + y).get());
23
24     System.out.print("Concat names: ");
25     System.out.println(Stream.of(names)
26       .reduce((x, y) -> x + y).get());
27
28     System.out.print("Total number of characters: ");
29     System.out.println(Stream.of(names)
30       .reduce((x, y) -> x + y).get().length());
31   }
32 }
```

```
The values are 3, 4, 1, 5, 20, 1, 3, 3, 4, 6
The result of multiplying all values is 259200
The values are 3, 4, 1, 5, 20, 1, 3, 3, 4, 6
The names are John, Peter, Susan, Kim, Jen, George, Alan, Stacy,
  Michelle, john
Concat names: JohnPeterSusanKimJenGeorgeAlanStacyMichellejohn
Total number of characters: 47
```

该程序创建了一个 int 值的数组（第 6 行）。流管道从这个 int 数组创建一个 IntStream 并调用 forEach 方法显示流中的每个整数（第 9 行）。

该程序为 int 数组创建一个并行流管道，并应用 reduce 方法来得到流中 int 值的乘积（第 12 行）。

mapToObj 方法从 IntStream 返回一个字符串对象的流（第 15 行）。可以不带标识调用 reduce 方法。在这种情况下，它返回一个 Optional<T> 对象。第 16 行中的 reduce 方法将流中字符串归约为一个复合字符串，该字符串由流中的逗号分隔所有字符串组成。

第 26 行的 reduce 方法将流中所有字符串结合为一个长字符串。Stream.of(names).reduce((x, y) -> x + y).get() 返回一个由流中所有字符串连接在一起的字符串，在字符串上调用 length() 方法返回字符串中的字符数（第 30 行）。

注意，第 30 行中 reduce((x, y) -> x + y) 可以用方法引用 reduce(String::concat) 来简化。

复习题

30.5.1 给出下列代码的输出：

```
int[] values = {1, 2, 3, 4};
System.out.println(IntStream.of(values)
  .reduce(0, (e1, e2) -> e1 + e2));
System.out.println(IntStream.of(values)
```

```
           .reduce(1, (e1, e2) -> e1 * e2));
System.out.println(IntStream.of(values).map(e -> e * e)
           .reduce(0, (e1, e2) -> e1 * e2));
System.out.println(IntStream.of(values).mapToObj(e -> "" + e)
           .reduce((e1, e2) -> e1 + " " + e2).get());
System.out.println(IntStream.of(values).mapToObj(e -> "" + e)
           .reduce((e1, e2) -> e1 + ", " + e2).get());
```

30.5.2 给出下列代码的输出：

```
int[][] m = {{1, 2}, {3, 4}, {5, 6}};
System.out.println(Stream.of(m)
   .map(e -> IntStream.of(e).reduce(1, (e1, e2) -> e1 * e2))
   .reduce(1, (e1, e2) -> e1 * e2));
```

30.5.3 给出下列代码的输出：

```
int[][] m = {{1, 2}, {3, 4}, {5, 6}, {1, 3}};
Stream.of(m).map(e -> IntStream.of(e))
   .reduce((e1, e2) -> IntStream.concat(e1, e2))
   .get().distinct()
   .forEach(e -> System.out.print(e + " "));
```

30.5.4 给出下列代码的输出：

```
int[][] m = {{1, 2}, {3, 4}, {5, 6}, {1, 3}};
System.out.println(
   Stream.of(m).map(e -> IntStream.of(e))
      .reduce((e1, e2) -> IntStream.concat(e1, e2))
      .get().distinct().mapToObj(e -> e + "")
      .reduce((e1, e2) -> e1 + ", " + e2).get());
```

30.6 使用 collect 方法进行流的归约

☞ 要点提示：可以使用 collect 方法将流中的元素归约在可变容器中。

在之前的例子中，Stream.of(names).reduce((x, y) -> x + y) 中的 reduce 方法用了 String 的 concat 方法。这一操作会导致在连接两个字符串时创建一个新的字符串，这是非常低效的。更好的方法是使用 StringBuilder 并将结果累加到 StringBuilder 中。这可以通过 collect 方法完成。

collect 方法把流中的元素收集在一个可变容器中，例如，使用如下句法的 Collection 对象：

```
<R> R collect(Supplier<R> supplier,
              BiConsumer<R, ? super T> accumulator,
              BiConsumer<R, R> combiner)
```

该方法传入三个函数式参数：一个用来构造结果容器的新实例的供应者函数，一个将来自流的元素并入结果容器的累加器函数，一个将结果容器的内容合并到另一个结果容器中的合并函数。

例如，要把字符串结合到 StringBuilder 中，可以编写如下使用 collect 方法的代码：

```
String[] names = {"John", "Peter", "Susan", "Kim", "Jen",
   "George", "Alan", "Stacy", "Michelle", "john"};
StringBuilder sb = Stream.of(names).collect(() -> new StringBuilder(),
   (c, e) -> c.append(e), (c1, c2) -> c1.append(c2));
```

lambda 表达式 () -> new StringBuilder() 创建了用来存储结果的 StringBuilder 对象，这可以通过使用方法引用 StringBuilder::new 来简化。lambda 表达式 (c, e) -> c.append(e) 将字符串 e 添加至 StringBuilder c 中，这可以通过使用方法引用 StringBuilder::append

来简化。lambda 表达式 (c1, c2) -> c1.append(c2) 将 c2 中的内容合并到 c1 中，这也可以通过使用方法引用 StringBuilder::append 来简化。所以，可以将之前的语句做如下简化：

```
StringBuilder sb = Stream.of(names).collect(StringBuilder::new,
  StringBuilder::append, StringBuilder::append);
```

针对这个 collect 方法的顺序 foreach 循环可实现如下：

```
StringBuilder sb = new StringBuilder();
for (String s: Stream.of(names)) {
  sb.append(s);
}
```

注意，在顺序实现中没有使用合并器函数 (c1, c2) -> c1.append(c2)。它在流管道并行执行时使用。当并行地执行 collect 方法时，创建了多个 StringBuilder 结果，之后再用合并器函数合并。所以，合并器函数的目的是进行并行处理。

这里有另一个从流中字符串创建 ArrayList 的例子：

```
ArrayList<String> list = Stream.of(names).collect(ArrayList::new,
  ArrayList::add, ArrayList::addAll);
```

供应器函数是 ArrayList 的构造方法。累加器是 add 方法，它将一个元素添加到 ArrayList 中。合并器函数将一个 ArrayList 合并到另一个 ArrayList 中。这三个参数——供应器、累加器、合并器——是紧密耦合的，并使用标准方法进行定义。为了简单起见，Java 提供了另一种 collect 方法，它读入 Collector 类型的参数，称为收集器（collector）。Collector 接口定义了返回供应器、累加器以及合并器的方法。可以使用 Collector 类中的静态工厂方法 toList() 来创建 Collector 接口的实例。所以，之前的语句可以用标准收集器做如下简化：

```
List<String> list = Stream.of(names).collect(Collectors.toList());
```

程序清单 30-5 给出了使用 collect 方法和 Collector 的工厂方法的示例。

程序清单 30-5 CollectDemo.java

```
 1  import java.util.ArrayList;
 2  import java.util.List;
 3  import java.util.Map;
 4  import java.util.Set;
 5  import java.util.stream.Collectors;
 6  import java.util.stream.Stream;
 7
 8  public class CollectDemo {
 9    public static void main(String[] args) {
10      String[] names = {"John", "Peter", "Susan", "Kim", "Jen",
11        "George", "Alan", "Stacy", "Michelle", "john"};
12      System.out.println("The number of characters for all names: " +
13        Stream.of(names).collect(StringBuilder::new,
14          StringBuilder::append, StringBuilder::append).length());
15
16      List<String> list = Stream.of(names).collect(ArrayList::new,
17        ArrayList::add, ArrayList::addAll);
18      System.out.println(list);
19
20      list = Stream.of(names).collect(Collectors.toList());
21      System.out.println(list);
22
23      Set<String> set = Stream.of(names).map(e -> e.toUpperCase()).
```

```
24      collect(Collectors.toSet());
25    System.out.println(set);
26
27    Map<String, Integer> map = Stream.of(names).collect(
28      Collectors.toMap(e -> e, e -> e.length()));
29    System.out.println(map);
30
31    System.out.println("The total number of characters is " +
32      Stream.of(names).
33        collect(Collectors.summingInt(e -> e.length())));
34
35    java.util.IntSummaryStatistics stats = Stream.of(names).
36      collect(Collectors.summarizingInt(e -> e.length()));
37    System.out.println("Max is " + stats.getMax());
38    System.out.println("Min is " + stats.getMin());
39    System.out.println("Average is " + stats.getAverage());
40  }
41 }
```

```
The number of characters for all names: 47
[John, Peter, Susan, Kim, Jen, George, Alan, Stacy, Michelle, john]
[John, Peter, Susan, Kim, Jen, George, Alan, Stacy, Michelle, john]
[JEN, GEORGE, ALAN, SUSAN, JOHN, PETER, MICHELLE, KIM, STACY]
{Michelle=8, Stacy=5, Jen=3, George=6, Susan=5, Alan=4, John=4,
  john=4, Peter=5, Kim=3}
The total number of characters is 47
Max is 8
Min is 3
Average is 4.7
```

该程序创建了一个字符串数组 names（第 10～11 行）。collect 方法（第 13 行）指定了一个用于创建 StringBuilder 的供应器（StringBuilder::new），用于向 StringBuilder 添加字符串的累加器（StringBuilder::append），以及用于组合两个 StringBuilder 的合并器（StringBuilder::append）（第 13～14 行）。流管道获取包含流中所有字符串的 StringBuilder。length() 方法返回 StringBuilder 中字符的长度。

collect 方法（第 16 行）指定了一个用来创建 ArrayList 的供应器（ArrayList::new），一个用于向 ArrayList 添加字符串的累加器（ArrayList::add），一个用于合并两个 ArrayList 的合并器（ArrayList::addAll）（第 16～17 行）。流管道获取一个包含流中所有字符串的 ArrayList。该语句通过使用针对线性表的标准收集器 Collectors.toList() 进行了简化（第 20 行）。

该程序创建一个字符串流，将每个字符串映射为大写字符（第 23 行），并使用 collect 方法和针对规则集的标准收集器 Collectors.toSet() 创建一个规则集（第 24 行）。流中有两个大写字符串 JOHN。由于一个规则集中不包含重复的元素，因此只有一个 JOHN 存储在集合中。

该程序创建一个字符串流，并使用 collect 方法和用于映射的标准收集器创建一个映射（第 28 行）。映射的键是字符串，值是字符串的长度。注意，键必须是唯一的。如果在流中两个字符串相同，则会发生运行异常。

Collectors 类还包含返回收集器的方法，这些收集器生成汇总信息。例如，Collectors.summingInt 会生成流中整数值的总和（第 33 行），而 Collectors.summarizingInt 会为流中的整数值生成一个 IntSummaryStatistics（第 35～36 行）。

复习题

30.6.1 给出下列代码的输出：

```
int[] values = {1, 2, 3, 4, 1};
List<Integer> list = IntStream.of(values).mapToObj(e -> e)
  .collect(Collectors.toList());
System.out.println(list);

Set<Integer> set = IntStream.of(values).mapToObj(e -> e)
  .collect(Collectors.toSet());
System.out.println(set);

Map<Integer, Integer> map = IntStream.of(values).distinct()
  .mapToObj(e -> e)
  .collect(Collectors.toMap(e -> e, e -> e.hashCode()));
System.out.println(map);

System.out.println(
  IntStream.of(values).mapToObj(e -> e)
    .collect(Collectors.summingInt(e -> e)));

System.out.println(
  IntStream.of(values).mapToObj(e -> e)
    .collect(Collectors.averagingDouble(e -> e)));
```

30.7 使用 groupingBy 收集器进行元素分组

> **要点提示**：可以使用 groupingBy 收集器和 collect 方法按组收集元素。

可以使用 groupingBy 收集器将流中的元素分组，再在每个组上使用聚合方法。例如，可以按照首字符对所有字符串进行分组，并获取每个组中元素的数量，如下所示：

```
String[] names = {"John", "Peter", "Susan", "Kim", "Jen",
  "George", "Alan", "Stacy", "Steve", "john"};
Map<Character, Long> map = Stream.of(names).collect(
  Collectors.groupingBy(e -> e.charAt(0), Collectors.counting()));
```

groupingBy 方法中的第一个参数指定了分组的标准，称为分类器（classifier）。第二个参数指定了组中元素的处理方式，称为组处理器（processor）。处理器通常是一个汇总收集器，比如 counting()。使用 groupingBy 收集器、collect 方法返回以分类器作为键的映射。也可以在 groupingBy 方法中指定供应器，如下所示：

```
Map<Character, Long> map = Stream.of(names).collect(
Collectors.groupingBy(e -> e.charAt(0),
  TreeMap::new, Collectors.counting()));
```

在这个例子中，使用了树映射来存储映射的条目。

程序清单 30-6 给出了一个使用 groupingBy 方法的示例。

程序清单 30-6 CollectGroupDemo.java

```
 1  import java.util.Map;
 2  import java.util.TreeMap;
 3  import java.util.stream.Collectors;
 4  import java.util.stream.IntStream;
 5  import java.util.stream.Stream;
 6
 7  public class CollectGroupDemo {
 8    public static void main(String[] args) {
 9      String[] names = {"John", "Peter", "Susan", "Kim", "Jen",
10        "George", "Alan", "Stacy", "Steve", "john"};
```

```java
11
12      Map<String, Long> map1 = Stream.of(names).
13        map(e -> e.toUpperCase()).collect(
14          Collectors.groupingBy(e -> e, Collectors.counting()));
15      System.out.println(map1);
16
17      Map<Character, Long> map2 = Stream.of(names).collect(
18        Collectors.groupingBy(e -> e.charAt(0), TreeMap::new,
19          Collectors.counting()));
20      System.out.println(map2);
21
22      int[] values = {2, 3, 4, 1, 2, 3, 2, 3, 4, 5, 1, 421};
23      IntStream.of(values).mapToObj(e -> e).collect(
24        Collectors.groupingBy(e -> e, TreeMap::new,
25          Collectors.counting())).
26        forEach((k, v) -> System.out.println(k + " occurs " + v +
27          (v > 1 ? " times " : " time ")));
28
29      MyStudent[] students = {new MyStudent("John", "Lu", "CS", 32, 78),
30        new MyStudent("Susan", "Yao", "Math", 31, 85.4),
31        new MyStudent("Kim", "Johnson", "CS", 30, 78.1)};
32
33      System.out.printf("%10s%10s\n", "Department", "Average");
34      Stream.of(students).collect(Collectors.
35        groupingBy(MyStudent::getMajor, TreeMap::new,
36          Collectors.averagingDouble(MyStudent::getScore))).
37        forEach((k, v) -> System.out.printf("%10s%10.2f\n", k, v));
38    }
39  }
40
41  class MyStudent {
42    private String firstName;
43    private String lastName;
44    private String major;
45    private int age;
46    private double score;
47
48    public MyStudent(String firstName, String lastName, String major,
49        int age, double score) {
50      this.firstName = firstName;
51      this.lastName = lastName;
52      this.major = major;
53      this.age = age;
54      this.score = score;
55    }
56
57    public String getFirstName() {
58      return firstName;
59    }
60
61    public String getLastName() {
62      return lastName;
63    }
64
65    public String getMajor() {
66      return major;
67    }
68
69    public int getAge() {
70      return age;
71    }
72
73    public double getScore() {
74      return score;
75    }
76  }
```

```
{JEN=1, ALAN=1, GEORGE=1, SUSAN=1, JOHN=2, STEVE=1, PETER=1, STACY=1,
  KIM=1}
{A=1, G=1, J=2, K=1, P=1, S=3, j=1}
1 occurs 2 times
2 occurs 3 times
3 occurs 3 times
4 occurs 2 times
5 occurs 1 time
421 occurs 1 time
Department    Average
        CS     78.05
      Math     85.40
```

该程序创建一个字符串流,将其元素映射为大写(第 12 ~ 13 行),并将元素收集到一个映射中,以字符串作为键,并将字符串出现的次数作为值(第 14 行)。注意,counting() 收集器使用 Long 类型的值。所以,Map 被声明为 Map<String, Long>。

该程序创建一个字符串流,并将元素收集到一个映射中,将字符串首字符作为键,并将首字符出现的次数作为值(第 17 ~ 18 行)。键和值的条目存储在一个 TreeMap 中。

该程序创建一个 int 数组(第 22 行)。从这个数组创建一个流,并使用 mapToObj 方法将这些元素映射到 Integer 对象(第 23 行)。collect 方法返回一个以 Integer 值作为键、以整数出现次数作为值的映射(第 24 ~ 25 行)。forEach 方法显示键和值的条目。注意,collect 方法是一个终止方法。在这种情况下,它返回一个 TreeMap 的实例。TreeMap 具有 forEach 方法,用于对集合中的每个元素执行操作。

该程序创建一个 Student 对象数组(第 29 ~ 31 行)。在第 41 ~ 76 行中定义了具有属性 firstName、lastName、major、age 和 score 的 Student 类。该程序为该数组创建了一个流,而 collect 方法将学生按他们的专业分组,然后返回一个以专业为键、以该组平均分数为值的映射(第 34 ~ 36 行)。方法引用 MyStudent::getMajor 用于通过分类器指定组。方法引用 TreeMap::new 指定了结果映射的供应器。方法引用 MyStudent::getScore 指定要计算平均值的值。

✓ 复习题

30.7.1 给出下列代码的输出:

```
int[] values = {1, 2, 2, 3, 4, 2, 1};
IntStream.of(values).mapToObj(e -> e).collect(
  Collectors.groupingBy(e -> e, TreeMap::new,
    Collectors.counting())).
  forEach((k, v) -> System.out.println(k + " occurs " + v
    + (v > 1 ?  " times " :  " time ")));

IntStream.of(values).mapToObj(e -> e).collect(
  Collectors.groupingBy(e -> e, TreeMap::new,
    Collectors.summingInt(e -> e))).
  forEach((k, v) -> System.out.println(k + ": " + v));

MyStudent[] students = {
  new MyStudent("John", "Johnson", "CS", 23, 89.2),
  new MyStudent("Susan", "Johnson", "Math", 21, 89.1),
  new MyStudent("John", "Peterson", "CS", 21, 92.3),
  new MyStudent("Kim", "Yao", "Math", 22, 87.3),
  new MyStudent("Jeff", "Johnson", "CS", 23, 78.5)};

Stream.of(students)
  .sorted(Comparator.comparing(MyStudent::getLastName)
    .thenComparing(MyStudent::getFirstName))
  .forEach(e -> System.out.println(e.getLastName() + ", " +
             e.getFirstName()));
```

```
Stream.of(students).collect(Collectors.
    groupingBy(MyStudent::getAge, TreeMap::new,
      Collectors.averagingDouble(MyStudent::getScore))).
  forEach((k, v) -> System.out.printf("%10s%10.2f\n", k, v));
```

30.8 示例学习

要点提示：许多处理数组和集合的程序现在都可以通过流的聚合操作简化并运行得更快。

也可以不用流写出程序，然而，使用流可以写出更简短的程序，并通过利用多个处理器更快地并行执行。在前面章节中，许多涉及数组和集合的程序都可以使用流来简化。本节将介绍几个示例。

30.8.1 示例学习：数字分析

7.3 节给出了一个程序，提示用户输入数值，得到它们的平均值，并显示大于平均值的数值。程序可以使用 DoubleStream 简化，如程序清单 30-7 所示。

程序清单 30-7 AnalyzeNumbersUsingStream.java

```java
 1  import java.util.stream.*;
 2
 3  public class AnalyzeNumbersUsingStream {
 4    public static void main(String[] args) {
 5      java.util.Scanner input = new java.util.Scanner(System.in);
 6      System.out.print("Enter the number of items: ");
 7      int n = input.nextInt();
 8      double[] numbers = new double[n];
 9      double sum = 0;
10
11      System.out.print("Enter the numbers: ");
12      for (int i = 0; i < n; i++) {
13        numbers[i] = input.nextDouble();
14      }
15
16      double average = DoubleStream.of(numbers).average().getAsDouble();
17      System.out.println("Average is " + average);
18      System.out.println("Number of elements above the average is "
19        + DoubleStream.of(numbers).filter(e -> e > average).count());
20    }
21  }
```

```
Enter the number of items: 10 ↵Enter
Enter the numbers: 3.4 5 6 1 6.5 7.8 3.5 8.5 6.3 9.5 ↵Enter
Average is 5.75
Number of elements above the average is 6
```

该程序从用户处获得输入并将其存储在一个数组中（第 8 ～ 14 行），使用流获得这些数值的平均值（第 16 行），并使用过滤器查找大于平均值的数值个数（第 19 行）。

30.8.2 示例学习：计算字母的出现次数

程序清单 7-4 给出了一个随机生成 100 个小写字母并计算每个字母出现次数的程序。该程序可以用 Stream 来简化，如程序清单 30-8 所示。

程序清单 30-8 CountLettersUsingStream.java

```java
 1  import java.util.Random;
 2  import java.util.TreeMap;
```

```
3   import java.util.stream.Collectors;
4   import java.util.stream.Stream;
5
6   public class CountLettersUsingStream {
7     private static int count = 0;
8
9     public static void main(String[] args) {
10      Random random = new Random();
11      Object[] chars = random.ints(100, (int)'a', (int)'z' + 1).
12        mapToObj(e -> (char)e).toArray();
13
14      System.out.println("The lowercase letters are:");
15      Stream.of(chars).forEach(e -> {
16        System.out.print(e + (++count % 20 == 0 ? "\n" : " "));
17      });
18
19      count = 0; // Reset the count for columns
20      System.out.println("\nThe occurrences of each letter are:");
21      Stream.of(chars).collect(Collectors.groupingBy(e -> e,
22        TreeMap::new, Collectors.counting())).forEach((k, v) -> {
23        System.out.print(v + " " + k
24          + (++count % 10 == 0 ? "\n" : " "));
25      });
26    }
27  }
```

```
The lowercase letters are:
e y l s r i b k j v j h a b z n w b t v
s c c k r d w a m p w v u n q a m p l o
a z g d e g f i n d x m z o u l o z j v
h w i w n t g x w c d o t x h y v z y z
q e a m f w p g u q t r e n n w f c r f

The occurrences of each letter are:
5 a 3 b 4 c 4 d 4 e 4 f 4 g 3 h 3 i 3 j
2 k 3 l 4 m 6 n 4 o 3 p 3 q 4 r 2 s 4 t
3 u 5 v 8 w 3 x 3 y 6 z
```

该程序产生一个 100 个随机整数的流。这些整数的范围在（char）'a' 和（char）'z' 之间（第 11 行）。它们是小写字母的 ASCII 码。mapToObj 方法将整数映射到它们相应的小写字母上（第 12 行）。toArray() 方法返回一个由这些小写字母组成的数组。

该程序创建一个小写字母流（第 15 行），并使用 forEach 方法显示每个字母（第 16 行）。每行显示 20 个字母。静态变量 count 用于对显示过的字母计数。

该程序将 count 值重置为 0（第 19 行），创建一个小写字母流（第 21 行），返回一个以小写字母作为键、每个字母出现次数作为值的映射（第 21 ~ 22 行），并调用 forEach 方法显示每个键和值，每行显示 10 个（第 23 ~ 24 行）。

程序清单 7-4 中有 66 行代码。新的代码只有 27 行，大大简化了编码。此外，使用流更加高效。

30.8.3 示例学习：计算字符串中每个字母的出现次数

前面的例子随机生成小写字母并计算每个字母的出现次数。这个示例将计算字符串中每个字母的出现次数。程序清单 30-9 中给出的程序提示用户输入字符串，将所有字母转换为大写，并显示字符串中每个字母的计数。

程序清单 30-9 CountOccurrenceOfLettersInAString.java

```
1   import java.util.*;
```

```java
 2  import java.util.stream.Stream;
 3  import java.util.stream.Collectors;
 4
 5  public class CountOccurrenceOfLettersInAString {
 6    private static int count = 0;
 7
 8    public static void main(String[] args) {
 9      Scanner input = new Scanner(System.in);
10      System.out.print("Enter a string: ");
11      String s = input.nextLine();
12
13      count = 0; // Reset the count for columns
14      System.out.println("The occurrences of each letter are:");
15      Stream.of(toCharacterArray(s.toCharArray()))
16        .filter(ch -> Character.isLetter(ch))
17        .map(ch -> Character.toUpperCase(ch))
18        .collect(Collectors.groupingBy(e -> e,
19          TreeMap::new, Collectors.counting()))
20        .forEach((k, v) -> { System.out.print(v + " " + k
21          + (++count %  10 == 0 ?  "\n" :  " "));
22        });
23    }
24
25    public static Character[] toCharacterArray(char[] list) {
26      Character[] result = new Character[list.length];
27      for (int i = 0; i < result.length; i++) {
28        result[i] = list[i];
29      }
30      return result;
31    }
32  }
```

```
Enter a string: Welcome to JavaAA  ↵Enter
The occurrences of each letter are:
4 A 1 C 2 E 1 J 1 L 1 M 2 O 1 T 1 V 1 W
```

该程序读取一个字符串 s，并通过调用 s.toCharArray() 从字符串获得一个 char 数组。要创建一个字符流，需要将 char[] 转换为 Character[]。因此，该程序定义了 toCharacterArray 方法（25～31行），用于从 char[] 获得一个 Character[]。

该程序创建了一个 Character 对象流（第15行），使用 filter 方法从流中消除了非字母字符（第16行），使用 map 方法将所有字母转换为大写（第17行），并使用 collect 方法获得 TreeMap（第18～19行）。在 TreeMap 中，键是字母，值是字母的计数。TreeMap 中的 forEach 方法（第20～21行）用于显示值和键。

30.8.4 示例学习：处理二维数组中的所有元素

可以从一维数组中创建一个流。可以创建一个流来处理二维数组吗？程序清单 30-10 给出了一个使用流处理二维数组的例子。

程序清单 30-10 TwoDimensionalArrayStream.java

```java
 1  import java.util.IntSummaryStatistics;
 2  import java.util.stream.IntStream;
 3  import java.util.stream.Stream;
 4
 5  public class TwoDimensionalArrayStream {
 6    private static int i = 0;
 7    public static void main(String[] args) {
 8      int[][] m = {{1, 2}, {3, 4}, {4, 5}, {1, 3}};
 9
```

```
10    int[] list = Stream.of(m).map(e -> IntStream.of(e)).
11      reduce((e1, e2) -> IntStream.concat(e1, e2)).get().toArray();
12
13    IntSummaryStatistics stats =
14      IntStream.of(list).summaryStatistics();
15    System.out.println("Max: " + stats.getMax());
16    System.out.println("Min: " + stats.getMin());
17    System.out.println("Sum: " + stats.getSum());
18    System.out.println("Average: " + stats.getAverage());
19
20    System.out.println("Sum of row ");
21    Stream.of(m).mapToInt(e -> IntStream.of(e).sum())
22      .forEach(e ->
23        System.out.println("Sum of row " + i++ + ": " + e));
24  }
25 }
```

```
Max: 5
Min: 1
Sum: 23
Average: 2.875
Sum of row
Sum of row 0: 3
Sum of row 1: 7
Sum of row 2: 9
Sum of row 3: 4
```

该程序创建了一个二维数组 m（第 8 行）。调用 Stream.of(m) 创建一个由行作为元素的流（第 10 行）。map 方法将每一行映射到一个 IntStream。reduce 方法将这些流连接成一个大的流（第 11 行）。这个大的流现在包含了 m 中所有的元素。从流中调用 toArray() 返回一个由流中所有整数组成的数组（第 11 行）。

该程序获得从数组创建的针对 IntStream 的统计汇总（第 13～14 行），并显示流中整数的最大值、最小值、总和以及平均值（第 15～18 行）。

最后，该程序从 m 中创建一个由 m 中的行组成的流（第 21 行）。每行都使用 mapToInt 方法映射到一个 int 值（第 21 行）。int 值是该行中元素的总和。forEach 方法显示每行的总和（第 22～23 行）。

30.8.5 示例学习：得到目录大小

程序清单 18-7 给出了一个递归程序，用于得到目录的大小。目录的大小是目录中所有文件大小的总和。该程序可以使用如程序清单 30-11 所示的流来实现。

程序清单 30-11 DirectorySizeStream.java

```
1  import java.io.File;
2  import java.nio.file.Files;
3  import java.util.Scanner;
4
5  public class DirectorySizeStream {
6    public static void main(String[] args) throws Exception {
7      // Prompt the user to enter a directory or a file
8      System.out.print("Enter a directory or a file: ");
9      Scanner input = new Scanner(System.in);
10     String directory = input.nextLine();
11
12     // Display the size
13     System.out.println(getSize(new File(directory)) + " bytes");
14   }
15
```

```
16    public static long getSize(File file) {
17      if (file.isFile()) {
18        return file.length();
19      }
20      else {
21        try {
22          return Files.list(file.toPath()).parallel().
23            mapToLong(e -> getSize(e.toFile())).sum();
24        } catch (Exception ex) {
25          return 0;
26        }
27      }
28    }
29  }
```

```
Enter a directory or a file: c:\book  ↵Enter
48619631 bytes
```

```
Enter a directory or a file: c:\book\Welcome.java  ↵Enter
172 bytes
```

```
Enter a directory or a file: c:\book\NonExistentFile  ↵Enter
0 bytes
```

该程序提示用户输入文件或目录名（第 8～10 行），并调用 getSize(File file) 方法返回文件或目录的大小（第 13 行）。File 对象可以是目录或文件。如果是文件，则 getSize 方法返回文件的大小（第 17～19 行）。如果是目录，则调用 Files.list(file.toPath()) 返回由 Path 对象组成的流。每个 Path 对象代表该目录中的一个子路径（第 22 行）。从 JDK 1.8 开始，一些现有的类中添加了新的方法来返回流。Files 类现在具有静态 list(Path) 方法，该方法返回路径中的子路径流。对于每个子路径 e，mapToLong 方法通过调用 getSize(e) 将 e 映射到 e 的大小。最后，终止方法 sum() 返回整个目录的大小。

程序清单 18-7 和程序清单 30-9 中的 getSize 方法都是递归的。但程序清单 30-9 中的 getSize 方法可以在并行流中执行。因此，程序清单 30-9 具有更好的性能。

30.8.6 示例学习：关键字计数

程序清单 21-7 给出了一个程序，用于计算 Java 源文件中关键字的个数。该程序从文本文件中读取单词并测试该单词是否是关键字。可以使用流重写代码，如程序清单 30-12 所示。

程序清单 30-12 CountKeywordStream.java

```
1   import java.util.*;
2   import java.io.*;
3   import java.nio.file.Files;
4   import java.util.stream.Stream;
5
6   public class CountKeywordStream {
7     public static void main(String[] args) throws Exception {
8       Scanner input = new Scanner(System.in);
9       System.out.print("Enter a Java source file: ");
10      String filename = input.nextLine();
11
12      File file = new File(filename);
13      if (file.exists()) {
14        System.out.println("The number of keywords in " + filename
15          + " is " + countKeywords(file));
16      }
```

```
17      else {
18        System.out.println("File " + filename + " does not exist");
19      }
20    }
21
22    public static long countKeywords(File file) throws Exception {
23      // Array of all Java keywords + true, false and null
24      String[] keywordString = {"abstract", "assert", "boolean",
25        "break", "byte", "case", "catch", "char", "class", "const",
26        "continue", "default", "do", "double", "else", "enum",
27        "extends", "for", "final", "finally", "float", "goto",
28        "if", "implements", "import", "instanceof", "int",
29        "interface", "long", "native", "new", "package", "private",
30        "protected", "public", "return", "short", "static",
31        "strictfp", "super", "switch", "synchronized", "this",
32        "throw", "throws", "transient", "try", "void", "volatile",
33        "while", "true", "false", "null"};
34
35      Set<String> keywordSet =
36        new HashSet<>(Arrays.asList(keywordString));
37
38      return Files.lines(file.toPath()).parallel().mapToLong(line ->
39        Stream.of(line.split("[\\s++]")).
40          filter(word -> keywordSet.contains(word)).count()).sum();
41    }
42  }
```

```
Enter a Java source file: c:\Welcome.java ↵Enter
The number of keywords in c:\Welcome.java is 5
```

```
Enter a Java source file: c:\TTT.java ↵Enter
File c:\TTT.java does not exist
```

该程序提示用户输入文件名（第9～10行）。如果文件存在，则调用conutKeywords(file)返回文件中关键字的数目（第15行）。关键字存储在规则集keywordSet中（第35～36行）。调用Files.lines(file.toPath())从文件中返回由文件中的行组成的流（第38行）。流中的每一行都被映射为一个long值，用于统计行中关键字的数量。使用line.split("[\\ s ++]")将该行分割为一个单词数组，并使用该数组创建一个流（第39行）。filter方法应用于从流中选择关键字。调用count()返回一行中关键字的数量。sum()方法返回所有行中关键字的总数（第40行）。

在这个例子中使用并行流的真正好处是提高性能（第38行）。

30.8.7 示例学习：单词出现次数

程序清单21-9给出了一个用于计算文本中单词出现次数的程序。可以使用流重写该代码，如程序清单30-13所示。

程序清单30-13 CountOccurrenceOfWordsStream.java

```
1  import java.util.*;
2  import java.util.stream.Collectors;
3  import java.util.stream.Stream;
4
5  public class CountOccurrenceOfWordsStream {
6    public static void main(String[] args) {
7      // Set text in a string
8      String text = "Good morning. Have a good class. "
9        + "Have a good visit. Have fun!";
10
```

```
11      Stream.of(text.split("[\\s+\\p{P}]")).parallel()
12        .filter(e -> e.length() > 0).collect(
13          Collectors.groupingBy(String::toLowerCase, TreeMap::new,
14            Collectors.counting()))
15        .forEach((k, v) -> System.out.println(k + " " + v));
16    }
17  }
```

```
a        2
class    1
fun      1
good     3
have     3
morning  1
visit    1
```

使用空格 \s 或标点 \p{P} 作为分隔符将文本分割成单词（第 11 行），并为单词创建一个流。filter 方法用于选择非空单词（第 12 行）。collect 方法通过将每个单词转换为小写，并返回一个以小写单词作为键、以它们的计数值作为值的 TreeMap 来对单词进行分组（第 13 ~ 14 行）。forEach 方法显示键和它的值（第 15 行）。

程序清单 30-13 中的代码行数约为程序清单 21-9 中的一半。新程序大大简化了编码并提高了性能。

✓ 复习题

30.8.1 可以用下列代码代替程序清单 30-7 中的第 19 行吗？

```
DoubleStream.of(numbers).filter(e -> e >
  DoubleStream.of(numbers).average()).count());
```

30.8.2 可以用下列代码代替程序清单 30-8 中的第 15 ~ 16 行吗？

```
Stream.of(chars).forEach(e -> {
  int count = 0;
  System.out.print(e + (++count % 20 == 0 ? "\n" : " ")); });
```

30.8.3 给出下列代码的输出：

```
String s = "ABC";
Stream.of(s.toCharArray()).forEach(ch ->
  System.out.println(ch));
```

30.8.4 给出下列代码的输出（toCharacterArray 方法在程序清单 30-9 中给出）。

```
String s = "ABC";
Stream.of(toCharacterArray(s.toCharArray())).forEach(ch ->
  System.out.println(ch));
```

30.8.5 编写代码，从二维字符串数组 matrix 中得到一维字符串数组线性表 list。

本章小结

1. Java 8 引入了集合流上的聚合操作以简化代码并提高性能。
2. 流管道从一个数据源创建一个流，包含零个或多个中间方法（skip、limit、filter、distinct、sorted、map、mapToInt）和一个最后的终止方法（count、sum、average、max、min、forEach、findFirst、firstAny、anyMatch、allMatch、noneMatch、toArray）。
3. 流的执行是惰性的，这意味着流中的方法在终止方法启动之前不会被执行。
4. 流是临时对象。一旦终止方法被执行，它们就被销毁。
5. Stream <T> 类为 T 类型的对象序列定义了流。IntStream、LongStream 和 DoubleStream 是针对基本数据类型 int、long 和 double 值序列的流。

6. 使用流的一个重要好处是性能。流可以利用多核架构在并行模式下执行。可以通过调用 parallel() 或 sequential() 方法将流切换为并行或顺序模式。
7. 可以使用 reduce 方法将流归约为单个值，并使用 collect 方法将流中的元素放入集合中。
8. 可以使用 groupingBy 收集器对流中的元素进行分组，并为组中的元素应用聚合方法。

测试题

回答位于本书配套网站上的本章测试题。

编程练习题

30.1 （成绩评定）使用流改写编程练习题 7.1。
30.2 （数字出现次数计数）使用流改写编程练习题 7.3。
30.3 （成绩分析）使用流改写编程练习题 7.4。
30.4 （打印不同的数字）使用流改写编程练习题 7.5，并以升序显示数字。
30.5 （单个数位计数）使用流改写编程练习题 7.7。
30.6 （计算数组平均值）使用流改写编程练习题 7.8。
30.7 （寻找最小元素）使用流改写编程练习题 7.9。
30.8 （消除重复）使用流改写编程练习题 7.15，并以升序把元素存储在新的数组中。
30.9 （对学生排序）使用流改写编程练习题 7.17。定义一个具有 name 和 score 数据域以及它们的 getter 方法的 Student 类。将每个学生存储在一个 Student 对象中。
30.10 （二进制转十进制）编写一个程序，提示用户以字符串形式输入一个二进制数字，并显示它的十进制值。使用 Stream 的 reduce 方法将二进制数字转化为十进制。
30.11 （十六进制转十进制）编写一个程序，提示用户以字符串形式输入一个十六进制数字，并显示它的十进制值。使用 Stream 的 reduce 方法将十六进制数字转化为十进制。
30.12 （对整数中的位数求和）使用流重写编程练习题 6.2。
30.13 （对字符串中的字母计数）使用流重写编程练习题 6.20。
30.14 （特定字符出现次数）使用流重写编程练习题 6.23。
30.15 （以字母表顺序升序显示单词）使用流重写编程练习题 20.1。
30.16 （不同的分数）使用流编写一个程序，显示 8.8 节中 scores 数组中不同的分数。以升序显示，分数间以一个空格隔开，5 个一行。
30.17 （辅音和元音计数）使用流重写编程练习题 21.4。
30.18 （文本文件中单词的出现次数）使用流重写编程练习题 21.8。
30.19 （汇总信息）假设文件 test.txt 中包含了以空格分隔的浮点数。编写程序，获得这些数字的总和、平均值、最大值以及最小值。

附录 A
Introduction to Java Programming and Data Structures, Comprehensive Version, Twelfth Edition

Java 关键字和保留字

Java 中的关键字有特殊含义，是语法的一部分。保留字是不能作为标识符的单词。关键字是保留字。下面 50 个关键字是 Java 语言保留使用的：

abstract	double	int	super
assert	else	interface	switch
boolean	enum	long	synchronized
break	extends	native	this
byte	final	new	throw
case	finally	package	throws
catch	float	private	transient
char	for	protected	try
class	goto	public	void
const	if	return	volatile
continue	implements	short	while
default	import	static	
do	instanceof	strictfp⊖	

关键字 goto 和 const 是 C++ 保留的关键字，目前并没有在 Java 中用到。如果它们出现在 Java 程序中，Java 编译器能够识别它们，并产生错误信息。

字面常量 true、false 和 null 是保留字，但不是关键字，不能将其用作标识符。

在代码清单中，我们对 true、false 和 null 使用了关键字的颜色，以和 Java IDE 中它们的颜色保持一致。

⊖ strictfp 关键字用于修饰方法或者类，使其能使用严格的浮点计算。浮点计算可以使用以下两种模式：严格的和非严格的。严格模式可以保证计算结果在所有的虚拟机实现中都是一样的。非严格模式允许计算的中间结果以一种扩展的格式存储，该格式不同于标准的 IEEE 浮点数格式。扩展格式是依赖于机器的，可以使代码执行更快。然而，当在不同的虚拟机上使用非严格模式执行代码时，可能不会总能精确地得到相同的结果。默认情况下，非严格模式被用于浮点数的计算。若在方法和类中使用严格模式，需要在方法或者类的声明中增 strictfp 关键字。严格的浮点数可能会比非严格浮点数具有略好的精确度，但这种区别仅影响某些应用。严格模式不会被继承，也就是说，在类或者接口的声明中使用 strictfp 不会使得继承的子类或接口也是严格模式。

附录 B

ASCII 字符集

表 B-1 和表 B-2 分别列出了 ASCII 字符及其相应的十进制和十六进制编码。字符的十进制或十六进制编码是其行下标和列下标的组合。例如，在表 B-1 中，字母 A 在第 6 行第 5 列，所以它的十进制代码为 65；在表 B-2 中，字母 A 在第 4 行第 1 列，所以它的十六进制代码为 41。

表 B-1 十进制编码的 ASCII 字符集

	0	1	2	3	4	5	6	7	8	9
0	nul	soh	stx	etx	eot	enq	ack	bel	bs	ht
1	nl	vt	ff	cr	so	si	dle	dcl	dc2	dc3
2	dc4	nak	syn	etb	can	em	sub	esc	fs	gs
3	rs	us	sp	!	"	#	$	%	&	'
4	(	)	*	+	,	-	.	/	0	1
5	2	3	4	5	6	7	8	9	:	;
6	<	=	>	?	@	A	B	C	D	E
7	F	G	H	I	J	K	L	M	N	O
8	P	Q	R	S	T	U	V	W	X	Y
9	Z	[	\	]	^	_	'	a	b	c
10	d	e	f	g	h	i	j	k	l	m
11	n	o	p	q	r	s	t	u	v	w
12	x	y	z	{	|	}	~	del		

表 B-2 十六进制编码的 ASCII 字符集

	0	1	2	3	4	5	6	7	8	9	A	B	C	D	E	F
0	nul	soh	stx	etx	eot	enq	ack	bel	bs	ht	nl	vt	ff	cr	so	si
1	dle	dcl	dc2	dc3	dc4	nak	syn	etb	can	em	sub	esc	fs	gs	rs	us
2	sp	!	"	#	$	%	&	'	(	)	*	+	,	-	.	/
3	0	1	2	3	4	5	6	7	8	9	:	;	<	=	>	?
4	@	A	B	C	D	E	F	G	H	I	J	K	L	M	N	O
5	P	Q	R	S	T	U	V	W	X	Y	Z	[	\	]	^	_
6	'	a	b	c	d	e	f	g	h	i	j	k	l	m	n	o
7	p	q	r	s	t	u	v	w	x	y	z	{	|	}	~	del

附录 C

Introduction to Java Programming and Data Structures, Comprehensive Version, Twelfth Edition

操作符优先级表

操作符按照优先级递减的顺序从上到下列出。同一栏中的操作符优先级相同，它们的结合方向如表中所示。

操作符	名称	结合方向	操作符	名称	结合方向
( )	圆括号	从左向右	>>>	用零扩展的右移	从左向右
( )	函数调用	从左向右	<	小于	从左向右
[ ]	数组下标	从左向右	<=	小于等于	从左向右
.	对象成员访问	从左向右	>	大于	从左向右
++	后置自增	从右向左	>=	大于等于	从左向右
--	后置自减	从右向左	instanceof	检测对象类型	从左向右
++	前置自增	从右向左	==	相等	从左向右
--	前置自减	从右向左	!=	不等	从左向右
+	一元加	从右向左	&	（无条件与）	从左向右
-	一元减	从右向左	^	（异或）	从左向右
!	一元逻辑非	从右向左	|	（无条件或）	从左向右
(type)	一元类型转换	从右向左	&&	条件与	从左向右
new	创建对象	从右向左	||	条件或	从左向右
*	乘法	从左向右	?:	三元条件	从右向左
/	除法	从左向右	=	赋值	从右向左
%	取模	从左向右	+=	加法赋值	从右向左
+	加法	从左向右	-=	减法赋值	从右向左
-	减法	从左向右	*=	乘法赋值	从右向左
<<	左移	从左向右	/=	除法赋值	从右向左
>>	用符号位扩展的右移	从左向右	%=	取模赋值	从右向左

附录 D
Introduction to Java Programming and Data Structures, Comprehensive Version, Twelfth Edition

Java 修饰符

修饰符用于类和类的成员（构造方法、方法、数据和类一级的块），但 final 修饰符也可以用在方法中的局部变量上。可以用在类上的修饰符称为类修饰符（class modifier）。可以用在方法上的修饰符称为方法修饰符（method modifier）。可以用在数据域上的修饰符称为数据修饰符（data modifier）。可以用在类一级块上的修饰符称为块修饰符（block modifier）。下表对 Java 修饰符进行了总结。

修饰符	类	构造方法	方法	数据	块	解释
空白①	√	√	√	√	√	类、构造方法、方法或数据域在所在的包中可见
public	√	√	√	√		类、构造方法、方法或数据域对于任何包中的任何程序都可见
private		√	√	√		构造方法、方法或数据域仅在所在类中可见
protected		√	√	√		构造方法、方法或数据域在所属包中可见，或者在任何包中该类的子类中可见
static			√	√	√	定义类方法、类数据域或静态初始化模块
final	√		√	√		final 类不能被继承。final 方法不能在子类中修改。终极数据域是常量
abstract	√		√			抽象类必须被继承。抽象方法必须在具体的子类中实现
native			√			用 native 修饰的方法表明它是用 Java 之外的语言实现的
synchronized			√		√	同一时间只有一个线程可以执行这个方法
strictfp	√		√			使用精确浮点数计算模式，保证在所有的 Java 虚拟机中计算结果都相同
transient				√		标记不可序列化的实例数据域

①（空白）表明没有使用任何修饰符，例如 class Test{}。

public、private 以及 protected 等修饰符称为可见或者可访问性修饰符，因为它们给定了类，以及类的成员是如何被访问的。

public、private、protected、static、final 以及 abstract 也可以用于内部类。

Java 8 引入了 default 修饰符，用于在接口中声明默认方法。默认方法为接口中的方法提供了一种默认实现。

附录 E
Introduction to Java Programming and Data Structures, Comprehensive Version, Twelfth Edition

特殊浮点值

整数除以零是非法的，会抛出异常 ArithmeticException，但是浮点值除以零不会引起异常。在浮点运算中，如果运算结果对 double 型或 float 型来说太大，则向上溢出为无穷大；如果运算结果对 double 型或 float 型来说太小，则向下溢出为零。Java 提供了特殊的浮点值 POSITIVE_INFINITY、NEGATIVE_INFINITY 和 NaN（Not a Number，非数值）来表示这些结果。这些值被定义为 Float 类和 Double 类中的特殊常量。

如果正浮点数除以零，则结果为 POSITIVE_INFINITY。如果负浮点数除以零，则结果为 NEGATIVE_INFINITY。如果浮点数零除以零，则结果为 NaN，表示这个结果在数学意义上没有定义。这三个值的字符串表示分别为 Infinity、-Infinity 和 NaN。例如，

```
System.out.print(1.0 / 0); // Print Infinity
System.out.print(-1.0 / 0); // Print -Infinity
System.out.print(0.0 / 0); // Print NaN
```

这些特殊值也可以在运算中用作操作数。例如，一个数除以 POSITIVE_INFINITY 得到零。表 E-1 总结了运算符 /、*、%、+ 和 - 的各种组合。

表 E-1 特殊的浮点值

x	y	x/y	x*y	x%y	x+y	x-y
Finite	±0.0	±infinity	±0.0	NaN	Finite	Finite
Finite	±infinity	±0.0	±0.0	x	±infinity	infinity
±0.0	±0.0	NaN	±0.0	NaN	±0.0	±0.0
±infinity	Finite	±infinity	±0.0	NaN	±infinity	±infinity
±infinity	±infinity	NaN	±0.0	NaN	±infinity	infinity
±0.0	±infinity	±0.0	NaN	±0.0	±infinity	±0.0
NaN	Any	NaN	NaN	NaN	NaN	NaN
Any	NaN	NaN	NaN	NaN	NaN	NaN

注意：如果操作数之一是 NaN，则结果为 NaN。

附录 F

数 系

F.1 引言

因为计算机被制作为天然是存储和处理 0 和 1 的,所以其内部使用的是二进制数。二进制数系只有两个数字:0 和 1。在计算机中,数字或字符是以由 0 和 1 组成的序列来存储的。每个 0 或 1 都称为一个比特(二进制数字)。

日常生活中,我们使用十进制数。当我们在程序中编写一个数字,如 20,它被假定为一个十进制数。在计算机内部,通常会用软件将十进制数转换成二进制数,反之亦然。

我们使用十进制数编写程序。然而,如果要与操作系统打交道,需要使用二进制数以达到底层的"机器级"。二进制数冗长烦琐,所以经常使用十六进制数简化二进制数,每个十六进制数可以表示四个二进制数。十六进制数系有十六个数:0~9、A~F,其中字母 A、B、C、D、E 和 F 对应十进制数 10、11、12、13、14 和 15。

十进制数系中的数字是 0、1、2、3、4、5、6、7、8 和 9。一个十进制数是用一个或多个这些数字所构成的一个序列来表示的。这个序列中每个数所表示的值和它的位置有关,序列中数的位置决定了 10 的幂次。例如,十进制数 7423 中的数 7、4、2 和 3 分别表示 7000、400、20 和 3,如下所示:

$$\boxed{7\ 4\ 2\ 3} = 7 \times 10^3 + 4 \times 10^2 + 2 \times 10^1 + 3 \times 10^0$$
$$10^3\ 10^2\ 10^1\ 10^0 = 7000 + 400 + 20 + 3 = 7423$$

十进制数系有十个数字,它们的位置值都是 10 的整数次幂。我们表达为 10 是十进制数系的基数。类似地,由于二进制数系有两个数,所以它的基数为 2;而十六进制数系有 16 个数,所以它的基数为 16。

如果 1101 是一个二进制数,那么数 1、1、0 和 1 分别表示:

$$\boxed{1\ 1\ 0\ 1} = 1 \times 2^3 + 1 \times 2^2 + 0 \times 2^1 + 1 \times 2^0$$
$$2^3\ 2^2\ 2^1\ 2^0 = 8 + 4 + 0 + 1 = 13$$

如果 7423 是一个十六进制数,那么数字 7、4、2 和 3 分别表示:

$$\boxed{7\ 4\ 3\ 2} = 7 \times 16^3 + 4 \times 16^2 + 2 \times 16^1 + 3 \times 16^0$$
$$16^3\ 16^2\ 16^1\ 16^0 = 28672 + 1024 + 32 + 3 = 29731$$

F.2 二进制数与十进制数之间的转换

给定一个二进制数 $b_n b_{n-1} b_{n-2} \cdots b_2 b_1 b_0$,其等价的十进制数为

$$b_n \times 2^n + b_{n-1} \times 2^{n-1} + b_{n-2} \times 2^{n-2} + \cdots + b_2 \times 2^2 + b_1 \times 2^1 + b_0 \times 2^0$$

下面是二进制数转换为十进制数的一些例子:

二进制	转换公式	十进制
10	$1 \times 2^1 + 0 \times 2^0$	2
1000	$1 \times 2^3 + 0 \times 2^2 + 0 \times 2^1 + 0 \times 2^0$	8
10101011	$1 \times 2^7 + 0 \times 2^6 + 1 \times 2^5 + 0 \times 2^4 + 1 \times 2^3 + 0 \times 2^2 + 1 \times 2^1 + 1 \times 2^0$	171

把一个十进制数 d 转换为二进制数，就是求满足

$$d = b_n \times 2^n + b_{n-1} \times 2^{n-1} + b_{n-2} \times 2^{n-2} + \cdots + b_2 \times 2^2 + b_1 \times 2^1 + b_0 \times 2^0$$

的比特 b_n，b_{n-1}，b_{n-2}，$\cdots$，b_2，b_1 和 b_0。

用 2 不断地除 d，直到商为 0 为止。余数即为所求的比特 b_0，b_1，$\cdots$，b_{n-2}，b_{n-1}，b_n。

例如，十进制数 123 为二进制数 1111011。转换过程如下：

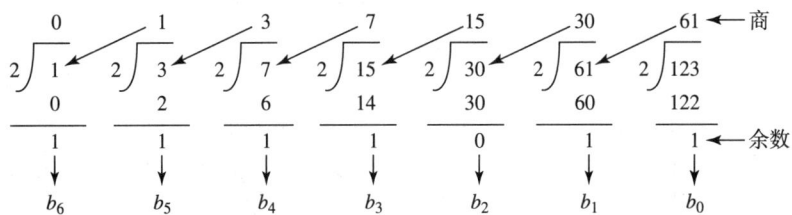

提示：Windows 操作系统所带的计算器是进行数制转换的一个有用的工具，如图 F-1 所示。要运行它，从 Start 按钮搜索 Calculator 并运行 Calculator，然后在 View 菜单下面选择 Scientific。

图 F-1　可以使用 Windows 的计算器进行数制转换

F.3　十六进制数与十进制数之间的转换

给定十六进制数 $h_n h_{n-1} h_{n-2} \cdots h_2 h_1 h_0$，其等价的十进制数为

$$h_n \times 16^n + h_{n-1} \times 16^{n-1} + h_{n-2} \times 16^{n-2} + \cdots + h_2 \times 16^2 + h_1 \times 16^1 + h_0 \times 16^0$$

下面是十六进制数转换为十进制数的例子：

十六进制	转换公式	十进制
7F	$7 \times 16^1 + 15 \times 16^0$	127
FFFF	$15 \times 16^3 + 15 \times 16^2 + 15 \times 16^1 + 15 \times 16^0$	65535
431	$4 \times 16^2 + 3 \times 16^1 + 1 \times 16^0$	1073

将一个十进制数 d 转换为十六进制数，就是求满足

$$d = h_n \times 16^n + h_{n-1} \times 16^{n-1} + h_{n-2} \times 16^{n-2} + \cdots + h_2 \times 16^2 + h_1 \times 16^1 + h_0 \times 16^0$$

的比特 $h_n, h_{n-1}, h_{n-2}, \cdots, h_2, h_1$ 和 h_0。用 16 不断地除 d，直到商为 0 为止。余数即为所求的比特 $h_0, h_1, \cdots, h_{n-2}, h_{n-1}, h_n$。

例如，十进制数 123 为十六进制数 7B。转换过程如下：

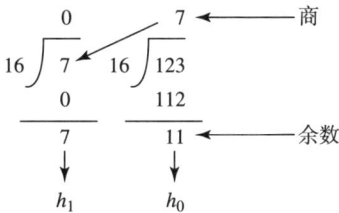

F.4 二进制数与十六进制数之间的转换

将一个十六进制数转换为二进制数，利用表 F-1，就可以简单地把十六进制数的每一位转换为四位二进制数。

例如，十六进制数 7B 转换为二进制是 1111011，其中 7 的二进制表示为 111，B 的二进制表示为 1011。

要将一个二进制数转换为十六进制数，从右向左将每四位二进制数转换为一位十六进制数。

例如，二进制数 001110001101 的十六进制表示是 38D，因为 1101 是 D，1000 是 8，0011 是 3，如下所示：

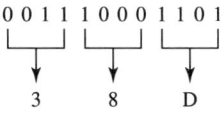

表 F-1 将十六进制数转换为二进制数

十六进制	二进制	十进制	十六进制	二进制	十进制
0	0000	0	8	1000	8
1	0001	1	9	1001	9
2	0010	2	A	1010	10
3	0011	3	B	1011	11
4	0100	4	C	1100	12
5	0101	5	D	1101	13
6	0110	6	E	1110	14
7	0111	7	F	1111	15

注意：八进制数也很有用。八进制数系有 0 到 7 共八个数。十进制数 8 在八进制数系中的作用就和十进制数系中的 10 一样。

复习题

F.1 将下列十进制数转换为十六进制数和二进制数。
 100；4340；2000

F.2 将下列二进制数转换为十六进制数和十进制数。
 1000011001；100000000；100111

F.3 将下列十六进制数转换为二进制数和十进制数。
 FEFA9；93；2000

位 操 作 符

用机器语言编写程序时，经常需要直接处理二进制数值，并在位级别上执行操作。Java 提供了位操作符和移位操作符，如表 G-1 所示。

表 G-1

操作符	名称	示例（例中使用字节）	描述
&	位与	10101110 & 10010010 得到 10000010	两个相应位上的比特如果都为 1，则执行与操作得到 1
|	位或	10101110 | 10010010 得到 10111110	两个相应位上的比特如果其中一个为 1，则执行或操作得到 1
^	位异或	10101110 ^ 10010010 得到 00111100	两个相应位上的比特如果相异，则执行异或操作得到 1
~	求反	~10101110 得到 01010001	操作符将每个比特进行从 0 到 1 以及从 1 到 0 的转换
<<	左移位	10101110 << 2 得到 10111000	操作符将第一个操作数按照第二个操作数指定的比特数进行左移位，右边补 0
>>	带符号位右移位	10101110 >> 2 得到 11101011 00101110 >> 2 得到 00001011	操作符将第一个操作数按照第二个操作数指定的比特数进行右移位，左边补上最高（符号）位
>>>	无符号位右移位	10101110 >>> 2 得到 00101011 00101110 >>> 2 得到 00001011	操作符将第一个操作数按照第二个操作数指定的位移数进行右移位，左边补 0

位操作符仅适用于整数类型（byte、short、int 和 long）。位操作涉及的字符将转换为整数。所有的位操作符都可以形成位赋值操作符，例如 ^=、|=、<<=、>>= 以及 >>>=。

注意：程序使用位操作符比使用算数操作符更高效。例如，要将一个 int 值 x 乘以 2，可以写成 x << 1，而不是 x * 2。

附录 H

Introduction to Java Programming and Data Structures, Comprehensive Version, Twelfth Edition

正则表达式

经常需要编写代码来验证用户输入，比如验证输入是否是一个数字，是否是一个全部小写的字符串，或者社会安全号。如何编写这类代码呢？一个简单而有效的做法是使用正则表达式来完成这个任务。

正则表达式（regular expression，简写为 regex）是一个字符串，用来描述匹配一个字符串集合的模式。对于字符串处理来说，正则表达式是一个强大的工具。可以使用正则表达式来匹配、替换和拆分字符串。

H.1 匹配字符串

让我们从 String 类中的 matches 方法开始。乍一看，matches 方法很类似 equals 方法。例如，以下两个语句都为 true。

```
"Java".matches("Java");
"Java".equals("Java");
```

然而，matches 方法功能更加强大。它不仅可以匹配一个固定的字符串，还可以匹配符合一个模式的字符串集。例如，以下语句结果都为 true。

```
"Java is fun".matches("Java.*")
"Java is cool".matches("Java.*")
"Java is powerful".matches("Java.*")
```

前面语句中的 "Java.*" 是一个正则表达式。它描述了一个字符串模式，以 Java 开始，后面跟 0 个或者多个字符串。这里，子字符串 .* 匹配 0 或者多个任意字符。

H.2 正则表达式语法

正则表达式由字面值字符和特殊符号组成。表 H-1 列出了一些正则表达式常用的语法。

表 H-1　常用的正则表达式

正则表达式	匹配	示例
x	指定字符 x	Java 匹配 Java
.	任意单个字符，除了换行符外	Java 匹配 J..a
(ab|cd)	ab 或者 cd	ten 匹配 t(en|im)
[abc]	a、b 或者 c	Java 匹配 Ja[uvwx]a
[^abc]	除了 a、b 或者 c 外的任意字符	Java 匹配 Ja[^ars]a
[a-z]	a 到 z	Java 匹配 [A-M]av[a-d]
[^a-z]	除了 a 到 z 外的任意字符	Java 匹配 Jav[^b-d]
[a-e[m-p]]	a 到 e 或 m 到 p	Java 匹配 [A-G[I-M]]av[a-d]
[a-e&&[c-p]]	a 到 e 与 c 到 p 的交集	Java 匹配 [A-P&&[I-M]]av[a-d]
\d	一位数字，等同于 [0-9]	Java2 匹配 "Java[\\d]"
\D	一位非数字	$Java 匹配 "[\\D][\\D]ava"

(续)

正则表达式	匹配	示例
\w	单词字符	Java1 匹配 "[\\w]ava[\\d]"
\W	非单词字符	$Java 匹配 "[\\W][\\w]ava"
\s	空白字符	"Java 2" 匹配 "Java\\s2"
\S	非空白字符	Java 匹配 "[\\S]ava"
p*	0 或者多次出现模式 p	aaaa 匹配 "a*" abab 匹配 "(ab)*"
p+	1 次或者多次出现模式 p	a 匹配 "a+b*" able 匹配 "(ab)+.*"
p?	0 或者 1 次出现模式 p	Java 匹配 "J?Java" ava 匹配 "J?ava"
p{n}	正好出现 n 次模式 p	Java 匹配 "Ja{1}.*" Java 不匹配 ".{2}"
p{n,}	至少出现 n 次模式 p	aaaa 匹配 "a{1,}" a 不匹配 "a{2,}"
p{n,m}	n 到 m (不包含) 次出现模式 p	aaaa 匹配 "a{1,9}" abb 不匹配 "a{2,9}bb"
\p{P}	一个标点字符 !"#$%&'()*+,-./:;<=>?@[\]^_`{|}~	J?a 匹配 "J\p{P}a" J?a. 不匹配 "J\p{P}a"

☞ **注意**：反斜杠是一个特殊的字符，在字符串中开始转义序列。因此 Java 中需要使用 \\ 来表示 \。

☞ **注意**：回顾一下，空白字符是 ' '、'\t'、'\n'、'\r' 或者 '\f'。因此，\s 和 [\t\n\r\f] 等同，\S 和 [^ \t\n\r\f] 等同。

☞ **注意**：单词字符是任何的字母，数字或者下划线字符。因此 \w 等同于 [a-z[A-Z][0-9]_] 或者简化为 [a-zA-Z0-9_]。\W 等同于 [^a-zA-Z0-9]。

☞ **注意**：表 H-1 中的 *、+、?、{n}、{n,} 以及 {n,m} 称为量词符 (quantifier)，用于指定量词符前面的模式可能重复的次数。例如，A* 匹配 0 或者多个 A，A+ 匹配 1 或者多个 A，A? 匹配 0 或者 1 个 A。A{3} 精确匹配 AAA，A{3,} 匹配至少 3 个 A，A{3,6} 匹配 3 到 6 个 A。* 等同于 {0,}，+ 等同于 {1,}，? 等同于 {0,1}。

☞ **警告**：不要在重复量词符中使用空白。例如，A{3,6} 不能写成逗号后面有一个空白符的 A{3, 6}。

☞ **注意**：可以使用括号来将模式进行分组。例如，(ab){3} 匹配 ababab，但是 ab{3} 匹配 abbb。

让我们用一些示例来演示如何构建正则表达式。

1. 示例 1

社会安全号的模式是 xxx-xx-xxxx，其中 x 是一位数字。社会安全号的正则表达式可以描述为

[\\d]{3}-[\\d]{2}-[\\d]{4}

例如

```
"111-22-3333".matches("[\\d]{3}-[\\d]{2}-[\\d]{4}")    返回    true
"11-22-3333".matches("[\\d]{3}-[\\d]{2}-[\\d]{4}")     返回    false
```

2. 示例 2

偶数以数字 0、2、4、6 或者 8 结尾。偶数的模式可以描述为

[\\d]*[02468]

例如，

```
"123".matches("[\\d]*[02468]")    返回    false
"122".matches("[\\d]*[02468]")    返回    true
```

3. 示例 3

电话号码的模式是 (xxx)xxx-xxxx，这里 x 是一位数字，并且第一位数字不能为 0。电话号码的正则表达式可以描述为

\\([1-9][\\d]{2}\\) [\\d]{3}-[\\d]{4}

注意：括符 (和) 在正则表达式中是特殊字符，用于对模式分组。为了在正则表达式中表示字面值（或者），必须使用 \\(和 \\)。

例如

```
"(912) 921-2728".matches("\\([1-9][\\d]{2}\\) [\\d]{3}-[\\d]{4}")
    返回    true
"921-2728".matches("\\([1-9][\\d]{2}\\) [\\d]{3}-[\\d]{4}")
    返回    false
```

4. 示例 4

假定姓由最多 25 个字母组成，并且第一个字母为大写形式。则姓的模式可以描述为

[A-Z][a-zA-Z]{1,24}

注意：不能随便放空白符到正则表达式中。如 [A-Z][a-Za-z]{1, 24} 将报错。例如：

```
"Smith".matches("[A-Z][a-zA-Z]{1,24}")      返回    true
"Jones123".matches("[A-Z][a-zA-Z]{1,24}")   返回    false
```

5. 示例 5

Java 标识符在 2.3 节中定义。

- 标识符必须以字母、下划线（_），或者美元符号（$）开始。不能以数字开头。
- 标识符是一个由字母、数字、下划线和美元符号组成的字符序列。

标识符的模式可以描述为

[a-zA-Z_$][\\w$]*

6. 示例 6

什么字符串匹配正则表达式 "Welcome to (Java|HTML)"？答案是 Welcome to Java 或者 Welcome to HTML。

7. 示例 7

什么字符串匹配正则表达式 ".*"？答案是任何字符串。

H.3 替换和拆分字符串

如果字符串匹配正则表达式，则 String 类的 matches 方法返回 true。String 类也包含 repleaceAll、replaceFirst 和 split 方法，用于替换和拆分字符串，如图 H-1 所示。

replaceAll 方法替换所有匹配的子字符串，replaceFirst 方法替换第一个匹配的子字

符串。例如，代码

```
System.out.println("Java Java Java".replaceAll("v\\w", "wi"));
```

显示

```
Jawi Jawi Jawi
```

代码

```
System.out.println("Java Java Java".replaceFirst("v\\w", "wi"));
```

显示

```
Jawi Java Java
```

java.lang.String	
+matches(regex: String): boolean	如果字符串匹配模式，则返回 true
+replaceAll(regex: String, replacement: String): String	将所有匹配的子字符串替换为 replacement 变量中的字符串，并返回新的字符串
+replaceFirst(regex: String, replacement: String): String	将匹配的第一个子字符串替换为 replacement 变量中的字符串，并返回新的字符串
+split(regex: String): String[]	返回一个字符串数组，包含被匹配模式的分隔符拆分的子字符串
+split(regex: String, limit: int): String[]	除使用了 limit 参数控制模式应用的次数外，与前面的拆分方法等同

图 H-1 String 类包含使用正则表达式来匹配、替换和拆分字符串的方法

有两个重载的 split 方法。split(regex) 方法使用匹配的分隔符将一个字符串拆分为子字符串。例如，以下语句

```
String[] tokens = "Java1HTML2Perl".split("\\d");
```

将字符串 "Java1HTML2Perl1" 拆分为 Java、HTML 以及 Perl 并且保存在 tokens[0]，tokens[1] 以及 tokens[2] 中。

在 split(regex,limit) 方法中，limit 参数确定模式匹配多少次。如果 limit <= 0，split(regex,limit) 等同于 split(regex)。如果 limit > 0，模式最多匹配 limit -1 次。下面是一些示例：

```
"Java1HTML2Perl".split("\\d", 0);    拆分为 Java, HTML, Perl
"Java1HTML2Perl".split("\\d", 1);    拆分为 Java1HTML2Perl
"Java1HTML2Perl".split("\\d", 2);    拆分为 Java, HTML2Perl
"Java1HTML2Perl".split("\\d", 3);    拆分为 Java, HTML, Perl
"Java1HTML2Perl".split("\\d", 4);    拆分为 Java, HTML, Perl
"Java1HTML2Perl".split("\\d", 5);    拆分为 Java, HTML, Perl
```

注意：默认情况下，所有的量词符都是"贪婪"的。这意味着它们会尽可能匹配最多次。比如，下面语句显示 JRvaa。因为第一个匹配成功的是 aaa。

```
System.out.println("Jaaavaa".replaceFirst("a+", "R"));
```

可以通过在后面添加问号（?）来改变量词符的默认行为。量词符变为"不情愿"或者"惰性"的，这意味着它将匹配尽可能少的次数。例如，下面的语句显示 JRaavaa，因为第一个匹配成功的是 a。

```
System.out.println("Jaaavaa".replaceFirst("a+?", "R"));
```

H.4 替换匹配的子字符串中的部分内容

有时，需要对匹配的子字符串中的部分内容进行替换。例如，假设有如下文本：

```
String text = "3 * (x - y) is in lines 12-56.";
```

我们希望将文本替换为

```
"3 * (x - y) is in lines 12 to 56.";
```

注意，如果符号"-"在两个数字中间且前面是单词"lines"的话，则将其替换为单词"to"。我们希望在这种情形下，将文本中所有出现的"-"都替换为"to"。为了实现这一点，我们将使用模式 [lines \\d+-\\d+] 找到匹配的子字符串，然后将模式中的"-"替换为单词"to"。可以使用 Pattern 类和 Matcher 类来实现。

Pattern 类代表编译好的正则表达式。可以使用 Pattern.compile(regex) 创建一个 Pattern 实例。产生的实例可以用于创建一个 Matcher 对象。例如，以下代码创建一个 Pattern 对象 p，并使用模式 p 为文本创建一个 Matcher 对象 m：

```
String regex = "lines \\d+-\\d+";
Pattern p = Pattern.compile(regex);
Matcher m = p.matcher(text);
```

现在可以使用 Matcher 类中的 find() 方法为模式找到一个匹配的子字符串，使用 group() 方法返回匹配的子字符串，并替换字符串中的"-"，然后使用 addReplacement 和 addTail 方法将文本及其替换部分加入一个 StringBuilder。

完整的代码在代码清单 H-1 中给出。

程序清单 H-1 PatternMatcherDemo.java

```
1   import java.util.regex.Matcher;
2   import java.util.regex.Pattern;
3
4   public class PatternMatcherDemo {
5     public static void main(String args[]) {
6       String text = "3 * (x - y) is in lines 12-56.";
7       String regex = "lines \\d+-\\d+";
8       Pattern p = Pattern.compile(regex);
9       Matcher m = p.matcher(text);
10
11      StringBuffer sb = new StringBuffer();
12      while (m.find()) {
13        String replacement = m.group();
14        replacement = replacement.replace("-", " to ");
15        m.appendReplacement(sb, replacement);
16      }
17
18      m.appendTail(sb);
19      System.out.println(sb.toString());
20    }
21  }
```

这是一个复杂的过程。调用 m.find（第 12 行）从起始位置扫描文本以找到下一个符合模式的匹配。开始时，起始位置位于下标 0 处。调用 m.group()（第 13 行）返回匹配的子字符串给 String replacement。String 的 replace 方法将"-"替换为"to"(第 14 行)。调用 m.appendReplace(sb, replacement) 将文本中当前没有匹配的内容追加到 sb 上,然后将 replacement 追加到 sb 上,这里的 sb 是一个 StringBuilder。注意,当前没有匹配的内容是已经被 m.find() 扫描过的子字符串,但不是匹配字符串的一部分。循环(第 12~16 行)继续寻找下一个匹配,得到匹配的子字符串,替代子字符串中的部分内容,然后追加没有匹配的内容和 replacement 到 sb 上。当无法找到更多匹配时循环结束。然后,程序调用 m.addTail(sb) 方法将文本中余下没有匹配的内容追加到 sb 上(第 18 行)。

注意,find()、group()、addReplacement 以及 addTail 方法一起用在一个寻找–替换–追加的循环中。当 m.find() 第一次被调用时,开始位置位于下标 0 处。当 m.find() 再次被调用时,它先重设起始位置以经过匹配子字符串的末尾。

附录 I

Introduction to Java Programming and Data Structures, Comprehensive Version, Twelfth Edition

枚 举 类 型

I.1 简单枚举类型

枚举类型定义了一个枚举值的列表。每个值是一个标识符。例如，下面的语句声明了一个类型，名为 MyFavoriteColor，依次具有 RED、BLUE、GREEN、YELLOW 值。

```
enum MyFavoriteColor {RED, BLUE, GREEN, YELLOW};
```

枚举类型的值类似于一个常量，因此，按惯例拼写都是使用大写字母。因此，前面的声明采用 RED，而不是 red。按惯例，枚举类型命名类似于一个类，每个单词的第一个字母大写。

一旦定义了一个类型，就可以声明这个类型的变量了：

```
MyFavoriteColor color;
```

变量 color 可以具有定义在枚举类型 MyFavoriteColor 中的一个值，或者 null，但是不能具有其他值。Java 的枚举类型是类型安全的，这意味着试图赋一个除枚举类型所列出的值或者 null 之外的值，都将导致编译错误。

枚举值可以使用下面的语法进行访问：

```
EnumeratedTypeName.valueName
```

例如，下面的语句将枚举值 BLUE 赋值给变量 color：

```
color = MyFavoriteColor.BLUE;
```

注意：必须使用枚举类型名称作为限定词来引用一个值，比如 BLUE。

如同其他类型一样，可以在一行语句中来声明和初始化一个变量：

```
MyFavoriteColor color = MyFavoriteColor.BLUE;
```

可将枚举类型看作一种特殊的类，因此，枚举类型的变量是引用变量。一个枚举类型是 Object 类和 Comparable 接口的子类型。因此，枚举类型继承了 Object 类中的所有方法，以及 Comparable 接口中的 compareTo 方法。另外，可以在一个枚举类型的对象上使用下面的方法：

- `public String name();`

  为对象返回名字值。

- `public int ordinal();`

  返回和枚举值关联的序号值。枚举类型中的第一个值具有序号数 0，第二个值具有序号值 1，第三个为 2，以此类推。

程序清单 I-1 给出了一个程序，演示了枚举类型的使用。

程序清单 I-1 EnumeratedTypeDemo.java

```java
1  public class EnumeratedTypeDemo {
2    static enum Day {SUNDAY, MONDAY, TUESDAY, WEDNESDAY, THURSDAY,
3      FRIDAY, SATURDAY};
4
5    public static void main(String[] args) {
6      Day day1 = Day.FRIDAY;
7      Day day2 = Day.THURSDAY;
8
9      System.out.println("day1's name is " + day1.name());
10     System.out.println("day2's name is " + day2.name());
11     System.out.println("day1's ordinal is " + day1.ordinal());
12     System.out.println("day2's ordinal is " + day2.ordinal());
13
14     System.out.println("day1.equals(day2) returns " +
15       day1.equals(day2));
16     System.out.println("day1.toString() returns " +
17       day1.toString());
18     System.out.println("day1.compareTo(day2) returns " +
19       day1.compareTo(day2));
20   }
21 }
```

```
day1's name is FRIDAY
day2's name is THURSDAY
day1's ordinal is 5
day2's ordinal is 4
day1.equals(day2) returns false
day1.toString() returns FRIDAY
day1.compareTo(day2) returns 1
```

在第 2 和 3 行定义了枚举类型 Day。变量 day1 和 day2 声明为 Day 类型，在第 6 和 7 行赋枚举值。由于 day1 的值为 FRIDAY，它的序号值为 5（第 11 行）。由于 day2 的值为 THURSDAY，它的序号值为 4（第 12 行）。

由于枚举类型是 Object 类和 Comparable 接口的子类。可以从一个枚举对象引用变量调用 equals，toString 以及 compareTo 方法（第 14 ~ 19 行）。如果 day1 和 day2 具有同样的序号数，day1.equals(day2) 返回真。day1.compareTo(day2) 返回 day1 的序号数到 day2 的序号数之间的差值。

也可以将程序清单 I-1 中的代码重新写为程序清单 I-2。

程序清单 I-2 StandaloneEnumTypeDemo.java

```java
1  public class StandaloneEnumTypeDemo {
2    public static void main(String[] args) {
3      Day day1 = Day.FRIDAY;
4      Day day2 = Day.THURSDAY;
5
6      System.out.println("day1's name is " + day1.name());
7      System.out.println("day2's name is " + day2.name());
8      System.out.println("day1's ordinal is " + day1.ordinal());
9      System.out.println("day2's ordinal is " + day2.ordinal());
10
11     System.out.println("day1.equals(day2) returns " +
12       day1.equals(day2));
13     System.out.println("day1.toString() returns " +
14       day1.toString());
15     System.out.println("day1.compareTo(day2) returns " +
```

```
16            day1.compareTo(day2));
17      }
18  }
19
20  enum Day {SUNDAY, MONDAY, TUESDAY, WEDNESDAY, THURSDAY,
21      FRIDAY, SATURDAY}
```

枚举类型可以在一个类中定义，如程序清单 I-1 中的第 2 和 3 行所示；或者单独定义，如程序清单 I-2 的第 20 和 21 行所示。在前一种情况下，枚举类型被作为内部类对待。程序编译后，将创建一个名为 EnumeratedTypeDemo$Day 的类。在后一种情况下，枚举类型作为一个独立的类来对待。程序编译后，将创建一个名为 Day.class 的类。

注意：当枚举类型在一个类中声明时，类型必须声明为类的一个成员，而不能在一个方法中声明。而且，类型总是 static 的。由于这个原因，程序清单 I-1 第 2 行的 static 关键字可以省略。可以用于内部类的可见性修饰符也可以应用在一个类中定义的枚举类型上。

提示：使用枚举值（例如，Day.MONDAY，Day.TUESDAY，等等）而不是字面量整数值（例如，0，1，等等）可以让程序更加易于阅读和维护。

I.2 通过枚举变量使用 if 或者 switch 语句

枚举变量具有一个值。程序经常需要根据取值来执行特定的动作。例如，如果值为 Day.MONDAY，则踢足球；如果值为 Day.TUESDAY，则学习钢琴课，等等。可以使用 if 语句或者 switch 语句来测试变量的值，如图 a）和 b）所示。

```
if (day.equals(Day.MONDAY)) {
  // process Monday
}
else if (day.equals(Day.TUESDAY)) {
  // process Tuesday
}
else
  ...
```
a)

等价于

```
switch (day) {
  case MONDAY:
    // process Monday
    break;
  case TUESDAY:
    // process Tuesday
    break;
  ...
}
```
b)

在 b 图的 switch 语句中，case 标签是一个无限定词的枚举值（即，MONDAY，而不是 Day.MONDAY）。

I.3 使用 foreach 循环处理枚举值

每个枚举类型都有一个静态方法 values()，可以以一个数组返回这个类型中所有的枚举值。例如，

```
Day[] days = Day.values();
```

可以使用如图 a 中所示的普通循环，或者图 b 中的 foreach 循环来处理数组中的所有值。

```
for (int i = 0; i < days.length; i++)
  System.out.println(days[i]);
```
a)

等价于

```
for (Day day: days)
  System.out.println(day);
```
b)

I.4 具有数据域、构造方法和方法的枚举类型

前面介绍的简单枚举类型定义了一个具有枚举值列表的类型。也可以定义一个具有数据域、构造方法和方法的枚举类型，如程序清单 I-3 所示。

程序清单 I-3 TrafficLight.java

```java
1  public enum TrafficLight {
2    RED ("Please stop"), GREEN ("Please go"),
3    YELLOW ("Please caution");
4
5    private String description;
6
7    private TrafficLight(String description) {
8      this.description = description;
9    }
10
11   public String getDescription() {
12     return description;
13   }
14 }
```

第 2 和 3 行定义了枚举值。值的声明必须是类型声明的第一条语句。第 5 行声明了一个名为 description 的数据域，用于描述一个枚举值。第 7～9 行声明了构造方法 TrafficLight。任何时候访问枚举值时，构造方法都将被调用。枚举值的参数将传递给构造方法，在构造方法中赋值给 description。

程序清单 I-4 给出了一个使用 TrafficLight 的测试程序。

程序清单 I-4 TestTrafficLight.java

```java
1  public class TestTrafficLight {
2    public static void main(String[] args) {
3      TrafficLight light = TrafficLight.RED;
4      System.out.println(light.getDescription());
5    }
6  }
```

一个枚举值 TrafficLight.RED 赋值给变量 light（第 3 行）。访问 TrafficLight.RED 引起 JVM 使用参数 "please stop" 调用构造方法。枚举类型中方法的调用和类中的方法是一样的。light.getDescription() 返回对枚举值的描述（第 4 行）。

> **注意**：Java 语法要求枚举类型的构造方法是私有的，避免被直接调用。私有修饰符可以省略。在这种情况下，默认为私有。

附录 J

Introduction to Java Programming and Data Structures, Comprehensive Version, Twelfth Edition

大 O、大 Ω 和大 Θ 表示法

第 22 章介绍了非专业用语中的大 O 表示法。在本附录中，我们给出大 O 表示法的精确数学定义。我们还将介绍大 Ω 和大 Θ 表示法。

J.1 大 O 表示法

大 O 表示法是一种渐近式表示法，用于描述函数的参数接近特定值或无穷大时的行为。设 $f(n)$ 和 $g(n)$ 为两个函数，如果对于常量 c（$c>0$）和值 m（$n \geq m$），有 $f(n) \leq c \times g(n)$，则我们说 $f(n)$ 为 $O(g(n))$。

例如，$f(n)=5n^3+8n^2$ 为 $O(n^3)$，因为可以找到 $c=13$ 以及 $m=1$，从而对于 $n \geq m$ 满足 $f(n) \leq cn^3$。$f(n)=6n\log n+n^2$ 为 $O(n^2)$，因为可以找到 $c=7$ 以及 $m=2$，从而对于 $n \geq m$ 满足 $f(n) \leq cn^2$。$f(n)= 6n\log n+400n$ 为 $O(n\log n)$，因为可以找到 $c=406$ 以及 $m=2$，从而对于 $n \geq m$ 满足 $f(n) \leq cn\log n$。$f(n)=n^2$ 为 $O(n^3)$，因为可以找到 $c=1$ 以及 $m=1$，从而对于 $n \geq m$ 满足 $f(n) \leq cn^3$。注意，有无数 c 和 m 的选择，从而对于 $n \geq m$ 满足 $f(n) \leq c \times g(n)$。

大 O 表示法表示函数 $f(n)$ 渐近小于或等于另一个函数 $g(n)$，从而可以通过忽略乘法常数并舍弃函数中的非主导项来简化函数。

J.2 大 Ω 表示法

大 Ω 表示法与大 O 表示法相反，它也是一种渐近表示法，表示函数 $f(n)$ 大于或等于另一个函数 $g(n)$。设 $f(n)$ 和 $g(n)$ 为两个函数，如果对于常量 c（$c>0$）和值 m（$n \geq m$），有 $f(n) \geq c \times g(n)$，则我们说 $f(n)$ 为 $\Omega(g(n))$。

例如，$f(n)=5n^3+8n^2$ 为 $\Omega(n^3)$，因为可以找到 $c=5$ 以及 $m=1$，从而对于 $n \geq m$ 满足 $f(n) \geq cn^3$。$f(n)=6n\log n+n^2$ 为 $\Omega(n^2)$，因为可以找到 $c=1$ 以及 $m=1$，从而对于 $n \geq m$ 满足 $f(n) \geq cn^2$。$f(n)= 6n\log n+400n$ 为 $\Omega(n\log n)$，因为可以找到 $c=6$ 以及 $m=1$，从而对于 $n \geq m$ 满足 $f(n) \geq cn\log n$。$f(n)=n^2$ 为 $\Omega(n)$，因为可以找到 $c=1$ 以及 $m=1$，从而对于 $n \geq m$ 满足 $f(n) \geq cn$。注意，有无数 c 和 m 的选择，从而对于 $n \geq m$ 满足 $f(n) \geq c \times g(n)$。

J.3 大 Θ 表示法

大 Θ 表示法表示两个函数渐近相同。设 $f(n)$ 和 $g(n)$ 为两个函数，如果 $f(n)$ 为 $O(g(n))$ 且 $f(n)$ 为 $\Omega(g(n))$，则我们说 $f(n)$ 为 $\Theta(g(n))$。

推荐阅读

 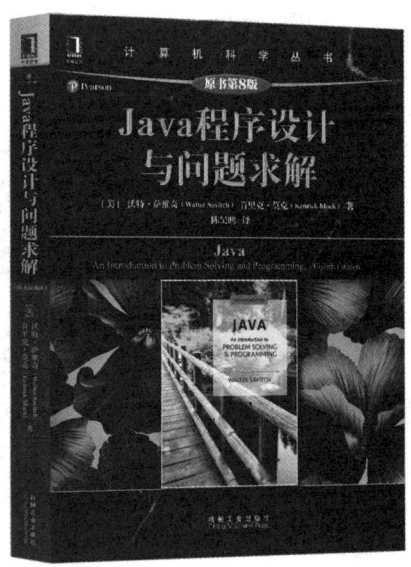

Java语言程序设计（基础篇）（原书第12版）

作者：[美] 梁勇（Y. Daniel Liang）著 译者：戴开宇
ISBN：978-7-111-66980-7 定价：139.00元

Java程序设计与问题求解（原书第8版）

作者：[美] 沃特·萨维奇 (Walter Savitch) 肯里克·莫克（Kenrick Mock）
译者：陈昊鹏 ISBN：978-7-111-62097-6 定价：139.00元

推荐阅读

Java语言程序设计（基础篇）（英文版·原书第11版）

作者：[美]梁勇（Y. Daniel Liang）著 ISBN：978-7-111-65517-6 定价：139.00元

Java语言程序设计与数据结构（进阶篇）（英文版·原书第11版）

作者：[美]梁勇（Y. Daniel Liang）著 ISBN：978-7-111-65515-2 定价：129.00元